EXPONENTS AND LOGARITHMS

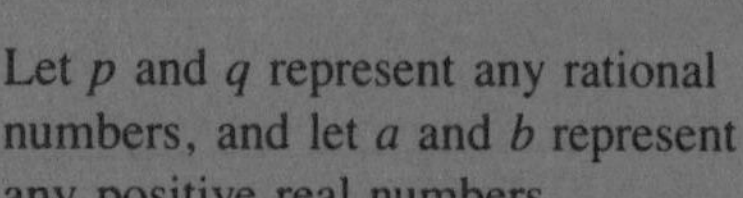

$y = \log_b x$ if and only if $x = b^y$ where $b > 0$, $b \neq 1$

If A and B are real

Let p and q represent any rational numbers, and let a and b represent any positive real numbers.

I. $\log_b AB = \log_b A + \log_b B$

The log of the product of two numbers is the sum of the logs of the numbers.

$$b^p b^q = b^{p+q}$$

II. $\log_b \dfrac{A}{B} = \log_b A - \log_b B$

The log of the quotient of two numbers is the log of the numerator minus the log of the denominator.

$$\frac{b^p}{b^q} = b^{p-q}$$

III. $\log_b A^n = n \log_b A$

The log of the nth power of a number is n times the log of the number.

$$(b^q)^p = b^{pq}$$

IV. $\log_b 1 = 0$

The log of one is zero for any base.

$$b^0 = 1$$

V. $\log_b \dfrac{1}{A} = -\log_b A$

The log of the reciprocal of a number is the negative of the log of the number.

$$b^{-1} = \frac{1}{b}$$

VI. $\log_b m = \log_b n$ is equivalent to $m = n$, if m and n are positive reals. If the logs are equal then the numbers are equal, and conversely if the numbers are equal then their logs are also equal.

VII. $\log_b A = \dfrac{\log A}{\log b}$

or $\dfrac{\ell n A}{\ell n b}$

The log of a number in any base is the quotient of the log (or ℓn) of the number divided by the log (or ℓn) of the base.

n	$b > 0$	$b < 0$	$b = 0$
even	$b^{1/n}$ is the positive nth real root of b $-b^{1/n}$ is the negative nth real root of b	not defined	0
odd	$b^{1/n}$ is the nth real root of b		

College Algebra

Third Edition

Other Brooks/Cole Titles by the Same Authors

Basic Mathematics for College Students
Karl J. Smith
Arithmetic for College Students
Karl J. Smith
Mathematics: Its Power and Utility
Karl J. Smith
The Nature of Mathematics, Fourth Edition
Karl J. Smith
Beginning Algebra for College Students, Third Edition
Karl J. Smith and Patrick J. Boyle
Study Guide for Beginning Algebra for College Students, Third Edition
Karl J. Smith and Patrick J. Boyle
Intermediate Algebra for College Students, Third Edition
Karl J. Smith and Patrick J. Boyle
Study Guide for Intermediate Algebra for College Students, Third Edition
Karl J. Smith and Patrick J. Boyle
Essentials of Trigonometry
Karl J. Smith
Trigonometry for College Students, Third Edition
Karl J. Smith
Precalculus Mathematics: A Functional Approach, Second Edition
Karl J. Smith
Finite Mathematics
Karl J. Smith

College Algebra

Third Edition

KARL J. SMITH
Santa Rosa Junior College

PATRICK J. BOYLE
Santa Rosa Junior College

Brooks/Cole Publishing Company
Monterey, California

Contemporary Undergraduate Mathematics Series
Robert J. Wisner, Editor

Brooks/Cole Publishing Company
A Division of Wadsworth, Inc.

Printed in the United States of America

10 9 8 7 6 5 4 3

Library of Congress Cataloging in Publication Data

Smith, Karl J.
College algebra

Includes index.
1. Algebra. I. Boyle, Patrick J.
II. Title.
QA154.2.S56 1984 512.9 84-11366
ISBN 0-534-03747-X

Subject Editor: Craig Barth
Manuscript Editor: Phyllis Niklas
Production: Joan Marsh
Interior and Cover Design: Vicki Van Deventer
Illustrations: Carl Brown
Typesetting: Graphic Typesetting Service

Cover Photo: Phillip A. Harrington/Fran Heyl Associates,
Courtesy of the Intel Corporation

About the Cover

This photograph was taken with an Olympus model BHS research microscope using a Nomarski system of top-light illumination. The Nomarski system is comprised of an intricate series of interrelated prisms and filters which color varying layers of subject materials.

Additional illumination was provided by fiber optic light guides. The film used was 35mm tungsten Kodachrome with an ASA rating of 40.

The mottled or stippled effect on the surface of the chip is caused by random light reflections on a transparent protective coating that has been applied to the chip in the manufacturing process.

Phillip A. Harrington is a well known photographer who specializes in photography through the microscope.

To Jack and Rosamond with love and affection
In loving memory of Frank and Maureen

Preface

Following the launching of *Sputnik* in the 1950s, mathematics went through dynamic changes that culminated in the "new math" of the 1960s. The goal of the new math was to create in the student a fundamental understanding of mathematics. But in so doing, we lost sight of the practicality of mathematics, and consequently the results of that grand experiment in the new math proved to be less than expected.

During the 1970s there was a slow but steady trend toward renewed emphasis on practical, solid mathematics. In the 1980s competency requirements are now being imposed on many students receiving high school diplomas, associate degrees, and even bachelor's degrees. These competency requirements are forcing schools to review their curricula and develop courses that not only provide basic skills, but also develop critical thinking and a fundamental understanding of mathematics. This "back to basics" movement of the 1980s does not seek to reestablish the methods used before *Sputnik*. Instead, there is a deep concern to present practical, solid mathematics while keeping the goal of giving our students a fundamental understanding of mathematics. It is not enough simply to present the superficial "monkey see, monkey do" type of exercises. The problem sets must be designed to develop the student's thought processes.

This *renewed mathematics* is the goal of our series of algebra textbooks. In this text we carefully and deliberately present exposition, examples, and problems that have a threefold purpose:

1. No-nonsense drill problems teach the students essential mathematical skills; this edition provides a *significant* number of new drill problems. The A and B drill problems are graded in difficulty and are presented as matched pairs or triplets of problems. The C problems give practice at a more difficult level. Each chapter has a summary and review problem set that includes over 100 problems for additional practice and study.
2. Applied problems teach the student some of the usefulness and practicality of mathematics; nearly every section of this edition has a sampling of applied problems so that the student can gain continued experience in dealing with applied problems. The applied problems are self-contained and do not require any outside knowledge.
3. Problems that actually *teach* the thought process to the student provide a fundamental understanding of mathematical processes in a way that will enhance the student's reasoning ability, not only in mathematics, but also in nonmathematical situations requiring a reasoned conclusion.

The exposition and examples in the text also reflect this goal. The first chapter provides a review of the real number system with exponents as the unifying theme. This chapter will review and refresh the student's manipulative skills. Chapters 2 and 3 review solving and graphing first- and second-degree equations and inequalities in one and two variables. The concept of a function is presented in Chapter 4 and is the unifying theme for the remaining chapters, which are relatively independent. Chapter 5 develops the theory of equations, including a discussion and graphing of both polynomial and rational functions. Chapter 6 focuses on the nature and applications of the exponential and logarithmic functions, rather than on their computational properties. In Chapter 7, we discuss matrices and their properties, the determinant function, and systems of equations. Sequences, series, and the Binomial Theorem are presented in Chapter 8, with combinatorics and probability in Chapter 9. Chapter 10 offers a unique introduction to higher mathematics and the nature of proof. The student is taught how to prove mathematical theorems in the framework of a college algebra course. This provides a smooth transition to more advanced mathematics.

The organization gives maximum flexibility in content, order, and emphasis in order to meet the wide-ranging mathematical needs of the students who enroll in a college algebra course. A logical, deductive approach is emphasized, beginning with a reasonable level of rigor; only those results that further the understanding of the topic are proved. Initially, many arguments are intuitive and many developments are inductive, but the notion of proof grows throughout the book. The final chapter is devoted to the nature of mathematical proof, but elements of this chapter may easily be included earlier as deemed appropriate. For example, the instructor may wish to cover the topic of mathematical induction in conjunction with the chapter on series and sequences.

Full advantage is taken of the availability of hand-held scientific calculators. We feel that the calculator is indispensable for topics such as exponential and logarithmic functions, but we have resisted adding calculator problems with "ugly numbers" simply for the sake of designating them "calculator problems." Calculators should be used as a natural extension of other available tools, such as a pencil and paper.

Answers to odd-numbered problems are included in the back of the book. An instructor's manual is available, which gives the answers to all the problems.

Our thanks go to the reviewers of this edition: William Anderson of East Texas State University, Elton Beougher of Fort Hays State University, Martin J. Brown of Jefferson Community College, J. S. Collins of Embry-Riddle University, Libby W. Holt of Florida Junior College, Harlan Koca of Washburn University, Fr. Micah E. Kozoil of Saint Vincent College, Helen Kriegsman of Pittsburg State University, and Wayne Mackey of Johnson County Community College. And our continued gratitude goes to the reviewers of the previous edition: James E. Arnold of the University of Wisconsin, David Barwick of Macon Junior College, Robert A. Chaffer of Central Michigan University, Jean Coover of West Virginia University, Allen Hesse of Rochester Community College, Joseph B. Hoffert of Drake University, William Perry of Texas A & M University, David Wend of Montana State University, and Charles R. Williams of Midwestern University.

Thanks also go to Phyllis Niklas, Joan Marsh, Carl Brown, Craig Barth, and Karen Sharp for all their contributions to the book. We especially appreciate the loving support of our wives, Linda and Theresa.

Karl J. Smith
Patrick J. Boyle

Contents

*Optional

*Optional

*Optional
*Section 10.5 does not require Sections 10.1–10.4.

CHAPTER 1

Numbers and Expressions

Gottlob Frege
(1848–1925)

The laws of number . . . are not the laws of nature . . . , they are the laws of the laws of nature.

Gottlob Frege,
Foundations of Science,
edited by N. Campbell, Dover, New York, 1957, p. 292.

Historical Note

Gottlob Frege was a German logician and mathematician who, in 1884, wrote a book on *The Fundamental Laws of Arithmetic* because he was not satisfied with its basic concepts. Have you ever tried to explain or define what we mean by the number five? Try it if you want an exercise in frustration. In his book, Frege attempted to put arithmetic on a sound logical basis and tried to answer questions such as "What is five?" His logical development depended heavily on the notion of "the set of all sets." After many years of work on this project, he received a letter from Bertrand Russell which contained Russell's famous "set of all sets paradox." This letter forced Frege to close his book with the statement: "A scientist can hardly encounter anything more undesirable than to have the foundation collapse just as the work is finished. I was put in this position by a letter from Mr. Bertrand Russell when the work was almost through the press."

The "set of all sets paradox" can be summarized by the following question: "The barber in a certain small town shaves those persons and only those persons who do not shave themselves. Who shaves the barber?"

Chapter Overview

This chapter will serve as a quick review for much of the material contained in a previous algebra course. The main thread tying the material together is the idea of exponents. Using positive integer exponents, we review polynomials, simplifying expressions, and factoring. Next, we review rational expressions by considering integral exponents. Finally, we allow exponents to be rational, which leads to a review of irrational expressions and radicals.

1.1 Sets of Numbers

You are probably familiar with the various sets of numbers used in mathematics, so they are simply summarized for you in Figure 1.1 and Table 1.1 to refresh your memory.

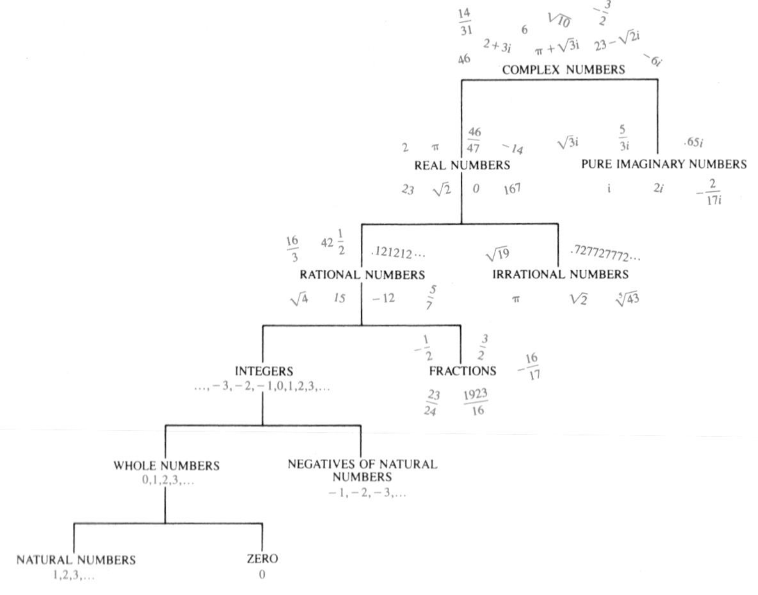

Figure 1.1

This course will focus attention on the set of real numbers, which is easily visualized by using a **coordinate system** called a **number line** (Figure 1.2). A **one-to-one correspondence** is established between all real numbers and all points on a number line.

Table 1.1
Sets of Numbers

Name	**Symbol**	**Set**
Counting numbers Natural numbers	**N**	{1, 2, 3, 4, . . .}
Whole numbers	**W**	{0, 1, 2, 3, 4, . . .}
Integers	**Z**	{. . . , −3, −2, −1, 0, 1, 2, 3, . . .}
Rational numbers	**Q**	Numbers that can be written in the form p/q, where p and q are integers with $q \neq 0$;* they are also characterized by numbers whose decimal representations either terminate or repeat
Irrational numbers	**Q′**	Numbers whose decimal representations do not terminate or repeat
Real numbers	**R**	Numbers that are either rational or irrational
Pure imaginary numbers	**I**	Numbers of the form bi, where b is a nonzero real number and $i^2 = -1$
Complex numbers	**C**	Numbers of the form $a + bi$, where a and b are real numbers and $i^2 = -1$

*Set-builder notation can be used for many of these definitions. For example, we can write

$$\{p/q \mid p \text{ and } q \in \mathbf{Z},\ q \neq 0\}$$

for the definition of the set of rational numbers. It is read "set of all numbers p/q such that p and q are integers and q does not equal 0." A review of sets and set notation is given in Appendix A.

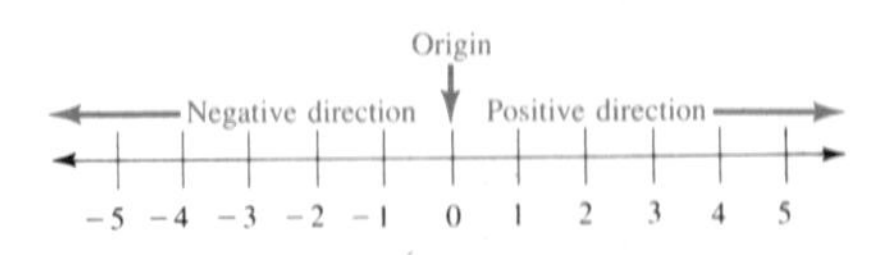

Figure 1.2

1. Every point on the line corresponds to precisely one real number.
2. For each real number, there corresponds one and only one point.

A point associated with a particular number is called the **graph** of that number. Numbers associated with points to the right of the origin are called **positive real numbers,** and those associated with points to the left are called **negative real numbers.**

Axioms concerning order of real numbers and formal definitions of inequality are considered in Chapter 10, but it is assumed that you are familiar with the following informal definitions:

$a = b$ Is read "a is equal to b" and means the graphs of both a and b are the same point.

$a < b$ Is read "a is less than b" and means the graph of a is to the left of the graph of b on a number line.

$a > b$ Is read "a is greater than b" and means $b < a$.

$a \leq b$ Is read "a is less than or equal to b" and means $a = b$ or $a < b$.

$a \geq b$ Is read "a is greater than or equal to b" and means $a = b$ or $a > b$.

$|a|$ Is read "the absolute value of a" and means the distance between the graph of a and the origin. For example:

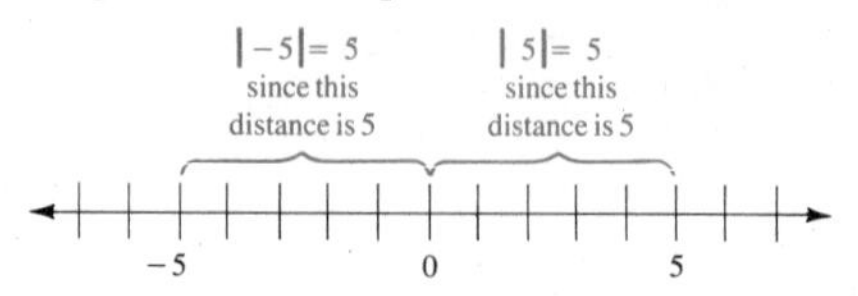

Example 1 **a.** $|2| = 2$ **b.** $|-2| = 2$ **c.** $|-\frac{2}{3}| = \frac{2}{3}$

d. $|\sqrt{5}| = \sqrt{5}$ **e.** $-|2| = -2$ **f.** $-|\pi| = -\pi$ □

Since we will want to use the notion of absolute value in a variety of contexts, we will now give an algebraic definition of absolute value that is equivalent to the geometric one using distances given above.

ABSOLUTE VALUE

The **absolute value** of a real number a is defined by

$$|a| = \begin{cases} a & \text{if } a \geq 0 \\ -a & \text{if } a < 0 \end{cases}$$

Example 2 Find the absolute values using the algebraic definition.

a. $|2| = 2$ Since $2 \geq 0$

b. $|-2| = -(-2) = 2$ Since $-2 < 0$

c. $-|5| = -5$ Since $5 \geq 0$

d. $-|-46| = -[-(-46)] = -46$ Since $-46 < 0$

e. $|\pi - 3| = \pi - 3$ Since $\pi - 3 \geq 0$

f. $|3 - \pi| = -(3 - \pi) = -3 + \pi = \pi - 3$ Since $3 - \pi < 0$

g. $|\sqrt{20} - 4| = \sqrt{20} - 4$ By definition, $\sqrt{20}$ is the positive square root of 20. We know that $4^2 = 16$ and $5^2 = 25$, so the positive square root of 20 is between 4 and 5; this means $\sqrt{20} - 4$ is positive.

h. $|\sqrt{20} - 5| = -(\sqrt{20} - 5) = 5 - \sqrt{20}$ Since $\sqrt{20} - 5$ is negative (see the reasoning in part g) □

Some properties of absolute value will be considered later (in Section 2.3).

To begin our study of algebra, it is necessary to assume certain properties of equality and inequality. The first is called the **trichotomy property,** or **property of comparison.**

PROPERTY OF COMPARISON

> Given any two real numbers a and b, exactly one of the following holds:
>
> 1. $a = b$ 2. $a < b$ 3. $a > b$

This property relates to any two real numbers and tells you that, if you are given *any* two real numbers, either they are equal or one of them is greater than the other. It establishes order on the number line. We need this property to derive a formula for the distance between two points on a number line. Let P_1 and P_2 be any points on a number line with coordinates x_1 and x_2, respectively. This is usually denoted by $P_1(x_1)$ and $P_2(x_2)$. Then, by the property of comparison, we know that $x_1 = x_2$, $x_1 < x_2$, or $x_1 > x_2$. Consider these possibilities one at a time:

$x_1 = x_2$ The distance between P_1 and P_2 is 0.

$x_1 < x_2$

The distance is $x_2 - x_1$.

$x_1 > x_2$

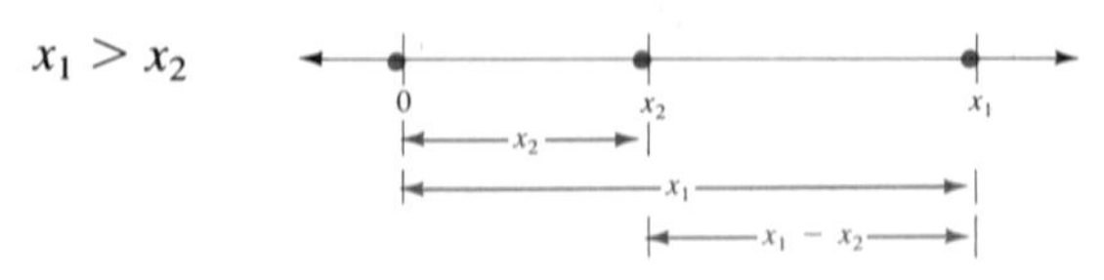

The distance is $x_1 - x_2$.

We can combine these different possibilities into one distance formula by using the idea of absolute value.

DISTANCE ON A NUMBER LINE

> The distance d between points $P_1(x_1)$ and $P_2(x_2)$ is
>
> $d = |x_2 - x_1|$

Example 3 Let A, B, C, D, and E be points with coordinates as shown on the number line:

a. Distance from D to E: $|5 - 2| = |3| = 3$

b. Distance from E to D: $|2 - 5| = |-3| = 3$

c. Distance from B to D: $|2 - (-1)| = |3| = 3$

d. Distance from B to A: $|-3 - (-1)| = |-2| = 2$

e. Distance from A to C: $|0 - (-3)| = |3| = 3$

f. Distance from A to D: $|2 - (-3)| = |5| = 5$

g. Distance from E to B: $|-1 - 5| = |-6| = 6$ □

When real numbers are added, the result is called the **sum** and the numbers are called **terms.** When real numbers are multiplied, the result is called the **product** and the numbers multiplied are called **factors.** The result from subtraction is called the **difference,** while that from division is called the **quotient.**

The real numbers, together with the relation of equality and the operations of addition and multiplication, satisfy the **field properties,** which are listed in the box.

FIELD PROPERTIES FOR THE SET **R** OF REAL NUMBERS

Let a, b, and c be real numbers.

	Addition Properties	**Multiplication Properties**
CLOSURE	$a + b$ is a unique real number	ab is a unique real number
COMMUTATIVE	$a + b = b + a$	$ab = ba$
ASSOCIATIVE	$(a + b) + c = a + (b + c)$	$(ab)c = a(bc)$
IDENTITY	There exists a unique real number 0 such that $a + 0 = 0 + a = a$	There exists a unique real number 1 such that $a \cdot 1 = 1 \cdot a = a$
INVERSE	For each real number a, there is a unique real number $-a$ such that $a + (-a) = (-a) + a = 0$	For each *nonzero* real number a, there is a unique real number $1/a$ such that $a\left(\frac{1}{a}\right) = \left(\frac{1}{a}\right)a = 1$
DISTRIBUTIVE	$a(b + c) = ab + ac$	

All these properties of real numbers will be important in our study of algebra, but for now we need to focus upon the commutative, associative, and distributive properties:

COMMUTATIVE	Properties of order
ASSOCIATIVE	Properties of grouping
DISTRIBUTIVE	Property that changes a product to a sum or changes a sum to a product

Example 4 Distinguish among the commutative, associative, and distributive properties.

a. $2 + (3 + 4) = (2 + 3) + 4$ Associative for addition

b. $2 + (3 + 4) = (3 + 4) + 2$ Commutative for addition

c. $2(3 + 4) = 2 \cdot 3 + 2 \cdot 4$ Distributive

d. $ba + a(b + c) = ba + ab + ac$ Distributive

e. $ba + a(b + c) = ba + a(c + b)$ Commutative for addition

f. $ba + a(b + c) = ba + (b + c)a$ Commutative for multiplication

g. $2\sqrt{3} + 5\sqrt{3} = (2 + 5)\sqrt{3}$ Distributive □

The remaining examples involve the other field properties.

Example 5 The set of natural numbers is closed for addition since

$$m + n \text{ is a natural number}$$

whenever m and n are natural numbers. On the other hand,

$$2 - 5 = -3$$

which is *not* a natural number, so the set of natural numbers is not closed for subtraction. □

Example 6 The set $\{0, 1\}$ is not closed for addition because

$$1 + 1 = 2$$

which is not a member of the set. On the other hand, it *is* closed for multiplication, because

$$\begin{array}{ll} 0 \cdot 0 = 0 & 1 \cdot 0 = 0 \\ 0 \cdot 1 = 0 & 1 \cdot 1 = 1 \end{array}$$

which shows that all possible products are members of the set. □

Example 7 Identify the field property being used.

a. $5 \cdot \left(6 \cdot \frac{1}{6}\right) = 5 \cdot 1$ Multiplicative inverse

b. $5 \cdot 1 = 5$ Multiplicative identity

c. $(a + 3)\left(\frac{1}{a + 3}\right) = 1, \quad a + 3 \neq 0$ Multiplicative inverse

d. $(5 \cdot 6) \cdot \frac{1}{6} = 5 \cdot \left(6 \cdot \frac{1}{6}\right)$ Associative for multiplication

e. $(a + 3) + [-(a + 3)] = 0$ Additive inverse

f. $5 + [x + (-x)] = 5$ Additive inverse and additive identity □

Problem Set 1.1

Classify each statement in Problems 1–18 as true or false.

A

1. 0 is a natural number.

2. $\sqrt{5}$ is a real number.

3. $-.333\ldots$ is a complex number.

4. $-\frac{1}{4}$ is a complex number.

5. 3.1416 is a rational number.

6. π is a rational number.

7. π is a real number.

8. π is a complex number.

9. .777. . . is a rational number.

10. .121221222. . . is an irrational number.

11. .606606660. . . is a complex number.

12. $\sqrt{-2}$ is a real number.

13. The set of integers is closed with respect to addition.

14. The set of rational numbers is closed with respect to multiplication.

15. The set of irrational numbers is closed with respect to multiplication.

16. The set of real numbers is closed with respect to multiplication.

17. The set $\{-1, 0, 1\}$ is closed with respect to addition.

18. The set $\{-1, 0, 1\}$ is closed with respect to multiplication.

19. Plot the numbers given in Problems 1, 3, 5, 7, and 9 on a number line.

20. Plot the numbers given in Problems 2, 4, 6, 8, and 10 on a number line.

Write each statement in Problems 21–40 without the absolute value symbol.

21. $|17|$
22. $-|23|$
23. $|-23|$
24. $-|-10|$
25. $|\pi - 2|$
26. $|2 - \pi|$
27. $-|\pi - 6|$
28. $|\pi - 12|$
29. $|\sqrt{20} - 4|$
30. $|\sqrt{20} - 5|$
31. $|\sqrt{30} - 5|$
32. $|\sqrt{30} - 6|$
33. $|2\pi - 5|$
34. $|2\pi - 7|$
35. $|x + 3|$ if $x \geqslant 3$
36. $|x + 3|$ if $x < -3$
37. $|y - 5|$ if $y < 5$
38. $|y - 5|$ if $y \geqslant 5$
39. $|5 - 2s|$ if $s > 10$
40. $|4 + 3t|$ if $t < -10$

Find the distance between the pairs of points whose coordinates are given in Problems 41–48.

41. $A(3)$ and $B(21)$

42. $C(-4)$ and $D(-12)$

43. $E(12)$ and $F(-8)$

44. $G(-5)$ and $H(9)$

45. $I(\pi)$ and $J(3)$

46. $K(\pi)$ and $L(4)$

47. $M(\sqrt{5})$ and $N(2)$

48. $P(\sqrt{5})$ and $Q(3)$

B *Identify the property illustrated by each statement in Problems 49–64.*

49. $5(6 + 2) = 5 \cdot 6 + 5 \cdot 2$

50. $4 + 8 = 8 + 4$

51. $15 + (85 + 23) = (85 + 23) + 15$

52. $35 \cdot 1 = 35$

53. $15 + (85 + 23) = (15 + 85) + 23$

54. $35 \cdot \frac{1}{35} = 1$

55. $(10 + a)b = 10b + ab$

56. $3\pi + 5\pi = (3 + 5)\pi$

57. $a + (10 + b) = (10 + b) + a$

58. $15 + [a + (-a)] = 15 + 0$

59. $a + (10 + b) = (a + 10) + b$

60. $(x + 9)\left(\dfrac{1}{x + 9}\right) = 1, \quad x + 9 \neq 0$

61. $(x + y) + [-(x + y)] = 0$

62. $xy + x(w + z) = xy + xw + xz$

63. $(a + b) \cdot \dfrac{1}{a + b} = 1, \quad a + b \neq 0$

64. $(x^2 + \pi) \cdot \dfrac{1}{x^2 + \pi} = \dfrac{1}{x^2 + \pi} \cdot (x^2 + \pi) = 1$

C **65.** B is 19 when $2 + 3 = 7$
B is 12 when $9 + 8 = 17$
B is 15 when $0 \cdot 1 = 1$
What is B if only one of the above statements is true?

66. A is 6 when $\{-1, 0, 1\}$ is closed for addition
A is 10 when $\{-1, 0, 1\}$ is closed for multiplication
A is 11 when $\{-1, 0, 1\}$ is closed for both addition and multiplication
What is A if only one of the above statements is true?

67. Which of the field properties are satisfied by the set **N** of natural numbers?

68. Which of the field properties are satisfied by the set **Z** of integers?

69. Which of the field properties are satisfied by the set **Q** of rational numbers?

70. Which of the field properties are satisfied by the set $H = \{1, -1, i, -i\}$, where i is a number such that $i \cdot i = -1$?

1.2 Exponents and Simplifying Expressions

An **algebraic expression** is a grouping of constants and variables obtained by applying a finite number of operations (such as addition, subtraction, multiplication, nonzero division, extraction of roots, or raising to integral or fractional powers). In the last section, we used **constants** (symbols with just one possible value, such as 2, 0, 63.4, or π) and **variables** (symbols such as x, y, s, or t representing an unspecified number selected from some set called the **domain**). In this book, if a domain for a variable is not specified, it will be assumed to be the set of all real numbers. Constants and variables may be used to generate a variety of algebraic expressions; for example,

$$(x + 5y)(y + 7) \qquad (a + b) + c \qquad a\left(\frac{1}{a}\right)$$

As you progress through this course, you will see additional types of algebraic expressions. In the last example, it is understood that $a = 0$ is excluded from the domain, because division by 0 is undefined. We will always assume that values that cause division by 0 are excluded from the domain.

DEFINITION OF EXPONENT

If b is any real number and n is any natural number, then

$$b^n = \underbrace{b \cdot b \cdot b \cdot \cdots \cdot b}_{n \text{ factors}}$$

Further, if $b \neq 0$, then

$$b^0 = 1 \qquad \text{and} \qquad b^{-n} = \frac{1}{b^n}$$

Here, b is called the **base,** n is the **exponent,** and b^n is the **power** of b.

In an algebraic expression of the form $A + B + C + \cdots$, where $A, B, C, \ldots$ are themselves algebraic expressions, A, B, and C are called the **terms** of the expression. For example, in

$6x^2 + 2x + 3$ the terms are $6x^2$, $2x$, and 3

whereas in

$6x^2 + (2x + 3)$ the terms are $6x^2$ and $(2x + 3)$

In this section we will discuss the simplification of algebraic expressions. Usually, we will simplify an expression before naming its terms.

If an algebraic expression is a finite sum of terms with **whole number exponents** on the variables, it is called a **polynomial.** For example,

$$6x, \quad 2x^2y + z^4, \quad \frac{2}{3}x + 4, \quad 0, \quad 86\tfrac{1}{2}, \quad \text{and} \quad \frac{x+3}{5}$$

are all polynomials, but

$$\frac{1}{x}, \quad 2 + \sqrt{x}, \quad \frac{6\sqrt{x+1}}{5}, \quad \text{and} \quad x^{2/3}$$

are not polynomials. These forms will be defined and discussed later in this book. The general form of a polynomial with a single variable will also be used later.

GENERAL FORM OF AN nTH-DEGREE POLYNOMIAL IN x

$$a_nx^n + a_{n-1}x^{n-1} + \cdots + a_2x^2 + a_1x + a_0 \qquad a_n \neq 0$$

Polynomials are frequently classified according to the number of terms. A **monomial** is a polynomial with one term, a **binomial** has two terms, and a **trinomial** has three terms.

The **degree of a monomial** is the number of variable factors of that monomial. For example, the degree of $5x^2$ is 2 because $5x^2 = 5xx$ (2 variable factors); 5^2x is degree 1 because there is 1 variable factor; and the degree of $5x^2y$ is 3. The expressions 7, 9^2, π, and π^3 are all of degree 0 since none of them contains a variable factor. A special case is the monomial 0, to which no degree is assigned. The **degree of a polynomial** is the same as the degree of its highest-degree term.

Some additional terminology concerning polynomials with which you should already be familiar is reviewed in Problems 1–20 of Problem Set 1.2. Be sure to work all these problems to refresh your memory, because we will assume you are familiar with that terminology.

From beginning algebra, recall the proper order of operations given in the box.

ORDER OF OPERATIONS AGREEMENT

When simplifying an algebraic expression:

1. Carry out all operations within parentheses.
2. Do exponents next.
3. Complete multiplications and divisions, working from left to right.
4. Finally, do additions and subtractions, working from left to right.

Example 1

$$\begin{aligned} 2 + 3 \cdot 4 &= 2 + 12 \\ &= 14 \end{aligned}$$ □

Example 2

$$\begin{aligned} 9 + 12 \div 3 + 4 \div 2 + 1 \cdot 2 &= 9 + 4 + 2 + 2 \\ &= 17 \end{aligned}$$ □

In algebra you rarely use the symbol ÷; instead, Example 2 would usually be written in the form

$$9 + \frac{12}{3} + \frac{4}{2} + 1 \cdot 2$$

The fractional bar used for division also functions as a grouping symbol:

$$6 + 4 \div 2 \quad \text{is written as} \quad 6 + \frac{4}{2}$$

whereas

$$(6 + 4) \div 2 \quad \text{is written as} \quad \frac{6 + 4}{2}$$

Example 3 In algebra, $6 + 4 \div 2 + (6 + 4) \div 2 - (6 + 4 \div 2)$ is written as

$$\begin{aligned} 6 + \frac{4}{2} + \frac{6 + 4}{2} - \left(6 + \frac{4}{2}\right) &= 6 + \frac{4}{2} + \frac{10}{2} - (6 + 2) \\ &= 6 + \frac{4}{2} + \frac{10}{2} - (8) \\ &= 6 + 2 + 5 - 8 \\ &= 13 - 8 \\ &= 5 \end{aligned}$$ □

Example 4 Find $[-(x+y)][-(s-t)]$, where $x=-2$, $y=-3$, $s=4$, and $t=7$.

$$\begin{aligned}\{-[(-2)+(-3)]\}[-(4-7)] &= \{-[-5]\}[-(-3)]\\ &= (5)(3)\\ &= 15\end{aligned}$$

□

We now turn to simplifying polynomials.

Addition and Subtraction of Polynomials

The distributive property shows you how to add polynomials by adding similar terms. Terms that are identical, except for their numerical coefficients, are called **similar terms.** That is, similar terms contain the same exponent (or exponents) on the variables. Since the variable parts are identical, the distributive property can be applied. For example:

$$\begin{aligned}5x+3x &= (5+3)x\\ &= 8x\end{aligned}$$

If you also freely use the commutative and associative properties, you can simplify more complicated expressions using the idea of similar terms.

$$(5x^2+2x+1)+(3x^3-4x^2+3x-2) = \underbrace{3x^3}_{\text{Third-degree term}} + \overbrace{\underbrace{5x^2+(-4)x^2}_{\text{Second-degree terms}}}^{\text{Similar terms}} + \overbrace{\underbrace{2x+3x}_{\text{First-degree terms}}}^{\text{Similar terms}} + \overbrace{\underbrace{1+(-2)}_{\text{Zero-degree terms}}}^{\text{Similar terms}}$$

$$= 3x^3+x^2+5x-1$$

Example 5

$$\begin{aligned}(4x-5)+(5x^2+2x+1) &= 5x^2+(4x+2x)+(-5+1)\\ &= 5x^2+6x-4\end{aligned}$$

□

Example 6

$$\begin{aligned}(4x-5)-(5x^2+2x+1) &= 4x-5-5x^2-2x-1\\ &= -5x^2+2x-6\end{aligned}$$

□

Compare Examples 5 and 6. In Example 5, you add similar terms. In Example 6, notice that when you subtract a polynomial you subtract *each* term. Another way of considering subtraction is to use the definition of subtraction,

$$\begin{aligned}a-b &= a+(-b)\\ &= a+(-1)b\end{aligned}$$

and then distribute the -1 as shown in Example 7.

Example 7

$$\begin{aligned}(5x^2 + 2x + 1) - (3x^3 - 4x^2 + 3x - 2) &= 5x^2 + 2x + 1 + (-1)(3x^3 - 4x^2 + 3x - 2)\\ &= 5x^2 + 2x + 1 + (-3)x^3 + 4x^2 + (-3)x + 2\\ &= -3x^3 + 9x^2 - x + 3\end{aligned}$$ □

Example 8

$$\begin{aligned}4(4x - 5) - 2(5x^2 + 2x + 1) &= 16x - 20 - 10x^2 - 4x - 2\\ &= -10x^2 + 12x - 22\end{aligned}$$ □

Multiplication of Polynomials

To understand the multiplication of polynomials, it is necessary to understand multiplication of monomials, as illustrated by Examples 9–11.

Example 9

$$\begin{aligned}x^2(x^3) &= xx(xxx)\\ &= x^5\end{aligned}$$ □

Example 10

$$\begin{aligned}(x^2)^3 &= (x^2)(x^2)(x^2)\\ &= (xx)(xx)(xx)\\ &= x^6\end{aligned}$$ □

Example 11

$$\begin{aligned}(x^2y^3)^4 &= (x^2y^3)(x^2y^3)(x^2y^3)(x^2y^3)\\ &= (x^2x^2x^2x^2)(y^3y^3y^3y^3)\\ &= x^8y^{12}\end{aligned}$$ □

These examples illustrate three laws of exponents that are used to simplify algebraic expressions.

LAWS OF EXPONENTS

Let a and b be any real numbers, let m and n be any integers, and assume that each expression is defined.

FIRST LAW	$b^m \cdot b^n = b^{m+n}$
SECOND LAW	$(b^n)^m = b^{mn}$
THIRD LAW	$(ab)^m = a^m b^m$

These laws of exponents are used with the distributive property when multiplying polynomials. Consider Examples 12–15.

Example 12

$$\begin{aligned}(4x-5)(5x^2+2x+1) &= (4x-5)5x^2+(4x-5)2x+(4x-5)\cdot 1\\ &= 20x^3-25x^2+8x^2-10x+4x-5\\ &= 20x^3-17x^2-6x-5\end{aligned}$$ □

Example 13

$$\begin{aligned}(2x-3)(x+4) &= (2x-3)x+(2x-3)\cdot 4\\ &= 2x^2-3x+8x-12\\ &= 2x^2+5x-12\end{aligned}$$ □

Example 14

$$\begin{aligned}(2x-3)(x^2-x-2) &= (2x-3)x^2+(2x-3)(-x)+(2x-3)(-2)\\ &= 2x^3-3x^2-2x^2+3x-4x+6\\ &= 2x^3-5x^2-x+6\end{aligned}$$ □

Example 15

$$\begin{aligned}(2x-3)(x+3)(3x-4) &= [(2x-3)x+(2x-3)\cdot 3](3x-4)\\ &= [2x^2-3x+6x-9](3x-4)\\ &= (2x^2+3x-9)(3x-4)\\ &= (2x^2+3x-9)3x+(2x^2+3x-9)(-4)\\ &= 6x^3+9x^2-27x-8x^2-12x+36\\ &= 6x^3+x^2-39x+36\end{aligned}$$ □

As you might have noticed from the preceding examples, it is frequently necessary to find products of binomials, and for this reason a special pattern, called **FOIL,** is observed. Four pairs of terms are multiplied:

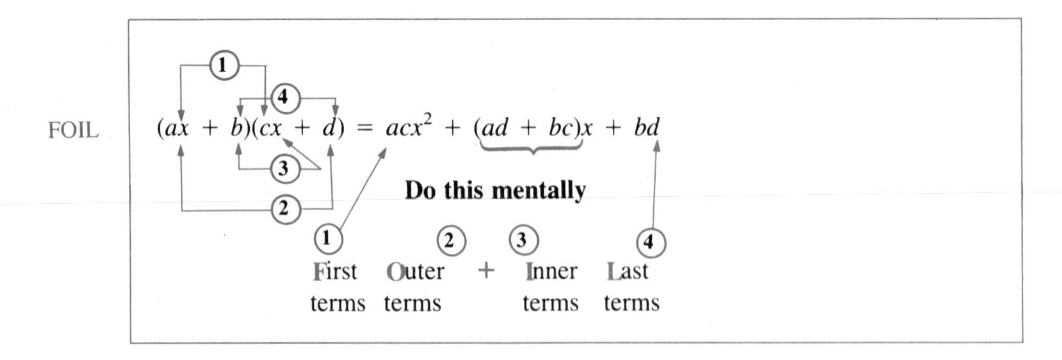

The FOIL method is used to multiply mentally in Examples 16–18.

Example 16 $(2x - 3)(x + 3) = 2x^2 + 3x - 9$

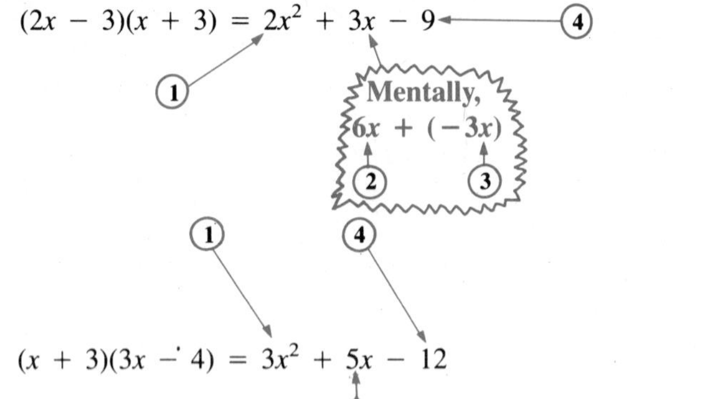

□

Example 17 $(x + 3)(3x - 4) = 3x^2 + 5x - 12$

② + ③

□

Example 18 $(5x - 2)(3x + 4) = 15x^2 + 14x - 8$ □

Some other special binomial products are given in the box:

SPECIAL PRODUCTS

DIFFERENCE OF SQUARES	$(x + y)(x - y) = x^2 - y^2$
PERFECT SQUARES	$(x + y)^2 = x^2 + 2xy + y^2$
	$(x - y)^2 = x^2 - 2xy + y^2$
PERFECT CUBES	$(x + y)^3 = x^3 + 3x^2y + 3xy^2 + y^3$
	$(x - y)^3 = x^3 - 3x^2y + 3xy^2 - y^3$

Example 19 $(6xy^2 + 3x^2y)(6xy^2 - 3x^2y) = 36x^2y^4 - 9x^4y^2$ Difference of squares □

Example 20 $(2a + 1)^2 = 4a^2 + 4a + 1$ Perfect square □

Calculator Comment If you plan to use a calculator in algebra, it is necessary that you know the type of logic used by your calculator. Three types are commonly available. The first is called **RPN** and is characterized by a [SAVE] or [ENTER] button. We will not illustrate the use of RPN logic in this text, so if you own an RPN calculator, you will need to consult your owner's manual. The other two types of logic are **arithmetic** and **algebraic,** and the difference is that algebraic calculators recognize the order of operations agreement, whereas arithmetic calculators do not. As a test problem for your calculator, use Example 1 and press:

[2] [+] [3] [×] [4] [=]

If the display shows *20.*, then your calculator worked from left to right, $(2 + 3) \cdot 4 = 20$, and you have an arithmetic logic calculator. If the display shows *14.*, then your calculator performed multiplication before addition (according to the order of operations agreement), $2 + (3 \cdot 4) = 14$, and you have an algebraic logic calculator. If you own an arithmetic logic calculator, you must remember to insert parentheses (available on many calculators) or remember to input the multiplication part first. To work Example 1 on an arithmetic calculator, you would press:

[3] [×] [4] [+] [2] [=] or [2] [+] [(] [3] [×] [4] [)] [=]

The calculator sequences shown in this book will assume an algebraic logic calculator with a 12 digit display.

Problem Set 1.2

A *In your previous algebra courses, you were introduced to a great deal of terminology involving polynomials. Problems 1–20 are designed to help you review these terms.*

In Problems 1–10 match the term at the left with its definition at the right.

1. Term
2. Degree of a term
3. Polynomial
4. Degree of a polynomial
5. Coefficient
6. Numerical coefficient
7. Monomial
8. Binomial
9. Trinomial
10. Real polynomial

A. A finite sum of terms with whole number exponents on the variables
B. B is an algebraic expression of the form $A + B + C + \cdots$
C. A polynomial with one term
D. A polynomial whose coefficients and variables are restricted to real numbers
E. A polynomial with three terms
F. The exponent of the variable or the sum of the exponents of the variables if more than one variable
G. A polynomial with two terms
H. The product of all the factors of a term except a certain one(s)
I. The numerical factor of a term
J. An expression with more than three terms
K. A polynomial with less than three terms
L. The degree of its highest-degree term

In Problems 11–20 match the description at the left with an example from the right.

11. A quadratic binomial in x and y
12. A quadratic trinomial over the rational numbers

M. $\frac{2}{3}$
N. $\frac{5}{2}x$
P. $3x + 2$

13. A monomial that is second degree in x

14. A linear monomial in x

15. A zero-degree polynomial

16. A linear polynomial in x over the integers

17. A second-degree monomial in x and y

18. A third-degree polynomial in x and y over the integers

19. A fifth-degree polynomial that contains a third-degree term

20. A fifth-degree polynomial over the integers

Q. $4x^3y^3$

R. $\sqrt{3}xy$

S. $2xy^2$

T. x^2y^2

U. $x^2 + y^2$

V. $\frac{2}{3}x^2 + 5x - \frac{1}{2}$

W. $x^2y^3 + 2xy^3 + y^4$

X. $\sqrt{5}x^2 - x + 1$

Y. $x^2y^3 + \frac{1}{2}x^2y^2 + xy^2$

Z. $\frac{1}{3}x^2 + y^3$

Simplify the expressions in Problems 21–37.

21. $3 + 2 \cdot 5$ **22.** $4 + 2 \cdot 3$ **23.** $5 + 3 \cdot 2$

24. $(-5)^2$ **25.** $(-6)^2$ **26.** $(-7)^2$

27. -5^2 **28.** -6^2 **29.** -7^2

30. $10 + 6(3 - 5)$ **31.** $8 - 5(2 - 7)$

32. $3 \cdot 5 - (-2)(-4)$ **33.** $6 \cdot 2 - (-5)(-4)$

34. $\dfrac{(-2) \cdot 6 + (-4)}{-4} + \dfrac{(-3)(-4)}{6}$ **35.** $(-2) + \dfrac{6}{-3} - 4 + \dfrac{-8}{-4}$

36. $(9 + 12) \div 3 + 4 \div (2 + 1 \cdot 2)$ **37.** $9 + (12 \div 3 + 4) \div 2 + 1 \cdot 2$

Let $s = -3$, $t = 2$, $u = -1$, and $v = -4$. Evaluate the expressions in Problems 38–53 and simplify.

38. s^2 **39.** $-s^2$ **40.** t^2

41. $-t^2$ **42.** v^2 **43.** $-v^2$

44. $s + t - u$ **45.** $s - (t - v)$ **46.** $(st)^2 + st^2 + u^2v$

47. $\dfrac{s - v^2}{u}$ **48.** $\dfrac{s - u}{t}$ **49.** $(s + t) - (-s + v)$

50. $5s - (4u + 3v)$ **51.** $s^2 - u^3 + v^2 - s^3$

52. $st - (t^4 - 3s)$ **53.** $s^2 - v^2(t^2 + u^2)$

In Problems 54–67 multiply mentally.

54. a. $(x + 3)(x + 2)$ **b.** $(y + 1)(y + 5)$

55. a. $(z - 2)(z + 6)$ **b.** $(s + 5)(s - 4)$

56. a. $(x + 1)(x - 2)$ **b.** $(y - 3)(y + 2)$

57. a. $(a - 5)(a - 3)$ **b.** $(b + 3)(b - 4)$

58. a. $(c + 1)(c - 7)$ **b.** $(z - 3)(z + 5)$

59. **a.** $(2x + 1)(x - 1)$ **b.** $(2x - 3)(x - 1)$

60. **a.** $(x + 1)(3x + 1)$ **b.** $(x + 1)(3x + 2)$

61. **a.** $(2a + 3)(3a - 2)$ **b.** $(2a + 3)(3a + 2)$

62. **a.** $(x + y)(x + y)$ **b.** $(x - y)(x - y)$

63. **a.** $(x + y)(x - y)$ **b.** $(a + b)(a - b)$

64. **a.** $(5x - 4)(5x + 4)$ **b.** $(3y - 2)(3y + 2)$

65. **a.** $(a + 2)^2$ **b.** $(b - 2)^2$

66. **a.** $(x + 4)^2$ **b.** $(y - 3)^2$

67. **a.** $(s + t)^2$ **b.** $(u - v)^2$

B *Simplify the expressions in Problems 68–83.*

68. **a.** $(x + 1)^3$ **b.** $(x - 1)^3$

69. $(3x^2 - 5x + 2) + (x^3 - 4x^2 + x - 4)$

70. $(5x + 1) + (x^3 - 4x^2 + x - 4)$

71. $(3x^2 - 5x + 2) - (5x + 1)$

72. $(x^3 - 4x^2 + x - 4) - (3x^2 - 5x + 2)$

73. $(5x + 1)(3x^2 - 5x + 2)$

74. $(5x + 1)(3x^2 - 5x + 2) - (x^3 - 4x^2 + x - 4)$

75. $3(3x^2 - 5x + 2) - 4(x^3 - 4x^2 + x - 4)$

76. $(3x - 1)(x^2 + 3x - 2)$

77. $(5x + 1)(x^3 - 2x^2 + 3x - 5)$

78. $(x + 1)(x - 3)(2x + 1)$

79. $(2x - 1)(x + 3)(3x + 1)$

80. $(x - 2)^2(x + 1)$

81. $(x - 2)(x + 1)^2$

82. $(2x - 1)(3x^4 - 2x^3 + 3x^2 - 5x + 12)$

83. $(x - 3)(2x^3 - 5x^2 + 4x - 7)$

C *Simplify the expressions in Problems 84–91.*

84. $(a + b + c)^2$

85. $(2x + 3y - z)^2$

86. $(2a + b + 3c)^3$

87. $(x - y - 2z)^3$

88. $(t^2 - 2tu + u^2)(t^2 + 3tu - u^2)$

89. $(3w^3 + 2w^2x + 5wx^2 + x^3)^2$

90. $(5u^3 - 4u^2v + 3uv^2 - 3v^3)^2$

91. $(3x^2 - 4xy + y^2)^2(2x^3 - x^2y - 2xy^2 + 3y^3)$

1.3 Factoring

A **factor** of a given algebraic expression is an algebraic expression (perhaps of some specified type) that divides evenly (that is, without a remainder) into the given expression. The process of **factoring** involves resolving the given expression into factors. The procedure we will use is to carry out a series of tests for different types of factors, as summarized in Table 1.2. These types are listed in the order in which you should check them when factoring an expression.

Example 1 $2x + 3xy = x(2 + 3y)$ Common factor x (Type 1) □

Example 2 $4 - 20z = 4(1 - 5z)$ Common factor 4 (Type 1) □

Example 3
$$9s^2 - 16t^2 = (3s)^2 - (4t)^2$$
$$= (3s - 4t)(3s + 4t)$$
No common factors (Type 1), so try to write it as a difference of squares (Type 2). □

Example 4
$$8p^3 + 1 = (2p)^3 + (1)^3$$
$$= (2p + 1)(4p^2 - 2p + 1)$$
No common factors (Type 1), not a difference of squares (Type 2), not a difference of cubes (Type 3), so try sum of cubes (Type 4). □

Example 5 $a^2 - 2ab + b^2 = (a - b)^2$ Perfect square (Type 5) □

Table 1.2 Factoring Types

Type	Form	Comments
1. Common factors	$ax + ay + az = a(x + y + z)$	This simply uses the distributive property. It can be applied with any number of terms.
2. Difference of squares	$x^2 - y^2 = (x - y)(x + y)$	Note that the *sum* of two squares cannot be factored in the set of real numbers.
3. Difference of cubes	$x^3 - y^3 = (x - y)(x^2 + xy + y^2)$	This type is similar to the difference of squares and can be proved by multiplying the factors.
4. Sum of cubes	$x^3 + y^3 = (x + y)(x^2 - xy + y^2)$	Unlike the sum of squares, the sum of cubes can be factored.
5. Perfect squares	$x^2 + 2xy + y^2 = (x + y)^2$ $x^2 - 2xy + y^2 = (x - y)^2$	The middle term is twice the product of x and y.
6. FOIL	$acx^2 + (ad + bc)xy + bdy^2$ $= (ax + by)(cx + dy)$	This trial-and-error procedure is used with trinomials. It should be used after Types 1–5 have been checked. See examples in this section.
7. Grouping	See examples that follow.	After Types 1–6 have been checked, you can factor some expressions by proper grouping.
8. Irreducible (not able to be factored)	Examples arise in every factoring situation. Expressions such as $x^2 + 4$ and $x^2 + y^2$ are irreducible.	When factoring, you are not through until all the factors are **irreducible** over some set of numbers (such as the integers).

Example 6 $2x(2a - b) + y(2a - b) = (2x + y)(2a - b)$ Common factor $(2a - b)$ □

Example 7 $3x^2 - 75 = 3(x^2 - 25)$ Common factor 3
$= 3(x - 5)(x + 5)$ Difference of squares □

Do not forget to determine whether the factors themselves can be factored further.

Example 8 $(x + 3y)^2 - 1 = [(x + 3y) - 1][(x + 3y) + 1]$ Difference of squares
$= (x + 3y - 1)(x + 3y + 1)$ Simplify □

Example 9 $x^2 - 8x + 15 = (x - 5)(x - 3)$ Try Types 1–5 (they do not apply). This is a trinomial, so try FOIL (a trial-and-error procedure) to factor it (if possible) into two binomials. □

Example 10 $6m^2 + m - 12 = (2m + 3)(3m - 4)$ FOIL □

Example 11 $6m^2 - 9m - 15 = 3(2m^2 - 3m - 5)$ Common factor 3
$= 3(2m - 5)(m + 1)$ FOIL □

Example 12 $4x^4 - 13x^2y^2 + 9y^4 = (x^2 - y^2)(4x^2 - 9y^2)$ FOIL
$= (x - y)(x + y)(2x - 3y)(2x + 3y)$ Difference of squares (twice) □

Example 13 $x^2 - 3$ Try Types 1–7 (they do not apply). This is irreducible over the set of integers. □

Factoring over the set of integers rules out factoring the expression in Example 13 as

$$x^2 - 3 = (x - \sqrt{3})(x + \sqrt{3})$$

since the factors do not have integer coefficients. In this book, an expression is called **completely factored** if all fractions are eliminated by common factoring and if no further factoring is possible *over the set of integers*. This means that all the numerical coefficients are integers. If the original problem involves fractions, write all the coefficients with a common denominator and then factor out the fractional part as a common factor, as shown in Example 14.

Example 14 $\frac{9}{4}a^2 - b^2 = \frac{9}{4}a^2 - \frac{4}{4}b^2$ Write $1 = \frac{4}{4}$
$= \frac{1}{4}(9a^2 - 4b^2)$ Common factor of $\frac{1}{4}$
$= \frac{1}{4}(3a - 2b)(3a + 2b)$ Difference of squares □

Example 15

$$\begin{aligned} &x^6 - 1 \\ &= (x^3)^2 - (1^3)^2 && \text{Difference of squares} \\ &= (x^3 - 1)(x^3 + 1) && \text{[\textit{Note:} You might also try difference of cubes, but the result would not lead you to a completely factored form.]} \\ &= (x - 1)(x^2 + x + 1)(x + 1)(x^2 - x + 1) && \text{Difference of cubes and sum of cubes} \end{aligned}$$

□

Example 16

$$\begin{aligned} &9x^3 + 18x^2 - x - 2 \\ &= (9x^3 + 18x^2) - (x + 2) && \text{Try Types 1–6 (they do not apply). Try to group the terms; some groupings may lead to a factorable form, but others may not.} \\ &= 9x^2(x + 2) - (x + 2) && \text{Common factor} \\ &= (9x^2 - 1)(x + 2) && \text{Common factor} \\ &= (3x - 1)(3x + 1)(x + 2) && \text{Difference of squares} \end{aligned}$$

□

Example 17

$$\begin{aligned} x^{3a} - x^{3b} &= (x^a)^3 - (x^b)^3 \\ &= (x^a - x^b)[(x^a)^2 + x^a x^b + (x^b)^2] \\ &= (x^a - x^b)(x^{2a} + x^{a+b} + x^{2b}) \end{aligned}$$

□

Example 18

$$\begin{aligned} x^{3+a} + x^{3+b} &= x^3 x^a + x^3 x^b \\ &= x^3(x^a + x^b) \end{aligned}$$

□

Problem Set 1.3

If possible, completely factor the expressions in Problems 1–75 over the set of integers.

A

1. $20xy - 12x$ **2.** $8st - 6t$ **3.** $6m - 2$

4. $5n + 5$ **5.** $xy + xz^2 + 3x$ **6.** $me + mi + my$

7. $a^2 - b^2$ **8.** $a^2 + b^2$ **9.** $a^3 - b^3$

10. $a^3 + b^3$ **11.** $s^2 + 2st + t^2$ **12.** $m^2 - 2mn + n^2$

13. $u^2 + 2uv - v^2$ **14.** $(a + b)(a^2 - ab + b^2) + (a + b)3ab$

15. $x^2y + xy^2$ **16.** $mn^2 + m^2n$

17. $x^2 - 2x - 35$ **18.** $2x^2 + 7x - 15$

19. $3x^2 - 5x - 2$ **20.** $6y^2 - 7y + 2$

21. $2x^2 - 7x - 15$ **22.** $3x^2 - x - 4$

23. $4x^2 - 21x - 18$ **24.** $3x^2 - 14x - 24$

25. $(a + b)x + (a + b)y$ **26.** $(4x - 1)x + (4x - 1) \cdot 3$

27. $x^2(3x + 1) - 4(3x + 1)$ **28.** $9(2x - 1) - x^2(2x - 1)$

29. $2s^2 - 10s - 48$ **30.** $12m^2 - 7m - 12$

31. $8a^2b + 10ab - 3b$ **32.** $9x^2y + 15xy - 14y$

33. $4y^3 + y^2 - 21y$ **34.** $12p^4 + 11p^3 - 15p^2$

B

35. $(x - y)^2 - 1$

36. $(2x + 3)^2 - 1$

37. $(5a - 2)^2 - 9$

38. $(3p - 2)^2 - 16$

39. $\frac{x^6}{4} - 16$

40. $\frac{x^2}{9} - \frac{y^2}{4}$

41. $(m - 2)^2 - (m + 1)^2$

42. $(a + b)^2 - (x + y)^2$

43. $x^{2+a} + x^{2+b}$

44. $x^{2n} - y^{2n}$

45. $x^{3n} - y^{3n}$

46. $x^{3n} + y^{3n}$

47. $x^{2n} - 2x^n y^n + y^{2n}$

48. $(x - 2)^3 - (x + 2)^3$

49. $(x - 2)^2 - (x - 2) - 6$

50. $(x + 3)^2 - (x + 3) - 6$

51. $4x^4 - 17x^2 + 4$

52. $4x^4 - 45x^2 + 81$

53. $z^6 - 64$

54. $x^8 - y^8$

55. $z^5 - 8z^2 - 4z^3 + 32$

56. $x^5 + 8x^2 - x^3 - 8$

57. $\left(x^3 + \frac{1}{27}\right)\left(x^3 - \frac{1}{8}\right)$

58. $\left(x^3 + \frac{1}{8}\right)\left(x^2 - \frac{1}{4}\right)$

59. $x^6 - 6x^3 - 16$

60. $x^6 + 9x^3 + 8$

C

61. $x^2 - 2xy + y^2 - a^2 - 2ab - b^2$

62. $x^2 + 2xy + y^2 - a^2 - 2ab - b^2$

63. $x^2 + y^2 - a^2 - b^2 - 2xy + 2ab$

64. $x^3 + 3x^2y + 3xy^2 + y^3 + a^3 + 3a^2b + 3ab^2 + b^3$

65. $(x + y + 2z)^2 - (x - y + 2z)^2$

66. $(x^2 - 3x - 6)^2 - 4$

67. $2(x + y)^2 - 5(x + y)(a + b) - 3(a + b)^2$

68. $2(s + t)^2 + 3(s + t)(s + 2t) - 2(s + 2t)^2$

69. $2x^2y + 2xz - xy - z$

70. $3x^2y - 3xz^2 + xy - z^2$

71. $\frac{4}{9}x^2 - (x + y)^2$

72. $\frac{4}{25}x^2 - (x + 2)^2$

73. $(2x - 3)(3)(1 - x)^2(-1) - (1 - x)^3(-3)$

74. $4(x + 5)^3(x^2 - 2)^3 + (x + 5)^4(3)(x^2 - 2)^2(2x)$

75. $5(x - 2)^4(x^2 + 1)^3 + (x - 2)^5(3)(x^2 + 1)^2(2x)$

1.4 Integral Exponents and Rational Expressions

In Section 1.2 negative exponents were defined by

$$b^{-n} = \frac{1}{b^n}$$

where b is a nonzero real number. If b is a nonzero integer, then negative exponents can be used to represent rational numbers:

$$\frac{2}{3} = 2 \cdot 3^{-1} \qquad \frac{5}{7} = 5 \cdot 7^{-1} \qquad \frac{x}{y} = x \cdot y^{-1}$$

If a variable is used in a denominator (as in xy^{-1}), then the expression is not called a polynomial.

RATIONAL EXPRESSION

> A **rational expression** is an expression that can be written as a polynomial divided by a polynomial. Any values that cause division by zero are excluded from the domain.

The fundamental property used to simplify rational expressions involves factoring both the numerator and denominator and then eliminating common factors according to the property

$$\frac{PK}{QK} = \frac{P}{Q} \qquad Q, K \neq 0$$

Some rational expressions are simplified in Examples 1–5. Do not forget, all values for variables that cause division by zero are excluded from the domain.

Example 1

$$\frac{3xyz}{x^2 + 3x} = \frac{x(3yz)}{x(x+3)} = \frac{3yz}{x+3}$$

□

Example 2

$$\frac{x-2}{x^2-4} = \frac{1(x-2)}{(x+2)(x-2)} = \frac{1}{x+2}$$

□

Sometimes the factors that are eliminated (as shown in color in Examples 1 and 2) are marked off in pairs as shown in Example 3. The slashes should be viewed as replacing the factor K by the number 1 since

$$\frac{PK}{QK} = \frac{P\not{K}}{Q\not{K}} = \frac{P \cdot 1}{Q \cdot 1} = \frac{P}{Q} \cdot 1 = \frac{P}{Q}$$

Example 3

$$\frac{(x-5)(x+2)(x+1)}{(x+1)(x-2)(x-5)} = \frac{\cancel{(x-5)}(x+2)\cancel{(x+1)}}{\cancel{(x+1)}(x-2)\cancel{(x-5)}} = \frac{x+2}{x-2}$$

□

Example 4

$$\frac{x^3-1}{x^2+x+1} = \frac{(x-1)\cancel{(x^2+x+1)}}{\cancel{x^2+x+1}} = x - 1$$

□

Example 5

$$\frac{6x^2 + 2x - 20}{30x^2 - 68x + 30} = \frac{2(3x^2 + x - 10)}{2(15x^2 - 34x + 15)}$$ Common factor first

$$= \frac{2(3x - 5)(x + 2)}{2(3x - 5)(5x - 3)}$$ Complete the factoring; then reduce

$$= \frac{x + 2}{5x - 3}$$ □

The laws of exponents can be extended to include rational expressions and negative exponents.

Example 6

$$\left(\frac{x}{y}\right)^3 = \left(\frac{x}{y}\right)\left(\frac{x}{y}\right)\left(\frac{x}{y}\right)$$

$$= \frac{xxx}{yyy}$$

$$= \frac{x^3}{y^3}$$ □

Example 7

$$\frac{x^5}{x^3} = \frac{xxxxx}{xxx}$$

$$= xx$$

$$= x^2$$ □

Example 8

$$\frac{x^3}{x^5} = \frac{xxx}{xxxxx}$$

$$= \frac{1}{x^2}$$

$$= x^{-2}$$ □

LAWS OF EXPONENTS

Let a *and* b be any real numbers, let m and n be any integers, and assume that each expression is defined.

FIRST LAW $b^m \cdot b^n = b^{m+n}$

SECOND LAW $(b^n)^m = b^{mn}$

THIRD LAW $(ab)^m = a^m b^m$

FOURTH LAW $\left(\frac{a}{b}\right)^m = \frac{a^m}{b^m}$

FIFTH LAW $\frac{b^m}{b^n} = b^{m-n}$

The first three laws are repeated here for your easy reference. To simplify expressions involving negative exponents, use the definition and laws of exponents.

Simplify the expressions in Examples 9–11. Assume that all values for variables that cause division by zero are excluded.

Example 9

$$(x^2y^{-3})^{-2} = x^{-4}y^6 = \frac{y^6}{x^4}$$ □

Example 10

$$\left(\frac{24x^3y^{-1}}{16xz^3}\right)^{-1} = \left(\frac{2^3 3x^3y^{-1}}{2^4xz^3}\right)^{-1}$$
$$= (2^{3-4}3x^{3-1}y^{-1}z^{-3})^{-1}$$
$$= (2^{-1}3x^2y^{-1}z^{-3})^{-1}$$
$$= 2 \cdot 3^{-1}x^{-2}yz^3$$
$$= \frac{2yz^3}{3x^2}$$ □

Example 11

$$\frac{(-2xy^2z^{-1})^3}{(8x^{-2}y^{-2}z^3)^2} = \frac{-8x^3y^6z^{-3}}{64x^{-4}y^{-4}z^6}$$
$$= \frac{-8}{64}x^{3-(-4)}y^{6-(-4)}z^{-3-6}$$
$$= \frac{-1}{8}x^7y^{10}z^{-9} \quad \text{or} \quad \frac{-x^7y^{10}}{8z^9}$$ □

One of the more common uses of negative exponents is with a base 10 in the representation of very small numbers. For example,

$$.00000\ 00038\ 7 = 3.87 \times 10^{-9}$$

To multiply numbers by different powers of 10, you simply move the decimal point. This easy method is exploited in what is called **scientific notation.**

SCIENTIFIC NOTATION

> A number is in **scientific notation** if it is a power of 10 or if it is written as a number between 1 and 10 times a power of 10.

Example 12 $93{,}000{,}000 = 9.3 \times 10^7$ □

Example 13 $.00000\ 0845 = 8.45 \times 10^{-7}$ □

Example 14 The numbers 1 and 10 are in scientific notation (they are powers of 10). □

Notice that a times sign (×) is customarily used with a scientific notation. Calculations with large and small numbers are often done by writing the numbers in scientific notation and then using the laws of exponents.

Example 15

$$\frac{(6{,}800{,}000) \times (50{,}000)^3}{.00034} = \frac{(6.8 \times 10^6) \times (5.0 \times 10^4)^3}{3.4 \times 10^{-4}}$$

$$= \frac{6.8 \times 10^6 \times 5.0^3 \times 10^{12}}{3.4 \times 10^{-4}}$$

$$= \frac{6.8 \times 5.0^3}{3.4} \times \frac{10^6 \times 10^{12}}{10^{-4}}$$

$$= \left(\frac{6.8}{3.4} \times 5.0^3\right) \times \frac{10^{6+12}}{10^{-4}}$$

$$= (2 \times 125) \times 10^{6+12-(-4)}$$

$$= 250 \times 10^{22}$$

$$= (2.5 \times 10^2) \times 10^{22}$$

$$= 2.5 \times 10^{24}$$

□

Calculator Comment Many calculators can work directly in scientific notation by separating the number between 1 and 10 and the exponent. For example, 3.4×10^{-4} can be entered as follows:

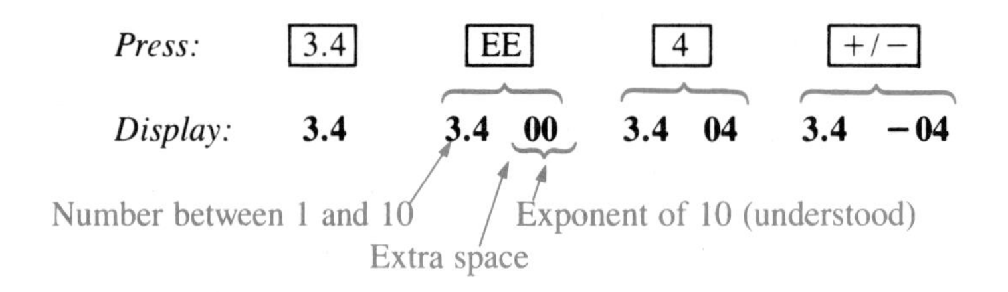

Your calculator may label the scientific notation key by SCI, EXP, EX, or EEx instead of EE. Check your owner's manual.

Operations with Rational Expressions

The procedures for operating on rational expressions are identical to the procedures you learned in arithmetic for operations with fractions. However, with rational expressions, the numerators and denominators are any polynomials (division by zero is excluded) rather than constants. The procedures for handling rational expressions are summarized in the box. Notice that the property for reducing rational expressions is repeated here for the sake of completeness.

PROPERTIES OF RATIONAL EXPRESSIONS

Let P, Q, R, S, and K be any polynomials such that all values of the variable that cause division by zero are excluded from the domain.

EQUALITY $\dfrac{P}{Q} = \dfrac{R}{S}$ if and only if $PS = QR$

FUNDAMENTAL PROPERTY $\dfrac{PK}{QK} = \dfrac{P}{Q}$

ADDITION $\dfrac{P}{Q} + \dfrac{R}{S} = \dfrac{PS + QR}{QS}$

SUBTRACTION $\dfrac{P}{Q} - \dfrac{R}{S} = \dfrac{PS - QR}{QS}$

MULTIPLICATION $\dfrac{P}{Q} \cdot \dfrac{R}{S} = \dfrac{PR}{QS}$

DIVISION $\dfrac{P}{Q} \div \dfrac{R}{S} = \dfrac{PS}{QR}$

Some operations on rational expressions are performed in Examples 16–20. All values of variables that could cause division by zero are excluded.

Example 16

$$\frac{13}{x-y} + \frac{2}{x-y} = \frac{15}{x-y}$$

Common denominator □

Example 17

$$\begin{aligned}\frac{13}{x-y} - \frac{2}{y-x} &= \frac{13}{x-y} + \frac{-2}{y-x} \\ &= \frac{13}{x-y} + \frac{-2}{y-x} \cdot \frac{-1}{-1} \\ &= \frac{13}{x-y} + \frac{2}{x-y} \\ &= \frac{15}{x-y}\end{aligned}$$

It is helpful to write subtractions as additions.

Multiply by 1 (written as $\frac{-1}{-1}$) in order to obtain a common denominator. □

Example 18

$$\begin{aligned}x^{-1}(x+1) - x^{-2}(x+2) + x^{-3}(x+3) &= \frac{x+1}{x} - \frac{x+2}{x^2} + \frac{x+3}{x^3} \\ &= \frac{x+1}{x} \cdot \frac{x^2}{x^2} + \frac{-(x+2)}{x^2} \cdot \frac{x}{x} + \frac{x+3}{x^3} \\ &= \frac{x^2(x+1) - x(x+2) + (x+3)}{x^3}\end{aligned}$$

$$= \frac{x^3 + x^2 - x^2 - 2x + x + 3}{x^3}$$

$$= \frac{x^3 - x + 3}{x^3} \quad \text{or} \quad x^{-3}(x^3 - x + 3)$$

□

Notice in Example 18 that an alternate form of answer is given. It is often customary to give the answer in the same form as the original question, so a negative exponent can be used. This example also illustrates the close tie between rational expressions and negative exponents.

If the denominators of two rational expressions to be added or subtracted have no common factors, it is often convenient just to use the definition of addition, as illustrated in Example 19.

Example 19

$$\frac{a + b}{a - b} + \frac{a - 2b}{2a + b} = \frac{(a + b)(2a + b) + (a - b)(a - 2b)}{(a - b)(2a + b)}$$

$$= \frac{2a^2 + 3ab + b^2 + a^2 - 3ab + 2b^2}{(a - b)(2a + b)}$$

$$= \frac{3a^2 + 3b^2}{(a - b)(2a + b)}$$

$$= \frac{3(a^2 + b^2)}{(a - b)(2a + b)}$$

□

Example 20

$$\left[\frac{x^2 + 5x + 6}{2x^2 - x - 1} \cdot \frac{2x^2 - 9x - 5}{x^2 + 7x + 12}\right] \div \frac{2x^2 - 13x + 15}{x^2 + 3x - 4}$$

$$= \left[\frac{(x + 2)(x + 3)}{(x - 1)(2x + 1)} \cdot \frac{(2x + 1)(x - 5)}{(x + 3)(x + 4)}\right] \div \frac{(x - 5)(2x - 3)}{(x - 1)(x + 4)}$$

$$= \frac{(x + 2)\cancel{(x + 3)}\cancel{(2x + 1)}\cancel{(x - 5)}\cancel{(x - 1)}\cancel{(x + 4)}}{\cancel{(x - 1)}\cancel{(2x + 1)}\cancel{(x + 3)}\cancel{(x + 4)}\cancel{(x - 5)}(2x - 3)}$$

$$= \frac{x + 2}{2x - 3}$$

□

The key to working problems involving multiplication and division of rational expressions is the proper factoring of each polynomial. However, it is sometimes difficult to factor either the numerator or the denominator. In such a case, you can use long division of polynomials, as illustrated by Example 21.

Example 21 Simplify: $\dfrac{9x^3 + 19x^2 - 4}{x + 2}$

Solution

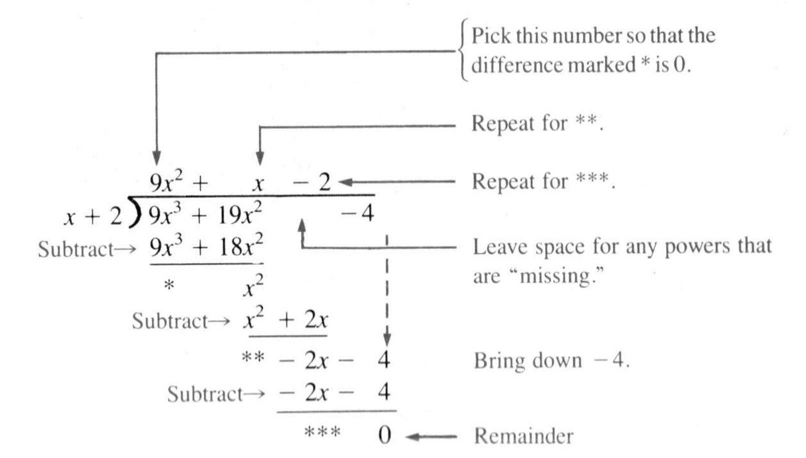

Thus,

$$\frac{9x^3 + 19x^2 - 4}{x + 2} = 9x^2 + x - 2$$ □

If the remainder from long division is 0 (as in Example 21), then the divisor is a factor of the numerator. Thus, from the long division in Example 21, we see that

$$(x + 2)(9x^2 + x - 2) = 9x^3 + 19x^2 - 4$$

so

$$\frac{9x^3 + 19x^2 - 4}{x + 2} = \frac{\cancel{(x + 2)}(9x^2 + x - 2)}{\cancel{x + 2}}$$
$$= 9x^2 + x - 2$$

Example 22 Simplify: $\dfrac{2x^4 - 13x^3 - x^2 + 5x + 1}{8x^2 - 2x - 3}$

Solution You could begin by long division with the given polynomials, but in this case it is easy to factor the denominator:

$$8x^2 - 2x - 3 = (2x + 1)(4x - 3)$$

The given rational expression can be simplified if division by $2x + 1$, $4x - 3$, or both, gives a zero remainder. If not, then it is simplified as given.

$$
\begin{array}{r}
x^3 - 7x^2 + 3x + 1 \\
2x + 1 \overline{\smash{)}\,2x^4 - 13x^3 - x^2 + 5x + 1} \\
\underline{2x^4 + x^3} \qquad\qquad\qquad\quad \\
-14x^3 - x^2 \qquad\qquad \\
\underline{-14x^3 - 7x^2} \qquad\qquad \\
6x^2 + 5x \qquad \\
\underline{6x^2 + 3x} \qquad \\
2x + 1 \\
\underline{2x + 1} \\
0
\end{array}
$$

Since a zero remainder was obtained, you can write

$$
\frac{2x^4 - 13x^3 - x^2 + 5x + 1}{8x^2 - 2x - 3} = \frac{(2x + 1)(x^3 - 7x^2 + 3x + 1)}{(2x + 1)(4x - 3)}
$$

$$
= \frac{x^3 - 7x^2 + 3x + 1}{4x - 3}
$$

You should also make certain that $x^3 - 7x^2 + 3x + 1$ does not have $4x - 3$ as a factor.

$$
\begin{array}{r}
\frac{1}{4}x^2 - \frac{25}{16}x - \frac{27}{64} \\
4x - 3 \overline{\smash{)}\, x^3 - 7x^2 + 3x + 1} \\
\underline{x^3 - \frac{3}{4}x^2} \qquad\qquad\quad \\
{}^{*} - \frac{25}{4}x^2 + \frac{48}{16}x \qquad \\
\underline{-\frac{25}{4}x^2 + \frac{75}{16}x} \qquad \\
-\frac{27}{16}x + \frac{64}{64} \\
\underline{-\frac{27}{16}x + \frac{81}{64}} \\
-\frac{17}{64}
\end{array}
$$

Fractional coefficients may be necessary to yield the proper coefficients for the subtraction step (* must be zero).

Since the remainder is not 0, the rational expression

$$
\frac{x^3 - 7x^2 + 3x + 1}{4x - 3}
$$

is simplified. □

Problem Set 1.4

A *In Problems 1–6 find the numerical value.*

1.	**a.** $(-2)^2$	**b.** -2^2	**c.** 2^{-2}	**d.** -2^{-2}	**e.** $(-2)^{-2}$
2.	**a.** $(-6)^2$	**b.** 6^{-2}	**c.** -6^2	**d.** -6^{-2}	**e.** $(-6)^{-2}$

3. **a.** $(-3)^2$ **b.** -3^2 **c.** 3^{-2} **d.** -3^{-2} **e.** $(-3)^{-2}$

4. **a.** -5^2 **b.** $(-5)^2$ **c.** 5^{-2} **d** $(-5)^{-2}$ **e.** -5^{-2}

5. **a.** $(5 \cdot 10^3)(20 \cdot 10^{-4})$ **b.** $\dfrac{12 \cdot 10^5}{2 \cdot 10^{-3}}$

6. **a.** $(5 \cdot 10^3)^{-1}(20 \cdot 10^{-4})$ **b.** $\dfrac{25 \cdot 10^{-4}}{5 \cdot 10^4}$

Write the answers to Problems 7–12 in scientific notation.

7. $\dfrac{(2 \times 10^3)(14 \times 10^5)}{(7 \times 10^{14})}$

8. $\dfrac{(8 \times 10^{-3})(2 \times 10^{-5})}{(4 \times 10^3)}$

9. $\dfrac{(1.32 \times 10^8)(2.8 \times 10^4)}{(7 \times 10^{-3})}$

10. $\dfrac{(3.14 \times 10^5)(2 \times 10^4)(8.61 \times 10^3)}{(1.722 \times 10^{-8})}$

11. $\dfrac{(6.14 \times 10^6)(4.9 \times 10^3)}{(2 \times 10^{-4})}$

12. $\dfrac{(5.2 \times 10^3)(3 \times 10^4)(4.75 \times 10^2)}{(5.2 \times 10^{-5})}$

Simplify the expressions in Problems 13–76. Values that cause division by zero are excluded. All rational expressions should be reduced, and negative exponents eliminated.

13. $(-2t^5)^2$ **14.** $(-3s^4)^3$ **15.** $(-2t^5)^{-2}$

16. $(-3s^4)^{-3}$ **17.** $(3xy^{-2}z)^4$ **18.** $(2x^{-1}y^2z)^3$

19. $(-2x^2y)^3\left(\dfrac{x}{8y^2}\right)^2$ **20.** $\dfrac{(8s^{-2}t)^2}{(6s^3t^{-1})^3}$ **21.** $\left(\dfrac{x^2y^{-3}}{z^2}\right)^{-1}$

22. $\left(\dfrac{x^3y^{-5}}{z^2}\right)^{-1}$ **23.** $\left(\dfrac{a^3b^4c^{-6}}{ab^{-2}c^2}\right)^{-2}$ **24.** $\left(\dfrac{10d^2e^{-1}}{25d^{-3}e^2}\right)^{-3}$

25. $\left(\dfrac{6m^{-3}n^2p}{18m^4n^{-8}p^3}\right)^{-5}$ **26.** $\left(\dfrac{4x^{-2}y^{-1}z}{12x^4y^{-3}z^{-2}}\right)^{-3}$ **27.** $2^{-1} + 3^{-1}$

28. $2^{-1} + 3^{-1} + 4^{-1}$ **29.** $x^{-1} + y^{-1}$ **30.** $2x^{-1} + 3y^{-1}$

31. $3x^{-1} - 2y^{-1}$ **32.** $(2x)^{-1} + (3y)^{-1}$ **33.** $(3x)^{-1} - (2y)^{-1}$

B **34.** $2(x + y)^{-1} + 3$ **35.** $3(x + y)^{-1} - 2$ **36.** $\dfrac{x^2 - y^2}{2x + 2y}$

37. $\dfrac{x^2 - y^2}{3x - 3y}$ **38.** $\dfrac{3x^2 - 4x - 4}{x^2 - 4}$ **39.** $\dfrac{y^2 + 3y - 18}{y^2 - 9}$

40. $\dfrac{3}{x + y} + \dfrac{5}{2x + 2y}$ **41.** $\dfrac{4}{x - y} - \dfrac{3}{2x - 2y}$

42. $[7^3 + 2^6(3^3 + 4^4)]^0$ **43.** $[9^3 + 3^5(5^3 + 8^3)]^0$

44. $(x^2 - 36) \cdot \dfrac{3x + 1}{x + 6}$ **45.** $(y^2 - 9) \div \dfrac{y + 3}{y - 3}$

46. $\dfrac{x + 3}{x} + \dfrac{3 - x}{x^2}$ **47.** $\dfrac{2}{x - y} + \dfrac{5}{y - x}$

48. $\dfrac{a}{a-1} + \dfrac{a-3}{1-a}$

49. $x^{-1}(x+1) + x^{-2}(2-x)$

50. $x^{-2}(2x+3) + x^{-1}(3-x)$

51. $x^{-3} + 2xy + x^2y^{-2}$

52. $xy^{-1} + 2 + x^{-1}y$

53. $x^{-3}y^{-2} + x^{-1}y + 2xy$

54. $x^{-2} + y^{-2} + 2$

55. $\frac{1}{2}y^{-1} + x^{-1} + \frac{1}{2}x^{-2}y$

56. $\frac{1}{3}x^{-1}y^{-2} + xy + x^{-3}y^{-1}$

57. $\dfrac{2s+t}{(s+t)^2} + \dfrac{s^2-2t^2}{(s+t)^3}$

58. $\dfrac{4x-12}{x^2-49} \div \dfrac{18-2x^2}{x^2-4x-21}$

59. $\dfrac{36-9y}{3y^2-48} \div \dfrac{15+13y+2y^2}{12+11y+2y^2}$

60. $\dfrac{x^3+5x^2-3x-7}{2x^2-x-3}$

C

61. $(x^2+1) - x^2(x^2+1)^{-1}$

62. $(x^2+1) + x^2(x^2+1)^{-2}$

63. $\dfrac{x^4-x^3-7x^2+x+6}{4x^2-13x+3}$

64. $\dfrac{2x^4-x^3-9x^2+11x-10}{2x^2+x-10}$

65. $\dfrac{3x^4-x^3-9x^2+9x-2}{3x^2+5x-2}$

66. $\dfrac{x^6-y^6}{x^2+xy+y^2} \cdot (x^2-xy+y^2)^{-1}$

67. $x\left(\frac{1}{3}\right)(x^2-3)^{-2}(2x) - (x^2-3)^{-1}$

68. $(x^2-4) - (x^2-x-2)(x^2-4)^{-1}$

69. $\dfrac{x^6-y^6}{x^2-y^2} \div (x^4+x^2y^2+y^4)$

70. $\dfrac{4x^2}{x^4-2x^3} + \dfrac{8}{4x-x^3} - \dfrac{-4}{x+2}$

71. $\dfrac{6x}{2x+1} - \dfrac{2x}{x-3} + \dfrac{4x^2}{2x^2-5x-3}$

72. $\dfrac{xy^{-1}+x^2y^{-2}}{yx^{-2}-x^{-1}+y^{-1}}$

73. $\dfrac{\dfrac{m}{n}+\dfrac{m^2}{n^2}}{\dfrac{m}{n^2}-m+\dfrac{1}{n}}$

74. $1 + \dfrac{x}{x+\dfrac{1}{1+x}}$

75. $x^{-2}(2x^2+x-1) + x^{-3}(x^3-2x^2+x-1)$

76. $(a+b)^{-2}(2a+b) + (a+b)^{-3}(a^2+2ab+b^2)$

1.5 Rational Exponents and Irrational Expressions

If n is a natural number greater than 1, then an nth root of a number b is a only if $a^n = b$. If $n = 2$, the root is called a **square root;** and if $n = 3$, it is called a **cube root.** In your past work, you have probably used the symbols $\sqrt{b}$ and $\sqrt[3]{b}$ to denote square roots and cube roots of b, respectively. These symbols will be used in the next section, but for the present we will extend the idea of exponents to encompass this idea of roots. For example, if

2^x is squared to obtain 2 then $(2^x)^2 = 2$
3^x is squared to obtain 3 $(3^x)^2 = 3$
5^x is squared to obtain 5 $(5^x)^2 = 5$

If the laws of exponents stated in Section 1.4 are to hold, then

$$2^{2x} = 2 \qquad 3^{2x} = 3 \qquad 5^{2x} = 5$$

and it would seem reasonable that $x = \frac{1}{2}$. Therefore, $b^{1/2}$ is defined to be the positive square root of b and $b^{1/n}$ to be an nth root of b according to the following definition:

*n*TH ROOT

If n is a natural number, $n > 1$, and b is a real number, then we define $b^{1/n}$ for each of the following cases:

n	$b > 0$	$b < 0$	$b = 0$
EVEN	$b^{1/n}$ is the positive real nth root of b $-b^{1/n}$ is the negative real nth root of b	Not defined	0
ODD	$b^{1/n}$ is the real nth root of b		0

Example 1

Relate to definition

	Number	n ↓	b ↓	Simplification
a.	$25^{1/2}$	2	25	$25^{1/2} = 5$ since $5^2 = 25$
b.	$-25^{1/2}$	2	25	$-25^{1/2} = -5$ since $5^2 = 25$, so $-(5)^2 = -25$
c.	$(-25)^{1/2}$	2	-25	Not defined since n is even and $b < 0$
d.	$-(-25^{1/2})$	2	25	$-(-25^{1/2}) = -(-5) = 5$
e.	$64^{1/3}$	3	64	$64^{1/3} = 4$ since $4^3 = 64$
f.	$-64^{1/3}$	3	64	$-64^{1/3} = -(64^{1/3}) = -4$
g.	$(-64)^{1/3}$	3	-64	$(-64)^{1/3} = -4$ since $(-4)^3 = -64$
h.	$0^{1/5}$	5	0	$0^{1/5} = 0$ since $0^5 = 0$

□

Given b^p it is now possible to extend the set of replacements for p to include all rational numbers. If m and n are natural numbers and $b^{1/n}$ is defined, then

$$b^{m/n} = (b^{1/n})^m \qquad \text{and} \qquad b^{-(m/n)} = \frac{1}{b^{m/n}}$$

Furthermore, if m/n is reduced, then

$$(b^{1/n})^m = (b^m)^{1/n}$$

The laws of exponents stated in Section 1.4 are now extended to include rational exponents.

EXTENDED LAWS OF EXPONENTS

Let p and q represent any rational numbers, and let a and b represent any positive real numbers.

FIRST LAW	$b^p b^q = b^{p+q}$	FOURTH LAW	$\left(\frac{a}{b}\right)^p = \frac{a^p}{b^p}$
SECOND LAW	$(b^q)^p = b^{pq}$		
THIRD LAW	$(ab)^p = a^p b^p$	FIFTH LAW	$\frac{b^p}{b^q} = b^{p-q}$

An example will show why it is necessary to let a and b represent only positive real numbers.

$$(1)\qquad [(-2)^2]^{1/2} = (-2)^{2/2} = (-2)^1 \qquad = -2$$

↕ Same number to start ↕ Different numbers

$$(2)\qquad [(-2)^2]^{1/2} = [4]^{1/2} = [2^2]^{1/2} = 2^{2/2} = 2$$

Since both lines begin with the same expression, the absurd conclusion $-2 = 2$ is reached. But what is "wrong" with these examples? In equation (1) we incorrectly applied the second law of exponents since the base is a negative number.

A more subtle example of the same fallacy is shown below.

$$(x^2)^{1/2} = x^{2/2} = x$$

If $x = -2$, then $(x^2)^{1/2}$ is certainly defined and exists. But as equation (2) shows, it equals the opposite of x, namely, $-(-2) = 2$. That is, for $x = -2$, you should have

$$(x^2)^{1/2} = -x$$

On the other hand, if $x = 2$,

$$(x^2)^{1/2} = x$$

as shown by equation (1). Looking at these together,

$$\begin{array}{rl} x &= -2 \\ [x^2]^{1/2} &= -x \\ \updownarrow\, x=-2 & \updownarrow\, x=-2 \\ [(-2)^2]^{1/2} &= -(-2) \\ &= 2 \end{array} \qquad \begin{array}{rl} x &= 2 \\ [x^2]^{1/2} &= x \\ \updownarrow\, x=2 & \updownarrow\, x=2 \\ [(2)^2]^{1/2} &= 2 \end{array}$$

This inconsistency can be resolved by using the idea of absolute value:

$$(x^2)^{1/2} = |x|$$

On the other hand,

$$(x^{1/2})^2 = x$$

since the variables used in this section are all restricted so that $x^{1/b}$ is defined. In the case of $b = 2$, this means that x must be positive.

At this point, you might be thinking of a general result

$$(x^n)^{1/n} = |x|$$

Unfortunately, this is not true either. For example, if $x = -2$ and $n = 3$ and you try to apply the above result, you get

$$[(-2)^3]^{1/3} = |-2| = 2$$

But this result is false because $(-2)^3 = -8$ and the cube root of -8 is *not* 2. The correct form is

$$\begin{aligned}[(-2)^3]^{1/3} &= (-8)^{1/3} \\ &= -2\end{aligned}$$

by definition of cube root. We can reconcile matters with the following statement:

> If x is a real number and n is a natural number, so that $x^{1/n}$ is defined, then
>
> $(x^{1/n})^n = x$ for all n
> $(x^n)^{1/n} = |x|$ for n even
> $(x^n)^{1/n} = x$ for n odd

Example 2

$$\begin{aligned}125^{2/3} &= (5^3)^{2/3} \\ &= 5^2 \\ &= 25\end{aligned}$$

□

Example 3

$$\begin{aligned}10{,}000^{3/2} &= (10^4)^{3/2} \\ &= 10^6 \\ &= 1{,}000{,}000\end{aligned}$$

□

Example 4

$$\begin{aligned}16^{-3/2} &= (2^4)^{-3/2} \\ &= 2^{-6} \\ &= \frac{1}{2^6} \\ &= \frac{1}{64}\end{aligned}$$

□

Example 5

$$\begin{aligned}-16^{3/2} &= -(2^4)^{3/2} \\ &= -2^6 \\ &= -64\end{aligned}$$

□

Example 6

$(-16)^{3/2}$ Notice the difference between this example and the one in Example 5. The negative symbol is included as part of the base in this example. This is the square root of a negative number, which is not defined in the set of real numbers. □

Since we are working in the set of real numbers, notice that there is an *implied* restriction in Example 7; namely $x \geq 0$ and $y \geq 0$.

Example 7 $(x^{1/2} + y^{1/2})(x^{1/2} - y^{1/2}) = x^{1/2}x^{1/2} - x^{1/2}y^{1/2} + x^{1/2}y^{1/2} - y^{1/2}y^{1/2}$
$= x - y$ □

Example 8 $4(x^{1/2})^2 = 4x$ □

Now, compare Example 8 with the following example, which has no implied restriction on the variables.

Example 9 $4(x^2)^{1/2} = 4|x|$ Since 2 is even, absolute values are necessary. □

Example 10
$$\begin{aligned}(4x^2)^{1/2} &= (2^2x^2)^{1/2}\\ &= 2^{2/2}x^{2/2}\\ &= 2|x|\end{aligned}$$
□

Example 11 $-(x^2)^{1/2} = -|x|$ □

Example 12
$$\begin{aligned}(-8x^3)^{1/3} &= [(-2)^3x^3]^{1/3}\\ &= (-2)^{3/3}x^{3/3}\\ &= -2x\end{aligned}$$
$x^{3/3} = x$ since 3 is odd □

Example 13
$$\begin{aligned}(x^2 - 2xy + y^2)^{1/2} &= [(x - y)^2]^{1/2}\\ &= |x - y|\end{aligned}$$
□

The result in Example 13 is consistent with what you recall from elementary algebra:

$$|x - y| = \begin{cases} x - y & \text{if} \quad x - y \geqslant 0 \\ -(x - y) & \text{if} \quad x - y < 0 \end{cases}$$

That is, $(x - y)^2 = x^2 - 2xy + y^2$, so $(x - y)$ is a square root of $x^2 - 2xy + y^2$, and since

$$-(x - y) = y - x$$

we check this:

$$\begin{aligned}(y - x)^2 &= y^2 - 2yx + x^2\\ &= x^2 - 2xy + y^2\end{aligned}$$

So, $(y - x)$ is also a square root of $x^2 - 2xy + y^2$.

Example 14 $[(x - y)^{1/2}]^2 = x - y$

Compare this with Example 13. The example given here has an implied restriction—namely, $x - y \geqslant 0$, so that $(x - y)^{1/2}$ is defined. Notice that this restriction is not necessary in Example 13 since $[(x - y)^2]^{1/2}$ is defined for all real numbers x and y. □

Problem Set 1.5

If defined, simplify each of the expressions in Problems 1–35.

A

1. $9^{1/2}$
2. $-9^{1/2}$
3. $(-9)^{1/2}$
4. $27^{1/3}$
5. $-27^{1/3}$
6. $(-27)^{1/3}$
7. $64^{1/6}$
8. $(-64)^{1/6}$
9. $(-32)^{1/5}$
10. $-32^{1/5}$
11. $0^{1/3}$
12. $-16^{1/2}$
13. $27^{2/3}$
14. $-27^{2/3}$
15. $-8^{2/3}$
16. $1000^{-2/3}$
17. $100^{-3/2}$
18. $7^{-2/3} \cdot 7^{-1/3}$
19. $8^{1/3} \cdot 8$
20. $9^{1/2} \cdot 9$

B

21. $2^{1/2} \cdot 32^{1/2}$
22. $9^{2/3} \cdot 9^{5/6}$
23. $4^{1/3} \cdot 2^{1/3}$
24. $2^{2/3} \cdot 4^{2/3} \cdot 8^{5/3}$
25. $\dfrac{9^{3/2}}{9}$
26. $\dfrac{4^{5/2}}{4}$
27. $\dfrac{3^{2/3}}{3^{-1/3}}$
28. $\dfrac{9^{2/3} \cdot 3^{1/6}}{27^{1/2}}$
29. $(2^{1/3} \cdot 2^{1/2})^6$
30. $(3^{2/3} \cdot 3^{1/2})^6$
31. $(5^{1/3} \cdot 5^{1/2})^6$
32. $(10^{2^2+1} \cdot 10^5)^{1/10}$
33. **a.** $[4^{1/2}]^2$ **b.** $[4^2]^{1/2}$ **c.** $[(-4)^{1/2}]^2$ **d.** $[(-4)^2]^{1/2}$ **e.** $[-4^{1/2}]^2$ **f.** $[-4^2]^{1/2}$
34. **a.** $[3^{1/2}]^2$ **b.** $[3^2]^{1/2}$ **c.** $[(-3)^{1/2}]^2$ **d.** $[(-3)^2]^{1/2}$ **e.** $[-3^{1/2}]^2$ **f.** $[-3^2]^{1/2}$
35. **a.** $[9^{1/2}]^2$ **b.** $[9^2]^{1/2}$ **c.** $[(-9)^{1/2}]^2$ **d.** $[(-9)^2]^{1/2}$ **e.** $[-9^{1/2}]^2$ **f.** $[-9^2]^{1/2}$

If defined, simplify each of the expressions in Problems 36–58. Assume that the domain for all variables is the set of positive real numbers.

36. $x^{3/7}x^{2/7}$
37. $x^{1/2}x^{1/3}$
38. $x^{2/3}x^{1/2}$
39. $(x^{2/3}y^{-1/3})^3$
40. $(x^{3/4}y^{-1/4})^4$
41. $(16x^4)^{-3/2}$
42. $(25x^2)^{-3/2}$
43. $(125x^6)^{-2/3}$
44. $\dfrac{x^{1/2}}{x^{2/3}}$
45. $\dfrac{x^{2/3}}{x^{1/2}}$
46. $\dfrac{y^{1/3}}{y^{1/2}}$
47. $\left(\dfrac{m^{-2}n}{m^3n^2}\right)^{-1}$
48. $\left(\dfrac{s^3t^3}{s^{-3}t^{-5}}\right)^{1/2}$
49. $\left(\dfrac{x^{-1/3}}{x^{1/2}}\right)^{-6}$
50. $x^{1/4}x^{1/3}$
51. $x^{2/3}x^{1/6}$
52. $y^{1/6}y^{1/15}$
53. $(x^{1/2} + y^{1/2})^2$
54. $(x^{1/2} - y^{1/2})^2$
55. $(x^{1/3} + y^{1/3})(x^{2/3} - x^{1/3}y^{1/3} + y^{2/3})$
56. $(x^{1/3} - y^{1/3})(x^{2/3} + x^{1/3}y^{1/3} + y^{2/3})$
57. $(x + 1)^{2/3}[(x + 1)^{1/3} + (x + 1)]$
58. $(x - y)^{-2/3}[(x - y)^{5/3} + (x - y)]$
59. Find the error in the following "proof":

$$
\begin{aligned}
2 &= 2 \\
&= 4^{1/2} && \text{Definition of square root} \\
&= [(-2)^2]^{1/2} && \text{Substitute } (-2)^2 = 4 \\
&= (-2)^{2(1/2)} && \text{Theorem } (b^q)^p = b^{pq}
\end{aligned}
$$

$= (-2)^1$ Since $2 \cdot (\frac{1}{2}) = 1$
$= -2$

Therefore, $2 = -2$.

60. The third law of exponents states $(ab)^p = a^p b^p$. For example,

$$\begin{aligned} 8 &= 64^{1/2} \\ &= (4 \cdot 16)^{1/2} \\ &= 4^{1/2} \cdot 16^{1/2} \\ &= 2 \cdot 4 \\ &= 8 \end{aligned}$$

However, where is the error in the following steps, which show that 8 is equal to a number that is not defined?

$$\begin{aligned} 8 &= 64^{1/2} \\ &= [(-4)(-16)]^{1/2} \\ &= (-4)^{1/2}(-16)^{1/2} \end{aligned}$$

C *Simplify the expressions in Problems 61–72, assuming that the domain for all variables is the set of real numbers.*

61. $(x^2)^{1/2}$ **62.** $(x^3)^{1/3}$ **63.** $(8x^3)^{1/3}$
64. $(-8x^3)^{1/3}$ **65.** $(16x^2)^{1/2}$ **66.** $(25x^4)^{1/2}$
67. $(16x^4y^6)^{1/2}$ **68.** $(8x^5y^6)^{1/3}$ **69.** $(x^2 - 4x + 4)^{1/2}$
70. $[(x + 3)^3]^{1/3}$ **71.** $[(x + 3)^4]^{1/2}$ **72.** $(2x^2 - 16x + 32)^{1/2}$

73. Supply the missing factor: $[2^{1/2} + 3^{1/2} + 5^{1/2}][\qquad] = 12$

1.6 Radicals

Radical Notation

In this section the more familiar radical notation is used.

RADICAL

The symbol $\sqrt[n]{b}$ is called a **radical,** n is called the **index,** and b is called the **base** or **radicand.** A radical is defined by

$$\sqrt[n]{b} = b^{1/n}$$

if:

1. n is an even natural number and b is nonnegative; or
2. n is an odd natural number and b is any real number.

If an algebraic expression contains a radical with b (the radicand) as a variable, or a variable expression, then it is called a **radical expression.** It also follows that if m is a natural number, then

$$b^{m/n} = (b^{1/n})^m = (\sqrt[n]{b})^m$$

The laws of exponents are used to write analogous laws of radicals, and these laws of radicals follow directly from the laws of exponents.

LAWS OF RADICALS

Let a and b be real numbers such that $\sqrt{a}$ and $\sqrt{b}$ exist and are real, and let m, n, and k be natural numbers greater than or equal to 2. Then the following hold:

FIRST LAW $\sqrt[n]{b^n} = b$

Proof $\sqrt[n]{b^n} = (b^n)^{1/n} = b^{n/n} = b$

SECOND LAW $\sqrt[n]{a}\sqrt[n]{b} = \sqrt[n]{ab}$

Proof $\sqrt[n]{a}\sqrt[n]{b} = a^{1/n}b^{1/n} = (ab)^{1/n} = \sqrt[n]{ab}$

THIRD LAW $\dfrac{\sqrt[n]{a}}{\sqrt[n]{b}} = \sqrt[n]{\dfrac{a}{b}}$

Proof $\dfrac{\sqrt[n]{a}}{\sqrt[n]{b}} = \dfrac{a^{1/n}}{b^{1/n}} = \left(\dfrac{a}{b}\right)^{1/n} = \sqrt[n]{\dfrac{a}{b}}$

FOURTH LAW $\sqrt[kn]{b^{km}} = \sqrt[n]{b^m}$

Proof $\sqrt[kn]{b^{km}} = (b^{km})^{1/kn} = b^{km/kn} = b^{m/n} = \sqrt[n]{b^m}$

Simplifying Radicals

These laws of radicals allow us to formulate the criteria for a radical expression to be in simplest form. All variables in this section satisfy the conditions set forth in the laws of radicals.

CONDITIONS FOR SIMPLIFYING RADICALS

A radical expression is simplified if all four of the following conditions are satisfied:

1. When the radicand is written in completely factored form there is no factor raised to a power greater than or equal to the index of the radical.
2. No radical appears in a denominator.
3. No fraction (or negative exponent) appears within a radical.
4. There is no common factor (other than 1) between the power of the radicand and the index of the radical.

Example 1

$$\sqrt[3]{1080} = \sqrt[3]{2^3 \cdot 3^3 \cdot 5} \quad \text{or} \quad (2^3 \cdot 3^3 \cdot 5)^{1/3}$$
$$= 2 \cdot 3 \cdot \sqrt[3]{5} \qquad = 2^{3/3}3^{3/3}5^{1/3}$$
$$= 6\sqrt[3]{5} \qquad = 6\sqrt[3]{5}$$

The first condition says that you first factor the expression under the radical symbol and eliminate any powers that are greater than or equal to the index of the radical. □

Example 2

$$\sqrt{32} = \sqrt{2^5} \quad \text{or} \quad 2^{5/2}$$
$$= \sqrt{2^4 \cdot 2} \qquad = 2^{2\frac{1}{2}}$$
$$= \sqrt{(2^2)^2 \cdot 2} \qquad = 2^{2+(1/2)}$$
$$= 2^2\sqrt{2} \qquad = 2^2 \cdot 2^{1/2}$$
$$= 4\sqrt{2} \qquad = 4\sqrt{2}$$

Use the first and second laws of radicals. □

Example 3

$$\frac{4x}{\sqrt[6]{32x^2}} = \frac{4x}{\sqrt[6]{2^5x^2}} \qquad x \neq 0$$
$$= \frac{4x}{\sqrt[6]{2^5x^2}} \cdot \frac{\sqrt[6]{2x^4}}{\sqrt[6]{2x^4}}$$
$$= \frac{4x\sqrt[6]{2x^4}}{\sqrt[6]{2^6x^6}}$$
$$= \frac{4x\sqrt[6]{2x^4}}{2|x|}$$
$$= 2\sqrt[6]{2x^4} \text{ if } x \text{ is positive} \quad \text{or} \quad -2\sqrt[6]{2x^4} \text{ if } x \text{ is negative}$$

The second condition says that no radical appears in a denominator. Multiply by 1 (shown in color) to eliminate the radical in the denominator. This process is called **rationalizing the denominator.** □

If the variable x in Example 3 had been restricted to $x > 0$, then the last step would have looked like this:

$$\frac{4x\sqrt[6]{2x^4}}{2x} = 2\sqrt[6]{2x^4}$$

Using exponents instead of radicals to show the solution, we have

$$\frac{4x}{\sqrt[6]{32x^2}} = \frac{4x}{(2^5x^2)^{1/6}}$$
$$= \frac{2^2x}{2^{5/6}x^{2/6}} \cdot \frac{2^{1/6}x^{4/6}}{2^{1/6}x^{4/6}}$$
$$= \frac{4x(2x^4)^{1/6}}{2^{6/6}x^{6/6}}$$
$$= \frac{4x(2x^4)^{1/6}}{2|x|}$$
$$= 2x\sqrt[6]{2x^4} \text{ if } x \text{ is positive} \quad \text{or} \quad -2x\sqrt[6]{2x^4} \text{ if } x \text{ is negative}$$

Multiply by 1 (in color) written so that the exponents on similar factors in the denominator add up to 1. □

Example 4

$$\sqrt[5]{128x^{-6}y^{12}} = \sqrt[5]{\frac{2^7y^{12}}{x^6}}$$

$$= \sqrt[5]{\frac{(2)^5(2)^2(y^2)^5y^2}{x^5 \cdot x} \cdot \frac{x^4}{x^4}}$$

$$= \sqrt[5]{\left(\frac{2y^2}{x^2}\right)^5 \cdot \frac{(2)^2y^2x^4}{1}}$$

$$= \frac{2y^2}{x^2}\sqrt[5]{4x^4y^2}$$

The third condition says that no fraction appears within a radical. This includes negative exponents. Notice the step (in color) which allows the radical to be simplified. □

Can you simplify this using exponents?

Example 5

$$\frac{\sqrt[6]{4x^{14}}}{\sqrt[6]{32x^2}} = \frac{2^{2/6}x^{14/6}}{2^{5/6}x^{2/6}}$$

$$= 2^{-3/6}x^{12/6}$$

$$= \left|\frac{x^2}{2^{1/2}}\right| \cdot \frac{2^{1/2}}{2^{1/2}}$$

$$= \frac{x^2}{2}\sqrt{2}$$

The fourth condition says the index must be as small as possible. Notice that for this condition, it is usually easier to work via exponents.

x^2 is nonnegative and $2^{1/2}$ is positive, so $\frac{x^2}{2^{1/2}} \geqslant 0$. Thus, by the definition of absolute value, $\left|\frac{x^2}{2^{1/2}}\right| = \frac{x^2}{2^{1/2}}$ □

Operations with Radicals

For adding and subtracting radicals, it is necessary to first simplify the given radicals and then use the distributive property to carry out the operations.

Example 6

$$\frac{\sqrt{x}}{x} + \sqrt[3]{\frac{2}{3}} + \frac{2\sqrt[3]{18}}{3} + \frac{1}{\sqrt{x}} = \frac{\sqrt{x}}{x} + \sqrt[3]{\frac{2}{3} \cdot \frac{3^2}{3^2}} + \frac{2\sqrt[3]{18}}{3} + \frac{1}{\sqrt{x}} \cdot \frac{\sqrt{x}}{\sqrt{x}}$$

$$= \frac{\sqrt{x}}{x} + \frac{\sqrt[3]{18}}{3} + \frac{2\sqrt[3]{18}}{3} + \frac{\sqrt{x}}{x}$$

$$= \left(\frac{\sqrt{x}}{x} + \frac{\sqrt{x}}{x}\right) + \left(\frac{\sqrt[3]{18}}{3} + \frac{2\sqrt[3]{18}}{3}\right)$$

$$= \frac{2\sqrt{x}}{x} + \frac{3\sqrt[3]{18}}{3}$$

$$= \frac{2}{x}\sqrt{x} + \sqrt[3]{18}$$

□

Notice that the domain for x in Example 6 is the set of positive real numbers in order for $\sqrt{x}$ and $1/\sqrt{x}$ to be defined.

For multiplication of radicals use

$$\sqrt[n]{a}\sqrt[n]{b} = \sqrt[n]{ab}$$

which means you can multiply radicals *provided* they have the same index.

Example 7 $\sqrt[4]{5}\sqrt[4]{25} = \sqrt[4]{125}$ □

Example 8

$$\begin{aligned}\sqrt{2}\sqrt[3]{5} &= 2^{1/2}\cdot 5^{1/3}\\ &= 2^{3/6}\cdot 5^{2/6}\\ &= (2^3\cdot 5^2)^{1/6}\\ &= \sqrt[6]{200}\end{aligned}$$

You cannot directly multiply these radicals since the index is not the same. However, you can rewrite them with the same index. For this example, the least common index is 6. □

Example 9

$$\begin{aligned}\sqrt{12x}(\sqrt{x} + \sqrt{3}) &= \sqrt{12x^2} + \sqrt{36x}\\ &= 2x\sqrt{3} + 6\sqrt{x}\end{aligned}$$ □

Example 10

$$\begin{aligned}(\sqrt{5} + \sqrt{2})(\sqrt{5} - \sqrt{2}) &= \sqrt{5}\sqrt{5} + (-\sqrt{10} + \sqrt{10}) - \sqrt{2}\sqrt{2}\\ &= 5 - 2\\ &= 3\end{aligned}$$ □

Example 11

$$\begin{aligned}(1 - \sqrt{3})(1 + \sqrt{3}) &= 1 - 3\\ &= -2\end{aligned}$$ □

Example 12 $(\sqrt{x} + y)(\sqrt{x} - y) = x - y^2$ □

Examples 10–12 involve the product of *conjugates*. In general, expressions of the form $a + b$ and $a - b$ are called **conjugates.** Notice the result when multiplying conjugates:

$$(a + b)(a - b) = a^2 - b^2$$

For division of radicals, it is helpful to rationalize the denominators. As shown in the examples below, it is sometimes necessary to multiply by the conjugate of the denominator in order to carry out the division.

Example 13

$$\begin{aligned}\frac{1 + \sqrt{3}}{1 - \sqrt{3}} &= \frac{1 + \sqrt{3}}{1 - \sqrt{3}}\cdot\frac{1 + \sqrt{3}}{1 + \sqrt{3}}\\ &= \frac{1 + 2\sqrt{3} + 3}{1 - 3}\\ &= \frac{4 + 2\sqrt{3}}{-2}\\ &= \frac{-2(-2 - \sqrt{3})}{-2}\\ &= -2 - \sqrt{3}\end{aligned}$$ □

Example 14

$$\frac{x\sqrt{5}}{\sqrt{2x+3}} = \frac{x\sqrt{5}}{\sqrt{2x+3}} \cdot \frac{\sqrt{2x+3}}{\sqrt{2x+3}} = \frac{x\sqrt{10x+15}}{2x+3}$$

□

Example 15

$$\frac{x\sqrt{5}}{\sqrt{2x}+3} = \frac{x\sqrt{5}}{\sqrt{2x}+3} \cdot \frac{\sqrt{2x}-3}{\sqrt{2x}-3} = \frac{x\sqrt{10x}-3x\sqrt{5}}{2x-9}$$

□

Notice the difference between Examples 14 and 15. In Example 14, you are dividing by a single radical. In Example 15, you are dividing by the *sum* of a radical and 3.

Problem Set 1.6

Express each of the algebraic expressions in Problems 1–78 in simplest form. Assume that the replacement set of the variables is the set of nonnegative real numbers.

A

1. $\sqrt{121}$ **2.** $-\sqrt{144}$ **3.** $\sqrt[3]{-8}$ **4.** $\sqrt{48}$

5. $-\sqrt{64}$ **6.** $-\sqrt[3]{-64}$ **7.** $\sqrt{(-8)^2}$ **8.** $\sqrt[3]{(-4)^3}$

9. $\sqrt[3]{-64}$ **10.** $\sqrt[3]{64}$ **11.** $4\sqrt{75}$ **12.** $\sqrt[5]{x^{-2}}$

13. $\sqrt[3]{-48x^4y^5z^6}$ **14.** $-3\sqrt[6]{625^{-1}}$ **15.** $\sqrt{\frac{3}{4}\sqrt{\frac{144}{3}}}$ **16.** $\sqrt[6]{9}$

17. $\sqrt[8]{4}$ **18.** $\frac{1}{\sqrt[3]{2}}$ **19.** $\frac{1}{\sqrt[3]{4}}$ **20.** $\sqrt[6]{512x^{-9}y^{12}}$

21. $\frac{\sqrt{8x^{12}}}{\sqrt{128x^2}}$ **22.** $\sqrt[5]{\left(\frac{x^{-10}}{128}\right)^{-2}}$ **23.** $\frac{4+6\sqrt{2}}{-2}$ **24.** $\frac{9+3\sqrt{x}}{-3}$

25. $3\sqrt[3]{4}+5\sqrt[3]{9}-(2\sqrt[3]{4}+5\sqrt[3]{9})$ **26.** $7\sqrt{x}-3\sqrt{y}+2\sqrt{y}-7\sqrt{x}$

27. $\sqrt{5}(3\sqrt{5}+2\sqrt{2})$ **28.** $\sqrt{3}(1-\sqrt{3})$

29. $(4+\sqrt{3})(4-\sqrt{3})$ **30.** $(3-\sqrt{5})(\sqrt{5}+3)$

31. $(\sqrt{xy}+\sqrt{w})(\sqrt{xy}-\sqrt{w})$ **32.** $(\sqrt{x}+5)(\sqrt{x}-5)$

33. $(\sqrt{y}+1)^2$ **34.** $(\sqrt{x}-1)^2$

B

35. $(2\sqrt{3}-3\sqrt{2})(3\sqrt{2}+2\sqrt{3})$ **36.** $(\sqrt{12}+\sqrt{3})(\sqrt{12}-2\sqrt{3})$

37. $(2-\sqrt{3})(\sqrt{3}+1)^2$ **38.** $\sqrt{3}(\sqrt{27}+\sqrt{48}-\sqrt{12})$

39. $(\sqrt{6}+\sqrt{3})(\sqrt{6}-2\sqrt{3})(\sqrt{8})$ **40.** $(2+\sqrt{2})(3-\sqrt{2})(4-\sqrt{2})$

41. $(3\sqrt{2}-\sqrt{8})(2\sqrt{2}-3\sqrt{8})$ **42.** $4\left(\frac{1}{2}+\frac{\sqrt{2}}{2}\right)\left(\frac{1}{2}-\frac{\sqrt{2}}{2}\right)$

43. $\sqrt{\frac{3}{5}} + \sqrt{60}$

44. $\sqrt{5} + \dfrac{1}{\sqrt{5}}$

45. $\dfrac{\sqrt{6} + \sqrt{8}}{\sqrt{2}}$

46. $\dfrac{xy - y\sqrt{x^2 y}}{xy}$

47. $\sqrt{\dfrac{1}{4} + \dfrac{1}{x}}$

48. $\sqrt{9^{-1} + x^{-1}}$

49. $\sqrt{(9 + x)^{-1}}$

50. $\sqrt{18} + \dfrac{2}{\sqrt{2}} - 4\sqrt{50}$

51. $\dfrac{1 - \sqrt{y}}{1 + \sqrt{y}}$

52. $\sqrt[3]{81} + \dfrac{3}{\sqrt[3]{3}} - 2\sqrt[3]{24}$

53. $\dfrac{1}{\sqrt{3} + 1}$

54. $\dfrac{2}{\sqrt{5} + 1}$

55. $\dfrac{(2\sqrt{3} - \sqrt{6})(3\sqrt{3} + 3\sqrt{6})}{\sqrt{2}}$

56. $\dfrac{2\sqrt{3} - 1}{\sqrt{3} + 2}(5\sqrt{3} + 8)$

57. $(\sqrt{2} - \sqrt{3})^2 + (\sqrt{2} + \sqrt{3})^2$

58. $(2 + \sqrt{6})(3 + \sqrt{2})(2 - \sqrt{6})(3 - \sqrt{2})$

59. $\sqrt{8 - 2\sqrt{7}}\,\sqrt{8 + 2\sqrt{7}}$

60. $2(3 + \sqrt{5})(2 - \sqrt{3})(3 - \sqrt{5})(\sqrt{3} + 2)$

61. $\dfrac{(2\sqrt{7} + 3)(2\sqrt{7} - 3)}{(5 - \sqrt{6})(5 + \sqrt{6})}$

62. $\dfrac{1}{6}(6 - \sqrt{6})(3\sqrt{2} - 2\sqrt{3})(4\sqrt{2} + 3\sqrt{3})$

63. $\dfrac{2 + \sqrt{x}}{\sqrt{x} - 1}$

64. $\dfrac{2\sqrt{t} - 5}{\sqrt{t} + 5}$

65. $\dfrac{3 + \sqrt{y}}{\sqrt{y} - 1}$

66. $\dfrac{1 - 3\sqrt{x}}{3\sqrt{x} + 2}$

67. $\dfrac{2}{\sqrt{x - 1}}$

68. $\dfrac{5}{\sqrt{t + 5}}$

69. $\dfrac{m + n}{3\sqrt{n + 2}}$

70. $\dfrac{2\sqrt{s} - 5}{\sqrt{s + 5}}$

71. $\dfrac{1}{\sqrt{x} + \sqrt{y}}$

72. $\dfrac{1}{\sqrt{x + y}}$

C **73.** $\dfrac{1}{x + \sqrt{y + z}}$

74. $\dfrac{1}{x + y + \sqrt{z}}$

75. $\dfrac{1}{\sqrt{x + y} + z}$

76. $\dfrac{1}{\sqrt{x + y + z}}$

77. $\dfrac{-5}{\sqrt[3]{2} - \sqrt[3]{7}}$

78. $\dfrac{x + 1}{\sqrt[3]{x} + 1}$

Express each of the algebraic expressions in Problems 79–88 in simplest form. The replacement set for the variables is the set of all real numbers.

79. $\sqrt{x^2}$

80. $\sqrt{(x+7)^2}$

81. $\sqrt[3]{(x+y)^3}$

82. $\sqrt{a^2+2ab+b^2}$

83. $\dfrac{\sqrt[6]{8x^{12}}}{\sqrt[6]{128x^2}}$

84. $\sqrt[3]{a^3+3a^2b+3ab^2+b^3}$

85. $\sqrt[4]{\dfrac{10y^6}{80y^2}}$

86. $\dfrac{5x}{\sqrt[4]{125x^2}}$

87. $\dfrac{7x}{\sqrt[3]{25x^2}}$

88. $(\sqrt{x}+5)(\sqrt{x}-5)$

89. Let: $X = \sqrt{169+15\sqrt{77}} - \sqrt{(5\sqrt{11}-13\sqrt{2})(3\sqrt{7}-13\sqrt{2})}$

a. Use a calculator to find an approximate value for X.

b. Find X exactly without using a calculator.

1.7 Complex Numbers

Throughout the previous parts of this chapter, it was necessary to require that variables under radicals be positive when the indexes were even. For example,

$$\sqrt{-4} = a$$

means, by definition, $-4 = a^2$, but no real number squared can be negative. Nevertheless, problems that give rise to the square root of a negative number are still occasionally encountered, so we will now investigate that possibility.

It is important to remember that *in the set of real numbers the square root of a negative number does not exist.* Therefore, a number that is *not a real number* must be introduced. This number is denoted by the symbol $\boldsymbol{i}$.

DEFINITION OF i, THE IMAGINARY UNIT

> The number i, called the **imaginary unit**, is defined as a number with the following properties:
>
> $i^2 = -1$ and $\sqrt{-a} = i\sqrt{a}$ if $a > 0$

With this number available, the square root of any negative number can be written as the product of a real number and the number i.

Example 1 $\sqrt{-1} = i\sqrt{1} = i$ This example shows that $i = \sqrt{-1}$. You should remember this. □

Example 2 $\sqrt{-4} = i\sqrt{4} = 2i$ *Check:* $(2i)^2 = 4i^2 = 4(-1) = -4$. So, by definition, $\sqrt{-4} = 2i$. □

Example 3 $\sqrt{-13} = i\sqrt{13}$ or $\sqrt{13}\,i$ If you write the numerical coefficient first, make sure the i is *not* included under the radical. □

Example 4 $\sqrt{-b} = i\sqrt{b}$ or $\sqrt{b}\,i$ $(b > 0)$ □

Now consider another set of numbers that includes the real numbers as a *subset.*

THE SET OF COMPLEX NUMBERS

> The set of numbers of the form
>
> $a + bi$
>
> where a and b are real numbers and i is the imaginary unit, is called the **set of complex numbers.**

If $b = 0$, then $a + 0i$ is a *real number.*
If $a = 0$ and $b = 1$, then $0 + 1i$ is the *imaginary unit.*
If $a = 0$ and $b \neq 0$, then $0 + bi$ is a *pure imaginary number.*

A complex number is *simplified* when it is written in the form $a + bi$.

In order to work with complex numbers you will need definitions for equality, addition, subtraction, and distributivity of two complex numbers $a + bi$ and $c + di$.

DEFINITION OF EQUALITY, ADDITION, SUBTRACTION, AND DISTRIBUTIVITY OF COMPLEX NUMBERS

> Let $a + bi$ and $c + di$ be any complex numbers.
>
> | | |
> |---|---|
> | EQUALITY | $a + bi = c + di$ if and only if $a = c$ and $b = d$ |
> | ADDITION | $(a + bi) + (c + di) = (a + c) + (b + d)i$ |
> | SUBTRACTION | $(a + bi) - (c + di) = (a - c) + (b - d)i$ |
> | DISTRIBUTIVITY | $c(a + bi) = ca + cbi$ |

In Examples 5–12 each of the complex numbers is rewritten in the form $a + bi$.

Example 5
$$\sqrt{-9} = i\sqrt{9}$$
$$= 3i$$
In the form $a + bi$, this can be written as $0 + 3i$; notice that this is the special case of a complex number called a **pure imaginary.** The form $3i$ is considered simplified and is the preferred form. □

Example 6 7 In the form $a + bi$, this can be written as $7 + 0i$; this is the special case of a complex number called a **real number.** The form 7 is considered simplified and is the preferred form. □

Example 7 $5 + 3i, \quad 3 - i, \quad \sqrt{2} - \sqrt{3}i$ are all in simplified form $a + bi$. □

Example 8
$$\frac{2 - \sqrt{-8}}{2} = \frac{1}{2}(2 - \sqrt{8}\,i)$$

$$= \frac{1}{2}(2 - 2\sqrt{2}\, i)$$
$$= 1 - \sqrt{2}\, i \qquad \square$$

Example 9 $(3 + 4i) + (2 + 3i) = 5 + 7i$ $\square$

Example 10 $(2 - i) - (3 - 2i) = -1 + i$ $\square$

Example 11 $(6 + 2i) + (2 - 2i) = 8$ In the form $a + bi$, this is $8 + 0i$. $\square$

Example 12 $(4 + 3i) - (4 - 2i) = 5i$ In the form $a + bi$, this is $0 + 5i$. $\square$

Notice that the operations of addition and subtraction of complex numbers conform to the usual way you handle the addition and subtraction of binomials. Multiplication is defined similarly. Recall that

$$i^2 = -1$$

which means that i^2 is replaced by -1 wherever it occurs. Now consider the complex numbers $a + bi$ and $c + di$. If multiplication of complex numbers is handled in the same manner as binomials, perform the operation as follows:

$$\begin{aligned} (a + bi)(c + di) &= ac + adi + bci + bdi^2 && \text{Using FOIL} \\ &= ac + adi + bci - bd && \text{Since } i^2 = -1 \\ &= (ac - bd) + (ad + bc)i \end{aligned}$$

This can be taken as the definition of multiplication of complex numbers.

MULTIPLICATION OF COMPLEX NUMBERS

> If $a + bi$ and $c + di$ are any two complex numbers, then
>
> $$(a + bi)(c + di) = (ac - bd) + (ad + bc)i$$

It is not necessary to memorize this definition since two complex numbers are handled as you would any binomials.

Example 13
$$\begin{aligned} (2 + 3i)(4 + 2i) &= 8 + 16i + 6i^2 \\ &= 8 + 16i - 6 \\ &= 2 + 16i \end{aligned}$$
$\square$

Example 14
$$\begin{aligned} (4 - 3i)(2 - i) &= 8 - 10i + 3i^2 \\ &= 5 - 10i \end{aligned}$$
$\square$

Example 15
$$\begin{aligned} (3 - 2i)(3 + 2i) &= 9 - 4i^2 \\ &= 13 \end{aligned}$$
$\square$

If a and b are both positive, then

$$\sqrt{a}\sqrt{b} = \sqrt{ab}$$

but if a and b are both negative, then

$$\sqrt{a}\sqrt{b} = -\sqrt{ab}$$

This means that when working with complex numbers, you should *first* write them in the form

$$a + bi$$

and *then* perform the arithmetic.

Example 16

$$\begin{aligned}(2 - \sqrt{-4})(4 + \sqrt{-9}) &= (2 - 2i)(4 + 3i)\\ &= 8 - 2i - 6i^2\\ &= 14 - 2i\end{aligned}$$

□

Example 17

$$\begin{aligned}(\sqrt{7} - \sqrt{-16})^2 &= (\sqrt{7} - 4i)^2\\ &= (\sqrt{7})^2 + 2(\sqrt{7})(-4i) + (-4i)^2\\ &= 7 - 8\sqrt{7}\,i + 16i^2\\ &= 7 - 8\sqrt{7}\,i - 16\\ &= -9 - 8\sqrt{7}\,i\end{aligned}$$

□

For division, use the process of rationalizing the denominator, which was developed in the last section. Notice from Example 15 that when you multiply a complex number by its conjugate, the result is a real number. In general,

$$\begin{aligned}(a + bi)(a - bi) &= a^2 - b^2i^2\\ &= a^2 + b^2\end{aligned}$$

which is real since a and b are real. Thus, for division,

$$\begin{aligned}\frac{a + bi}{c + di} \cdot \frac{c - di}{c - di} &= \frac{(ac + bd) + (bc - ad)i}{c^2 - d^2i^2}\\ &= \frac{ac + bd}{c^2 + d^2} + \frac{bc - ad}{c^2 + d^2}i\end{aligned}$$

Conjugate of the denominator

DIVISION OF COMPLEX NUMBERS

If $a + bi$ and $c + di$ are any two complex numbers (c and d are not both zero), then

$$\frac{a + bi}{c + di} = \frac{ac + bd}{c^2 + d^2} + \frac{bc - ad}{c^2 + d^2}i$$

Notice that the result is in the form of a complex number, but instead of memorizing this definition, simply remember: To divide complex numbers, you multiply both the numerator and the denominator by the conjugate of the denominator.

Example 18

$$\frac{7+i}{2+i} = \frac{7+i}{2+i} \cdot \frac{2-i}{2-i} = \frac{14-5i-i^2}{4-i^2} = \frac{15-5i}{5} = 3-i$$

□

Example 19

$$\frac{15-\sqrt{-25}}{2-\sqrt{-1}} = \frac{15-5i}{2-i} \cdot \frac{2+i}{2+i} = \frac{30+5i-5i^2}{4-i^2} = \frac{35+5i}{5} = 7+i$$

□

Problem Set 1.7

Simplify each of the expressions in Problems 1–57.

A

1. $\sqrt{-25}$ **2.** $\sqrt{-100}$ **3.** $\sqrt{-16}$ **4.** $\sqrt{-81}$

5. $\sqrt{-8}$ **6.** $\sqrt{-20}$ **7.** $4\sqrt{-36}$ **8.** $2\sqrt{-49}$

9. $-3\sqrt{-144}$ **10.** $-6\sqrt{-4}$ **11.** $\frac{3\sqrt{-49}}{7}$ **12.** $\frac{-2\sqrt{-24}}{8}$

13. $2+\sqrt{2}-4+\sqrt{-2}$ **14.** $6-\sqrt{3}-8+\sqrt{-3}$

15. $(3+\sqrt{3})+(5-\sqrt{-9})$ **16.** $(5-\sqrt{2})+(2-\sqrt{-4})$

17. $i(3+5i)$ **18.** $i(4-6i)$ **19.** $-7(5-2i)$

20. $6i-(2-5i)$ **21.** $5(2+4i)$ **22.** $-2(-5+5i)$

23. $(4-i)(2+i)$ **24.** $(5-i)(5+i)$ **25.** $(8-2i)(8+2i)$

26. $(5+3i)+(2+6i)$ **27.** $(2-7i)-(3-2i)$ **28.** $(5-4i)-(5-9i)$

29. $-i^2$ **30.** $-i^3$ **31.** $-i^5$

32. i^7 **33.** $(1-3i)^2$ **34.** $(6-2i)^2$

35. $(3+4i)^2$ **36.** $(\sqrt{2}+3i)^2$

B

37. $(\sqrt{5}-3i)^2$ **38.** $(6-\sqrt{-1})(2+\sqrt{-1})$

39. $(2-\sqrt{-4})(3-\sqrt{-4})$ **40.** $(3-\sqrt{-3})(3+\sqrt{-3})$

41. $(2-\sqrt{-3})(2+\sqrt{-2})$ **42.** $(2-\sqrt{-2})(3-\sqrt{-3})$

43. $\frac{-2}{1+i}$ **44.** $\frac{5}{1-2i}$ **45.** $\frac{9-7i}{3+i}$ **46.** $\frac{3-i}{2+i}$

47. $\frac{3 - i}{1 - i}$ **48.** $\frac{-1 + 4i}{1 + 2i}$ **49.** $\frac{2 + 3i}{4 - i}$ **50.** $\frac{1}{i}$

51. $\frac{3}{i}$ **52.** i^{-3} **53.** $\frac{\sqrt{-1} + 1}{\sqrt{-1} - 1}$ **54.** $\frac{10 - \sqrt{-25}}{2 - \sqrt{-1}}$

55. $\frac{4 - \sqrt{-4}}{1 + \sqrt{-1}}$ **56.** $\frac{2 - \sqrt{-9}}{1 + \sqrt{-1}}$ **57.** $\left(\frac{2 + \sqrt{3}\, i}{\sqrt{2}}\right)^2$

58. Show that $5i$ is a root of $x^2 + 25 = 0$.

59. Show that $\sqrt{3}\, i$ is a root of $x^2 + 3 = 0$.

60. Show that $1 + 2i$ is a root of $x^2 - 2x + 5 = 0$.

C **61.** Show that $1 + \frac{1}{2}\sqrt{6}\, i$ is a root of $2x^2 - 4x + 5 = 0$.

62. Show that $1 - \frac{1}{2}\sqrt{6}\, i$ is a root of $2x^2 - 4x + 5 = 0$.

63. Simplify: $[(2 + \sqrt{3}) + (4 - \sqrt{2})i][(2 + \sqrt{3}) - (4 - \sqrt{2})i]$

64. Simplify: $[(4 - \sqrt{2}) + (3 - \sqrt{2})i][(1 + \sqrt{2}) + (2 - \sqrt{2})i]$

65. Show that $x = 4$ and $x = \sqrt[3]{2 + \sqrt{-121}} + \sqrt[3]{2 - \sqrt{-121}}$ are roots of $x^3 = 15x + 4$.

66. What is wrong with the following "proof"?

$\sqrt{-1} = \sqrt{-1}$ Property of equality

$\sqrt{\frac{1}{-1}} = \sqrt{\frac{-1}{1}}$ $-1 = \frac{1}{-1}$ and $-1 = \frac{-1}{1}$

$\frac{\sqrt{1}}{\sqrt{-1}} = \frac{\sqrt{-1}}{\sqrt{1}}$ Property of radicals

$\frac{1}{i} = \frac{i}{1}$ $\sqrt{1} = 1$ and $\sqrt{-1} = i$

$\frac{1}{i} \cdot \frac{i}{i} = i$ Multiplication by 1 and $\frac{i}{1} = i$

$\frac{i}{i^2} = i$ $\frac{1}{i} \cdot \frac{i}{i} = \frac{i}{i^2}$

$\frac{i}{-1} = i$ $i^2 = -1$

$-i = i$ Which is impossible!

67. Show that the field properties are satisfied for the operations of addition and multiplication over the set of complex numbers.

1.8 Summary and Review

In this chapter, we have laid the foundation for a study of college algebra. The set of complex numbers was developed along with the operations and properties of its various subsets: natural numbers, integers, rational numbers, irrational numbers, and real numbers. It is this foundation upon which the course will be built. You are probably familiar with much of the content of this chapter, and it should serve mostly as a review of your previous mathematics courses.

The last section of each chapter in this book is a review intended to help you understand the material of that chapter. We suggest that you review the list of chapter objectives first. Then, take the fill-in self-test to review the concepts and check your answers (all answers are provided in the back of the book); problems missed indicate areas of weakness and sections to be reviewed further. Next, work selected review problems to gain additional practice. The sections to be reviewed when problems are missed are indicated in brackets in the margin.

CHAPTER OBJECTIVES

After studying this chapter, you should be able to:

1. Distinguish among the following sets of real numbers: natural numbers, whole numbers, integers, rational numbers, irrational numbers, real numbers, pure imaginary numbers, and complex numbers
2. State and use the definition of absolute value
3. Find the distance between points on a number line
4. Identify examples of each of the following properties for a given operation: closure, commutative, associative, identity, and inverse
5. State and use the distributive property
6. Define b^n
7. Carry out operations with polynomials (including proper use of the order of operations agreement)
8. Completely factor a given expression
9. State and use the five laws of exponents
10. Carry out operations with rational expressions and simplify the result
11. Define b^n for n a rational number and b a real number
12. Simplify expressions involving rational exponents
13. State and use the four laws of radicals
14. State the four conditions for a radical to be in simplified form
15. Simplify radical expressions
16. Simplify expressions involving complex numbers

CONCEPT PROBLEMS

Problem Set 1.8

Fill in the word or words to make the statements in Problems 1–16 complete and correct.

[1.1] **1.** The elements of the set of ______ numbers can be characterized as repeating or terminating decimals.

[1.1] **2.** If $a \in \mathbf{R}$, then

$$|a| = \begin{cases} \_\_\_\_\_\_ & \text{if } a \geqslant 0 \\ \_\_\_\_\_\_ & \text{if } a < 0 \end{cases}$$

[1.1] **3.** Given any two real numbers a and b, exactly one of the following holds: (1) $a = b$, (2) ______ , (3) ______ .

[1.1] **4.** The distance between the points (x_1) and (x_2) on a number line is ______ .

[1.2] **5.** In the notation b^n, b is called the ______ , n is called the ______ , and b^n is called a(n) ______ .

[1.2] **6.** When simplifying an algebraic expression, the proper order of operations is:

a. First, ________________________

b. ________________________

c. ________________________

d. Finally, ________________________

[1.3] **7.** **a.** $a^2 - b^2 = (a - b)(\_\_\_\_\_\_)$

b. $a^3 - b^3 = (a - b)(\_\_\_\_\_\_)$

c. $a^3 + b^3 = (a + b)(\_\_\_\_\_\_)$

[1.4] **8.** **a.** $b^m \cdot b^n =$ ______

b. $(ab)^m =$ ______

c. $\dfrac{b^m}{b^n} =$ ______

d. $(a^n)^m =$ ______

e. $\left(\dfrac{a}{b}\right)^n =$ ______

[1.4] **9.** In Problem 8, the necessary restrictions for the variables are: a is a(n) ______ ; b is a(n) ______ ; m is a(n) ______ ; and n is a(n) ______ .

[1.4] **10.** A rational expression is an expression that can be written ______ .

[1.4] **11.** Let P, Q, R, S, and K be any polynomials such that all values of the variables that cause division by zero are excluded from the domain.

a. $\dfrac{P}{Q} = \dfrac{R}{S}$ if and only if ______

b. $\dfrac{PK}{QK} =$ ______

c. $\dfrac{P}{Q} + \dfrac{R}{S} =$ ______

d. $\dfrac{P}{Q} - \dfrac{R}{S} =$ ______

e. $\dfrac{P}{Q} \cdot \dfrac{R}{S} =$ ______

f. $\dfrac{P}{Q} \div \dfrac{R}{S} =$ ______

[1.5] **12.** **a.** $(x^{1/n})^n =$ ______ for any n.

b. $(x^n)^{1/n} =$ ______ if n is odd.

c. $(x^n)^{1/n} =$ ______ if n is even.

[1.6] **13.** A radical expression is simplified if all four of the following conditions are satisfied:

a. When the radicand is written in completely factored form ______________ ______________.

b. ______________.

c. ______________.

d. There is no common factor (other than 1) between ______________.

[1.7] **14.** The number i is defined as a number with the following properties:

a. ______________ **b.** ______________ if ______________

[1.7] **15.** A complex number is ______________.

[1.7] **16.** If $(a + bi)$ and $(c + di)$ are any complex numbers, then:

a. $a + bi = c + di$ if and only if ______________.

b. $(a + bi) + (c + di) =$ ______________.

c. $(a + bi) - (c + di) =$ ______________.

d. $(a + bi)(c + di) =$ ______________.

e. $\dfrac{a + bi}{c + di} =$ ______________.

REVIEW PROBLEMS [1.1] *Write each statement in Problems 17–22 without the absolute value symbol.*

17. $-|\sqrt{5}|$ **18.** $|2\sqrt{2} - 3|$ **19.** $|8 - 2\pi|$

20. $|x|$ **21.** $|x^2 + 5|$ **22.** $|3 - 5t|$ if $t > 1$

[1.1] *Find the distance between the pairs of points whose coordinates are given in Problems 23–26.*

23. $A(-8)$ and $B(3)$ **24.** $C(14)$ and $D(-12)$

25. $E(\sqrt{6})$ and $F(9)$ **26.** $G(\sqrt{5})$ and $H(2)$

[1.1] *Identify the property illustrated by each statement in Problems 27–37.*

27. $a(b + c) = a(c + b)$ **28.** $a(b + c) = (b + c)a$

29. $a(b + c) = ab + ac$ **30.** $a(b + c) = (c + b)a$

31. $a + (b + c) = (a + b) + c$ **32.** $a + (b + c) = a + (c + b)$

33. $\dfrac{x + 2}{x + 1} = \dfrac{x + 2}{x + 1} \cdot 1$

34. $(a + b)(a^2 + 2ab + b^2) = (a + b)a^2 + (a + b)(2ab) + (a + b)b^2$

35. $2x + (-3x^2) + 4 = -3x^2 + 2x + 4$

36. $5\sqrt{2} + 3\sqrt{2} = (5 + 3)\sqrt{2}$

37. $x \cdot \left(5 \cdot \frac{1}{5}\right) - 7 \cdot \frac{1}{5} = x \cdot 1 - 7 \cdot \frac{1}{5}$

[1.2] *Simplify the expressions in Problems 38–48.*

38. **a.** -8^2 **b.** $(-8)^2$ **c.** $-(-8)^2$ **d.** -1^4

39. $\dfrac{6 + 4(-3)}{-2}$ **40.** $6 + 5 \cdot 2 - 8 \div 4 + 2$

41. $5x - (3x - 5)$ **42.** $(x + 2y) - (3x + 5y)$

43. $(3x - 2)(5x + 3)$

44. $(2y - 1)(4 + 4y)$

45. $(2x - 1)^2$

46. $(3x - 2)^2$

47. $(x - 1)^3$

48. $(x + 2)(2x^2 + 3x - 2)$

[1.3] *Completely factor each expression in Problems 49–63.*

49. $36x - 6$

50. $x^2y + 3xy^2$

51. $m^2 - n^2$

52. $2p^2 - 8$

53. $36x^2 - 31x + 3$

54. $6x^2 - 23x - 4$

55. $(3x - 1)x^2 - (3x - 1)$

56. $8(x + 1)^3 + x^3(x + 1)^3$

57. $18x^3 + 9x^2 - 36x$

58. $10 - 24x - 18x^2$

59. $s^{2x} - t^{2x}$

60. $5x^3 - 625$

61. $(m - n)^2 - 1$

62. $\dfrac{k^2}{16} - h^2$

63. $u^{3t} + v^{3t}$

[1.4] *Simplify the expressions in Problems 64–78. Values that cause division by zero are excluded. All rational expressions should be reduced, and negative exponents should be eliminated.*

64. $-3^{-2}x^2$

65. $-3(x^2)^{-2}$

66. $(-3x^2)^{-2}$

67. $\left(\dfrac{x^2y^{-3}}{z^4}\right)^{-1}$

68. $\left(\dfrac{m^{-3}n^{-2}}{q^{-4}}\right)^{-3}$

69. $\left(\dfrac{3st^3}{12s^4}\right)^{-2}$

70. $2x^{-1} + (3y)^{-1}$

71. $3(x + y)^{-1} - 2$

72. $(x^2 - y^2)^{-1}(x + y)$

73. $\dfrac{6x^2 + 5x - 4}{4 + 3x}$

74. $\dfrac{7}{2x - 1} - \dfrac{13}{4 + 3x}$

75. $\dfrac{x}{x - 2} - \dfrac{x + 2}{2 - x}$

76. $y + 2x^{-1}y^2 + x^{-2}y^3$

77. $\dfrac{2x^4 + 7x^3 + 4x^2 + 2x - 3}{2x^2 + 5x - 3}$

78. $\dfrac{6x^4 + 7x^3 + 11x^2 - x + 4}{3x^2 - x + 1}$

[1.5] *Simplify each expression in Problems 79–90. Assume that the domain for all variables is the set of positive real numbers.*

79. $16^{1/2}$

80. $-16^{1/2}$

81. $(-16)^{1/2}$

82. $-16^{-1/2}$

83. $(x^{1/2}x^{1/3})^6$

84. $x^{1/2}x^{1/3}$

85. $\dfrac{x^{1/2}}{x^{1/3}}$

86. $\dfrac{y^{2/3}}{y^{1/2}}$

87. $\left(\dfrac{x^3y^4}{y^{-2}}\right)^{1/3}$

88. $\left(\dfrac{x^{-2}}{y^{-4}}\right)^{1/2}$

89. $(x^{1/2} + y^{1/2})^2$

90. $(x^{1/2} + y^{1/2})(x^{1/2} - y^{1/2})$

[1.6] *Simplify the expressions in Problems 91–108. Assume that the domain for all variables is the set of nonnegative real numbers.*

91. $\sqrt{50}$

92. $\sqrt[3]{72}$

93. $\sqrt[5]{64x^5y^7}$

94. $\sqrt[6]{4}$

95. $\sqrt[4]{32x^4y^{-5}}$

96. $\dfrac{4 + 12\sqrt{3}}{-4}$

97. $\dfrac{8 - 12\sqrt{5}}{20}$

98. $\sqrt{3}(6\sqrt{3} - 2\sqrt{2})$

99. $(2 - \sqrt{5})(3 - 2\sqrt{5})$

100. $(\sqrt{x} + \sqrt{y})^2$

101. $\sqrt{\dfrac{4}{5}} + \sqrt{20}$

102. $\sqrt{\dfrac{1}{9} + \dfrac{1}{y}}$

103. $\dfrac{1}{4 + \sqrt{3}}$

104. $\dfrac{-4}{2\sqrt{3} - 1}$

105. $\dfrac{3}{\sqrt{5} - \sqrt{4}}$

106. $\dfrac{x}{2 + \sqrt{x}}$

107. $\dfrac{x}{\sqrt{2 + x}}$

108. $\dfrac{y}{\sqrt{3y - 1}}$

[1.7] *Simplify each of the expressions in Problems 109–120.*

109. $\sqrt{-121}$

110. $-\sqrt{-1}$

111. $\dfrac{3\sqrt{-16}}{4}$

112. i^{11}

113. $\dfrac{6}{i}$

114. $(2i - 1) + (4 + 3i^2)$

115. $\dfrac{1}{4 + \sqrt{-3}}$

116. $(2 + \sqrt{-5})(1 - \sqrt{-5})$

117. $(\sqrt{2} + i)^2$

118. $\dfrac{-5}{2 - i}$

119. $\dfrac{3 - i}{2 + i}$

120. $\dfrac{2 - \sqrt{-25}}{1 + \sqrt{-25}}$

CHAPTER 2

Solving Equations and Inequalities

William Kingdon Clifford (1845–1879)

We may always depend upon it that algebra which cannot be translated into good English and sound common sense is bad algebra.

W. K. Clifford,
Common Sense in the Exact Sciences, London, 1885

Historical Note

W. K. Clifford was a witty, charming man of many talents. He was England's most promising mathematician when his early death from tuberculosis at 34 prevented the maturing of his genius. He was one of only a handful of writers who suggested notions of relativity prior to Einstein. His *Common Sense in the Exact Sciences,* a work on the foundations of mathematics, is still a classic.

Clifford was an incredibly clever youth. There is an account of his being shown a spherical puzzle from India, perhaps the Rubik's cube of his day. He looked it over without touching it, and thought quietly. Finally, when he did pick it up, he was able to solve it immediately. A contemporary of Lewis Carroll, he had a similar affection for young children, which prompted him to write *The Little People,* a collection of fairy tales, and many poems. He was also a gifted speaker, reputedly the only mathematician in England that could address a general audience and be clearly understood. He had an unusual aptitude for languages. Part of his motivation was to allow him to read foreign mathematics journals. He learned French, Spanish, German, Greek, Arabic, and hieroglyphics. He tackled a language as if it were a problem or a riddle, and even found Morse code and shorthand of interest.

Chapter Overview

Solving equations and inequalities is fundamental to most of what you do in algebra. In this chapter those ideas and skills are reviewed, extended, and used with applied problems. Linear, quadratic, rational, and radical equations and inequalities are each reviewed. Particular attention is given to absolute value and extraneous results that arise in rational and radical equations. The work is concluded with an intensive review of nonlinear inequalities, where the expressions may be factored.

2.1 Equivalent Equations and Inequalities

Any statement that two expressions are equal is called an **equation.** If the two expressions are constants, you can easily determine whether the statement is true or false. However, if any of the expressions contain a variable, the truth of the statement may depend on the value of the variable. Such a statement is called a **conditional equation.** To solve a conditional equation, the values of the variable that satisfy the equation must be identified. These values are called **solutions,** or **roots.** The set of all solutions to an equation is called the **solution set** of the equation. If a conditional equation is true for all values of the variable, it is called an **identity.** When a conditional equation is never true, its solution set is empty and the equation is called a **contradiction.** Two equations with the same solution set are called **equivalent equations.** The process of solving an equation involves finding a sequence of equivalent equations; it ends when the solution or solutions are obvious. There are a few operations that produce equivalent equations, some of which we summarize in the box.

PROPERTIES OF EQUIVALENT EQUATIONS

If P and Q are algebraic expressions, and k is a real number, then each of the following is equivalent to $P = Q$:

1. $P + k = Q + k$
2. $P - k = Q - k$
3. $kP = kQ \quad k \neq 0$
4. $\dfrac{P}{k} = \dfrac{Q}{k} \quad k \neq 0$

Briefly, these properties allow you to obtain an equivalent equation by adding any real number to or subtracting any real number from each side of an equation. You can also multiply or divide both sides of an equation by a nonzero real number to obtain an equivalent equation.

Example 1

$$
\begin{aligned}
3x + 5 &= x - 3 \\
3x + 5 - x &= x - x - 3 && \text{Subtract } x. \\
2x + 5 &= -3 \\
2x + 5 - 5 &= -3 - 5 && \text{Subtract 5.} \\
2x &= -8 \\
\frac{2x}{2} &= \frac{-8}{2} && \text{Divide by 2.} \\
\mathbf{x} &= \mathbf{-4}
\end{aligned}
$$

□

The statement of the solution set $\{-4\}$ is often omitted and the solution left in terms of a very simple equation or inequality.

Example 2

$$
\begin{aligned}
(x-3)(2x+5) &= 2x(x-2) \\
2x^2 - x - 15 &= 2x^2 - 4x \\
2x^2 - 2x^2 - x - 15 &= 2x^2 - 2x^2 - 4x \\
-x - 15 &= -4x \\
-x + x - 15 &= -4x + x \\
-15 &= -3x \\
\frac{-15}{-3} &= \frac{-3x}{-3} \\
\mathbf{5} &= \mathbf{x}
\end{aligned}
$$

□

Example 3

$$
\begin{aligned}
2x - 5(x-2) &= 3(3-x) \\
2x - 5x + 10 &= 9 - 3x \\
-3x + 10 &= 9 - 3x \\
3x - 3x + 10 &= 9 - 3x + 3x \\
10 &= 9 && \text{False equation}
\end{aligned}
$$

This is a contradiction and, since it is equivalent to the original equation, **the solution set is empty.** The empty or null set is written as $\{\ \ \}$ or denoted by the symbol $\varnothing$.

□

Example 4

$$
\begin{aligned}
2x - (7 - x) &= x + 1 - 2(4 - x) \\
2x + x - 7 &= x + 1 - 8 + 2x \\
3x - 7 &= 3x - 7 \\
3x &= 3x \\
0 &= 0 && \text{True equation}
\end{aligned}
$$

This is an identity; because it is equivalent to the original equation, **the solution set is all real numbers.** □

Another type of statement is an **inequality.** Inequalities can be sentences that are always true (for example, $x - 1 < x$ or $5 < 7$), called **absolute inequalities;** always false (for example, $x > x$), called **contradictions;** or sometimes true and sometimes false (for example, $x > 2$), called **conditional inequalities.** The latter can be solved by using a set of properties similar to those for equations.

PROPERTIES OF EQUIVALENT INEQUALITIES

If P and Q are algebraic expressions, and k is a real number, then each of the following is equivalent to $P < Q$:

1. $P + k < Q + k$
2. $P - k < Q - k$
3. $kP < kQ$ if $k > 0$
   $kP > kQ$ if $k < 0$
4. $\frac{P}{k} < \frac{Q}{k}$ if $k>0$
   $\frac{P}{k} > \frac{Q}{k}$ if $k<0$

These properties also hold for $\leqslant$, $>$, and $\geqslant$.

Essentially, these properties allow any operation that is allowed for equations, except that multiplying or dividing by a negative number reverses the sense of the inequality.

Example 5

$$5 - 2(3x - 4) < 4x - 7$$
$$5 - 6x + 8 < 4x - 7$$
$$-6x + 13 < 4x - 7 \quad \text{Simplify.}$$
$$-6x - 4x + 13 < 4x - 4x - 7 \quad \text{Subtract } 4x.$$
$$-10x + 13 < -7$$
$$-10x + 13 - 13 < -7 - 13 \quad \text{Subtract 13.}$$
$$-10x < -20$$
$$\frac{-10x}{-10} > \frac{-20}{-10} \quad \text{Divide by } -10\text{. Note that the sense of the inequality is changed.}$$
$$\mathbf{x > 2}$$

The solution set is graphed as

where the open point indicates that $x = 2$ is excluded. □

Example 6

$$5(2x - 3) - 4(x - 2) \leqslant 3(x - 3)$$
$$10x - 15 - 4x + 8 \leqslant 3x - 9$$
$$6x - 7 \leqslant 3x - 9$$
$$3x - 7 \leqslant -9$$
$$3x \leqslant -2$$
$$\mathbf{x \leqslant -\frac{2}{3}}$$

The solution set is graphed as

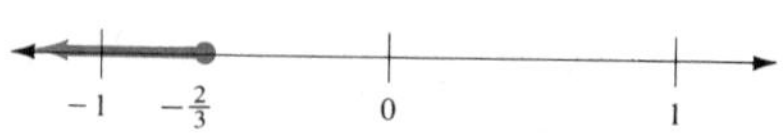

where the closed point indicates that $x = -\frac{2}{3}$ is included. □

A "string of inequalities" may be used to show the order of three or more quantities. For instance, $2 < x < 5$ states that x is a number *between* 2 and 5. The statement is a **compound inequality,** equivalent to $x > 2$ *and* $x < 5$, and is graphed as the interval on the number line between 2 and 5:

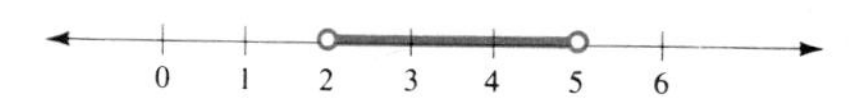

Such inequalities may be solved in a similar way to that used for other inequalities. Note that what is done to one member of the string in each of the following inequalities is done to all three members.

Example 7

$$-3 \leq 2x - 5 \leq 7$$

$$-3 + 5 \leq 2x - 5 + 5 \leq 7 + 5 \qquad \text{Add 5.}$$

$$2 \leq 2x \leq 12$$

$$\frac{2}{2} \leq \frac{2x}{2} \leq \frac{12}{2} \qquad \text{Divide by 2.}$$

$$\mathbf{1 \leq x \leq 6}$$

□

Example 8

$$-2 < 1 - 3x < 7$$

$$-3 < -3x < 6 \qquad \text{Subtract 1.}$$

$$1 > x > -2 \qquad \text{Divide by } -3\text{, reversing the sense of both inequalities.}$$

or

$$\mathbf{-2 < x < 1}$$

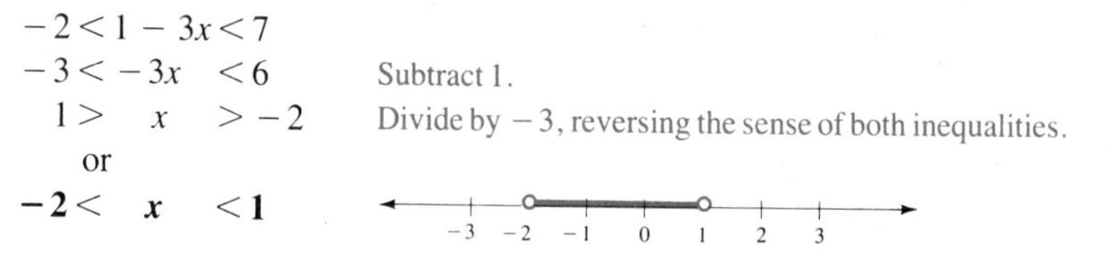

□

The string of inequalities $1 < 2 < 3$ is equivalent to $3 > 2 > 1$. The first states the order of the three integers from smallest to largest, and the second from largest to smallest. It is standard practice to use the ascending order from smallest to largest, so inequalities are stated with the less than, $<$, relation whenever possible. In Example 8, notice that $1 > x > -2$ is rewritten as $-2 < x < 1$.

Problem Set 2.1

A *Solve each equation in Problems 1–20.*

1. $3x = 5x - 4$

2. $2x = 9 - x$

3. $7x + 10 = 5x$

4. $9x - 8 = 5x$

5. $3x + 22 = 1 - 4x$

6. $5x - 4 = 3x + 8$

7. $2x - 13 = 7x + 2$

8. $3x + 32 = 18 - 4x$

9. $7x + 18 = -2x$

10. $6x - 1 = 13 - x$

11. $8x - 3 = 15 - x$

12. $-5x = 9 - 2x$

13. $2(x - 1) - 1 = 7 - 3x$

14. $11 - x = 2 - 3(x - 1)$

15. $5 - 2x = 1 + 3(x - 2)$

16. $1 - 2(x - 3) = 2x - 1$

17. $6 - 2[x - 2(3 - x)] = 0$

18. $5 - x = 2 - [2 - 3(x - 1)]$

19. $3 - [2(1 - x) - x] = 4$

20. $[x - 3(x - 2)] - 3 = x$

Solve each inequality in Problems 21–30 and graph the solution set on a number line.

21. $3x - 2 \leqslant 7$

22. $2x - 1 \geqslant 9$

23. $4 - 5x > 29$

24. $7 - 4x < 3$

25. $9x + 7 \geqslant 5x - 9$

26. $3x + 7 > 7x - 5$

27. $2(3 - 4x) < 30$

28. $3(2x - 5) \leqslant 9$

29. $2x - [(x - 3) - 5] < 0$

30. $[7 - 3x - 2(1 - x)] + x \geqslant 11 - 2x$

B *Solve each equation or inequality in Problems 31–50 and then graph the solution set of each inequality on a number line.*

31. $2(m - 3) - 5m = 3(1 - 2m)$

32. $3(2n + 7) + 11 = 5(2 - n)$

33. $5(x - 1) + 3(2 - 4x) > 8$

34. $4(1 - y) - 7(2y - 5) < 3$

35. $4(a - 2) + 1 \leqslant 3(a + 1)$

36. $6(2b - 1) \geqslant 3(b + 4)$

37. $5h + 3(h + 2) + (h + 4) + 17 = 0$

38. $6(k - 2) - 4(k + 3) + 42 = 10$

39. $2(r + 5) - 7 = 6 + 2(r + 1)$

40. $5(s - 3) + 29 = 4 + 5(s + 2)$

41. $2(4 - 3x) > 4(3 - x)$

42. $3(2 - 5y) \leqslant 5(y + 2) + 36$

43. $2(3 - 7a) \geqslant -4 - (5 - a)$

44. $3(7 - b) < 2(b - 2)$

45. $3(1 - x) - 5(x - 2) = 5$

46. $5(5 - 3y) - (y - 8) = 1$

47. $2[3 - 2(m - 5)] < 5 - m$

48. $2(2n - 9) \geqslant 3[(2 - n) - 1]$

49. $2(2x + 7) \geqslant 2(x - 3) - 3(x + 5)$

50. $5(y + 7) - 3(2y - 4) < 4(y - 2)$

Solve the compound inequalities in Problems 51–60 and graph each solution on a number line.

51. $7 < x + 2 < 11$

52. $12 < 5 + y < 14$

53. $-2 < x - 1 < 3$

54. $-5 \leqslant y - 2 \leqslant 4$

55. $-8 \leqslant 2x - 4 \leqslant 6$

56. $-14 < 3y - 5 < 4$

57. $-4 \leqslant 2x - 3 < 2$

58. $3 < 3y + 5 \leqslant 6$

59. $9 < 1 - 2x < 15$

60. $-3 \leqslant 1 - 2y \leqslant 7$

C *Solve each equation or inequality in Problems 61–68 and then graph the solution set of each inequality on a number line.*

61. $(2a - 1)(a + 3) - (a + 1)(a + 7) = (a - 1)(a + 5)$

62. $(2b + 5)(b + 3) + (b + 3)(b + 7) = 3(b + 1)^2$

63. $(4y - 3)(y - 2) - (3y + 1)(y - 2) \leq (y - 3)^2$

64. $(2x - 1)(2x + 3) \geq (x - 1)(2x + 1) + 2(x + 1)^2$

65. $(z + 2)^2 - (z - 2)^2 = 2$

66. $(w - 3)^2 - 3 = (w + 3)^2$

67. $u(u - 3) - 2u - 3(2 - u) - 3 = (u - 2)^2$

68. $2v(2 + v) - 5(v + 4) - 2[2 - v(3 - v) - 1] + 1 = 0$

69. The *arithmetic mean* of two numbers is one-half their sum. The *harmonic mean* of two numbers is twice their product divided by their sum. Given two unequal positive numbers a and b, prove that their arithmetic mean is always greater than their harmonic mean.

2.2 Applications, Word Problems

You must approach word problems with the realization that algebra will not solve them—*you will.* If an equation or inequality can be found to represent accurately and fully the relationships presented in the problem, then that equation or inequality can be solved, and the solutions can be interpreted as an answer to the problem. However, the direct application of algebra to the words of the problem will not yield an equation; that is your contribution. This step can be accomplished only if you understand the problem well enough to state its relationships algebraically. You must, therefore, thoroughly *understand* the problem *before* you can solve it. The following suggestions give a strategy that is helpful in attacking such problems:

STRATEGY OF SOLVING APPLIED PROBLEMS

1. Read the problem. Determine what it is all about. Be certain you know what information is given and what must be found. You must understand a problem if you are to solve it.
2. Analyze the problem. Focus on processes rather than numbers. Recall what you know about the subject of the problem. Display this information to help clarify the facts and relationships.
3. Identify the variable and write the equation or inequality that evolves from an analysis of the problem after defining a variable. The discovery of this equation will be the result of your understanding of the problem, not your algebraic skill.
4. Solve the equation or inequality. Check the result in the original problem to make certain that you discovered the right equation. That is, your solution should make sense.
5. State the answer or answers to the problem. Pay attention to units of measure and other details of the problem.

Current studies of learning theories support such a strategy. The following is quoted from a recent address at a national convocation called by the National Academy of Science and the National Academy of Engineering: *"Good problem-solvers do not rush in to apply a formula or an equation. Instead, they try to understand the problem situation; they consider alternative representations and relations among variables. Only when satisfied that they understand the situation and all the variables in a qualitative way do they start to apply the quantification."** You are well-advised to begin to develop a strategy with which you are comfortable. If the strategy makes sense, your confidence will provide greater success.

Applied problems are often concerned with length, area, and volume. Since these are largely geometric ideas, begin whenever possible with a sketch. Watch carefully in the following examples as the strategy is applied to help solve the problems.

Example 1 Two rectangles have the same width, but one is 40 m^2 larger in area. The larger rectangle is 6 m longer than it is wide. The other is only 1 m longer than its width. What are the dimensions of the larger rectangle?

Analysis The *areas* differ by 40, and each area is the product of length and width:

$$\begin{pmatrix}\text{AREA OF}\\ \text{LARGER}\end{pmatrix} - \begin{pmatrix}\text{AREA OF}\\ \text{SMALLER}\end{pmatrix} = 40$$

$$\begin{pmatrix}\text{LENGTH}\\ \text{OF LARGER}\end{pmatrix}(\text{WIDTH}) - \begin{pmatrix}\text{LENGTH}\\ \text{OF SMALLER}\end{pmatrix}(\text{WIDTH}) = 40$$

The width is the same in each case, but is related differently in each figure to its length:

$$(\text{WIDTH} + 6)(\text{WIDTH}) - (\text{WIDTH} + 1)(\text{WIDTH}) = 40$$

Solution Let W be the width of each rectangle. Then

$$\begin{aligned} (W + 6)(W) - (W + 1)(W) &= 40 \\ W^2 + 6W - W^2 - W &= 40 \\ 5W &= 40 \\ W &= 8 \qquad \text{and} \qquad W + 6 = 14 \end{aligned}$$

The larger rectangle is 8 m wide by 14 m long. □

Alternate Analysis A sketch will frequently simplify a problem. In this case, since the widths are the same, the difference in area can be seen in the larger rectangle:

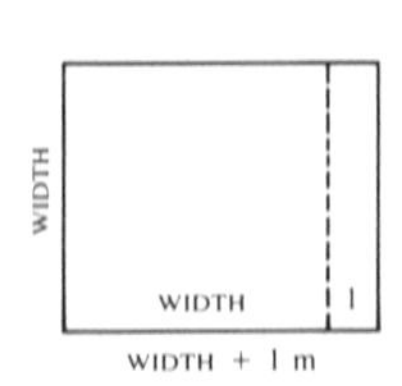

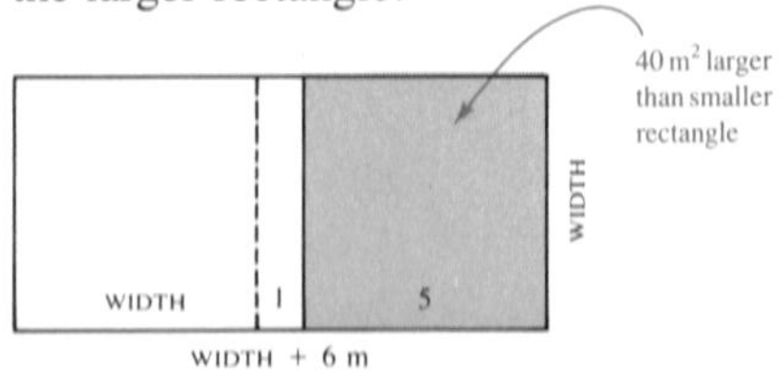

*An address by Lauren B. Resnick, National Convocation on Precollege Education in Mathematics, National Academy of Science and National Academy of Engineering, Washington, D.C., May 1982. (Quoted in *Science*, vol. 220, no. 4596, April 29, 1983, p. 29.)

The difference is shown as the shaded region. The area of this region is stated in the problem:

$$\begin{pmatrix}\text{DIFFERENCE}\\ \text{IN AREA}\end{pmatrix} = 40$$

$$(5)(\text{WIDTH}) = 40$$

Alternate Solution Let W be the width of each rectangle. Then

$$\begin{aligned} 5W &= 40 \\ W &= 8 \quad \text{and} \quad W + 6 = 14 \end{aligned}$$

The larger rectangle is 8 m wide by 14 m long. □

A dictionary definition of *rate* is "the quantity of a thing in relation to the units of something else." This definition is quite general, yet rate is too often limited to rate–time–distance relationships. A typing rate is the number of words typed per unit of time, and most prices are based on cost per unit. The following example deals with the rate of interest earned as a percent of the principal invested.

Example 2 A total of \$1400 is invested for a year, part at 5% and the rest at $6\frac{1}{2}\%$. If \$76 is collected in interest, how much is invested at 5%?

Analysis First, interest is earned from two separate investments, and since interest is the product of the principal and the rate, we have

$$\begin{pmatrix}\text{INTEREST}\\ \text{AT } 5\%\end{pmatrix} + \begin{pmatrix}\text{INTEREST}\\ \text{AT } 6\frac{1}{2}\%\end{pmatrix} = \begin{pmatrix}\text{TOTAL}\\ \text{INTEREST}\end{pmatrix}$$

$$\begin{pmatrix}\text{PRINCIPAL}\\ \text{AT } 5\%\end{pmatrix}(\text{RATE}) + \begin{pmatrix}\text{PRINCIPAL}\\ \text{AT } 6\frac{1}{2}\%\end{pmatrix}(\text{RATE}) = \begin{pmatrix}\text{TOTAL}\\ \text{INTEREST}\end{pmatrix}$$

Now some of the known quantities may be replaced by their values and since the total investment is known, one principal may be written in terms of the other:

$$\begin{pmatrix}\text{PRINCIPAL}\\ \text{AT } 5\%\end{pmatrix}(.05) + \left(1400 - \begin{matrix}\text{PRINCIPAL}\\ \text{AT } 5\%\end{matrix}\right)(.065) = 76$$

Solution Let P be the principal invested at 5%. Then

$$\begin{aligned} .05P \;\; .065(1400 - P) &= 76 \\ .05P + 91 - .065P &= 76 \\ 91 - .015P &= 76 \\ -.015P &= -15 \\ P &= 1000 \end{aligned}$$

The amount invested at 5% is \$1000. □

Example 3 An office worker takes 55 minutes to return from the job each day. This person rides a bus that averages 30 mph and walks the rest of the way at 4 mph. If the total distance is 21 miles from office to home, what is the distance walked each day?

Analysis The total trip is composed of two distinct distances. These distances are the product of rate and time, so once again *rate* is an important ingredient of the analysis.

BUS WALK TOTAL

$$\begin{pmatrix}\text{BUS}\\ \text{DISTANCE}\end{pmatrix} + \begin{pmatrix}\text{WALK}\\ \text{DISTANCE}\end{pmatrix} = \begin{pmatrix}\text{TOTAL}\\ \text{DISTANCE}\end{pmatrix}$$

$$\begin{pmatrix}\text{BUS}\\ \text{RATE}\end{pmatrix}\begin{pmatrix}\text{BUS}\\ \text{TIME}\end{pmatrix} + \begin{pmatrix}\text{WALK}\\ \text{RATE}\end{pmatrix}\begin{pmatrix}\text{WALK}\\ \text{TIME}\end{pmatrix} = \begin{pmatrix}\text{TOTAL}\\ \text{DISTANCE}\end{pmatrix}$$

The total time is known; therefore, the bus time may be expressed in terms of the walk time. Be careful to express the time in *hours* since the rate is given in miles per *hour:*

$$(30)\left(\frac{55}{60} - \begin{matrix}\text{WALK}\\ \text{TIME}\end{matrix}\right) + (4)\begin{pmatrix}\text{WALK}\\ \text{TIME}\end{pmatrix} = 21$$

Solution Let W be the time walked, in hours. Then

$$\begin{aligned} 30\left(\frac{55}{60} - W\right) + 4W &= 21 \\ \frac{55}{2} - 30W + 4W &= 21 \\ 27.5 - 26W &= 21 \\ -26W &= -6.5 \\ W &= .25 \end{aligned}$$

The worker walks .25 hour, but the problem asks for the distance, so returning to our original analysis, we find

$$\begin{aligned} \begin{pmatrix}\text{WALK}\\ \text{DISTANCE}\end{pmatrix} &= \begin{pmatrix}\text{WALK}\\ \text{RATE}\end{pmatrix}\begin{pmatrix}\text{WALK}\\ \text{TIME}\end{pmatrix} \\ &= 4(.25) \\ &= 1 \end{aligned}$$

Be certain that the solution answers the problem posed.

The worker walks 1 mile daily. □

Example 4 Milk containing 10% butterfat and cream with 80% butterfat are mixed to produce half-and-half, which is 50% butterfat. How many gallons of each must be mixed to produce 140 gallons of half-and-half?

Analysis In mixture problems of this sort, focus on the *amount* of each substance:

$$\begin{pmatrix}\text{AMT OF}\\ \text{BUTTERFAT}\\ \text{IN MILK}\end{pmatrix} + \begin{pmatrix}\text{AMT OF}\\ \text{BUTTERFAT}\\ \text{IN CREAM}\end{pmatrix} = \begin{pmatrix}\text{AMT OF}\\ \text{BUTTERFAT IN}\\ \text{HALF-AND-HALF}\end{pmatrix}$$

Each amount is given as a percentage (or rate) of its volume, so we may write the amounts as products:

$$\begin{pmatrix}\text{PCT IN}\\ \text{MILK}\end{pmatrix}\begin{pmatrix}\text{VOL OF}\\ \text{MILK}\end{pmatrix} + \begin{pmatrix}\text{PCT IN}\\ \text{CREAM}\end{pmatrix}\begin{pmatrix}\text{VOL OF}\\ \text{CREAM}\end{pmatrix}$$

$$= \begin{pmatrix}\text{PCT IN}\\ \text{HALF-AND-HALF}\end{pmatrix}\begin{pmatrix}\text{VOL OF}\\ \text{HALF-AND-HALF}\end{pmatrix}$$

$$.10\begin{pmatrix}\text{VOL OF}\\ \text{MILK}\end{pmatrix} + .80\left(140 - \begin{matrix}\text{VOL OF}\\ \text{MILK}\end{matrix}\right) = .50(140)$$

Solution Let M be the volume of milk used, in gallons. Then

$$\begin{aligned} .10M + .80(140 - M) &= .50(140) \\ .10M + 112 - .80M &= 70 \\ -.70M &= -42 \\ M &= 60 \qquad \text{and} \qquad 140 - M = 80 \end{aligned}$$

The half-and-half contains 60 gal of milk and 80 gal of cream. □

Problem Set 2.2

A *For Problems 1–48 carefully analyze the problem, evolve and solve an equation, and state an answer. Watch for rates involved in the problems.*

1. Find two consecutive numbers whose sum is 125.
2. Find three consecutive odd integers whose sum is 57.
3. Find the first of three consecutive odd integers if their sum is 99.
4. What is the largest of three consecutive integers if their sum is 84?
5. The product of two consecutive integers is 62 less than the product of the next two integers. What is the second of the four integers?
6. The product of two consecutive odd integers is 208 more than the product of the preceding two odd integers. What is the third of these odd integers?
7. Two rectangles have the same width, but one is 20 ft^2 larger in area. The larger rectangle is 6 ft longer than it is wide. The other is only 2 ft longer than its width. What are the base and height of the larger rectangle?
8. Two rectangles have the same length, but one is 24 in.^2 larger in area. The larger rectangle is only 4 in. shorter than it is long. The other is 6 in. shorter than its length. What are the base and height of the larger rectangle?
9. Two triangles have the same height. The base of the larger one is 3 cm greater than its height. The base of the other is 1 cm greater than its height. If the areas differ by 3 cm^2, then find the dimensions of the smaller figure.

10. Two triangles have the same base. The height of the larger one is 2 yd shorter than its base. The height of the other is 4 yd shorter than its base. If the areas differ by 6 yd^2, then find the dimensions of the smaller figure.

B **11.** A rectangle is 2 ft longer than it is wide. If you increase the length by a foot and reduce the width the same, the area is reduced by 3 ft^2. Find the width of the new figure.

12. The length of a rectangle is 3 m more than its width. The width is increased by 2 m, and the length is shortened by a meter. If the two figures have the same perimeter, what is it?

13. *Business* A total of \$1000 was invested for one year, part of it at 10% and the rest at 12.5%. If \$110 interest was earned, how much was invested at 10%?

14. *Business* A total of \$1200 was invested for one year, part of it at 10% and the rest at 12%. If \$130 interest was earned, how much was invested at 12%?

15. *Business* A total of \$8000 is invested for one year, part at 9% and the remainder at 8%. If \$665 interest is earned, how much is invested at 9%?

16. *Business* A total of \$12,000 is invested for one year, part at $9\frac{1}{2}$% and the remainder at 11%. If \$1275 interest is earned, how much is invested at 11%?

17. *Business* Part of \$1500 is invested at $9\frac{1}{2}$% and the remainder is invested at 14%. The combined investments will yield \$183 interest the first year. How much is invested at each rate?

18. *Business* Part of \$9000 is invested at 7.5% and the remainder is invested at 11.5%. The combined investments will yield \$804 interest the first year. How much is invested at each rate?

19. *Business* One part of \$20,000 is invested at $10\frac{1}{4}$% and the rest at $11\frac{1}{4}$%. A total of \$2165 will be earned from the first year's interest. How much is invested at the lower rate?

20. *Business* One part of \$12,400 is invested at 9.75% and the rest at 10.75%. A total of \$1256 will be earned from the first year's interest. How much is invested at the higher rate?

21. A trip is made by train and bus. The train averages 72 mph and the bus only 39 mph. The total trip of 405 mi takes 7 hr. What distance is traveled by train?

22. A trip is made by ship and train. The ship averages 32 mph and the train 52 mph. The total trip of 660 mi takes 20 hr. How many hours are spent on the ship?

23. A businesswoman logs time in an airliner and a rental car to reach her destination. The total trip is 1100 mi, the plane averaging 600 mph and the car 50 mph. How long is spent in the automobile if the trip took a total of $5\frac{1}{2}$ hr?

24. Stu hitchhikes back to campus from home, which is 82 mi away. He makes 4 mph walking, until he gets a ride. In the car, he makes 48 mph. If the trip took 4 hr, how far did Stu walk?

25. A commuter takes an hour to get to work each day. She takes rapid transit that averages 65 mph and a shuttle bus the rest of the way at 25 mph. If the total distance is 61 mi from home to work, how far does she ride the bus to get to work each day?

26. A commuter rides a train into the city and catches a taxi to the office, traveling 50 mi daily. The train averages 60 mph and the cab 20 mph. If his trip takes a total of 1 hr, then how much time is spent on the train each morning?

27. Two joggers set out at the same time from their homes 21 mi apart. They agree to meet at a point in between in an hour and a half. If the rate of one is 2 mph faster than the other, find the rate of each.

28. Two joggers set out at the same time but in opposite directions. If they were to maintain their normal rates for 4 hr, they could be 68 mi apart. If the rate of one is 1.5 mph faster than the other, find the rate of each.

29. Cori commutes 30 mi to work each day, partly on a highway and partly in city traffic. On the highway she doubles her city speed for just 15 minutes. If the entire trip takes an hour, how fast is she able to average in city traffic?

30. Chris drives 56 mi to his job in the city every day, part on the highway and part on city streets. Off the highway, traffic crawls along at a third of his highway speed for 10 minutes to complete the journey. If the trip takes an hour, how fast is Chris able to average in the city?

31. Milk containing 12% butterfat and cream containing 69% butterfat are blended to produce half-and-half, which is 50% butterfat. How many gallons of each must be mixed to make 150 gallons of half and half?

32. Milk containing 8% butterfat and cream with 64% butterfat are mixed to produce half-and-half, which is 50% butterfat. How many gallons of each must be mixed to make 200 gallons of half-and-half?

33. *Chemistry* A chemist has two solutions of sulfuric acid. One is a 72% solution, and the other is a 45% solution. How many liters of each does the chemist mix to get 4.5 liters of a 60% solution?

34. *Chemistry* A chemist has two solutions of hydrochloric acid. One is a 50% solution, and the other is a 32% solution. How many milliliters of each must be mixed to get 75 ml of 44% solution?

35. *Chemistry* How much water must be added to a 35% acid solution to obtain 100 cc of a 21% solution?

36. *Metallurgy* How much pure silver must be alloyed with a 36% silver alloy to obtain 100 g of a 52% alloy?

37. An aftershave lotion is 50% alcohol. If you have 8 fluid ounces of the lotion, how much water must be added to reduce the mixture to 20% alcohol?

38. If you have 8 fluid ounces of an aftershave lotion that is 20% alcohol, how much alcohol must be added to enhance the mixture to 50% alcohol?

39. You have 8 oz of an aftershave lotion that is 50% alcohol. How much of a 15% alcohol lotion must be added to reduce the mixture to 25% alcohol?

40. If 8 oz of an aftershave lotion is 15% alcohol, how much of a 50% alcohol lotion must be added to produce a 30% alcohol lotion?

C **41.** A radiator contains 8 quarts of a 40% antifreeze mixture. Some of the mixture must be drained and replaced by pure antifreeze to increase the amount of antifreeze in the mixture to 60% antifreeze. How much pure antifreeze must be added?

42. A radiator contains 8 quarts of a 75% antifreeze solution. Some of the solution is to be drained for another purpose and replaced by water to decrease the solution to 60% antifreeze. How much of the original solution should be replaced by water?

43. *Consumer issue* An electric company charges \$40 for the first 1000 kilowatt-hours or less, 6¢ per kilowatt-hour for the next 1000 kilowatt-hours, and 8¢ per kilowatt-hour for anything beyond. If the charge is \$119.20, how many kilowatt-hours were used?

44. Current postal regulations do not permit a package to be mailed if the combined length, width, and height exceed 72 in. What are the dimensions of the largest permissible package with length twice the size of its square end?

45. Two brothers leave home at the same time and walk in opposite directions, one at 3.5 mph and the other at 4 mph. How long will it be before the boys are 15 mi apart? How far must each boy walk?

46. Two trains leave towns 84 mi apart at the same time and travel toward one another. If one travels at 25 mph and the other at 31 mph, in how many hours will they meet? How far must each train travel?

47. A girl riding a motorcycle at 35 mph left home $2\frac{1}{2}$ hours after her brother, who was riding a bicycle at 10 mph. How long will it take the girl on the motorcycle to overtake her brother if they are traveling in the same direction?

48. An airplane flying 120 mph leaves Newport 4 hours after a yacht has sailed and overtakes it in 1 hour. What is the speed of the yacht, and how far had it sailed before being overtaken?

2.3 Absolute Value Equations and Inequalities

Recall the definition of absolute value given in Section 1.1:

ABSOLUTE VALUE

$$|n| = \begin{cases} n & \text{if } n \geq 0 \\ -n & \text{if } n < 0 \end{cases}$$

This definition, together with the usual properties of equality and order, can be applied to develop several useful properties of absolute value (see Section 10.3). For completeness, they are listed in Table 2.1 on page 70.

Consider Property 7 in Table 2.1. If $|a| = |b|$ and the definition of absolute value states that $|a| = \pm a$ and $|b| = \pm b$, then $\pm a = \pm b$. Reversing this reasoning, if $\pm a = \pm b$, then $|a| = |b|$. That is, numbers with equal absolute values differ only in sign. A careful proof is given in Section 10.3 (see Theorem 13 on page 420).

This property may be applied to an equation of the form $|x| = k$, where k is a positive constant. The equation is equivalent to the compound statement $x = k$ *or* $x = -k$, and each of these equations may be solved in the usual way. But first you should restate the equation without absolute value.

Table 2.1 Properties of Absolute Value

Let a be a real number.

Property	Comment
1. $\|a\| \geq 0$	1. Absolute value is nonnegative.
2. $\|-a\| = \|a\|$	2. The absolute value of a number and the absolute value of its opposite are equal.
3. $\|a\|^2 = a^2$	3. If an absolute value is squared, the absolute value sign can be dropped.
4. $\|ab\| = \|a\|\,\|b\|$	4. This property tells how to multiply absolute values. The absolute value of a product is the product of the absolute values.
5. $\left\|\frac{a}{b}\right\| = \frac{\|a\|}{\|b\|}$ $b \neq 0$	5. This property tells how to divide absolute values. The absolute value of a quotient is the quotient of the absolute values.
6. $-\|a\| \leq a \leq \|a\|$	6. Any number a is between + and − the absolute value of that number.
7. $\|a\| = \|b\|$ if and only if $a = \pm b$	7. This property tells how to solve absolute value equations.
8. $\|a\| < b$ if and only if $-b < a < b$ for $b > 0$	8. & 9. These properties tell how to solve absolute value inequalities. Property 8 also holds for $\leq$ and Property 9 holds for $\geq$. The condition $b > 0$ is necessary because of Property 1.
9. $\|a\| > b$ if and only if either $a > b$ or $a < -b$ for $b > 0$	
10. $\|a + b\| \leq \|a\| + \|b\|$	10. This property is called the **triangle inequality** and is an important property of absolute value.
11. $\|a - b\| \geq \|a\| - \|b\|$	11. This is a property of absolute value of a difference. Also, $\|a - b\| \leq \|a\| + \|b\|$.

Example 1

$|x| = 2$

$x = 2$ or $x = -2$ Restate without absolute value.

$\{2, -2\}$ State the solution set. □

Example 2

$|x - 3| = 5$

$x - 3 = 5$ or $x - 3 = -5$ Restate.

$x = 8$ $x = -2$ Solve separately.

$\{8, -2\}$ State the solution set. □

Example 3

$|2x + 1| = 7$

$2x + 1 = 7$ or $2x + 1 = -7$

$2x = 6$ $2x = -8$

$x = 3$ $x = -4$

$\{3, -4\}$ □

Example 4

$$|3x - 4| = 9$$

$$3x - 4 = 9 \quad \text{or} \quad 3x - 4 = -9$$

$$3x = 13 \qquad 3x = -5$$

$$x = \frac{13}{3} \qquad x = -\frac{5}{3}$$

$$\left\{\frac{13}{3}, -\frac{5}{3}\right\}$$

□

Example 5

$$|5 - 2x| = -3$$

No solution since the absolute value is always nonnegative. (See Property 1 in Table 2.1.) □

The absolute value inequality is actually encountered more often than the absolute value equation. It is a common means of specifying conditions, intervals, tolerances, errors, or restrictions of various sorts. Note that Properties 8 and 9 in Table 2.1 deal with absolute value inequalities. A careful proof of Property 8 is given in Section 10.3 (see Theorem 14 on page 420). Consider a special case: $|x| < k$, where k is a positive constant.

The inequality $|x| < k$ involves two conditions,

$$x < k \quad \text{if} \quad x \geq 0 \qquad \text{or} \qquad -x < k \quad \text{if} \quad x < 0$$

Since $-x < k$ is equivalent to $-k < x$, these two conditions can be written as

$$0 \leq x < k \qquad \text{or} \qquad -k < x < 0$$

or more simply,

$$-k < x < k$$

In general, $|x| < k$ is equivalent to $-k < x < k$.

Example 6

$$|x - 2| < 1$$

$$-1 < x - 2 < 1 \qquad \text{Rewrite without absolute value.}$$

$$\mathbf{1 < x < 3} \qquad \text{Add 2 to each member of the inequality.}$$

□

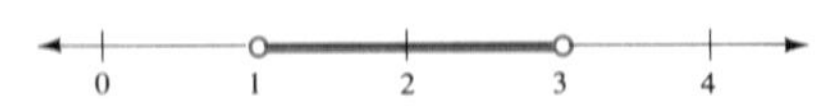

Example 7

$$|2x + 1| \leq 5$$

$$-5 \leq 2x + 1 \leq 5 \qquad \text{Rewrite.}$$

$$-6 \leq 2x \leq 4 \qquad \text{Subtract 1 from each member.}$$

$$\mathbf{-3 \leq x \leq 2} \qquad \text{Divide each member by 2.}$$

□

Example 8

$$|x + 1| < 0$$

No solution since the absolute value is always nonnegative. (See Example 5.) □

Example 9

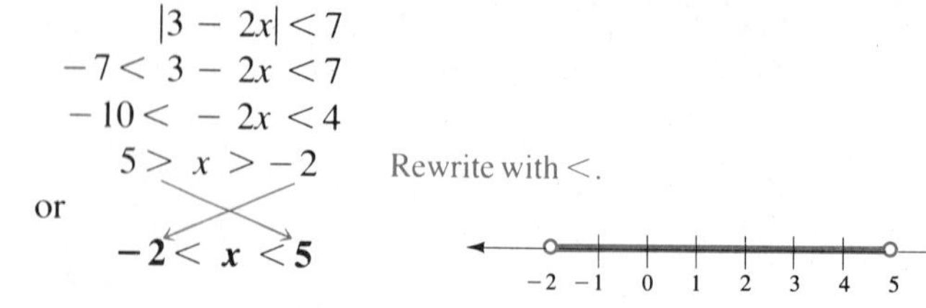

$$
\begin{aligned}
|3 - 2x| &< 7 \\
-7 < 3 - 2x &< 7 \\
-10 < -2x &< 4 \\
5 > x &> -2 \quad \text{Rewrite with} <.
\end{aligned}
$$

or

$$\mathbf{-2 < x < 5}$$

□

The variable is customarily written first in expressions such as $|3 - 2x|$, since $|3 - 2x| = |2x - 3|$. Note further that if we write $|2x - 3|$, we avoid having to rewrite the solution with $<$, since we have

$$
\begin{aligned}
|2x - 3| &< 7 \\
-7 < 2x - 3 &< 7 \\
-4 < 2x &< 10 \\
-2 < x &< 5
\end{aligned}
$$

Finally, consider the "greater than" statement as shown in Property 9. Again consider a special case: $|x| > k$, where k is a positive constant. Using the definition of absolute value, $|x| > k$ may be written as

$$x > k \quad \text{if} \quad x \geq 0 \qquad \text{or} \qquad -x > k \quad \text{if} \quad x < 0$$

and $-x > k$ is equivalent to $x < -k$. Thus,

$$x > k \qquad \text{or} \qquad x < -k$$

Thus, as in the other cases, the statement may be restated without absolute value and solved in the usual way.

Example 10

$$
\begin{aligned}
&|x + 3| > 4 \\
&x + 3 > 4 \quad \text{or} \quad x + 3 < -4 \\
&\mathbf{x > 1} \quad \textbf{or} \quad \mathbf{x < -7}
\end{aligned}
$$

□

Example 11

$$
\begin{aligned}
&|2x - 5| > 3 \\
&2x - 5 > 3 \quad \text{or} \quad 2x - 5 < -3 \\
&2x > 8 \quad \text{or} \quad 2x < 2 \\
&\mathbf{x > 4} \quad \textbf{or} \quad \mathbf{x < 1}
\end{aligned}
$$

□

Example 12 $|3 - 4x| > -5$

Notice that the absolute value is always nonnegative, so **this inequality is always true**. However, solving will verify this. (See Examples 5 and 8.)

$$\begin{aligned} 3 - 4x &> -5 \quad \text{or} \quad 3 - 4x < 5 \\ -4x &> -8 \quad \text{or} \quad -4x < 2 \\ x &< 2 \quad \text{or} \quad x > -\frac{1}{2} \end{aligned}$$

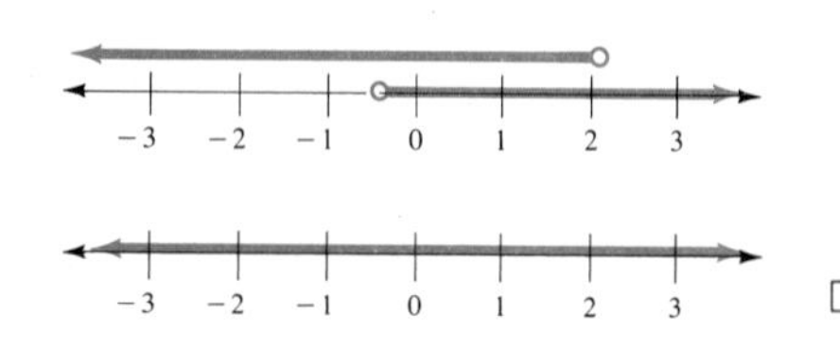

□

Example 13

$$|1 - 3x| \geqslant 2$$

$$\begin{aligned} 1 - 3x &\geqslant 2 \quad \text{or} \quad 1 - 3x \leqslant -2 \\ -3x &\geqslant 1 \quad \text{or} \quad -3x \leqslant -3 \\ \boldsymbol{x} &\leqslant \boldsymbol{-\frac{1}{3}} \quad \textbf{or} \quad \boldsymbol{x \geqslant 1} \end{aligned}$$

□

Recall that $|x| \leqslant k$ is equivalent to $|x| \not> k$. That is, the solution of $|x| > k$ is all the points that do not satisfy $|x| \leqslant k$. We call $|x| \leqslant k$ the *complement* of $|x| > k$. This provides an alternate approach to the case of $|x| > k$. Reconsider Example 10. To solve $|x + 3| > 4$, use its complement:

$$\begin{aligned} |x + 3| &\leqslant 4 \\ -4 \leqslant x + 3 &\leqslant 4 \\ -7 \leqslant x &\leqslant 1 \end{aligned}$$

Now sketch the complement of this portion of the number line to obtain the graph of $|x + 3| > 4$:

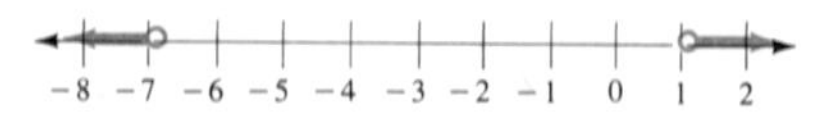

REWRITING ABSOLUTE VALUE STATEMENTS

To rewrite the absolute value statement, use the following, where k is a positive number:

$|x| < k$ $-k < x < k$	$|x| = k$ $x = k$ or $x = -k$	$|x| > k$ $x < -k$ or $x > k$

Thus, the first step in dealing with an absolute value equation or inequality is to rewrite the statement without the absolute value. The usual properties of equality or order can then be applied.

Distance Revisited

Absolute value was interpreted as a distance in Section 1.1, where $d = |x_2 - x_1|$ is the distance between x_1 and x_2 on the number line. The difference $x_2 - x_1$ is the change in x and is denoted by Δx (read delta x). The distance can be written in terms of Δx:

$$d = \sqrt{\Delta x^2}$$ Note that Δx is one symbol, so Δx^2 means $(\Delta x)^2$.

since $\sqrt{\Delta x^2} = |\Delta x| = |x_2 - x_1|$.

If a is some fixed value, then $|x - a|$ is the distance between x and a on the number line:

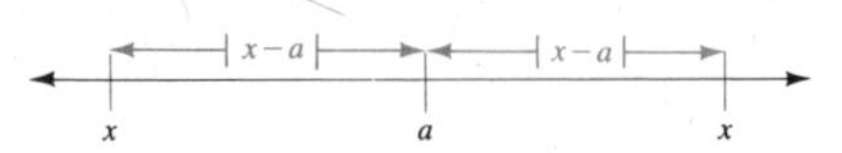

If the distance is fixed, as in $|x - a| = b$, there are two values of x that are the given distance from a. That is, $|x - a| = b$ states that x is a number b units from a. For example, $|x - 10| = 3$ states that x is 3 units from 10. Thus, x is either 7 or 13:

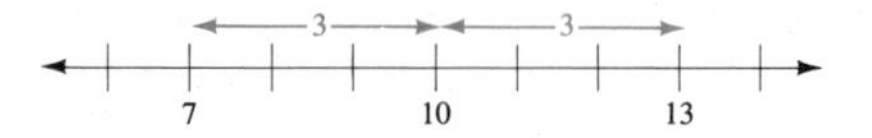

Algebraically, we verify this conclusion by solving the equation for x:

$$|x - 10| = 3$$
$$x - 10 = 3 \quad \text{or} \quad x - 10 = -3$$
$$x = 13 \qquad\qquad x = 7$$

The inequality $|x - 10| < 3$ may then be interpreted as any distance between x and 10 being less than 3. That is, x is any number less than 3 units from 10, which is any number between 7 and 13.

Suppose a 90 pound bag of cement is purchased. It will not be exactly 90 pounds. The material must be measured, and measurement is approximate. Some bags will be a little over 90 pounds, and some will be a little under 90 pounds, perhaps by a couple of pounds. If so, the bag could weigh as much as 92 pounds and as little as 88 pounds. This can be stated as an inequality. Let w be the weight of the bag in pounds. Then

$$88 \leq w \leq 92$$
$$88 - 90 \leq w - 90 \leq 92 - 90 \qquad \text{Subtract 90 from each member, since this is the average of the extremes, 88 and 92.}$$
$$-2 \leq w - 90 \leq 2$$

Equivalently,

$$|w - 90| \leq 2$$

To write the inequality in absolute value form, 90 was subtracted from each member above because 90 is the average value of the extremes, 88 and 92. In general, this average value, or distance, can be used to achieve the form $-k < x < k$, which is equivalent to $|x| < k$.

Example 14 Rewrite $-37 < x < 73$ using absolute value.

Solution First, find the average of -37 and 73:

$$\frac{-37 + 73}{2} = \frac{36}{2} = 18$$

Then subtract this average from each member:

$$-37 < x < 73$$
$$-37 - 18 < x - 18 < 73 - 18$$
$$-55 < x - 18 < 55$$

Finally, write as the equivalent absolute value inequality:

$$|x - 18| < 55$$ □

The final expression in Example 14 states that x is within 55 units of 18. This difference, when applied to measurement, is called **tolerance.** Tolerance is an allowable deviation from a standard. The example of the cement bag might be described as "90 pounds plus or minus 2 pounds."

Another case might involve temperature. Suppose a certain substance is unstable at or near room temperature. When storing the substance, you would want to avoid a temperature of or near 70°F. A safety factor would probably be in order, so you want to avoid temperatures between 60° and 80°F, inclusive. The distance from 70°F to a safe temperature must be more than 10°, so you may use absolute value when stating the situation algebraically:

$$|t - 70| > 10 \quad \text{Where } t \text{ is the temperature}$$
$$t - 70 > 10 \quad \text{or} \quad t - 70 < -10$$
$$t > 80 \quad \text{or} \quad t < 60$$

60 70 80

To summarize, the three cases of linear absolute value statements may be interpreted as distances.

ABSOLUTE VALUE AS A DISTANCE

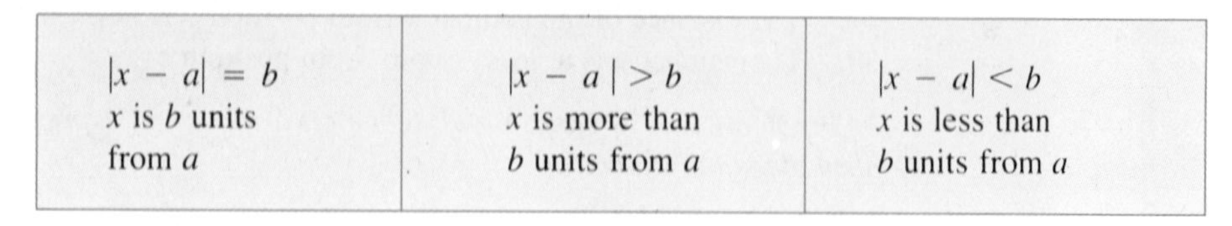

$\lvert x - a\rvert = b$	$\lvert x - a\rvert > b$	$\lvert x - a\rvert < b$
x is b units from a	x is more than b units from a	x is less than b units from a

In absolute value expressions containing a variable, the variable term is customarily written first. Although $|x - a| = |a - x|$, the first expression is preferred. That is, $|3n - 2|$ is preferred over $|2 - 3n|$.

Problem Set 2.3

A *Use the definition of* $|x|$ *in Problems 1–12 to rewrite each equation as two linear equations; then solve.*

1. $|x| = 4$ **2.** $|x| = 6$ **3.** $|x - 2| = 3$

4. $|x - 1| = 4$ **5.** $|a + 1| = 5$ **6.** $|b + 1| = 3$

7. $|2c - 1| = 7$ **8.** $|2d - 3| = 9$ **9.** $|3 - 4e| = 9$

10. $|2 - 3f| = 7$ **11.** $|3h + 1| = 10$ **12.** $|5k + 3| = 12$

B *Use the definition of $|x|$ to rewrite each inequality in Problems 13–40 without the absolute value; then solve and graph the solution set.*

13. $|y| < 2$
14. $|y| \leq 8$
15. $|x| \geq 5$
16. $|x| > 7$
17. $|x| < 1$
18. $|x| \leq 4$
19. $|2z| < 3$
20. $|3z| > 1$
21. $|5z| \leq 2$
22. $|b - 1| \geq 9$
23. $|d - 3| > 7$
24. $|h - 5| < 12$
25. $|k - 7| \leq 10$
26. $|m + 2| < 5$
27. $|n + 3| \geq 4$
28. $|2x - 5| \leq 11$
29. $|2x - 7| \leq 13$
30. $|2y - 11| < 9$
31. $|11 - 2p| < 5$
32. $|5 - 2m| > 15$
33. $|13 - 2n| < 17$
34. $|3x + 1| < 10$
35. $|5y + 2| > 1$
36. $|4z + 5| \geq 9$
37. $|2h + 31| \geq 19$
38. $|2k + 19| \geq 31$
39. $|2m + 25| \leq 33$
40. $|3n - 2| \leq 7$

Express Problems 41–50 as absolute value statements.

41. The number x is five units from seventeen.
42. The number y is seven units from twenty.
43. The value of r is at least four units from eleven.
44. The value of t remains within ten units of sixteen.
45. The number five is d units from forty.
46. Eighteen is within b units of twenty-seven.
47. The value of h is more than nine units from negative one.
48. The number k is at least three units from negative four.
49. The distance of the number x from position p is never more than t units.
50. The number y is at least s units from position r.

In Problems 51–60 choose and define a variable and express the information in an absolute value statement.

51. *Manufacturing* A machined part is to be 3.7 cm, with a tolerance of .04 cm.
52. *Consumer issue* A survey predicts a market of 5.2 million, with an allowable error of 0.3 million.
53. *Manufacturing* A manufactured item must measure between 9 in. and 9.3 in.
54. *Manufacturing* A certain product must weigh between 32 lb and 36 lb.
55. *Business* A salesperson travels no more than 150 mi per day and sometimes as little as 90 mi.
56. *Social science* A library has had a circulation count of as many as 10,000 volumes and as few as 6400 volumes.
57. *Economics* An economist estimates the price index will grow by 11%, give or take a percent.
58. *Consumer issue* A mechanic estimates repairs will run about \$250, \$30 more or less.
59. *Space science* Splash down for the astronauts is expected to be 10 mi from the carrier, plus or minus 3 mi.
60. *Aviation* An air traffic controller requires you to stay in an air corridor between 5000 ft and 8000 ft.

C *Solve and graph the solution set in Problems 61–70.*

61. $\left|\frac{1}{2}x - 3\right| = \frac{1}{4}$

62. $\left|\frac{1}{3}x + 2\right| = \frac{1}{3}$

63. $\left|\frac{1}{2}x - 3\right| = \frac{1}{2}$

64. $\left|\frac{1}{2}x + 5\right| > \frac{1}{2}$

65. $\left|1 - \frac{1}{4}x\right| < 2$

66. $\left|3 - \frac{1}{4}x\right| \leq 3$

67. $\left|\frac{1}{2}x + \frac{1}{3}\right| \leq \frac{1}{4}$

68. $\left|\frac{1}{3}x - \frac{1}{4}\right| > \frac{1}{2}$

69. $\left|\frac{3x}{4} - \frac{5}{8}\right| > 0$

70. $\left|\frac{4x}{7} + \frac{3}{5}\right| \leq 0$

2.4 Quadratic Equations

A second-degree polynomial is called a **quadratic polynomial,** or simply a **quadratic.** A **quadratic equation** is an equation of the form $ax^2 + bx + c = 0$. Quadratic equations are solved by several means. The simplest solution occurs if the quadratic expression $ax^2 + bx + c = 0$ is factorable over the integers. The solution depends on an important property of zero:

PROPERTY OF ZERO

$AB = 0$ if and only if $A = 0$ or $B = 0$

Thus, if the product of two factors is zero, then at least one of the factors is zero. If a quadratic is factorable, this property provides a method of solution.

Example 1

$$x^2 = 2x + 15$$
$$x^2 - 2x - 15 = 0$$
$$(x + 3)(x - 5) = 0$$
$$x + 3 = 0 \quad \text{or} \quad x - 5 = 0$$
$$x = -3 \qquad\qquad x = 5$$

Since the product is zero, one of the factors must be zero.

$\{-3, 5\}$ □

SOLUTION OF QUADRATIC EQUATIONS BY FACTORING

To solve a quadratic equation that can be expressed as a product of linear factors:

1. Rewrite all nonzero terms on one side of the equation.
2. Factor the expression.
3. Set each of the factors equal to zero.
4. Solve each of the linear equations.
5. Write the solution set which is the union of the solution sets of the linear equations.

Completing the Square

When the quadratic is not factorable, other methods must be employed. One such method depends on the square root property.

SQUARE ROOT PROPERTY

If $P^2 = Q$, then $P = \pm\sqrt{Q}$.

The equation $x^2 = 4$ could be rewritten as $x^2 - 4 = 0$, factored and solved. However, the square root property can be used instead, as illustrated below.

Using Square Root Property:	**Using Factoring:**
$x^2 = 4$	$x^2 - 4 = 0$
$x = \pm\sqrt{4}$	$(x + 2)(x - 2) = 0$
$x = \pm 2$	$x = -2$ or $x = 2$

The importance of this property, however, is that it can be applied to any quadratic! This is because every quadratic may be expressed in the form $P^2 = Q$ by isolating the variable terms and **completing the square,** as illustrated in the following examples.

Example 2

$$x^2 = 2x + 15$$

$$x^2 - 2x = 15 \qquad \text{Isolate the variable.}$$

$$x^2 - 2x + 1 = 15 + 1 \qquad \text{Complete the square by adding the square of half the coefficient of } x \text{ to both sides.}$$

$$(x - 1)^2 = 16 \qquad \text{Factor.}$$

$$x - 1 = \pm 4 \qquad \text{Use the square root property.}$$

$$x = 1 \pm 4 \qquad \text{Solve for } x.$$

$$x = 1 + 4 \quad \text{or} \quad x = 1 - 4$$

$$x = 5 \qquad x = -3$$

$\{5, -3\}$ □

Example 3

$$4x^2 - 4x - 7 = 0$$

$$4x^2 - 4x = 7 \qquad \text{Isolate the variable.}$$

$$x^2 - x = \frac{7}{4} \qquad \text{Divide by 4 to obtain 1 as coefficient of } x^2.$$

$$x^2 - x + \left(-\frac{1}{2}\right)^2 = \left(-\frac{1}{2}\right)^2 + \frac{7}{4} \qquad \text{Complete the square by adding the square of half the coefficient of } x \text{ to both sides.}$$

$$\left(x - \frac{1}{2}\right)^2 = 2 \qquad \text{Factor.}$$

$$x - \frac{1}{2} = \pm\sqrt{2} \qquad \text{Remember } \pm.$$

$$x = \frac{1 \pm 2\sqrt{2}}{2} \qquad \text{Solve for } x.$$

$$\left\{\frac{1 + 2\sqrt{2}}{2}, \frac{1 - 2\sqrt{2}}{2}\right\}$$ □

The Quadratic Formula

The process of completing the square is often cumbersome. However, if *any* quadratic $ax^2 + bx + c = 0$, $a \neq 0$, is considered, completing the square can be used to derive a formula for x in terms of the coefficients a, b, and c. The formula can then be used to solve all quadratics, even those that are nonfactorable.

$$ax^2 + bx + c = 0$$

$$ax^2 + bx = -c \qquad \text{Isolate the variable.}$$

$$x^2 + \frac{b}{a}x = -\frac{c}{a} \qquad \text{Divide by } a.$$

$$x^2 + \frac{b}{a}x + \left(\frac{b}{2a}\right)^2 = \left(\frac{b^2}{4a^2}\right) - \frac{c}{a} \qquad \text{Complete the square; } \tfrac{1}{2} \text{ of } \frac{b}{a} \text{ is } \frac{b}{2a}.$$

$$\left(x + \frac{b}{2a}\right)^2 = \frac{b^2 - 4ac}{4a^2} \qquad \text{Factor.}$$

$$x + \frac{b}{2a} = \pm\sqrt{\frac{b^2 - 4ac}{4a^2}} \qquad \text{Use the square root property.}$$

$$x + \frac{b}{2a} = \pm\frac{\sqrt{b^2 - 4ac}}{2a}$$

$$x = -\frac{b}{2a} \pm \frac{\sqrt{b^2 - 4ac}}{2a} \qquad \text{Solve for } x.$$

$$x = \frac{-b \pm \sqrt{b^2 - 4ac}}{2a}$$

QUADRATIC FORMULA

> If $ax^2 + bx + c = 0$, $a \neq 0$, then
>
> $$x = \frac{-b \pm \sqrt{b^2 - 4ac}}{2a}$$

Example 4 Solve: $4x^2 - 4x - 7 = 0$

Solution

↑	↑	↑
a	b	c
4	-4	-7

$$x = \frac{-(-4) \pm \sqrt{(-4)^2 - 4(4)(-7)}}{2(4)}$$

$$x = \frac{4 \pm \sqrt{128}}{8}$$

$$x = \frac{4 \pm 8\sqrt{2}}{8}$$

$$x = \frac{4(1 \pm 2\sqrt{2})}{4 \cdot 2}$$

$$x = \frac{1 \pm 2\sqrt{2}}{2}$$

$$\left\{ \frac{\mathbf{1 + 2\sqrt{2}}}{\mathbf{2}}, \frac{\mathbf{1 - 2\sqrt{2}}}{\mathbf{2}} \right\}$$ Compare with Example 3. □

Example 5 Solve: $2x^2 - 2x + 5 = 0$

Solution

$$\begin{array}{ccc} \uparrow & \uparrow & \uparrow \\ a & b & c \\ 2 & -2 & 5 \end{array}$$

$$x = \frac{-(-2) \pm \sqrt{(-2)^2 - 4(2)(5)}}{2(2)}$$

$$x = \frac{2 \pm \sqrt{-36}}{4}$$

$$x = \frac{2 \pm 6i}{4}$$

$$x = \frac{1 \pm 3i}{2}$$

$$\left\{ \frac{\mathbf{1 + 3i}}{\mathbf{2}}, \frac{\mathbf{1 - 3i}}{\mathbf{2}} \right\}$$ □

Since the quadratic formula contains a radical, the sign of the radicand will determine whether the roots will be real or nonreal. This radicand is called the **discriminant** of the quadratic, and its properties are summarized in the box.

THE DISCRIMINANT OF THE QUADRATIC

> If $ax^2 + bx + c = 0$, $a \neq 0$, then $b^2 - 4ac$ is called the *discriminant.*
> If $b^2 - 4ac < 0$, there are *no real solutions.*
> If $b^2 - 4ac = 0$, there is *one real solution.*
> If $b^2 - 4ac > 0$, there are *two real solutions.*

Find the discriminant of each quadratic in Examples 6–8, and specify the number of real roots.

Example 6

$$x^2 = 2x + 15$$

$$1x^2 - 2x - 15 = 0$$ Put in standard form.

$$\begin{aligned} b^2 - 4ac &= (-2)^2 - 4(1)(-15) \\ &= 4 + 60 \\ &= \mathbf{64} \end{aligned}$$

Since $64 > 0$, there are two real solutions. □

Example 7

$$4x^2 - 4x + 9 = 0$$

$$\begin{aligned} b^2 - 4ac &= 16 - 144 \\ &= \mathbf{-128} \end{aligned}$$

Since $-128 < 0$, there are no real solutions. □

Example 8

$$9x^2 + 6x + 1 = 0$$

$$\begin{aligned} b^2 - 4ac &= 36 - 36 \\ &= \mathbf{0} \end{aligned}$$

Since the discriminant is zero, there is one real solution. □

Calculator Comment To utilize a hand calculator efficiently in evaluating the roots of the quadratic equations $ax^2 + bx + c = 0$, first evaluate $\sqrt{b^2 - 4ac}$ and store this value. It may then be recalled for the evaluation of each of the two roots of the equation, as shown in Figure 2.1.

Note that the suggested key scheme is divided into three distinct segments: $\sqrt{b^2 - 4ac}$, first root, and second root. If the quadratic has no real roots, that will be identified in the first segment when the calculator "refuses" to take the square root of a negative discriminant.

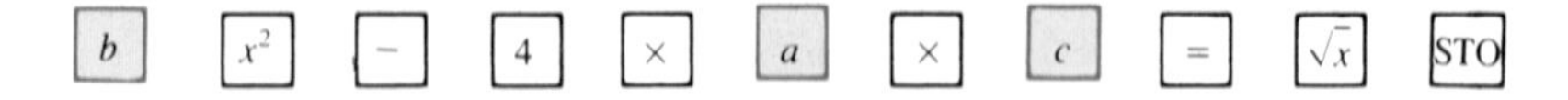

[b] [−/+] [−] [RCL] [=] [÷] [2] [÷] [a] [=]

The display now shows the first root. ↑

[b] [+/−] [+] [RCL] [=] [÷] [2] [÷] [a] [=]

The display now shows the second root. ↑

Figure 2.1

Example 9 Solve and state the solution set with roots correct to the nearest hundredth.

Solution

$$(3x - 5)(x - 1) = (x - 2)^2$$

$$3x^2 - 8x + 5 = x^2 - 4x + 4 \quad \text{Expand the products.}$$

$$2x^2 - 4x + 1 = 0 \quad \text{Isolate the terms.}$$

$$x = \frac{4 \pm \sqrt{16 - 8}}{4} \quad \text{Apply the quadratic formula.}$$

$$x = \frac{4 \pm 2\sqrt{2}}{4} \qquad \text{Simplify.}$$

$$x = \frac{2 \pm \sqrt{2}}{2}$$

$$x = \frac{2 + \sqrt{2}}{2} \quad \text{or} \quad x = \frac{2 - \sqrt{2}}{2}$$

$$x \approx 1.7071067 \qquad x \approx .29289321$$

See the Calculator comment above and Figure 2.1.

{1.71, .29} □

Example 10 The areas of a square and a rectangle are equal. Find the side of the square if the width of a rectangle is 3 cm and its length is 60 cm longer than the side of the square.

Analysis

$$\begin{pmatrix}\text{AREA OF}\\ \text{SQUARE}\end{pmatrix} = \begin{pmatrix}\text{AREA OF}\\ \text{RECTANGLE}\end{pmatrix}$$

$$\begin{pmatrix}\text{SIDE OF}\\ \text{SQUARE}\end{pmatrix}^2 = \begin{pmatrix}\text{WIDTH OF}\\ \text{RECTANGLE}\end{pmatrix}\begin{pmatrix}\text{LENGTH OF}\\ \text{RECTANGLE}\end{pmatrix}$$

$$\begin{pmatrix}\text{SIDE OF}\\ \text{SQUARE}\end{pmatrix}^2 = (3)\left(\begin{matrix}\text{SIDE OF}\\ \text{SQUARE}\end{matrix} + 60\right)$$

Solution Let s = SIDE OF SQUARE. Then

$$s^2 = 3(s + 60)$$
$$s^2 = 3s + 180$$
$$s^2 - 3s - 180 = 0$$
$$(s + 12)(s - 15) = 0$$
$$s = -12 \quad \text{or} \quad s = 15$$

The length of the side of the square must be positive, so $s = -12$ is rejected.

The square is 15 cm on a side. □

Problem Set 2.4

A *In Problems 1–10 find the discriminant of each and specify the number of real roots. Do not solve.*

1. $u^2 + 7u + 2 = 0$

2. $u^2 + 11u + 5 = 0$

3. $2v^2 + 3 = 4v$

4. $v^2 = 5 - 2v$

5. $w^2 = 6w - 9$

6. $3w^2 - 2w = 1$

7. $5x^2 = 3x$

8. $3x^2 + 5 = 0$

9. $\frac{1}{2}y = \sqrt{2} - 2y^2$

10. $\sqrt{3} + 3y^2 = \frac{3}{2}y$

Solve Problems 11–30.

11. $(a - 2)(a + 1) = 0$

12. $(a + 3)(a + 4) = 0$

13. $(b - 2)(b - 3) = 0$

14. $(b + 1)(b - 3) = 0$

15. $(c + 3)(2c - 1) = 0$

16. $(c - 2)(3c + 1) = 0$

17. $(2x - 1)(3x - 2) = 0$

18. $(3x + 2)(5x - 3) = 0$

19. $(y - 3)^2 = 0$

20. $(y + 2)^2 = 0$

21. $a^2 + 2a - 3 = 0$

22. $a^2 - 3a - 4 = 0$

23. $b^2 - b - 6 = 0$

24. $b^2 + b - 12 = 0$

25. $3c^2 - c - 4 = 0$

26. $2c^2 + c - 6 = 0$

27. $2h^2 = 5h - 3$

28. $3h^2 = 10h + 8$

29. $6k^2 + 19k = -15$

30. $9k^2 + 9k = 10$

B *Solve Problems 31–50.*

31. $w^2 - 4w + 1 = 0$

32. $x^2 - 6x + 7 = 0$

33. $y^2 + 2y = 4$

34. $z^2 + 2z = 6$

35. $9p^2 = 2(3p + 1)$

36. $3q(4 + 3q) = 13$

37. $13 = r(4 - r)$

38. $s^2 + 2(3s + 5) = 0$

39. $2s^2 + 1 = 2s$

40. $9t^2 + 2 = 6t$

41. $4t(t - 1) = 9$

42. $4u(2 - u) = 1$

43. $9v^2 - 18v + 1 = 0$

44. $4w^2 - 8w - 25 = 0$

45. $2x^2 = 6x - 17$

46. $9y^2 = 12y - 13$

47. $2m^2 + 3m = 4$

48. $3n^2 = 10n + 4$

49. $4p(p + 4) = -19$

50. $3q(2 - 3q) = 19$

For Problems 51–62 design an equation that can be used to answer each question, solve the equation, and state an answer.

51. The length of a rectangle is 7 cm more than twice its width. If the area is 114 cm^2, what are the dimensions of the figure?

52. The altitude of a triangle is 1 cm less than three times the base. If the area is 92 cm^2, what are the dimensions of the figure?

53. The length of a rectangle is 2 in. more than three times the width. If the area is 85 in.^2, what are the dimensions of the figure?

54. The altitude of a triangle is 1 in. more than twice the base. If the area is 39 in.^2, what are the dimensions of the figure?

55. A polygon of n sides has $\frac{1}{2}n(n - 3)$ diagonals. How many sides has a polygon with 135 diagonals?

56. A polygon of n sides has $\frac{1}{2}n(n - 3)$ diagonals. How many sides has a polygon with 90 diagonals?

57. A side of one square is 8 mm longer than that of a second square. The sum of their areas is 424 mm.^2. Find the side of the larger square.

58. Two rectangles with the same width differ in area by 6 cm^2. The larger one is three times longer than it is wide. Find the dimensions of the smaller figure if it is 1 cm longer than it is wide.

59. If one side of a square is increased by 2 ft and an adjacent side is decreased by 2 ft, the area of the resulting rectangle is 32 ft^2. Find the side of the square.

60. One base of a trapezoid is twice as long as the altitude, and the other base is 17 ft longer than the altitude. Find the length of the bases if the area of the trapezoid is 45 ft^2.

C 61. One leg of a right triangle is 7 cm longer than the other. The hypotenuse is 8 cm longer than the longer leg. What are the dimensions of the triangle to the nearest centimeter?

62. A swimming pool is 15 ft by 30 ft and an average of 5 ft in depth. It takes 25 minutes longer to fill than to drain the pool. If it can be drained at a rate of 15 cu ft/min faster than it can be filled, what is the drainage rate?

Solve Problems 63–68 and state each solution set with roots correct to the nearest hundredth.

63. $3w = 75(w - 2)(w + 2)$

64. $2(3 - 4z)(2z + 1) = (6z - 1)(2z - 1)$

65. $(3x + 2)(2x + 1) = 2(2x - 1)(9x + 8)$

66. $2(3 - 4z)(2z + 1) = (6z - 1)(2z - 11)$

67. $x^3 - (x - 2)^3 = 92$

68. $(x - 1)(2x - 1)(3x - 1) = 6x^3 - 82$

69. For what value of k do the equations $x^2 + kx - 3 = 0$ and $x^2 - 3x + k = 0$ have exactly one root in common?

70. Reconsider the equation $ax^2 + bx + c = 0$, $a \neq 0$. Isolate the variable terms, *multiply both sides by* $4a$, complete the square on the resulting terms, and solve for x. Use this procedure to prove the *quadratic formula*.

2.5 Extraneous Solutions

Solving equations with radical expressions requires the following property of powers:

PROPERTY OF POWERS

If P and Q are algebraic expressions in a variable x, and n is any positive integer, then the solution set of

$$P = Q$$

is a subset of the solution set of

$$P^n = Q^n$$

The equation $P^n = Q^n$ is called a **derived equation** of $P = Q$.

This property implies that not all the solutions of the derived equations may be solutions of the original equation. Whenever both sides of an equation are raised to a power, the solution must be checked in the original equation. Solutions that do not check are called **extraneous solutions.** To use the property, a radical expression is first isolated on one side of the equation. This ensures that raising each side to a power will eliminate that radical. The following examples illustrate the procedure.

Example 1

$$\sqrt{3-x}+1=x$$ First, isolate the radical.

$$\sqrt{3-x}=x-1$$

$$(\sqrt{3-x})^2=(x-1)^2$$ Square both sides.

$$3-x=x^2-2x+1$$

$$0=x^2-x-2$$

$$0=(x+1)(x-2)$$ Simplify and solve.

$$x=-1 \quad \text{or} \quad x=2$$

Check $x = -1$:

$$\sqrt{3-(-1)}+1 \stackrel{?}{=} -1$$

$$\sqrt{4}+1 \stackrel{?}{=} -1$$

$$3 \neq -1$$

$x = -1$ is extraneous

Check $x = 2$:

$$\sqrt{3-2}+1 \stackrel{?}{=} 2$$

$$\sqrt{1}+1 \stackrel{?}{=} 2$$

$$2 = 2$$

$x = 2$ is a solution

Check the roots. State the solution set.

The solution set is {2}. □

Example 2

$$2\sqrt{x+6}+3\sqrt{x+1}=0$$

$$2\sqrt{x+6}=-3\sqrt{x+1}$$

$$4(x+6)=9(x+1)$$ Square both sides.

$$5x=15$$

$$x=3$$

Check: $2\sqrt{9}+3\sqrt{4} \stackrel{?}{=} 0$

$$6+6 \stackrel{?}{=} 0$$

$$12=0$$

$x = 3$ does not check and is extraneous

The solution set is ∅. State the solution set. □

Example 3

$$\sqrt[3]{x^2+2}=3$$

$$x^2+2=27$$ Cube both sides.

$$x^2=25$$

$$x=5 \quad \text{or} \quad x=-5$$

Check: $\sqrt[3]{5^2 + 2} \stackrel{?}{=} 3$ $\qquad$ $\sqrt[3]{(-5)^2 + 2} = 3$

$\sqrt[3]{27} \stackrel{?}{=} 3$ $\qquad$ $\sqrt[3]{27} = 3$

$3 = 3$ $\qquad$ $3 = 3$

$x = 5$ is a solution $\qquad$ $x = -5$ is a solution

The solution set is {5, −5}. □

Example 4

$$\begin{aligned} \sqrt{2x+5} + \sqrt{x+2} &= 1 && \text{Isolate } \textit{one} \text{ radical.} \\ \sqrt{2x+5} &= 1 - \sqrt{x+2} && \text{Square and simplify.} \\ 2x + 5 &= 1 - 2\sqrt{x+2} + x + 2 && \text{Isolate remaining radical.} \\ x + 2 &= -2\sqrt{x+2} && \textit{Square again.} \\ x^2 + 4x + 4 &= 4x + 8 && \text{Simplify and solve.} \\ x^2 &= 4 \\ x = -2 \quad &\text{or} \quad x = 2 \end{aligned}$$

Check: $\sqrt{-4+5} + \sqrt{-2+2} \stackrel{?}{=} 1$ $\qquad$ $\sqrt{4+5} + \sqrt{2+2} \stackrel{?}{=} 1$

$1 + 0 \stackrel{?}{=} 1$ $\qquad$ $3 + 2 \stackrel{?}{=} 1$

$1 = 1$ $\qquad$ $5 \neq 1$

$x = -2$ is a solution $\qquad$ $x = 2$ does not check

The solution set is {−2}. □

Rational equations can also give rise to extraneous results. The multiplication property for equality specifies that both sides of an equation may be multiplied by any *nonzero* number. However, if you multiply by a variable or an expression containing a variable, then there may be values for which the expression is *zero*. Again, you must check to see that you have not introduced an extraneous result. Actually, it is sufficient to note those values of the variable that are excluded from its domain.

As with radical equations, a simple method applies in the solution of rational equations. To solve a rational equation, multiply by some expression to eliminate the fractions. The least common denominator (LCD) is used since it has as a factor each denominator of the expression.

Example 5

$$\frac{1}{x} - \frac{1}{4} = \frac{x-2}{4x} \qquad x \neq 0$$

Note the restriction. It is implied by the equation, but must be stated as part of the solution.

$$(4x)\frac{1}{x} - (4x)\frac{1}{4} = (4x)\frac{x-2}{4x}$$

Multiply by the LCD.

$$\begin{aligned} 4 - x &= x - 2 && \text{Simplify and solve.} \\ 6 &= 2x \\ 3 &= x \end{aligned}$$

By noting the restriction $x \neq 0$, or checking, $x = 3$ is verified as a solution.

The solution set is {3}. □

Example 6

$$\frac{2x}{x-3} - 3 = \frac{2x-12}{3-x} \qquad x \neq 3 \qquad \text{Note the restriction.}$$

$$(x-3)\frac{2x}{x-3} - 3(x-3) = (x-3)\frac{2x-12}{3-x} \qquad \text{Multiply by the LCD}$$

$$2x - 3(x-3) = (-1)(2x-12) \qquad \text{Simplify and solve.}$$

$$2x - 3x + 9 = -2x + 12$$

$$-x + 9 = -2x + 12$$

$$x = 3$$

But $x \neq 3$, so the solution set is empty. Check.

The solution set is ∅. State the solution set. □

Example 7

$$\frac{1}{t+3} + \frac{1}{t+1} = 1 \qquad t \neq -3, -1$$

$$(t+3)(t+1)\left(\frac{1}{t+3}\right) + (t+3)(t+1)\left(\frac{1}{t+1}\right) = (t+3)(t+1)(1)$$

$$(t+1) + (t+3) = t^2 + 4t + 3$$

$$0 = t^2 + 2t - 1$$

$$t = \frac{-2 \pm \sqrt{8}}{2}$$

$$t = \frac{-2 \pm 2\sqrt{2}}{2}$$

$$t = -1 \pm \sqrt{2}$$

The original restrictions were $t \neq -3$ or -1, so both irrational results are solutions.

The solution set is $\{-1 + \sqrt{2}, \quad -1 - \sqrt{2}\}$. □

Lost Solutions

If extra roots can be generated by some kinds of procedures, it is natural to ask if roots may be lost during the solution of an equation. An extraneous solution can be introduced by multiplying by a variable factor, and a solution may be eliminated, or lost, when dividing by such a factor.

Example 8

$$(2r-1)(r+3) = (r+3)(r-2)$$

$$2r^2 + 5r - 3 = r^2 + r - 6$$

$$r^2 + 4r + 3 = 0$$

$$(r+1)(r+3) = 0$$

$$r = -1 \quad \text{or} \quad r = -3$$

However, note that in the original equation both sides had a factor of $r + 3$. Solve again, first by dividing by $r + 3$:

$$\frac{(2r - 1)(r + 3)}{r + 3} = \frac{(r + 3)(r - 2)}{r + 3}$$
$$2r - 1 = r - 2$$
$$\mathbf{r = -1}$$

□

The root $r = -3$ was lost in Example 8 because the division was valid for all r except -3, which would have caused division by zero. Division by variable factors should be avoided since the process may produce **lost solutions.**

Many situations that give rise to radical equations involve geometric concerns. Very often the geometry involves the **Pythagorean theorem,** which is stated in the box as a reminder:

PYTHAGOREAN THEOREM

If a right triangle has sides a and b with hypotenuse c (the side opposite the right angle), then

$$a^2 + b^2 = c^2$$

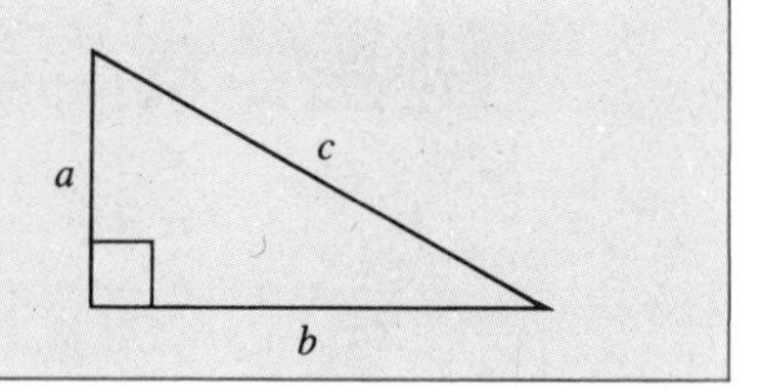

Example 9 A 1 mile length of pipeline connects two pumping stations. Special joints must be used along the line to provide for expansion and contraction due to changes in temperature. However, if the pipeline were actually one continuous length of pipe fixed at each end by the stations, then expansion would cause the pipe to bow. Approximately how high would the middle of the pipe rise if the expansion was just 1 inch over the mile?

Analysis Assume that the pipe bows in a circular arc, as shown in the figure. A triangle would produce a reasonable approximation since the distance x should be quite small compared to the total length.

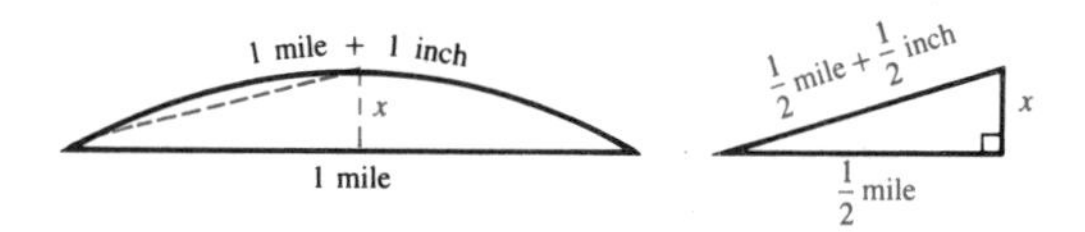

Since a right triangle is used to model the situation, the Pythagorean theorem may be employed.

Solution Let x be the height of the middle of the pipe. Then

$$\left(\tfrac{1}{2}\text{mile} + \tfrac{1}{2}\text{inch}\right)^2 = x^2 + \left(\tfrac{1}{2}\text{mile}\right)^2$$

$$(31{,}680.5)^2 = x^2 + (31{,}680)^2 \qquad \text{Convert to inches.}$$

$$x^2 = (31{,}680.5)^2 - (31{,}680)^2$$

$$x = \sqrt{(31{,}680.5)^2 - (31{,}680)^2} \qquad \text{The positive value is taken since } x \text{ is a distance.}$$

$$\mathbf{x \approx 177.99}$$

The solution 177.99 in. is approximately 14.8 ft. This is an extraordinary result if you consider what your estimate would have been before you worked the problem.

The pipe would bow approximately 14.8 ft at the middle. □

Problem Set 2.5

Solve the equations for real number solutions in Problems 1–54. State the solution sets, restrictions for rational equations, and all extraneous roots found.

A

1. $\frac{x^2}{12} - \frac{x}{3} = \frac{3}{4}$

2. $.09x^2 - .21x + .1 = 0$

3. $\frac{3}{4y} + \frac{7}{16} = \frac{4}{3y}$

4. $\frac{7}{4y} + \frac{5}{18} = \frac{1}{36}$

5. $\frac{4}{y} - \frac{3(2-y)}{y-1} = 3$

6. $\frac{6}{y} - \frac{2(y-5)}{y+1} = 3$

7. $\frac{z-1}{z} = \frac{z-1}{6}$

8. $\frac{3z-2}{z-1} = \frac{2z+1}{z+1}$

9. $\frac{1}{x+2} - \frac{1}{2-x} = \frac{3x+8}{x^2-4}$

10. $\frac{16}{x+5} + \frac{4}{5-x} = \frac{5-3x}{x^2-25}$

11. $\frac{x+2}{3x-1} - \frac{1}{x} = \frac{x+1}{3x^2-x}$

12. $\frac{x}{2x-1} - \frac{1}{x} = \frac{x-1}{2x^2-x}$

13. $\frac{a+1}{a-1} - \frac{a-1}{a+1} = \frac{5}{6}$

14. $\frac{5-a}{a+1} + \frac{a-3}{a-1} = \frac{8}{15}$

15. $\frac{b-1}{b-3} + \frac{b+2}{b+3} = \frac{3}{4}$

16. $\frac{b-2}{b-3} - \frac{b+2}{b+3} = \frac{5}{8}$

17. $2\sqrt{c} = c + 1$

18. $\sqrt{d} = d - 6$

19. $\sqrt{h+2} = 3$

20. $\sqrt{2k+3} = 3$

21. $\sqrt[3]{4x+4} = 2$

22. $\sqrt[3]{3y+1} = 1$

23. $m - \sqrt{m} - 2 = 0$

24. $n + \sqrt{n+8} = -2$

25. $r = 6 - 3\sqrt{r-2}$

26. $s - \sqrt{4s-11} - 4 = 0$

27. $\sqrt{x-3} = \sqrt{4x-5}$

28. $x\sqrt{6} = \sqrt{x+2}$

29. $\sqrt{y^2+4y-5} = \sqrt{2-2y}$

30. $\sqrt{y+1} = \sqrt{y^2+3y+2}$

B

31. $\frac{3}{2x+1} + \frac{2x+1}{1-2x} = 1 - \frac{8x^2}{4x^2-1}$

32. $\frac{2y+1}{y+3} + \frac{3y-y}{2-y} = \frac{9-3y-y^2}{y^2+y-6}$

33. $\frac{3r-5}{5r-5} + \frac{5r-1}{7r-7} - \frac{r-4}{1-r} = 2$

34. $\frac{2s+1}{s+2} - \frac{s+2}{s+1} = -1$

35. $\frac{x-2}{x+3} - \frac{1}{x-2} = \frac{x-4}{x^2+x-6}$

36. $\frac{x-2}{x+2} + \frac{1}{x-1} = \frac{4x-1}{x^2+x-2}$

37. $\frac{x-1}{x-2} + \frac{x+4}{2x+1} = \frac{1}{2x^2-3x-2}$

38. $\frac{x-1}{2x-1} + \frac{4-x}{x+1} = \frac{3x}{2x^2+x-1}$

39. $\frac{3}{x+2} + \frac{x-1}{x+5} = \frac{5x+20}{6x+24}$

40. $\frac{x-3}{x-2} + \frac{x-1}{x} = \frac{22x-110}{3x^2-15x}$

41. $2 - \sqrt{3b+1} = \sqrt{b-1}$

42. $\sqrt{3k+1} - \sqrt{2k-1} = 1$

43. $\sqrt{d+3} + \sqrt{d} = \sqrt{3}$

44. $1 + \sqrt{c+2} = \sqrt{c}$

45. $\sqrt{4x+1} - \sqrt{2x+1} = 2$

46. $\sqrt{3x+1} - \sqrt{2x-2} = 2$

47. $\sqrt{u+1} - \sqrt[4]{u+1} = 0$

48. $\sqrt{v-2} = \sqrt[4]{v^2-6v+1}$

49. $\sqrt{2a+3} + 3 = 3\sqrt{a+1}$

50. $\sqrt{2h-1} + \sqrt{h+4} = 6$

C **51.** $\frac{1}{\sqrt{3}} = \frac{\sqrt{2w+4}}{\sqrt{w}} - \frac{\sqrt{3w+4}}{\sqrt{3w}}$

52. $\frac{\sqrt{w}-3}{\sqrt{w}} - \frac{5-\sqrt{w}}{4} = 0$

53. $\sqrt{2x-1} = \sqrt{7x+2} - \sqrt{x+3}$

54. $\sqrt{3(y+2)} + \sqrt{y+4} = \sqrt{7y+1}$

Problems 55–57 refer to the fact that the distance d, in miles, one can see to the horizon may be approximated by taking the square root of three-halves of one's height h, in feet, above the surface of the earth.

P is the observation point.

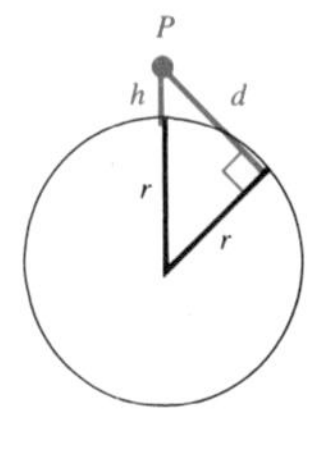

$d = \sqrt{\frac{3}{2}h}$

55. *Earth science* For what height will the number of feet in height be equal to the number of miles seen to the horizon?

56. If the number of miles seen to the horizon is one-half of the number of feet in height, what is the height?

57. Consider the approximation used in Problems 55 and 56. Show how this is obtained. Recall that the earth's radius r is approximately 4000 mi and that a mile is 5280 ft. (Note that h will normally be very small in comparison to r.)

58. If $x = \sqrt{7+\sqrt{48}} + \sqrt{7-\sqrt{48}}$, then find the simplified value of x without using a calculator.

For Problems 59 and 60, see Problem 69 in Problem Set 2.1.

59. The *arithmetic mean* of two numbers is one-half their sum. The *geometric mean* is the positive square root of their product. Given two unequal positive numbers a and b, prove that their arithmetic mean is always greater than their geometric mean.

60. Given two unequal positive numbers a and b, prove that their *geometric mean* is always greater than their *harmonic mean.*

2.6 Nonlinear Inequalities

A restatement of the rules of signs for multiplication proves very useful in developing a method for solving polynomial and rational inequalities that can be factored.

PROPERTY OF PRODUCTS

If $P \cdot Q = 0$, then either $P = 0$ or $Q = 0$.	If $P \cdot Q < 0$, then P and Q have opposite signs.	If $P \cdot Q > 0$, then P and Q have the same sign.

To apply the rules of signs to the solution of inequalities, look first at the associated equality and decide for what values of the variable the expression is zero. These values are called the **critical values** for the inequality. Then, the inequality must be greater than or less than zero everywhere else. The signs of the factors can then be examined on the intervals between critical values. Finally, the interval or intervals that satisfy the inequality can be selected.

Example 1 Solve: $x(x - 1) > 0$

Solution First find the critical values by determining where $x(x - 1) = 0$. The critical values are 0 and 1, as indicated on the number line:

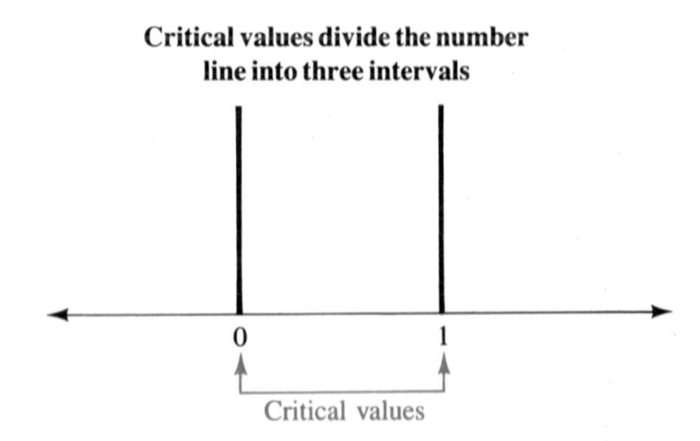

Now determine the signs of the factors on the intervals between critical values. **Choose any value in an interval** and evaluate each factor for that value. Indicate the signs of the factors on a chart:

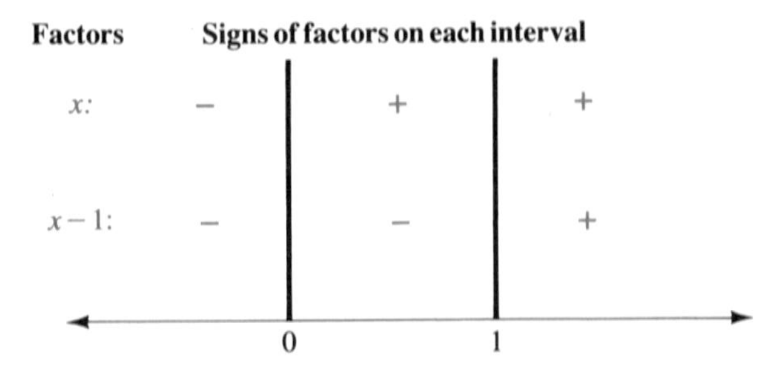

Finally, select the interval or intervals on which the product has the desired sign. In this case, the chosen intervals are those in which the signs of the factors are the same, because the product must be positive.

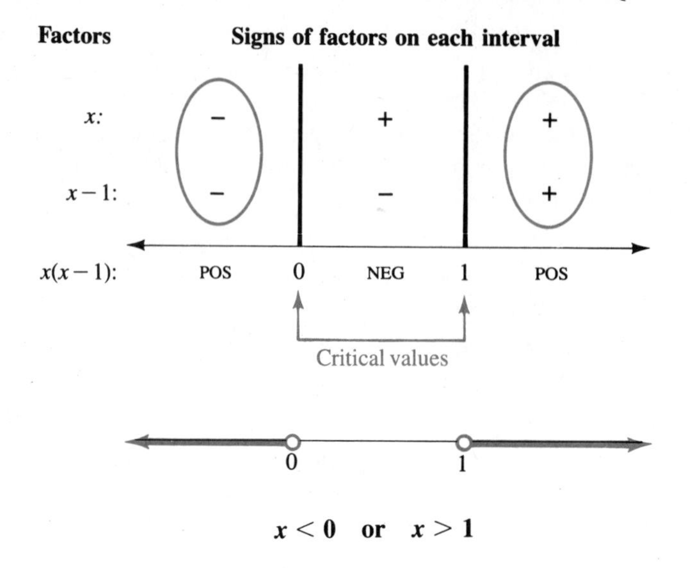

$$x < 0 \quad \text{or} \quad x > 1$$

□

SOLUTION OF POLYNOMIAL INEQUALITIES BY FACTORING

To solve an inequality that can be expressed as a product less than or greater than zero:

1. Rewrite all nonzero terms on one side of the inequality.
2. Factor the expression.
3. Determine the critical values.
4. Determine the signs of the factors on the intervals between critical values.
5. Select the interval or intervals on which the product has the desired sign.

Example 2 Solve: $x^2 - x - 6 < 0$

Solution $(x + 2)(x - 3) < 0$ The critical values are -2 and 3.

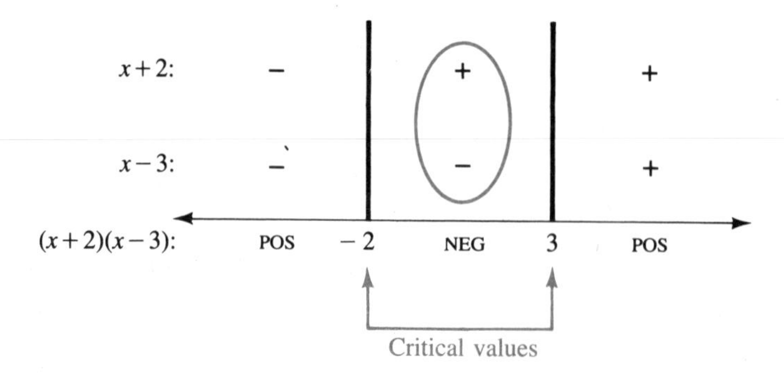

$$-2 < x < 3$$

□

Example 3 Solve: $5 + 4x - x^2 \geqslant 0$

Solution $(5 - x)(1 + x) \geqslant 0$ The critical values are 5 and -1. Notice that in this example the critical values are included in the solution since equality is included.

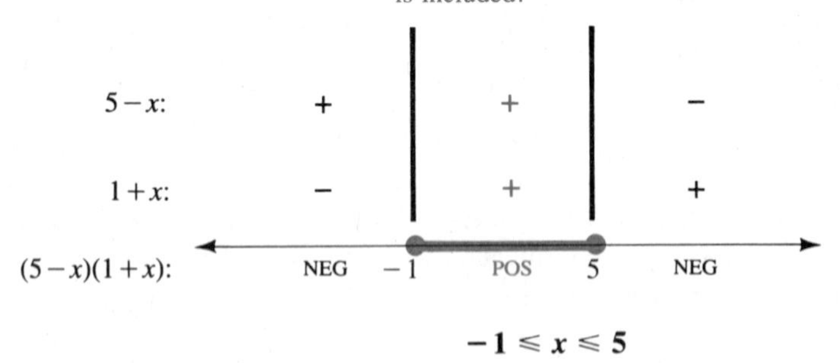

$$-1 \leqslant x \leqslant 5$$

□

Example 4 Solve: $(y + 1)(2y + 3)(5 - 3y) \geqslant 0$

Solution The critical values are -1, $-\frac{3}{2}$, and $\frac{5}{3}$.

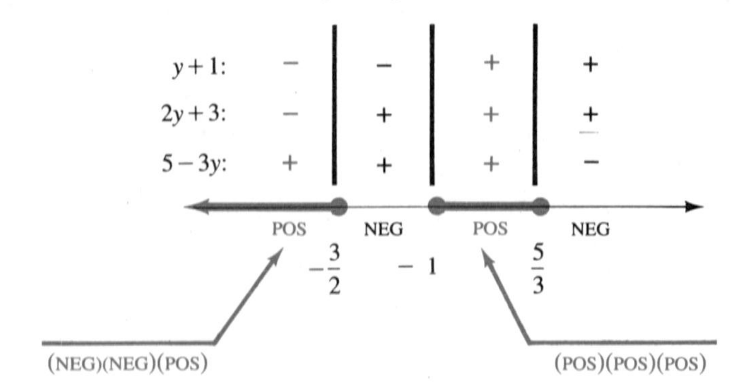

$$y \leqslant -\frac{3}{2} \quad \text{or} \quad -1 \leqslant y \leqslant \frac{5}{3}$$

□

Example 5 Solve: $6z^2(z + 2) > 35z + z^2$

Solution

$$6z^3 + 12z^2 > 35z + z^2$$

$6z^3 + 11z^2 - 35z > 0$ Bring nonzero terms to one side.

$z(6z^2 + 11z - 35) > 0$ Factor.

$z(2z + 7)(3z - 5) > 0$ Determine the critical values.

z:	−	−	+	+
$2z+7$:	−	+	+	+
$3z-5$:	−	−	−	+
	NEG	POS	NEG	POS

Critical values: $-7/2$, 0, $5/3$

$$-\frac{7}{2} < z < 0 \quad \text{or} \quad z > \frac{5}{3}$$

□

Certain inequalities involving rational expressions may be solved with an approach similar to that employed with polynomial inequalities. If this approach is to apply, you must be able to factor the numerator and the denominator of the rational expression. As with equations involving rational expressions, certain values must be excluded to avoid division by zero. Otherwise, the method of solution is very much the same.

Example 6 Solve: $\dfrac{x(x-3)}{x+2} \leqslant 0$

Solution The signs of the factors can be examined as in multiplication, because the rules of signs for division are the same as those for multiplication, except that division by zero is excluded. Thus, critical values that result in division by zero are excluded from the solution. The critical values are 0, 3, and -2, but -2 is an excluded value.

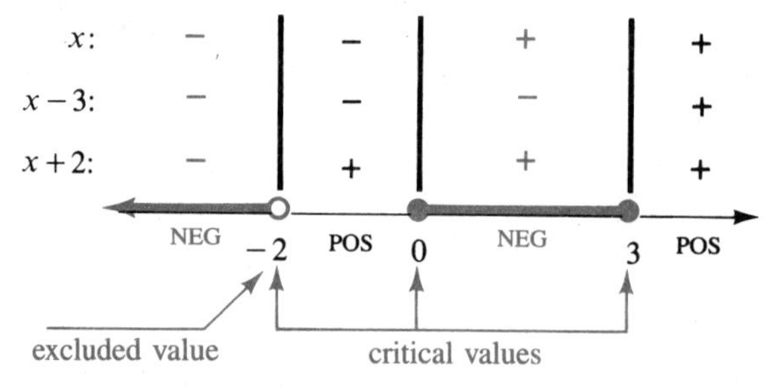

$$\mathbf{x < -2 \quad \text{or} \quad 0 \leqslant x \leqslant 3}$$ □

Example 7 Solve: $\dfrac{y-5}{y^2+4y} > 0$

Solution First factor and identify the critical values:

$$\frac{y-5}{y(y+4)} > 0$$

The critical values are 5, 0, and -4. The excluded values are 0 and -4.

$y-5$:	−	−	−	+
y:	−	−	+	+
$y+4$:	−	+	+	+
	NEG	POS	NEG	POS

$-4 \qquad 0 \qquad 5$

When selecting the intervals containing the correct combination of signs, notice that the excluded values need to be considered only if equality is included, as in $\leqslant$ or $\geqslant$.

$$\mathbf{-4 < y < 0 \quad \text{or} \quad y > 5}$$ □

Example 8 Solve: $\dfrac{2z^2-4z-6}{z^3+z^2-2z} \leqslant 0$

Solution
$$\frac{2(z^2 - 2z - 3)}{z(z^2 + z - 2)} \leqslant 0$$
$$\frac{2(z + 1)(z - 3)}{z(z - 1)(z + 2)} \leqslant 0$$

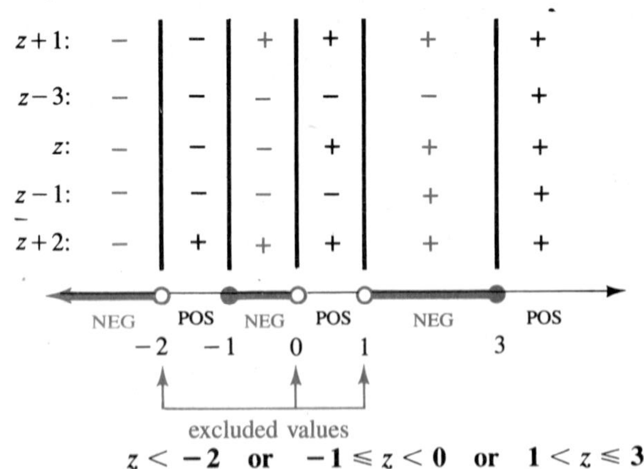

$z < -2$ **or** $-1 \leqslant z < 0$ **or** $1 < z \leqslant 3$ □

Problem Set 2.6

Solve each inequality in Problems 1–60, state the solution, and sketch the solution set on a number line.

A

1. $x(x + 1) < 0$

2. $x(x - 1) \geqslant 0$

3. $(x - 5)(x - 2) \geqslant 0$

4. $(y + 2)(y - 3) \leqslant 0$

5. $(x - 3)(x + 7) < 0$

6. $(x + 1)(x + 5) > 0$

7. $(y + 2)(2y - 1) \leqslant 0$

8. $(y - 2)(2y + 1) \leqslant 0$

9. $(3z + 2)(z - 3) > 0$

10. $(3z - 2)(z + 2) > 0$

11. $(x + 2)(3 - x) \leqslant 0$

12. $(2 - x)(x + 3) \geqslant 0$

13. $(1 - 3x)(x - 4) < 0$

14. $(2y + 1)(3 - y) > 0$

15. $y(y - 3)(y + 4) \leqslant 0$

16. $y(y + 3)(y - 4) \geqslant 0$

17. $(y - 2)(y + 3)(y - 4) \geqslant 0$

18. $(y + 1)(y - 2)(y + 3) < 0$

19. $(z + 1)(2z + 5)(7 - 3z) > 0$

20. $(z - 2)(3z + 2)(3 - 2z) < 0$

21. $\dfrac{x + 2}{x} < 0$

22. $\dfrac{x}{x + 2} < 0$

23. $\dfrac{x}{x - 3} > 0$

24. $\dfrac{x - 3}{x} > 0$

25. $\dfrac{x - 2}{x + 3} \leqslant 0$

26. $\dfrac{x + 1}{4 - x} \geqslant 0$

27. $\dfrac{y(2y - 1)}{3 - y} > 0$

28. $\dfrac{y}{(2y + 3)(y - 2)} < 0$

29. $\dfrac{1}{z(z - 3)(z + 2)} \leqslant 0$

30. $\dfrac{1}{z(z + 1)(5 - z)} \geqslant 0$

B

31. $x^2 - 3x + 2 > 0$
32. $x^2 + 3x + 2 < 0$
33. $y^2 - y - 2 \leqslant 0$
34. $y^2 + y - 2 \geqslant 0$
35. $z^2 - 2z + 1 \geqslant 0$
36. $z^2 + 2z + 1 < 0$
37. $5r^2 - 4r - 1 < 0$
38. $2s^2 - 3s - 35 > 0$
39. $t^2 + 5 \geqslant 6t$
40. $2r^2 + r \leqslant 6$
41. $2u^2 + 5u < 12$
42. $4v^2 + 10 > 13v$
43. $6w^2 + 6 < 13w$
44. $4x^2 + 23x + 15 > 0$
45. $10(y^2 + 1) \geqslant 29y$
46. $25z < 12(1 + z^2)$
47. $\dfrac{x^2 - 2x}{x + 1} > 0$
48. $\dfrac{x - 2}{x^2 + x} < 0$
49. $\dfrac{x^2 - 4}{x} \leqslant 0$
50. $\dfrac{x}{x^2 - 1} \geqslant 0$
51. $\dfrac{x^2 + 3x - 10}{x + 5} > 0$
52. $\dfrac{x - 7}{x^2 - 4x - 21} < 0$
53. $\dfrac{x - 4}{x^2 - 5x + 6} \leqslant 0$
54. $\dfrac{x^2 + 4x + 3}{x + 2} \geqslant 0$
55. $\dfrac{x^2 - 2x - 8}{x^2 + 2x - 8} > 0$
56. $\dfrac{x^2 - 6x + 8}{x^2 + 6x + 8} < 0$
57. $\dfrac{y^3 - 4y}{y^2 - 2y - 15} \leqslant 0$
58. $\dfrac{y^2 + 3y - 18}{y^3 - 9y} > 0$
59. $\dfrac{y^2 - 9}{y^3 - y^2 - 2y} \geqslant 0$
60. $\dfrac{y^3 - y^2 - 6y}{y^2 - 4} \leqslant 0$

C *For Problems 61–66 design an inequality that can be used to answer each question, solve the inequality, and state an answer.*

61. The product of two numbers is at least 340. One of the numbers is three less than the other. What are the possible values of the larger number?
62. The product of two numbers is no larger than 300. One number is five larger than the other. What are the possible values of the smaller number?
63. The quotient of two numbers is positive. The divisor is three larger than the dividend. What are the possibilities for the smaller number?
64. Two numbers have a negative quotient. What are the possibilities for the dividend if it is five larger than the divisor?
65. A rectangular area is to be fenced. If the space is twice as long as it is wide, for what dimensions is the area numerically greater than the perimeter?
66. A rectangular area three times as long as it is wide is to be fenced. For what dimensions is the perimeter numerically greater than the area?
67. Given two unequal positive numbers, prove that the arithmetic mean of their harmonic mean and their geometric mean is less than their arithmetic mean. See Problem 69 in Problem Set 2.1 and Problems 59 and 60 in Problem Set 2.5.

2.7 Summary and Review

This chapter recapped methods for solving linear and quadratic equations and inequalities, certain equations and inequalities involving absolute value, and certain rational and radical equations. These are the essential methods used for problem-solving throughout the remainder of the course. Much of this material has been a review of methods encountered in other courses. Notice, however, that whenever possible equal emphasis was given to equations and inequalities; this approach may differ from your previous work.

CHAPTER OBJECTIVES

After studying this chapter, you should be able to:

1. Distinguish between an identity, a contradiction, and a conditional equation or inequality
2. Solve linear and quadratic equations
3. Use the discriminant of the quadratic to classify the roots without solving
4. Solve certain absolute value equations and inequalities involving one absolute value
5. Interpret absolute value as a distance
6. Solve linear and quadratic inequalities and display their solution sets on the number line
7. Solve certain equations containing radicals, checking for extraneous roots
8. Solve certain equations containing rational expressions, noting restrictions on the variable
9. Design and solve an equation or inequality that describes the relationships of certain applied problems, and interpret the solution set as an answer to the problem

Problem Set 2.7

CONCEPT PROBLEMS

Fill in the word or words necessary to make the statements in Problems 1–25 complete and correct.

[2.1] **1.** A statement that says two expressions are equal is called a(n) ______ .

[2.1] **2.** Values that satisfy an equation are called ______ or ______ .

[2.1] **3.** A(n) ______ is a conditional equation that is true for all values of the variable.

[2.1] **4.** If a conditional equation is never true, it is called a(n) ______ .

[2.1] **5.** Two equations with the same solution set are called ______ equations.

[2.1] **6.** Any real number may be ______ to or ______ from each side of an equation to obtain an equivalent equation.

[2.1] **7.** Both sides of an equation may be multiplied or divided by a(n) ______ real number to obtain a(n) ______ equation.

[2.1] **8.** A(n) ______ inequality is sometimes true and sometimes false.

[2.1] **9.** The properties of equivalent inequalities allow any operation that is allowed for equations except that ______ or ______ by a(n) ______ reverses the ______ of the ______ .

[2.3] **10.** The absolute value of a number is equal to the ______ or its ______ .

[2.3] **11.** If $|x| < a$ and $a > 0$, then ______ $<$ ______ $<$ ______ .

[2.3] **12.** $|x - a|$ is the ______ between a and x.

[2.3] **13.** $|x - a| = b$ may be interpreted as: x is ______ units from ______ .

[2.3] **14.** $|x + 5| < 2$ may be interpreted as the distance between ______ and ______ being less than ______ .

[2.4] **15.** A(n) ______ polynomial is called a *quadratic polynomial.*

[2.4] **16.** The product of two factors is zero if and only if one of the factors is ______ .

[2.4] **17.** If $ax^2 + bx + c = 0$ and $a \neq 0$, then $x =$ ______ (formula).

[2.4] **18.** ______ is called the *discriminant* of $ax^2 + bx + c = 0$, $a \neq 0$.

[2.4] **19.** If the discriminant is negative, the quadratic has no ______ solutions.

[2.5] **20.** If P and Q are algebraic expressions in x, and n is a positive ______ , then the solutions of $P = Q$ are a(n) ______ of the solutions of $P^n = Q^n$.

[2.5] **21.** Since the solution of radical equations may generate ______ solutions, all solutions must be ______ .

[2.5] **22.** Before solving a radical equation, note any ______ on the variable.

[2.6] **23.** To solve a rational equation, multiply by the ______ to eliminate the fractions.

[2.6] **24.** The ______ of an inequality are the values of the variable for which the expression is zero.

[2.6] **25.** To solve a polynomial or rational inequality, determine the ______ of the factors on the ______ between the critical values.

REVIEW PROBLEMS

[2.1] *Solve each equation and inequality in Problems 26–40. Display the solution of each inequality on a number line.*

26. $2x = 7x - 10$

27. $5x + 28 = 1 - 4x$

28. $2(x + 2) = 6 - x$

29. $3y + 8 = 13 + 2(5 - y)$

30. $4z \leqslant z + 6$

31. $5 - 2z > 4z - 1$

32. $-6 < 3x < 9$

33. $4 \leqslant 1 - x \leqslant 7$

34. $2(x - 4) = 23 + 3(2 - x)$

35. $17 - 2(3x + 1) = 3(x + 2)$

36. $2(y - 1)^2 = (y + 3)(2y - 1)$

37. $3(y + 2)^2 = (3y - 1)(y + 2)$

38. $3(y + 1) - 2 \geqslant y + 2(y + 1)$

39. $2[3z - 2(z - 9)] > 5(z + 3)$

40. $2[z + 2(z + 3)] = 3 - 3[z - 3(z + 1)]$

[2.3] *Solve each absolute value statement in Problems 41–60. Display the solution of each inequality on a number line.*

41. $|x| = 5$

42. $|x| = 11$

43. $|x| < 2$

44. $|x| > 13$

45. $|2y| = 3$

46. $|3y| < 2$

47. $|y - 1| = 3$

48. $|y - 3| = 1$

49. $|x + 1| = 3$

50. $|x + 3| = 1$

51. $|y + 4| \leqslant 7$

52. $|y + 5| > 11$

53. $|z - 6| \geqslant 4$

54. $|z - 4| < 6$

55. $|3z + 2| = 8$

56. $|2z + 3| = 5$

57. $|5z - 2| = 3$

58. $|3z - 1| < 2$

59. $|5z - 3| > 2$

60. $|4z + 3| > 5$

[2.4, 2.6] *Solve each quadratic in Problems 61–80, and display the solution for each inequality on a number line.*

61. $(x - 5)(x + 9) = 0$

62. $(3x - 1)(2x + 5) = 0$

63. $y^2 - 2y - 3 = 0$

64. $3y^2 - 2y - 1 = 0$

65. $x^2 + 7 = 6x$

66. $2x^2 + 3x = 9$

67. $4y(y - 2) = 5$

68. $y = 3(5 - 2y^2)$

69. $(x - 5)(x - 7) < 0$

70. $(2x + 1)(2x - 3) > 0$

71. $(y + 5)(1 - 2y) \geqslant 0$

72. $(2 - 5y)(y - 2) \leqslant 0$

73. $z(z - 2)(2 - 3z) < 0$

74. $(z + 1)(z - 2)(z + 3) > 0$

75. $x^2 - 4x \geqslant 12$

76. $6y^2 > 11y + 10$

77. $\dfrac{y}{y^2 + 3y + 2} \geqslant 0$

78. $\dfrac{y^2 - 2y - 3}{y - 1} \geqslant 0$

79. $(3z + 4)^2 - (2z + 3)^2 \geqslant 0$

80. $(3z + 4)^2(2z + 3) \geqslant 0$

[2.5] *Solve each equation in Problems 81–94 for real solutions and state extraneous roots, if any are encountered.*

81. $\dfrac{7}{12} - \dfrac{1}{x} = \dfrac{3}{4x}$

82. $\dfrac{3}{2x} - \dfrac{1}{3x} = \dfrac{7}{18}$

83. $\dfrac{y + 1}{y + 3} = \dfrac{2y - 1}{2y + 1}$

84. $\dfrac{y + 1}{2y + 1} = \dfrac{2y - 1}{y + 3}$

85. $\dfrac{5}{x} - \dfrac{x - 5}{x - 3} = \dfrac{x - 6}{2x - 12}$

86. $\dfrac{8}{x + 6} - \dfrac{1}{x - 4} = \dfrac{1}{3}$

87. $\sqrt{2x - 3} = 5$

88. $2 = \sqrt{3x + 5}$

89. $z - \sqrt{3 - z} = 3$

90. $x - \sqrt{2x - 3} = 1$

91. $\sqrt{x + 5} = \sqrt{x} + 1$

92. $\sqrt{x - 9} = \sqrt{x} - 1$

93. $\sqrt{z + 2} = \sqrt{z + 3} - 1$

94. $\sqrt{x + 3} - \sqrt{x - 2} = 5$

[2.2] **95.** Two rectangles have the same width, but one is 48 m^2 larger in area. The larger rectangle is 8 m longer than it is wide. The other is only 4 m longer than it is wide. What are the dimensions of the larger rectangle?

[2.2] **96.** A rectangle is 5 cm longer than it is wide. If you increase the length by 2 cm and reduce the width by a centimeter, the area is unchanged. Find the dimensions of the new figure.

[2.2] **97.** A total of $1600 is invested for a year, part at 8% and the rest at $9\frac{1}{2}$%. If $134 is collected in interest, how much is invested at 8%?

[2.2] **98.** Milk containing 10% butterfat and cream with 70% butterfat are mixed to produce half-and-half, which is 50% butterfat. How many gallons of each must be mixed to produce 15 gallons of half-and-half?

[2.2] **99.** A traveler takes 9 hours to reach his destination. He rides a train that averages 65 mph and a bus the rest of the way at 36 mph. If the total distance is 498 miles, what is the distance traveled on the bus?

[2.2] **100.** A trip is made by plane and car. The plane averages 440 mph and the car 35 mph. The total trip of 800 mi takes $5\frac{1}{2}$ hr. How many hours are spent in the car?

CHAPTER 3

Graphing Equations and Inequalities

Hypatia
(370–415)

On the lecture slate
The circle rounded under female hands
With flawless demonstration.

Alfred Lord Tennyson
The Princess, II, 1847

Historical Note

Hypatia was the last important mathematician of the Alexandrian period and the first woman noted in the history of mathematics. She wrote scholarly commentaries on the works of Diophantus and Apollonius. One of her treatises was on the conic sections which are introduced in this chapter.

A professor of mathematics, philosophy, and astronomy at Alexandria, Hypatia was an important personality. She had considerable influence in the affairs of the day, particularly with the Roman prefect Orestes, as well as in the Christian community even though she followed the traditions of pagan Hellenism. In fact, her most notable student was Synesius of Cyrene, who later became the bishop of Ptolemais. Synesius called her his "mother, sister, and reverend teacher."

Her connections in high places ultimately made her a tragic heroine. She was accused among the Christians of having incited Orestes against the bishop Cyrillus. A mob sought her out, and she was pulled from her carriage, brutally tortured, and murdered. It was with her death that the long and glorious age of Greek mathematics came to an end.

Chapter Overview

Graphing of linear and quadratic equations and inequalities is thoroughly reviewed and extended in this chapter. The rectangular, or Cartesian, coordinate system is recalled and the graphs of lines, parabolas, circles, ellipses, and hyperbolas are developed. In each case the graphs of the corresponding inequalities are discussed, as well as graphs of systems of inequalities. Graphs of equations will later serve to illuminate the features of relationships with special clarity.

3.1 Coordinates and Graphs

If the order in which two numbers x and y are considered is important, then the pair is denoted by (x, y) and is called an **ordered pair.** Although single numbers are plotted on a number line, a **coordinate plane,** as shown in Figure 3.1, is used to plot ordered pairs. The first number listed is called the first component, or *abscissa,* and the second number listed is called the second component, or *ordinate*.

It is assumed that you are familiar with plotting points and some of the associated terminology: ***x*-axis, *y*-axis, origin,** and **quadrants I, II, III, and IV,** as shown in Figure 3.1.

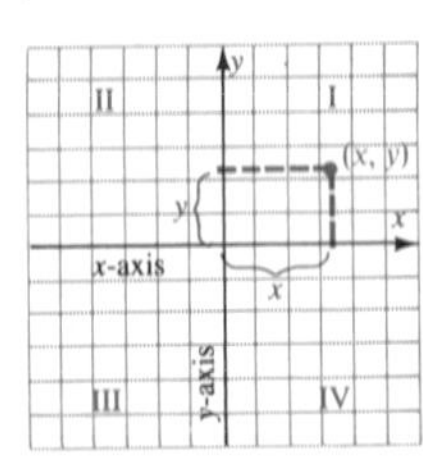

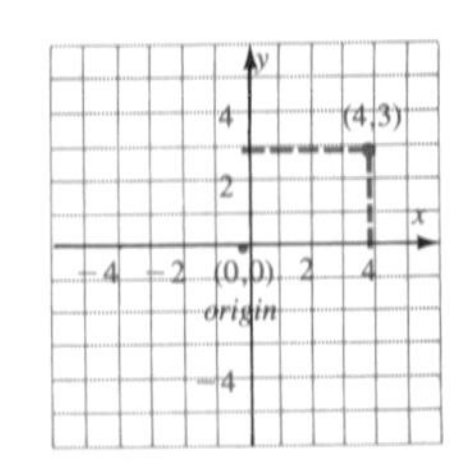

Figure 3.1
Cartesian Coordinate System

The idea of graphing, or plotting, a point—first introduced in Section 1.1—is now extended to graphing points specified by ordered pairs. One of the most common methods for specifying a graph is to give an equation involving two variables (usually x and y). The procedure for graphing an unfamiliar relationship is to compute a sufficient number of ordered pairs **satisfying** the equation, plot the points represented by the ordered pairs, and then sketch the line or curve through the points. The **domain** is the set of all replacements for the first component (usually x). The domain is *not* x, but rather a set from which x is chosen.

Example 1 Graph: $y = |2x - 1|$

Solution Complete the graph by plotting the points.

x	
-3	7
-2	5
-1	3
0	1
$\frac{1}{2}$	0
1	1
2	3
3	5

Let $x = -3$.
Then $y = |2(-3) - 1|$
$= |-7|$
$= 7$
Now plot $(-3, 7)$,
and find and plot the other points.

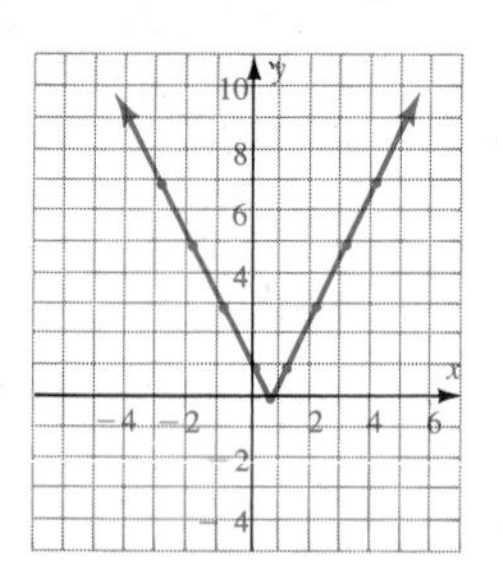

Example 2 Graph: $y = x^3 + 2x^2 - 5x - 6$

Solution

x	y
-4	-18
-3	0
-2	4
-1	0
0	-6
1	-8
2	0
3	24

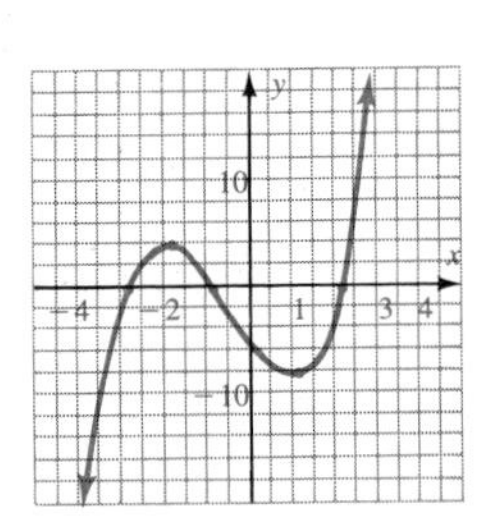

In later sections of this chapter, we will concentrate on the graphs of particular types of equations and find out much more efficient methods for graphing than plotting points. However, before beginning, it is important that you clearly understand the relationship between an equation and its graph. This relationship is very explicit; there is a one-to-one correspondence between ordered pairs satisfying an equation and coordinates of points on the graph.

RELATIONSHIP BETWEEN AN EQUATION AND ITS GRAPH

The **graph of an equation,** or the **equation of a graph,** means:

1. Every point on the graph has coordinates that satisfy the equation.
2. Every ordered pair satisfying the equation has coordinates that lie on the graph.

Distance

The **distance** between points in the coordinate plane can be found by using the Pythagorean theorem, since the axes are perpendicular. Notice, in Figure 3.2, that the distance between two points is the hypotenuse of a right triangle whose sides are vertical and horizontal segments. The difference $x_2 - x_1$ is called the **horizontal change,** or Δx, and $|\Delta x|$ is the length of the horizontal side of the triangle. Similarly, the difference in the vertical coordinates, $y_2 - y_1$, is the **vertical change,** or Δy, and $|\Delta y|$ is the length of the vertical side of the triangle. Thus,

$$d^2 = \Delta x^2 + \Delta y^2$$
$$d = \sqrt{\Delta x^2 + \Delta y^2}$$
or $$d = \sqrt{(x_2 - x_1)^2 + (y_2 - y_1)^2}$$

Note that $|\Delta x|^2 = (\Delta x)^2$ and $|\Delta y|^2 = (\Delta y)^2$, so the formula may be written without absolute values. Also, Δx is one symbol, so Δx^2 means $(\Delta x)^2$.

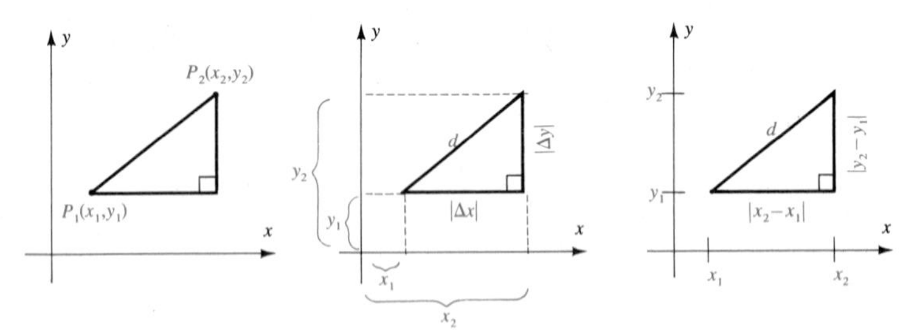

Figure 3.2
Distance Formula

DISTANCE FORMULA

The distance d between any two points (x_1, y_1) and (x_2, y_2) is given by

$$d = \sqrt{\Delta x^2 + \Delta y^2}$$

where $\Delta x = x_2 - x_1$ and $\Delta y = y_2 - y_1$.

Example 3 Find the distance between the points $(-5, 1)$ and $(4, -2)$.

Solution

$\Delta x = 4 - (-5) = 9 \qquad \Delta y = -2 - 1 = -3$

$d = \sqrt{9^2 + (-3)^2}$

$d = \sqrt{90}$

$\mathbf{d = 3\sqrt{10}}$

Midpoint

Another useful concept involving the coordinates of points is the *midpoint* of a segment determined by two points. The **midpoint** M of a segment determined by $P_1(x_1, y_1)$, and $P_2(x_2, y_2)$ can be found by considering the changes, Δx and Δy, used to determine distance.

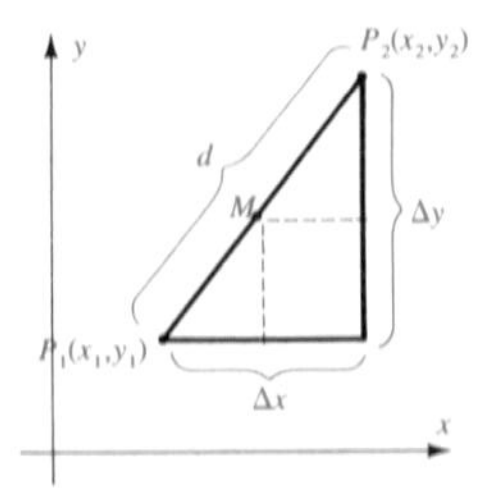

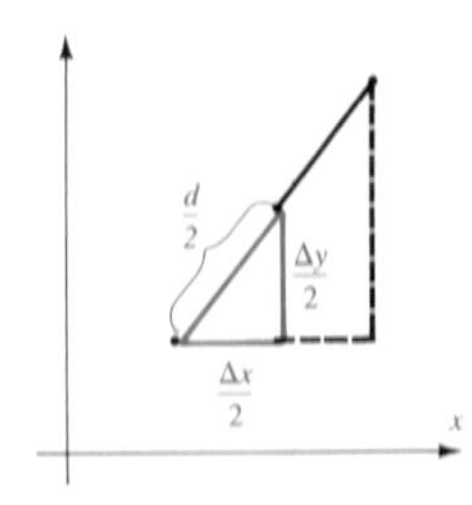

Figure 3.3
Midpoint Formula

In Figure 3.3, note that if M is half the distance from P_1 to P_2, then the changes in x and y should be just half as great as Δx and Δy. If these changes are added to the coordinates of the initial point P_1, the midpoint should be

$$\begin{aligned} M &= \left(x_1 + \frac{\Delta x}{2},\ y_1 + \frac{\Delta y}{2}\right) \\ &= \left(x_1 + \frac{x_2 - x_1}{2},\ y_1 + \frac{y_2 - y_1}{2}\right) \\ &= \left(\frac{2x_1 + x_2 - x_1}{2},\ \frac{2y_1 + y_2 - y_1}{2}\right) \\ &= \left(\frac{x_1 + x_2}{2},\ \frac{y_1 + y_2}{2}\right) \end{aligned}$$

Basically, you simply average the coordinates of the two end points, adding corresponding coordinates and dividing by 2 to find the coordinates of the midpoint.

MIDPOINT FORMULA

The midpoint M between points (x_1, y_1) and (x_2, y_2) is given by

$$\mathbf{M = \left(\frac{x_1 + x_2}{2},\ \frac{y_1 + y_2}{2}\right)}$$

Example 4 Find the midpoint between $(-5, 4)$ and $(1, 3)$.

Solution

$$M = \left(\frac{-5 + 1}{2},\ \frac{4 + 3}{2}\right)$$

$$\mathbf{M = \left(-2, \frac{7}{2}\right)}$$

□

Slope

Another useful concept involving the coordinates of points and the horizontal and vertical changes Δx and Δy is the **slope** of a line. A line is uniquely determined by any two of its points. A line can be described as *steep* or *flat*, referring to its inclination from the horizontal. The slope of a line is a measure of this inclination; it is the ratio of the vertical change to the horizontal change (see Figure 3.4). The horizontal change is sometimes called the *run*, and the vertical change is sometimes called the *rise*. The slope is then referred to as the *rise over the run*.

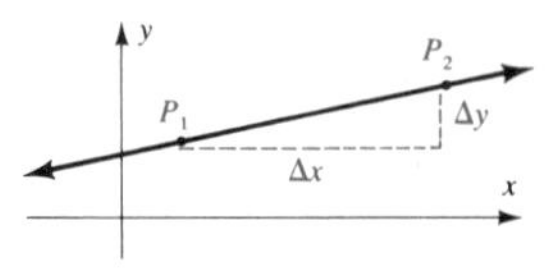

Figure 3.4
Slope

SLOPE FORMULA

> The slope m of the line passing through the points (x_1, y_1) and (x_2, y_2) is given by
>
> $$\mathbf{m = \frac{\Delta y}{\Delta x}}$$
>
> where $\Delta x = x_2 - x_1$ and $\Delta y = y_2 - y_1$.

A line for which $\Delta x = 0$ has undefined slope. We will consider this more completely in the next section.

Example 5 Find the slope of the line passing through the points $(-2, 3)$ and $(4, -6)$.

Solution

$$m = \frac{-6 - 3}{4 - (-2)} = \frac{-9}{6} = -\frac{3}{2}$$

□

In the next section you will see in what ways the slope of a line is an integral part of the equation representing the line. Slope is essentially a rate; it is also an important concept in the study of calculus. One theorem involving slope states a condition for parallel and perpendicular lines.

CONDITION FOR PARALLEL LINES

CONDITION FOR PERPENDICULAR LINES

> Let L_1 and L_2 be two nonvertical lines with slopes m_1 and m_2, respectively. Then
>
> 1. $L_1 \| L_2$ if and only if
>
> $$m_1 = m_2$$
>
> The lines are parallel if and only if the slopes are equal.
>
> 2. $L_1 \perp L_2$ if and only if
>
> $$m_1 m_2 = -1 \quad \text{or} \quad m_1 = -\frac{1}{m_2}$$
>
> The lines are perpendicular if and only if the product of the slopes is -1 (or the slopes are negative reciprocals).

Example 6 Determine whether $\triangle ABC$ is a right triangle if $A(-2, -3)$, $B(-5, 1)$, and $C(6, 3)$ are the vertices.

Solution The lengths of the sides, $\overline{AB}$, $\overline{BC}$, and $\overline{AC}$, may be found by using the distance formula:

$$\overline{AB} = \sqrt{\Delta x^2 + \Delta y^2}$$
$$\Delta x = -5 - (-2) = -3$$
$$\Delta y = 1 - (-3) = 4$$
$$\overline{AB} = \sqrt{(-3)^2 + 4^2} = \sqrt{25} = 5$$

$$\overline{BC} = \sqrt{\Delta x^2 + \Delta y^2}$$
$$\Delta x = 6 - (-5) = 11$$
$$\Delta y = 3 - 1 = 2$$
$$\overline{BC} = \sqrt{11^2 + 2^2} = \sqrt{125} = 5\sqrt{5}$$

$$\overline{AC} = \sqrt{\Delta x^2 + \Delta y^2}$$
$$\Delta x = 6 - (-2) = 8$$
$$\Delta y = 3 - (-3) = 6$$
$$\overline{AC} = \sqrt{8^2 + 6^2} = \sqrt{100} = 10$$

Then, trying the Pythagorean theorem, we have

$$5^2 + 10^2 \stackrel{?}{=} (5\sqrt{5})^2$$
$$25 + 100 = 125 \checkmark$$

Thus, $\triangle ABC$ is a right triangle.

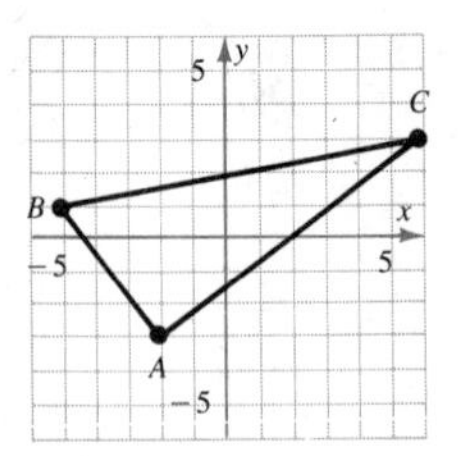

□

Alternate Solution The slopes of the sides of the triangle, m_{AB}, m_{BC}, and m_{AC}, may be found:

$$m_{AB} = \frac{\Delta y}{\Delta x} = \frac{4}{-3} = -\frac{4}{3} \qquad m_{BC} = \frac{\Delta y}{\Delta x} = \frac{2}{11} \qquad m_{AC} = \frac{\Delta y}{\Delta x} = \frac{6}{8} = \frac{3}{4}$$

Since $\left(-\frac{4}{3}\right)\left(\frac{3}{4}\right) = -1$, two sides are perpendicular and $\triangle ABC$ is a right triangle. □

Example 7 Show that $(0, 4)$, $(5, 5)$, $(1, -1)$, and $(-4, -2)$ are the vertices of a parallelogram.

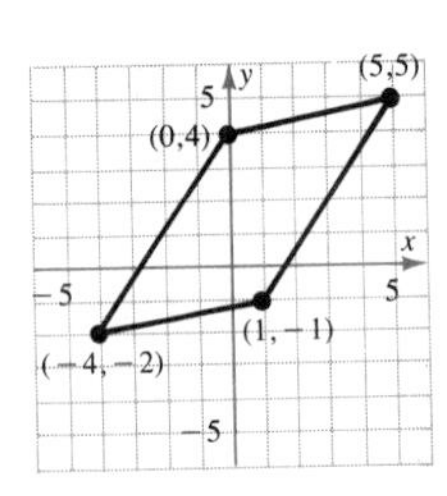

Solution Since opposite sides of a parallelogram are parallel, you must show that the lines containing the opposite sides have the same slope.

For $(0, 4), (5, 5)$: $\quad m_1 = \frac{5-4}{5-0} = \frac{1}{5}$

For $(5, 5), (1, -1)$: $\quad m_2 = \frac{-1-5}{1-5} = \frac{-6}{-4} = \frac{3}{2}$

For $(1, -1), (-4, -2)$: $\quad m_3 = \frac{-2+1}{-4-1} = \frac{-1}{-5} = \frac{1}{5}$

For $(-4, -2), (0, 4)$: $\quad m_4 = \frac{4+2}{0+4} = \frac{6}{4} = \frac{3}{2}$

$m_1 = m_3$ and $m_2 = m_4$, so the given points are the vertices of a parallelogram. □

Problem Set 3.1

A *Plot all points in Problems 1 and 2 on the same coordinate plane.*

1. $(-7,7), (13,-1), (1,-2), \left(-\frac{5}{3}, 1\right)$

2. $(-11,-3), (9-11), (4,6), \left(5, -\frac{5}{3}\right)$

In Problems 3–11 find points that satisfy the given equation, plot the points, and sketch the graph.

3. $y = |3x - 4|$
4. $y = |2x + 5|$
5. $x + y = 3$
6. $x + 2y = 4$
7. $2x = 10 - 5y$
8. $3x = 14 - 7y$
9. $y = x^2 + 3$
10. $y = x^2 - 4$
11. $y = (x - 1)^2$

Find the distance between the points given in Problems 12–26.

12. $(-5, 7), \quad (-2, 3)$
13. $(4, -6), \quad (1, -2)$
14. $(2, 1), \quad (7, 13)$
15. $(1, 3), \quad (16, 11)$
16. $(3, -4), \quad (8, 1)$
17. $(-5, -3), \quad (7, 9)$
18. $(5, 1), \quad (-2, 2)$
19. $(2, 7), \quad (5, -3)$
20. $(3, -5), \quad (-2, 7)$
21. $(\sqrt{3}, \sqrt{24}), \quad (\sqrt{12}, \sqrt{6})$
22. $(\sqrt{2}, \sqrt{28}), \quad (\sqrt{8}, \sqrt{7})$
23. $\left(\frac{\sqrt{3}}{2}, -\frac{1}{2}\right), \quad \left(\frac{1}{2}, \frac{\sqrt{3}}{2}\right)$
24. $(a, b), \quad (b, a)$
25. $(a, b), \quad (a + b, b - a)$
26. $(a, a + b), \quad (a + b, b)$

Find the midpoint between the points given in Problems 27–38.

27. $(4, 5), \quad (8,10)$
28. $(3,7), \quad (6, 14)$
29. $(-2, 1), \quad (4, 7)$
30. $(3, -4), \quad (-5, 8)$
31. $(1, 3), \quad (-5, 4)$
32. $(6, 1), \quad (4, -6)$
33. $(3, -7), \quad (-3, 4)$
34. $(8, -3), \quad (-5, -2)$
35. $(r, s), \quad (t, s)$
36. $(a, b), \quad (b, a)$
37. $(a, b), \quad (a + b, b - a)$
38. $(a, a + b), \quad (a + b, b)$

Find the slope of the line passing through the points given in Problems 39–50.

39. $(1, 2), \quad (5, 8)$
40. $(3, 1), \quad (6, 7)$
41. $(2, -3), \quad (-4, 1)$
42. $(-2, -5), \quad (4, 5)$
43. $(10, -11), \quad (-5, 4)$
44. $(12, -9), \quad (-4, 9)$
45. $\left(\frac{3}{4}, \sqrt{3}\right), \quad \left(-\frac{1}{4}, 2\sqrt{3}\right)$
46. $(\sqrt{2}, 1.8), \quad (1 + \sqrt{2}, 1.6)$
47. $(p, q), \quad (r, s)$
48. $(h, k), \quad (k, h)$
49. $(h, k), \quad (h + a, k + a)$
50. $(x, y), \quad (x + \Delta x, y + \Delta y)$

B *In Problems 51–60 construct a table of values that satisfy the given equation, plot the points, and sketch the graph.*

51. $y = 5 - x^2$
52. $y = \frac{1}{2}x^2 + 3$

53. $3y = |2x + 9|$

54. $2y = -|6 - x|$

55. $y = x^2 - 6x + 10$

56. $y = x^2 + 8x + 10$

57. $y = x^3 + 3x^2 - 9x - 22$

58. $y = x^3 - 6x^2 - 15x + 46$

59. $y = |x^2 - 6x + 5|$

60. $y = |3 + 2x - x^2|$

C **61.** Show that $P(2, 6)$, $A(-4, -2)$, and $T(6, 3)$ are the vertices of a right triangle, if possible.

62. Show that $M(2, 2)$, $I(6, -4)$, and $C(-7, -4)$ are the vertices of a right triangle, if possible.

63. Show that $A(2, 0)$, $B(4, 4)$, $C(0, 6)$, and $D(-2, 2)$ are the vertices of a square, if possible.

64. Show that $R(-2, -2)$, $B(3, -1)$, $U(4, 4)$, and $S(-1, 3)$ are the vertices of a rhombus, if possible. (A rhombus is an equilateral parallelogram.)

65. Show that $T(0, 0)$, $R(\sqrt{3}, 1)$, and $I(\sqrt{3}, -1)$ are the vertices of an equilateral triangle, if possible.

66. Show that $G(3, \sqrt{3})$, $L(1, -\sqrt{3})$, and $E(-1, \sqrt{3})$ are the vertices of an equilateral triangle, if possible.

67. Show that $A(6, 0)$, $B(4, 4)$, $C(-6, -1)$, and $D(-2, -4)$ are the vertices of a trapezoid with two right angles, if possible. (A trapezoid is a quadrilateral with two sides parallel.)

68. Show that $E(4, 1)$, $F(6, 6)$, $G(-4, 2)$, and $H(-6, -3)$ are the vertices of a parallelogram, if possible.

69. Show that $I(0, 0)$, $J(1, 0)$, $K(1 + a, b)$, and $L(a, b)$ are the vertices of a parallelogram for any nonzero real numbers a and b.

70. What type of quadrilateral has vertices $A(-a, 0)$, $B(b, 0)$, $C(a + b, c)$, and $D(0, c)$? Are any restrictions on the variables necessary?

3.2 Lines and Half-Planes

Consider a line with slope m passing through the point $Q(h, k)$. Let $P(x, y)$ be any other point on the line, as shown in Figure 3.5. Then

$$\frac{\Delta y}{\Delta x} = m$$

$$\frac{y - k}{x - h} = m$$

$$y - k = m(x - h)$$

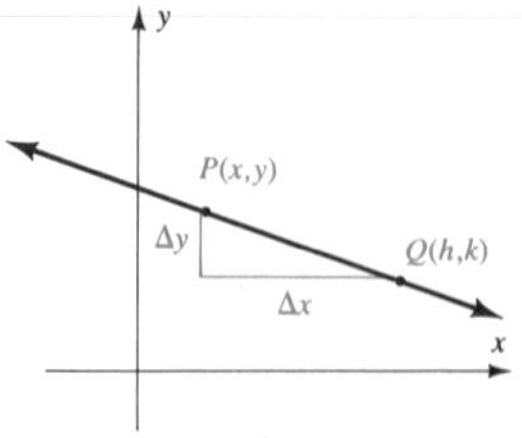

Figure 3.5

This is the equation of the line through (h, k) with slope m, and it is called the **point-slope form** of the equation of a line. The solution of this equation is the graph of the described line.

Example 1 Find the equation of the line through $(3, -2)$ with slope $-\frac{1}{2}$.

Solution

$$y - k = m(x - h)$$
$$y - (-2) = -\frac{1}{2}(x - 3)$$
$$y + 2 = -\frac{1}{2}x + \frac{3}{2}$$
$$2y + 4 = -x + 3$$
$$\mathbf{x + 2y + 1 = 0}$$

□

Example 2 Find the equation of the line through $(1, 5)$ and $(-2, -1)$.

Solution

$$m = \frac{-1 - 5}{-2 - 1} \qquad \text{First, find the slope.}$$
$$= \frac{-6}{-3}$$
$$= 2$$

$$y - k = m(x - h) \qquad \text{You may use either point as } (h, k).$$
$$y - 5 = 2(x - 1)$$
$$y = 2x + 3$$

or $\mathbf{2x - y + 3 = 0}$

□

When graphing lines, or curves in general, the points where the graph crosses the coordinate axes are usually easy to find, and they are often used to help sketch the curve. A **y-intercept** is a point where a graph crosses the y-axis and consequently is a point whose first component is 0. An **x-intercept** is a point where a graph crosses the x-axis; it has a second component of 0. When graphing lines, the y-intercept is used more often than the x-intercept. Every nonvertical line has exactly one y-intercept, so we speak about *the* y-intercept of a line. This point is almost universally denoted by $(0, b)$. Since a y-intercept always has a first component of 0, this notation is often shortened by simply saying *the y-intercept is b*. Remember, this means the line crosses the y-axis at the point $(0, b)$.

To derive another useful form, called the **slope–intercept form,** of the equation of a nonvertical line, we start with the point–slope form and let (h, k)—the given point—be the y-intercept $(0, b)$:

$$y - k = m(x - h)$$
$$y - b = m(x - 0)$$
$$y - b = mx - 0$$
$$\mathbf{y = mx + b}$$

This is the slope–intercept form of the equation of a line with given slope m and y-intercept b.

Example 3 Find the equation of a line with y-intercept 3 and passing through $(2, -5)$.

Solution Since $y = mx + b$ and $b = 3$, we have

$$y = mx + 3$$

The point $(2, -5)$ satisfies this equation, so we can substitute these values to find m:

$$\begin{aligned} -5 &= m(2) + 3 \\ -8 &= 2m \\ -4 &= m \end{aligned}$$

Now substitute in the known values for b and m to obtain the equation of the line:

$$\begin{aligned} y &= mx + b \\ y &= -4x + 3 \quad \text{or} \quad \mathbf{4x + y - 3 = 0} \end{aligned}$$

□

Every first-degree equation in two variables may be written in the form

$$Ax + By + C = 0$$

From this we may write

$$\begin{aligned} By &= -Ax - C \\ y &= -\frac{A}{B}x - \frac{C}{B} \qquad B \neq 0 \end{aligned}$$

which is of the form $y = mx + b$ with

$$m = -\frac{A}{B} \quad \text{and} \quad b = -\frac{C}{B}$$

If $B = 0$, then m is not defined and the line is vertical. Thus, every first-degree equation in two variables is linear; that is, its graph is a straight line. For this reason, first-degree equations in two variables are often called **linear equations.**

If you review the examples of this section, you will notice that the final equation in each case is the standard form of the equation of a line. The equation might be encountered in any of several forms. If you are able to recognize a form or put an equation into a given form, you will be able to sketch the graphs of lines more quickly and accurately. A summary of the forms of an equation of a line is given in Table 3.1.

Example 4 Sketch the graph: $2x - 3y + 6 = 0$

Solution For most graphs, the most efficient method is to solve the equation for y and then determine the y-intercept and the slope by inspection. Next, graph the line by plotting the y-intercept and a *slope point*. A slope point is found by starting at a known point (usually the y-intercept) and counting out the rise and run found from the slope.

Solve for y:

$$y = \tfrac{2}{3}x + 2$$

Table 3.1 Forms of the Equation of a Line

Form	Equation	Variables Defined	When Used
Standard form	$Ax + By + C = 0$	A, B, and C are any constants, A and B not both zero (x, y) is any point on the line	A common form in which equations are often represented; used to standardize the variety of available forms and is the final form you will use when finding equations of lines
Slope–intercept form	$y = mx + b$	m is the slope b is the y-intercept, so $(0, b)$ is on the line	Used when graphing a line and when finding the equation of a line in which you are given the slope and intercept
Point–slope form	$y - k = m(x - h)$	(h, k) is a given point on the line	Used when finding the equation of a line in which you are given the slope and a point, or two points
Horizontal line	$y = k$	Horizontal line passing through a point with a second component of k	Used whenever dealing with horizontal lines or lines with slopes of 0
Vertical line	$x = h$	Vertical line passing through a point with a first component of h	Used whenever dealing with vertical lines or lines with no slope (that is, slopes that are undefined)

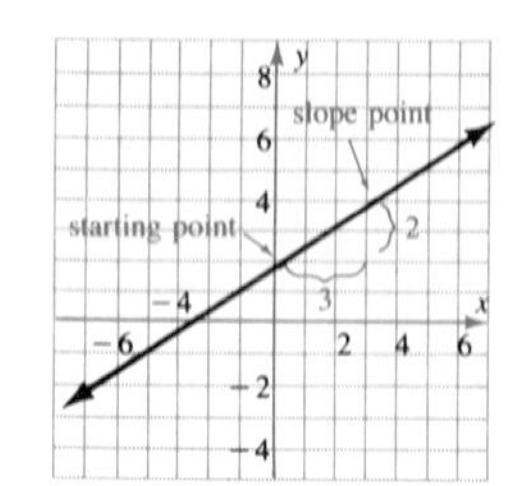

Thus, $m = \frac{2}{3}$ and $b = 2$. Plot the y-intercept at $(0, 2)$. Since

$$\frac{\Delta y}{\Delta x} = \frac{2}{3}$$

move up 2 and over 3; plot another point (a slope point) and draw the line. □

Example 5 Sketch the graph: $y - 3 = \frac{1}{4}(x + 2)$

Solution This equation is already in the point–slope form, so by inspection, we can see that it goes through $(-2, 3)$ with slope $\frac{1}{4}$. Count rise $= 1$ and run $= 4$ to find a slope point.

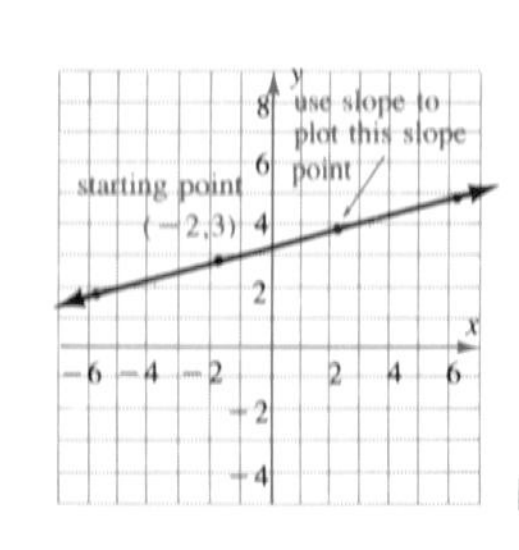

□

Example 6 Sketch the graph: $7x + y + 28 = 0$

Solution Since $y = -7x - 28$, $m = -7$ and $b = -28$. The y-intercept is a little awkward to plot, so find some other point to use. If $x = -4$, then

$$\begin{aligned} y &= -7(-4) - 28 \\ &= 28 - 28 \\ &= 0 \end{aligned}$$

and $(-4, 0)$ is a point on the line. Write the slope as $m = \frac{7}{-1}$ and count rise $= 7$ and run $= -1$ to find a slope point. Another slope point using $m = \frac{-7}{1}$ is also shown.

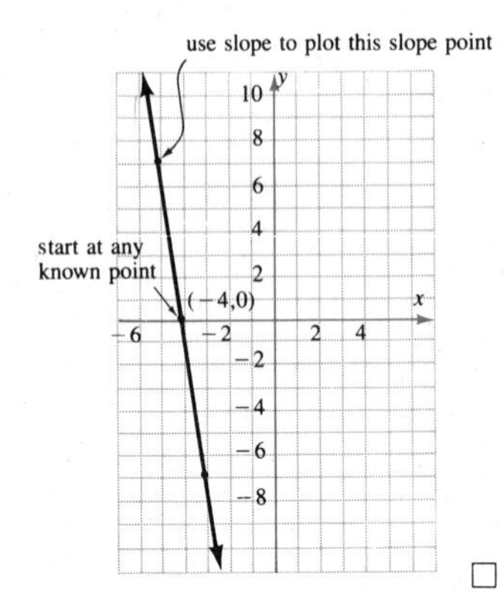

□

In Example 6 note that $m = \frac{7}{-1}$ means "up 7 and back 1," and $m = \frac{-7}{1}$ means "down 7 and over 1." A **positive** slope (as in Example 5) indicates that the line *increases* from left to right, and a **negative** slope (as in Example 6) indicates the line *decreases* from left to right.

Example 7 Sketch the graph: $y = 5$

Solution

$$y = 0x + 5$$
$$m = 0 \qquad b = 5$$

This is a horizontal line with y-intercept 5. It is horizontal because the vertical change must be 0 if the slope is 0, so every y-coordinate is the same.

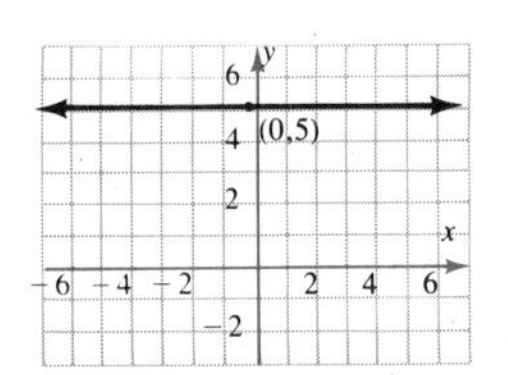

□

Example 8 Sketch the graph: $x + 4 = 0$

Solution

$$x = -4$$

This is a vertical line with x-intercept -4. It is vertical since every x-coordinate is the same, -4.

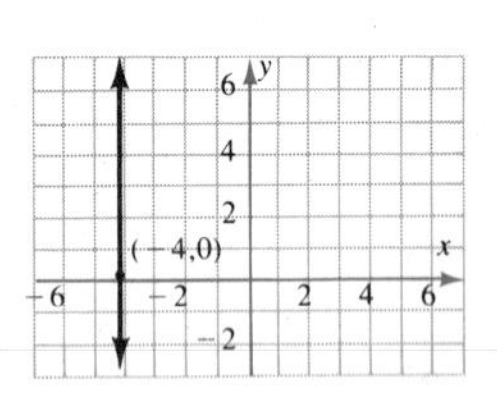

□

It is unusual to see a linear application without some restriction on the variables. For example, you might have an equation that represents the relationship between time x and temperature y, with the temperature remaining between 40°F and 100°F. This constraint or restriction could be written as $|y - 70| < 30$, or as

$$\begin{aligned} |y - 70| &< 30 \\ -30 < y - 70 &< 30 \\ 40 < y &< 100 \end{aligned}$$

Examples 9 and 10 show how to graph linear equations with restrictions. The symbol : used to indicate the restrictions means *such that*.

Example 9 $y = -\frac{2}{5}x + 5 : 1 \leq y \leq 5$

Solution Sketch the line segment that is part of the line with slope $-\frac{2}{5}$ and y-intercept 5, *but* with y between 1 and 5, inclusive. This is the segment shown in color.

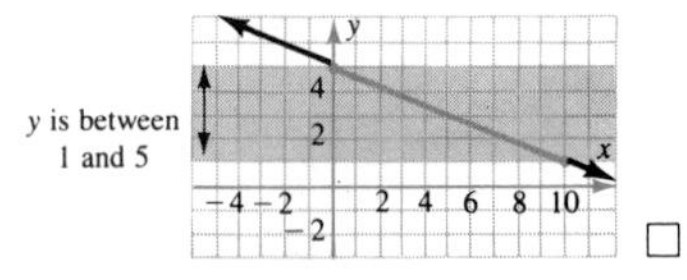

□

Example 10 $y + 2 = -3(x - 2) : |x - 1| < 2$

Solution From the properties of absolute value,

$$\begin{aligned} |x - 1| &< 2 \\ -2 < x - 1 &< 2 \\ -1 < x &< 3 \end{aligned}$$

Notice that the end points are not included. This segment is said to be *open* at both ends. Sketch the segment through $(2, -2)$ with slope -3, *but* so that $-1 < x < 3$.

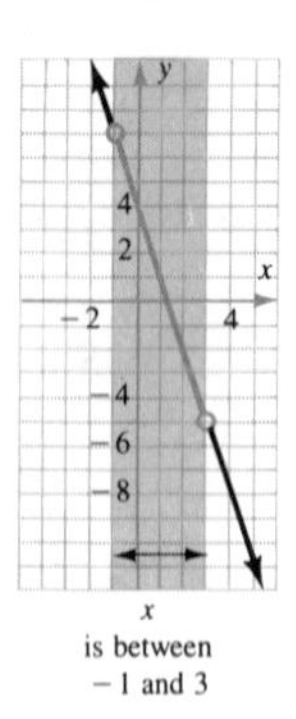

□

Now that you are able to graph linear equations, such as $y = 2x + 1$, it is natural to wonder about the graphs of $y > 2x + 1$ and $y < 2x + 1$. Consider a vertical line at $x = 1$, on which every point has coordinates $(1, y)$, as shown in Figure 3.6.

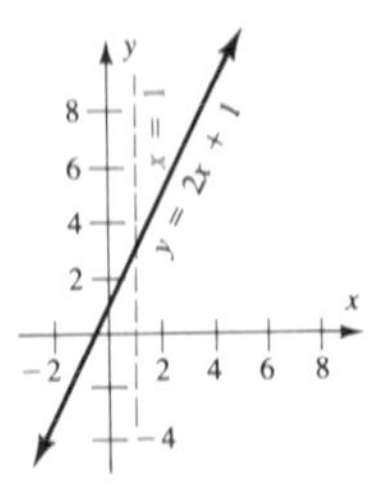

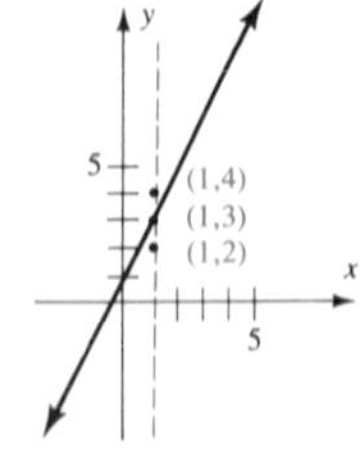

Figure 3.6

If $x = 1$ and $y = 2x + 1$, then $(1, 3)$ is *on* the line $y = 2x + 1$.
If $x = 1$ and $y > 2x + 1$, then

$$\begin{aligned} y &> 2(1) + 1 \\ y &> 3 \end{aligned}$$

Then the points $(1, y)$ will be *above* $(1, 3)$ on the line $y = 2x + 1$. Similarly for $x = 1$ and $y < 2x + 1$, the points $(1, y)$ will be *below* $(1, 3)$. Thus, $y > 2x + 1$ will be the set of points above the line $y = 2x + 1$, and $y < 2x + 1$ will be all those below.

In general, each line divides the coordinate plane into three parts: the line itself and two **half-planes.** The half-plane is **closed** if it includes the boundary line and **open** if it does not (see Figure 3.7). A summary of these concepts is given in Table 3.2.

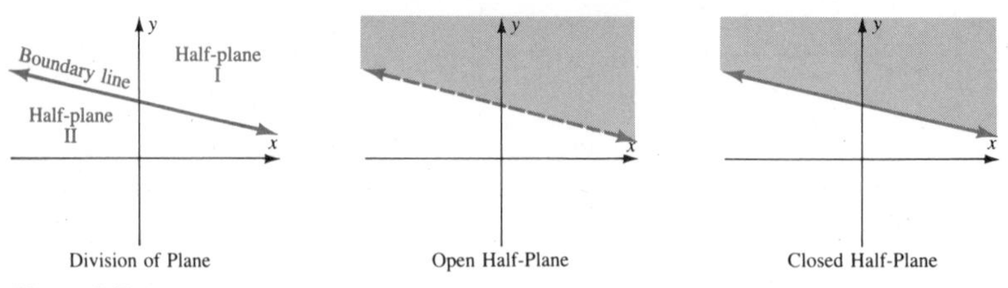

Figure 3.7

Sometimes it is difficult to refer to, or remember, the information given in Table 3.2. There is an easier and more general procedure. First, sketch the boundary; use a *solid line* if the boundary is included (closed) and a *dashed line* if it is excluded (open), as shown in Figure 3.7. Next, choose any test point (a, b) not on the boundary. Substitute the coordinates of the test point into the given inequality ($x = a$ and $y = b$). If it satisfies the inequality, then the test point is in the appropriate half-plane; if it does not satisfy the inequality, then the solution is the other half-plane.

Example 11 Sketch: $2x + 5y + 10 \geqslant 0$

Solution Solve for y:

$$y \geqslant -\frac{2}{5}x - 2$$

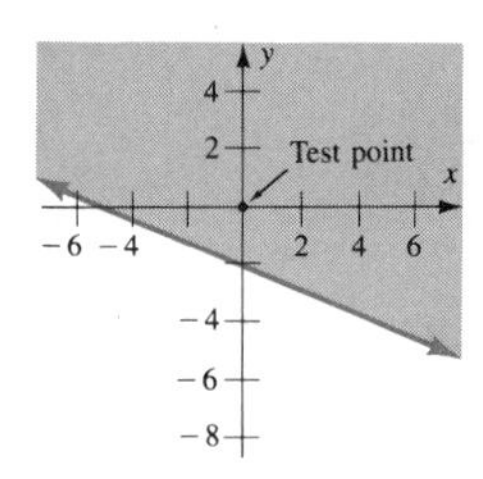

Draw the boundary (solid since it is included).
Test point $(0, 0)$ in the given inequality:

$$2(0) + 5(0) + 10 \geqslant 0$$

This is true, so shade in the portion of the plane including this test point. □

Example 12 Sketch: $3x - y - 6 > 0$

Solution Solve for y:

$$y < 3x - 6$$

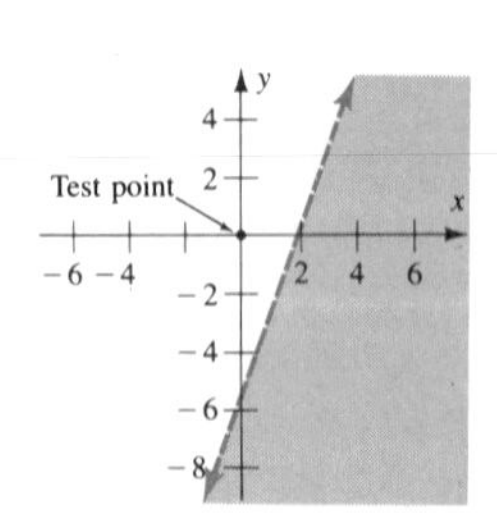

Draw the boundary (dashed since it is excluded).
Test point $(0, 0)$:

$$3(0) - 0 - 6 > 0$$

This is false, so shade in the half-plane that excludes this test point. □

Example 13 Sketch: $2x - 3y < 0$

Solution Solve for y:

$$y > \frac{2}{3}x$$

Table 3.2

The graph of:		
$y < mx + b$ is the *open half-plane* below $y = mx + b$	$y = mx + b$ is the *line* with slope m and y-intercept b	$y > mx + b$ is the *open half-plane* above $y = mx + b$
$y \leqslant mx + b$ is the *closed half-plane* below $y = mx + b$		$y \geqslant mx + b$ is the *closed half-plane* above $y = mx + b$

Draw the boundary (dashed).
The test point cannot be (0, 0) since the test point cannot be on the boundary. Remember, you can choose *any* point *not* on the boundary. In this case, we choose (6, 0):

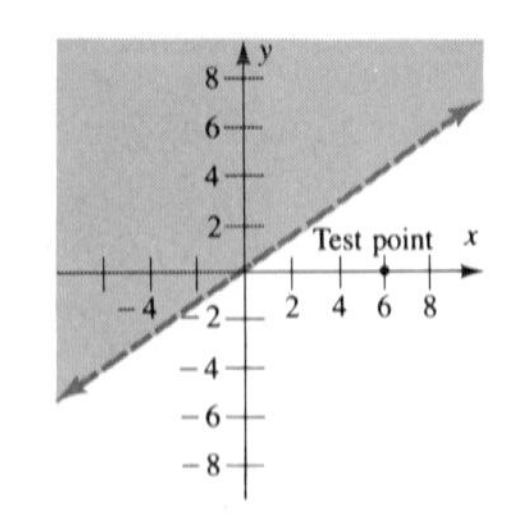

$$2(6) - 3(0) < 0$$
$$12 < 0$$

This is false, so shade the half-plane that does not include this test point. □

Now reconsider lines with restrictions as discussed earlier in this section. For example,

$$y = \frac{1}{2}x + 3 : 1 \leqslant x \leqslant 7$$

is actually three conditions imposed simultaneously on a set of points:

$$\begin{cases} y = \frac{1}{2}x + 3 \\ 1 \leqslant x \leqslant 7 \end{cases} \quad \text{or} \quad \begin{cases} y = \frac{1}{2}x + 3 \\ x \geqslant 1 \\ x \leqslant 7 \end{cases}$$

Using the symbol { here is the same as inserting the word *and* between the members of the system.

A solution to such a **system** of equalities or inequalities is one that satisfies each member of the system. Graphically speaking, the solution set of the system is the *intersection* of the solution sets of the members of the system. We will limit ourselves here to graphic solutions of systems of inequalities. To find the solution set of a system, graph each of the inequalities in the system and find the intersection of the graphs.

Systems of inequalities are frequently used to examine a number of conditions applied to a process. In business, the price of a product may be tied to the number of items produced in several ways. The demand for the product is usually dependent on the price charged. Similarly, the supply, the capacity of the production facilities, the availability of materials, and other factors are conditions that may be expressed as inequalities. A process called *linear programming* can be used to locate the point or points that provide the optimum results.

Example 14 Solve graphically:

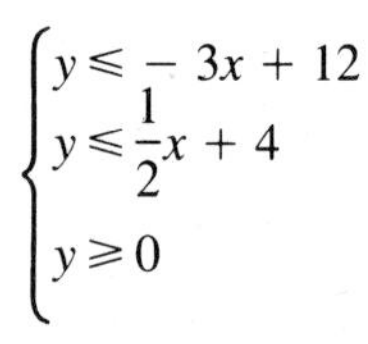

$$\begin{cases} y \leq -3x + 12 \\ y \leq \frac{1}{2}x + 4 \\ y \geq 0 \end{cases}$$

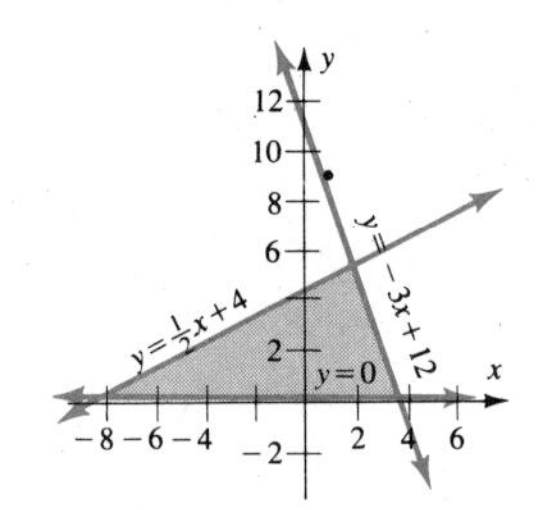

□

Example 15 Solve graphically:

$$\begin{cases} y \geq \frac{1}{4}x + 2 \\ y < 2x + 6 \\ x \leq 8 \\ y < 10 \end{cases}$$

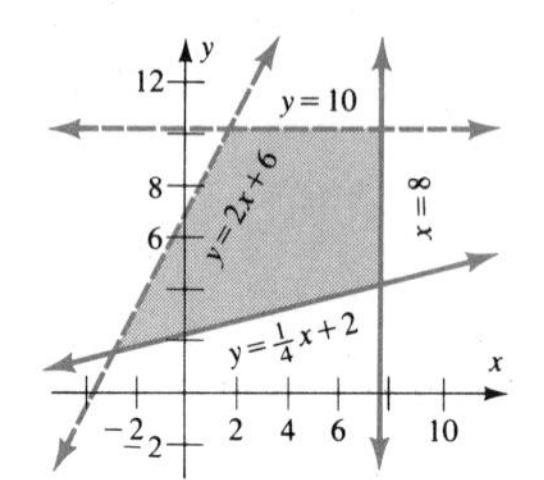

□

Problem Set 3.2

A *In Problems 1–16 find the equation of the line from the given information and then rewrite it in standard form.*

1. Slope 2; passing through (2, 3)

2. Slope $\frac{2}{5}$; passing through (5, −2)

3. Passing through (2, 4) and (4, −3)

4. Passing through (3, −5) and (5, 1)

5. y-intercept 6; slope 8

6. y-intercept −3; slope −2

7. y-intercept 4; passing through (−3, 5)

8. x-intercept −2; y-intercept 3

9. Zero slope; passing through (7, −2)

10. No slope; passing through (−7, 2)

11. Passing through (−1, 5) and (4, 5)

12. Passing through (5, 4) and (5, −1)

13. Intercepts (0, 2) and (−5, 0)

14. Intercepts (−7, 0) and (0, −3)

15. Passing through (3, −7) and (−2, 3)

16. Passing through (−6, −5) and (−1, 5

Sketch the lines given by the equations in Problems 17–30.

17. $y = 2x + 6$

18. $y = -x + 5$

19. $y = \frac{2}{3}x + 6$

20. $y = -\frac{3}{4}x + 6$

21. $2x + y = 3$
22. $2x + 3y + 6 = 0$
23. $x - 3y + 6 = 0$
24. $y = 5x$
25. $x = -3y$
26. $x = -10$
27. $y = 30$
28. $2x - 5y - 1200 = 0$
29. $100x - y = 0$
30. $50x + y = 300$

Sketch the graphs of the inequalities in Problems 31–38.

31. $y \geqslant \frac{5}{9}x - 5$
32. $y \leqslant -\frac{7}{3}x + 7$
33. $y > \frac{3}{5}x - 3$
34. $2x + 3y - 6 < 0$
35. $5x - 2y + 10 > 0$
36. $4x - 3y - 12 < 0$
37. $4x - 3y \geqslant 6$
38. $6y - 9x < 18$

B *Sketch the line segments in Problems 29–48.*

39. $x - 2y - 10 = 0 : 0 < x < 14$
40. $4x - 3y + 36 = 0 : -4 \leqslant x \leqslant 4$
41. $y + 7 = \frac{4}{7}(x - 6) : -1 \leqslant x \leqslant 13$
42. $y + 10 = -\frac{1}{5}(x + 6) : -11 \leqslant x \leqslant -1$
43. $x + 11 = 0 : -9 \leqslant y \leqslant -3$
44. $x = 13 : -6 \leqslant 2y < 3$
45. $y = -\frac{3}{2}x + 23 : 8 \leqslant x \leqslant 14$
46. $y = -\frac{3}{2}x - 5 : -6 \leqslant x \leqslant 0$
47. $y = \frac{1}{2}x + 7 : -6 \leqslant x \leqslant 8$
48. $x + 1 = 0 : 22 > -2y > 7$

In Problems 49–60 sketch the solution to each system of inequalities.

49. $\begin{cases} -10 \leqslant x \leqslant 6 \\ 2 \leqslant y \leqslant 5 \end{cases}$
50. $\begin{cases} -4 \leqslant x \leqslant -2 \\ -5 \leqslant y \leqslant 9 \end{cases}$
51. $\begin{cases} -3 \leqslant x \leqslant 3 \\ 10 \leqslant 3y \leqslant 15 \end{cases}$

52. $\begin{cases} -7 \leqslant 3x \leqslant 19 \\ -10 \leqslant 3y \leqslant 16 \end{cases}$
53. $\begin{cases} 2y \geqslant -x + 2 \\ x \leqslant 3 \\ y \leqslant 3 \end{cases}$
54. $\begin{cases} 2x + 3y \geqslant 12 \\ x \leqslant 6 \\ y \leqslant 6 \end{cases}$

55. $\begin{cases} y \geqslant \frac{3}{4}x - 4 \\ y \leqslant -\frac{3}{4}x + 11 \\ x \geqslant 6 \end{cases}$
56. $\begin{cases} y \geqslant \frac{3}{2}x + 3 \\ y \geqslant \frac{3}{2}x + 6 \\ 3 \leqslant y \leqslant 6 \end{cases}$
57. $\begin{cases} 5x + 2y \leqslant 30 \\ 5x + 2y \geqslant 20 \\ x \geqslant 0 \\ y \geqslant 0 \end{cases}$

58. $\begin{cases} 5x - 2y + 30 \geqslant 0 \\ 5x - 2y + 20 \leqslant 0 \\ x \leqslant 0 \\ y \geqslant 0 \end{cases}$
59. $\begin{cases} 8x + 3y \leqslant 9 \\ y - 4 \geqslant -\frac{8}{3}(x + 2) \\ -5 \leqslant y \leqslant 3 \end{cases}$
60. $\begin{cases} x + y - 9 \leqslant 0 \\ x + y + 3 \geqslant 0 \\ x - y \leqslant 7 \\ y - x \leqslant 5 \end{cases}$

Graph the line segments in Problems 61–68 by sketching the lines and noting restrictions on either x or y.

C **61.** $5x - 2y + 49 = 0 : |x + 9| < 2$
62. $2x + 5y + 37 = 0 : |y + 7| < 4$
63. $x - 12y - 25 = 0 : |x - 7| \leqslant 6$
64. $4x - 9y + 17 = 0 : |y + 1| \leqslant 2$

65. $2x + 5y - 21 = 0 : |x - 3| < 10$

66. $9x + 8y + 7 = 0 : \; |x + 3| \leq 4$

67. $5x - 2y - 67 = 0 : |y + 6| < 5$

68. $9x + 4y - 37 = 0 : |x - 7| \leq 2$

3.3 Quadratics and Parabolas

Many phenomena are linear, while others require more advanced equations and curves. The parabola, for instance, describes the shape of the path of a projectile—or of water, if it is squirted horizontally—as well as the path of certain comets. A parabolic mirror is the reflector in a flashlight, an auto headlamp, or a telescope; the parabola is an important curve with many applications.

The principles of analytic geometry allow us to prove that the graph of the *quadratic* $y = ax^2 + bx + c$, $a \neq 0$, is a **parabola.*** The equation

$$y + ax^2 + bx + c$$

can be written in the form

$$y - k = a(x - h)^2$$

by completing the square in x, as shown in the following example.

Example 1

$y = 2x^2 - 4x + 5$	
$y - 5 = 2x^2 - 4x$	Isolate the x terms.
$y - 5 = 2(x^2 - 2x)$	Factor out the coefficient of x^2.
$y - 5 + 2 = 2(x^2 - 2x + 1)$	Complete the square in x.
$\uparrow \quad \downarrow \qquad\qquad \downarrow$	Notice that adding 1 inside the parentheses is actually adding 2(1) to both sides.
$\mathbf{y - 3 = 2(x - 1)^2}$	Rewrite in factored form.

Now $(h, k) = (1, 3)$ and $a = 2$ in $y - k = a(x - h)^2$. □

Consider the parabola

$$y - k = a(x - h)^2 \qquad \text{with} \qquad a > 0$$

Since $a > 0$, by multiplying both sides by the nonnegative number $(x - h)^2$, we get $a(x - h)^2 \geq 0$. Also, $y - k \geq 0$, since $y - k = a(x - h)^2$ and $a(x - h)^2 \geq 0$. Thus, $y \geq k$. So k is the **minimum value** of y on the graph, and (h, k) is the lowest point of the parabola. This parabola opens *upward*, as shown in Figure 3.8. Similarly, for $a < 0$, k is the **maximum value** of y, (h, k) is the highest point of the parabola, and the parabola opens *downward*. In either case, the point (h, k) is called the **vertex** of the parabola.

In practice, knowing the vertex allows you to complete the sketch of the parabola by plotting a few points near the vertex.

---

*A formal definition is given in Problem 69, Problem Set 3.3.

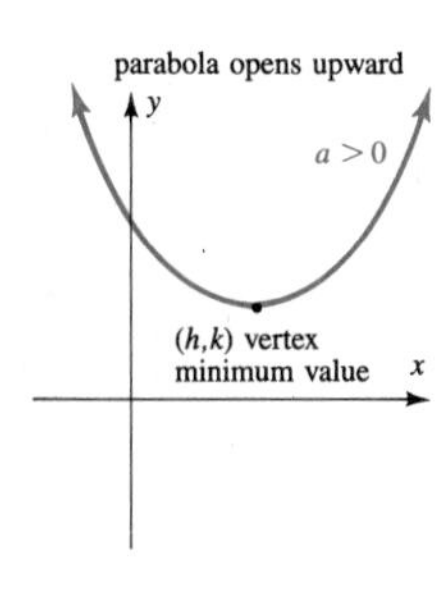

y
x
maximum value
(h,k) vertex
a < 0
parabola opens downward

Figure 3.8

Example 2 Find the vertex and sketch the graph of $y = -2x^2 + 6x - 5$.

Solution First, complete the square:

$$y + 5 = -2(x^2 - 3x \qquad)$$

$$y + 5 - \frac{9}{2} = -2\left(x^2 - 3x + \frac{9}{4}\right)$$

Caution: Adding $\frac{9}{4}$ inside the parentheses is actually adding $-2\left(\frac{9}{4}\right) = -\frac{9}{2}$ to both sides.

$$y + \frac{1}{2} \quad -2\left(x - \frac{3}{2}\right)^2$$

Now plot a few points and sketch the graph; the vertex is $\left(\frac{3}{2}, -\frac{1}{2}\right)$, and the parabola opens downward.

x	y
3	-5
2	-1
1	-1
0	-5

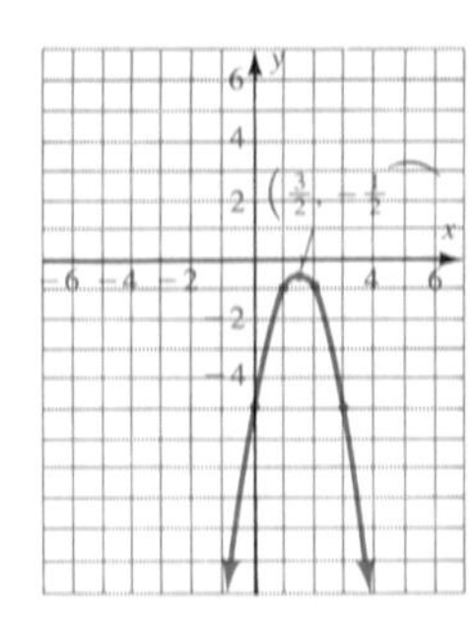

□

Example 3 Find the vertex and sketch the graph of $6y = 3x^2 + 6x + 5$.

Solution

$$6y - 5 = 3(x^2 + 2x \qquad)$$

$$y - \frac{5}{6} = \frac{3}{6}(x^2 + 2x \qquad)$$

Coefficient of first-degree variable must also be 1, so both sides are divided by 6.

$$y - \frac{5}{6} + \frac{1}{2} = \frac{1}{2}(x^2 + 2x + 1)$$

Add $\frac{1}{2}$ to both sides; that is, adding 1 inside the parentheses is actually adding $\frac{1}{2}(1)$.

$$y - \frac{1}{3} = \frac{1}{2}(x + 1)^2$$

The vertex is $\left(-1, \frac{1}{3}\right)$; the parabola opens upward.

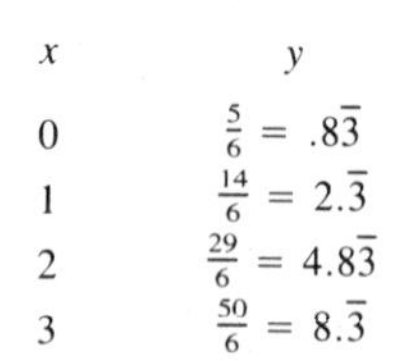

x	y
0	$\frac{5}{6} = .8\overline{3}$
1	$\frac{14}{6} = 2.\overline{3}$
2	$\frac{29}{6} = 4.8\overline{3}$
3	$\frac{50}{6} = 8.\overline{3}$

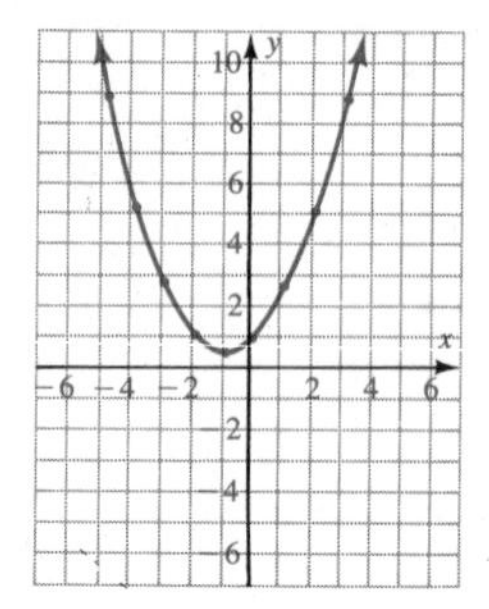

□

We now summarize the results about the graph of the parabola.

STANDARD FORM OF THE EQUATION OF A PARABOLA THAT OPENS UPWARD OR DOWNWARD

> The graph of every quadratic equation
>
> $$y = ax^2 + bx + c \qquad a \neq 0$$
>
> is a **parabola** and can be written in standard form
>
> $$y - k = a(x - h)^2$$
>
> where the **vertex is at (h, k)**. The parabola **opens upward if $a > 0$** and **downward if $a < 0$.**

Example 4 An arch is 18 ft at its highest point and spans 54 ft at its base. If the arch is parabolic, find its height at 9 ft intervals.

Solution A sketch can be made (Figure 3.9) and transferred to the coordinate system (Figure 3.10).

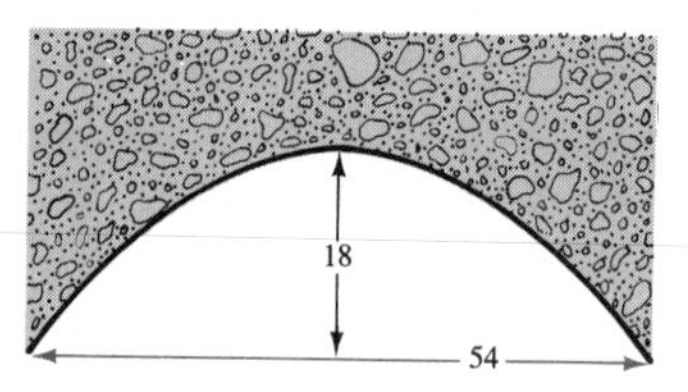

Figure 3.9

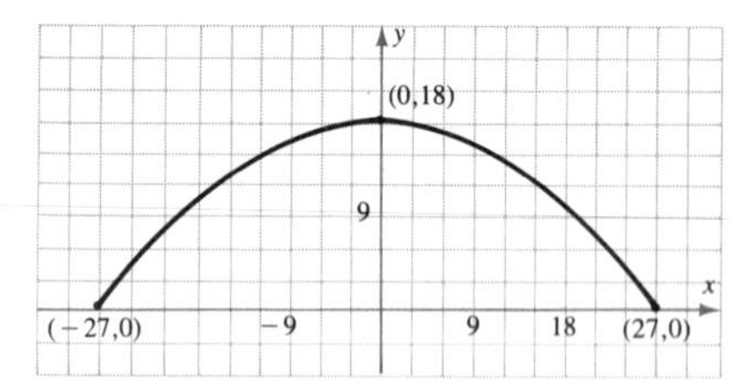

Figure 3.10

The parabola will have the equation $y - k = a(x - h)^2$, and with $(h, k) = (0, 18)$,

$$y - 18 = ax^2$$

The point (27, 0) must satisfy the equation, so

$$0 - 18 = a(27)^2$$
$$a = -\frac{18}{729}$$
$$a = -\frac{2}{81}$$

Thus, $y - 18 = -\frac{2}{81}x^2$ is the equation of the parabola.

Now to find the desired heights, let $x = \pm 9, \pm 18$.

$$y = -\frac{2}{81}(9)^2 + 18 \qquad y = -\frac{2}{81}(18)^2 + 18$$
$$y = 16 \qquad y = 10$$

Note that we will get the same values for -9 and -18.

The arch will be 10, 16, 18, 16, and 10 ft at the 9 ft intervals. □

Consider the points (x, y) and $(x, -y)$, and the x-axis (Figure 3.11a). Notice that the midpoint of the segment determined by the points is on the x-axis and the segment is perpendicular to the x-axis. The points are said to be *symmetric with respect to the x-axis*. That is, they are equidistant and on opposite sides of the x-axis. Similarly, (x, y) and $(-x, y)$ are *symmetric with respect to the y-axis* (Figure 3.11b), and (x, y) and $(-x, -y)$ are *symmetric with respect to the origin* (Figure 3.11c). If every point (x, y) of a curve were replaced by $(x, -y)$, the resulting curve would be symmetric with respect to the x-axis since every point on one curve would be symmetric to a corresponding point on the other curve. Similarly, replacements of $(-x, y)$ and $(-x, -y)$ would produce symmetries with respect to the y-axis and the origin, respectively.

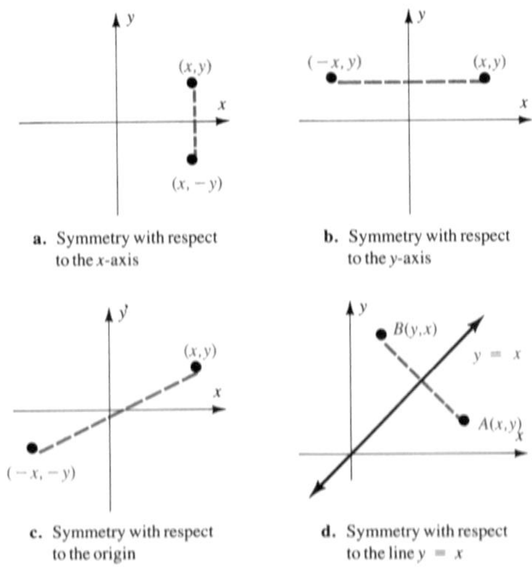

Figure 3.11
Symmetry

However, there is another symmetry which holds special importance for the parabola. Consider the points $A(x, y)$ and $B(y, x)$ and the line $y = x$ (Figure 3.11d). The midpoint formula may be used to find C, the midpoint of segment AB:

$$C = \left(\frac{x + y}{2}, \frac{y + x}{2}\right) = \left(\frac{x + y}{2}, \frac{x + y}{2}\right)$$

which lies on the line $y = x$, since the first and second components are equal. The slope of the line containing the segment AB is

$$m = \frac{y - x}{x - y} = -1$$

Since the slope of $y = x$ is 1, the lines are perpendicular. Thus, (x, y) and (y, x) are *symmetric with respect to the line* $y = x$. If every point (x, y) of a curve were replaced by (y, x), the resulting curve would be symmetric to the original curve with respect to the line $y = x$. In particular, $y = ax^2 + bx + c$ and $x = ay^2 + by + c$ are symmetric with respect to $y = x$. Their symmetry allows us to reach the following conclusion:

STANDARD FORM OF THE EQUATION OF A PARABOLA THAT OPENS TO THE LEFT OR RIGHT

The graph of every quadratic equation

$$x = ay^2 + by + c \qquad a \neq 0$$

is a **parabola** and can be written in standard form

$$\mathbf{x - h = a(y - k)^2}$$

where the **vertex is at (h, k).** The parabola **opens to the right if $a > 0$ and to the left if $a < 0$.**

Example 5 Write $2x + 12y = 3y^2 + 2$ in standard form, find its vertex, and sketch its graph.

Solution

$$2x + 12y = 3y^2 + 2$$
$$2x - 2 = 3(y^2 - 4y \qquad)$$
$$x - 1 + 6 = \frac{3}{2}(y^2 - 4y + 4)$$
$$x + 5 = \frac{3}{2}(y - 2)^2$$

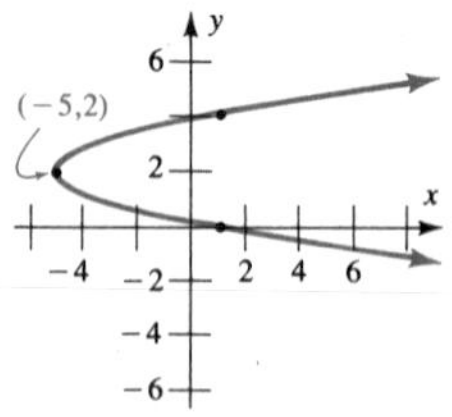

The vertex is $(-5, 2)$, and the parabola opens to the right. □

Every parabola divides the Cartesian plane into three parts; that is, the parabola is a *boundary* of two separate regions. The solution of a parabolic inequality is one of those two regions. While the boundary may or may not be included in the solution, the most direct way of determining the region described by a parabolic inequality is to take a convenient test point not on the boundary and check the inequality. The method is illustrated by Examples 6 and 7.

Example 6 Sketch the graph: $y > x^2 + 2x + 3$

Solution **a.** Consider the associated equation,

$$y = x^2 + 2x + 3$$

or $y - 2 = (x + 1)^2$, which is a parabola that has vertex at $(-1, 2)$ and opens upward.

b. Sketch this boundary equation. In this case, the boundary is dashed because it is not included.

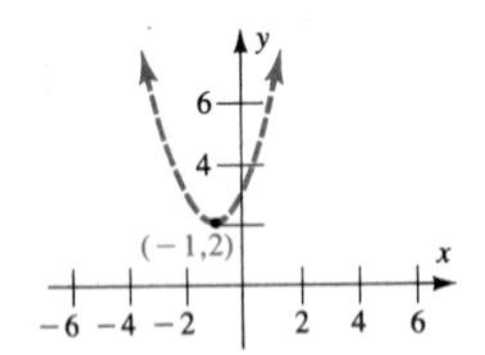

c. Take a convenient point not on the boundary. Try the origin, $(0, 0)$:

$$y > x^2 + 2x + 3$$
$$0 > 0^2 + 2 \cdot 0 + 3$$

Because $0 > 3$ is not true, the origin is not in the solution set.

d. Shade the appropriate region on the graph.

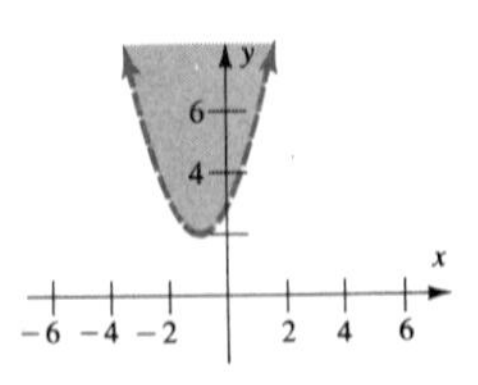

□

Example 7 Sketch the graph: $\begin{cases} y - 1 \geq (x - 3)^2 \\ x - 1 > \frac{1}{3}(y - 5)^2 \end{cases}$

Solution As with linear inequalities, first sketch the associated equations. Second, determine the individual graphs and their intersection. The intersection of the shaded portions (shown in darker color) is the solution.

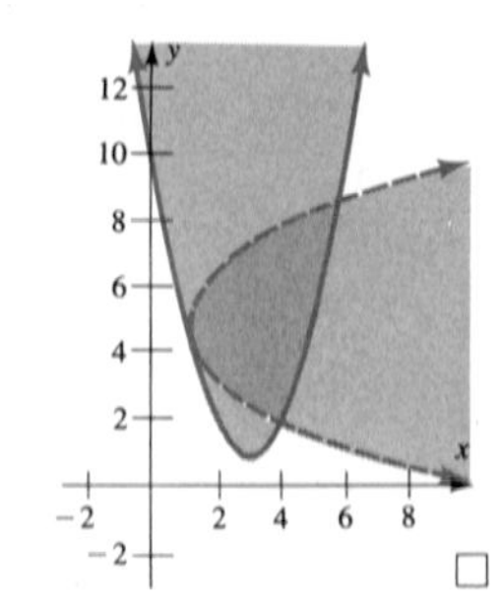

□

Problem Set 3.3

A *In Problems 1–20 rewrite the given quadratics in standard form. Give the vertex and tell in which direction the parabola opens.*

1. $y = x^2 + 4x + 7$

2. $y = x^2 + 4x - 3$

3. $x = y^2 - 6y - 1$

4. $x = y^2 - 2y - 2$

5. $y = x^2 + 10x + 10$

6. $y = x^2 - 12x + 12$

7. $2y = 2x^2 + 12x + 19$

8. $4y = 4x^2 - 4x - 11$

9. $3y = 3x^2 + 12x + 8$

10. $2y = 2x^2 - 16x + 21$

11. $x = 2y^2 - 20y + 52$

12. $x = -3y^2 + 24y - 55$

13. $y = -3x^2 + 18x - 20$
14. $y = 2x^2 + 4x + 1$
15. $6y = 4x^2 + 4x + 7$
16. $3y + 4x^2 = 24x - 38$
17. $3y^2 + 12y + 2x + 2 = 0$
18. $3x^2 + 4y + 216 = 48x$
19. $2x^2 + 8x - 3y + 11 = 0$
20. $4x^2 + 8x + 3y - 2 = 0$

Sketch the graph in Problems 21–40.

21. $y = x^2 - 2$
22. $y = (x - 2)^2$
23. $y - 1 = (x + 2)^2$
24. $x + 2 = -(y - 1)^2$
25. $x - 3 = \frac{1}{2}(y + 3)^2$
26. $y + 5 = -\frac{1}{3}(x - 4)^2$
27. $y - 2 = \frac{1}{4}(x - 2)^2$
28. $y - 3 = \frac{1}{2}(x + 5)^2$
29. $y - 10 = -2\left(x - \frac{1}{2}\right)^2$
30. $x + 9 = 3\left(y - \frac{3}{2}\right)^2$
31. $y > (x - 1)^2$
32. $y + 1 > x^2$
33. $x - 2 \leqslant (y + 3)^2$
34. $y + 3 < (x - 1)^2$
35. $y + 5 < -\frac{1}{2}(x - 2)^2$
36. $x - 4 \geqslant \frac{1}{4}(y + 1)^2$
37. $y + 2 \geqslant \frac{4}{3}(x + 5)^2$
38. $y - 6 \leqslant -\frac{5}{2}(x - 2)^2$
39. $x + \frac{1}{2} < -\frac{1}{4}(y - 1)^2$
40. $x - 3 > \frac{1}{6}\left(y + \frac{5}{2}\right)^2$

B *Sketch the graphs in Problems 41–56.*

41. $y = x^2 + 4x + 7$
42. $x = y^2 - 2y - 2$
43. $2y = 2x^2 + 12x + 19$
44. $4y = 4x^2 - 4x - 11$
45. $x = 2y^2 - 20y + 52$
46. $x = -3y^2 + 24y - 55$
47. $6y = 4x^2 + 4x + 7$
48. $3y + 4x^2 = 24x - 38$
49. $3y = x^2 + 21 : y \leqslant 10$
50. $3y + 16x = x^2 + 85 : y \leqslant 10$
51. $3y + x^2 = 2x + 25 : -1 \leqslant x \leqslant 2$
52. $3y + x^2 + 23 = 14x : 6 \leqslant x \leqslant 9$
53. $3x^2 + 29 = 2y + 18x : y \leqslant 7$
54. $4y + 6x + 3 = x^2 : |x - 3| \leqslant 4$
55. $4y + 6x = x^2 - 23 : y \leqslant 1$
56. $18y + x^2 = 6x + 261 : y \geqslant 14$
57. One side of a storage yard is against a building. The other three sides of the rectangular yard are to be fenced with 36 ft of fencing. How long should the sides be to produce the greatest area with the given length of fence? What is the area obtained?
58. The sum of the length and the width of a rectangular area is 50 m. Find the greatest area possible, and find the dimensions of the figure.
59. *Engineering* An arch has the shape of a parabola. If it is 72 ft wide at the base and 18 ft high at the center, how tall is it 12 ft from the center of the base?
60. A bridge is supported by a suspension cable. The cable is attached to towers at each end and to the roadway at midbridge. The towers are 100 ft above the level roadway and 1400 ft apart. How high is the cable above the roadway 140 ft from the center of the bridge if the cable is parabolic?

C *In Problems 61–68 sketch the systems of inequalities.*

61. $\begin{cases} y \geqslant x^2 \\ 5x - 4y + 26 \geqslant 0 \end{cases}$

62. $\begin{cases} y < 4 - x^2 \\ y > \frac{1}{2}x \end{cases}$

63. $\begin{cases} y \geqslant \frac{1}{2}(x - 2)^2 \\ y - 4 < \frac{1}{4}(x - 2)^2 \end{cases}$

64. $\begin{cases} y - 2 \geqslant \frac{1}{3}(x - 3)^2 \\ y - 4 \leqslant -\frac{1}{3}(x - 3)^2 \end{cases}$

65. $\begin{cases} x \geqslant (y + 2)^2 \\ y + 1 \geqslant \frac{1}{2}\left(x - \frac{3}{2}\right)^2 \end{cases}$

66. $\begin{cases} x + 1 \geqslant (y - 5)^2 \\ y - 3 \geqslant (x + 1)^2 \end{cases}$

67. $\begin{cases} y - 4 \leqslant -(x + 1)^2 \\ x + 2 \geqslant (y - 1)^2 \\ y - 1 > -(x - 1)^2 \end{cases}$

68. $\begin{cases} -1 \leqslant x - y \leqslant 1 \\ y > x^2 + 3 \end{cases}$

69. A parabola is defined as *the set of points equidistant from a point and a line*. Take the point to be $(0, c)$, the line as $y = -c$, and show that you obtain an equation of the form $y = ax^2$ from the definition.

70. The parabola $y - k = a(x - h)^2$ is symmetric to the vertical line $x = h$. Prove this assertion.

71. Prove that the rectangle with a given perimeter has greatest area only if it is a square.

72. Prove that the graph of an equation is symmetric with respect to the y-axis if and only if the replacement of x with $-x$ results in an equivalent equation.

73. Prove that the graph of an equation is symmetric with respect to the origin if and only if the replacement of x with $-x$ and y with $-y$ results in an equivalent equation.

74. Show that a graph symmetric with respect to both the x-axis and the y-axis is symmetric to the origin.

75. Show that a graph symmetric with respect to the origin need not be symmetric to either axis.

3.4 Circles

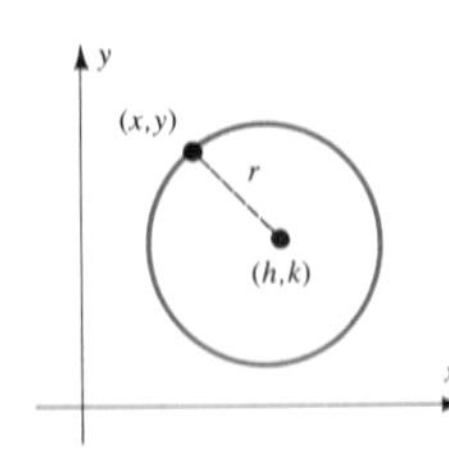

A **circle** is defined as the set of points equidistant from a fixed point, called its **center.** That distance is the **radius** of the circle, and the distance formula can be used to develop an equation for a circle. Let the center be the point (h, k), let the radius be r, and let any point on the circle be (x, y).

Recall the distance formula,

$$\sqrt{\Delta x^2 + \Delta y^2} = d$$

where in this case $\Delta x = x - h$, $\Delta y = y - k$, and $r = d$. Then

$$\sqrt{(x-h)^2+(y-k)^2} = r$$

or $\quad (x-h)^2+(y-k)^2 = r^2$

represents all the points of the circle.

STANDARD FORM OF THE EQUATION OF A CIRCLE

> The equation of the circle with center at (h, k) and radius r is
>
> $$(x-h)^2+(y-k)^2 = r^2$$

Example 1 Find the center and radius of the circle represented by $(x-2)^2+(y+5)^2 = 16$ and sketch the graph.

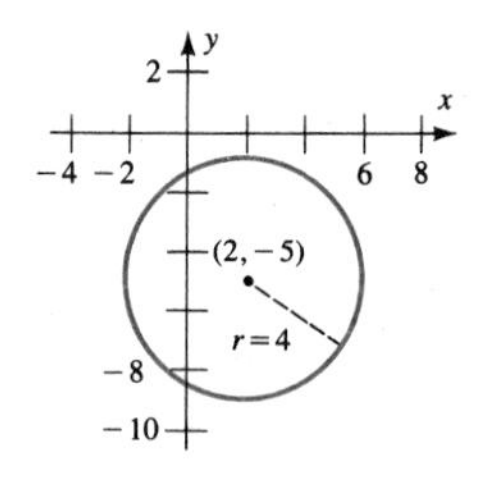

Solution $\quad h = 2 \qquad k = -5 \qquad r = \sqrt{16} = 4$

The center is $(2, -5)$. The radius is 4. □

Example 2 Find the center and radius, and sketch the graph:

$$\left(x+\frac{1}{2}\right)^2+\left(y-\frac{2}{3}\right)^2 = \frac{2}{9}$$

Solution $\quad h = \frac{-1}{2} \qquad k = \frac{2}{3} \qquad r = \frac{\sqrt{2}}{3}$

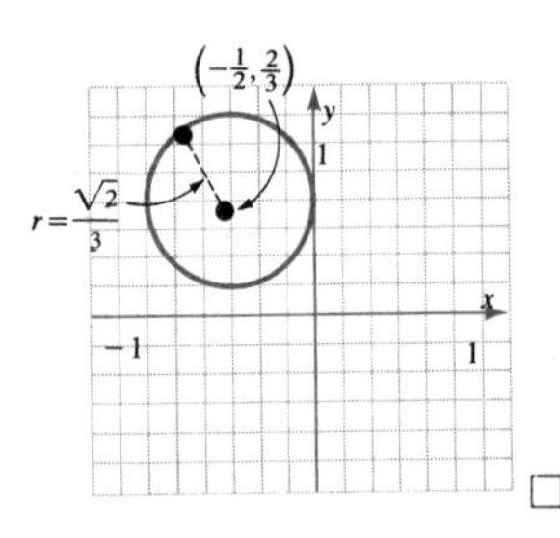

Do not let the scale bother you; simply choose an appropriate scale on your graph. For this problem, you might choose 6 squares per unit. □

GENERAL SECOND-DEGREE EQUATION IN x AND y

> The general form of a second-degree equation with two variables is
>
> $$Ax^2+Bxy+Cy^2+Dx+Ey+F = 0$$

If $B = 0$ and $A = C$ in this general equation, then the graph is a circle. Given such an equation, you should be able to rewrite it in standard form, from which the center and radius can be read. To determine this form, you complete the square on *both* variables.

Example 3 Complete the square for both x and y for $x^2+y^2+4x-10y+26 = 0$.

Solution Isolate and regroup variable terms:

$$(x^2+4x \qquad)+(y^2-10y \qquad) = -26$$

Complete the square on x and on y:

$$(x^2 + 4x + 4) + (y^2 - 10y + 25) = -26 + 4 + 25$$

Recall: Add the squares of half the coefficients of x and y to complete the square.

Rewrite in factored form:

$$(x + 2)^2 + (y - 5)^2 = 3$$

The final equation is in standard form, and the center and radius can easily be read from the equation. The center is at $(-2, 5)$ and the radius is $\sqrt{3}$. □

Example 4 Write the equation $4x^2 + 4y^2 + 24x - 4y + 21 = 0$ in standard form. Give the center and radius of the circle and sketch its graph.

Solution

$$4x^2 + 24x + 4y^2 - 4y = -21$$

$$x^2 + 6x + y^2 - y = -\frac{21}{4}$$

If you divide both sides by 4, the coefficients of x^2 and y^2 are 1 and can be handled as before.

$$(x^2 + 6x + 9) + \left(y^2 - y + \frac{1}{4}\right) = 9 + \frac{1}{4} - \frac{21}{4}$$

$$(x + 3)^2 + \left(y - \frac{1}{2}\right)^2 = 4$$

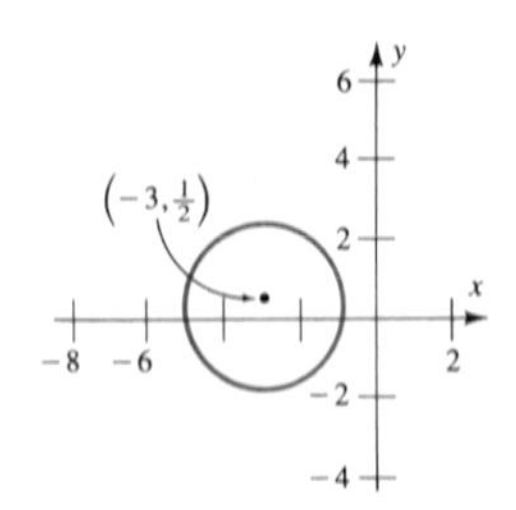

The center is at $\left(-3, \frac{1}{2}\right)$ and the radius is 2. □

Example 5 Graph: $x^2 + y^2 + 2x \leq 6y + 15$

Solution

a. Consider $x^2 + y^2 + 2x = 6y + 15$; this is the boundary:

$$x^2 + 2x + 1 + y^2 - 6y + 9 = 15 + 1 + 9$$
$$(x + 1)^2 + (y - 3)^2 = 25$$

b. Sketch the boundary. Notice that since it is $\leq$, the graph includes the boundary.

c. Check a point. For (0, 0),
$0^2 + 0^2 + 2(0) \leq 6(0) + 15$, or $0 \leq 15$;
true, so (0, 0) is on the graph.

d. Shade all points on or inside the circle. □

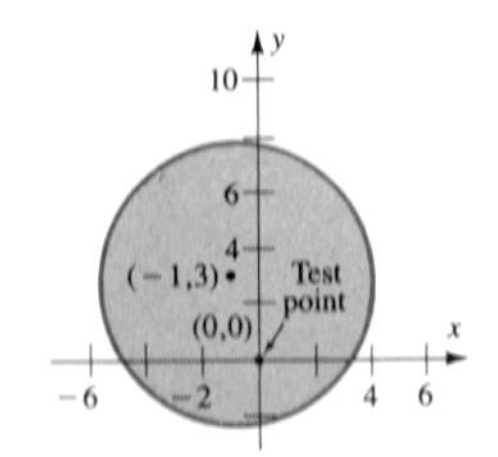

Example 6 Sketch: $\begin{cases} (x-2)^2 + (y+1)^2 \leq 9 \\ x - 3 > -(y+1)^2 \end{cases}$

Solution Circle: center $(2, -1)$, radius 3.
Parabola: vertex $(3, -1)$, opens left.
The intersection is the points inside the circle and to the right of the parabola.
Check by choosing some points within the region and outside.

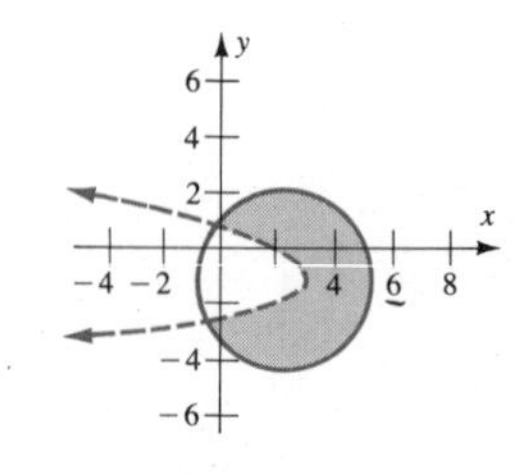

□

Problem Set 3.4

A *Rewrite the equations of circles in Problems 1–10 in standard form. Give the center and radius of each.*

1. $x^2 + y^2 - 4x - 10y + 28 = 0$
2. $x^2 + y^2 + 14x + 2y + 41 = 0$
3. $x^2 + y^2 + 2x - y + 1 = 0$
4. $x^2 + y^2 - 3x + 4y = 0$
5. $3x^2 + 3y^2 + 12y + 12 = 2x$
6. $4x^2 - 8y + 4y^2 + 1 = 4x$
7. $x^2 + y^2 - 4x - 5 = 0$
8. $x^2 + y^2 + 6y + 3 = 0$
9. $x^2 + y^2 + 6x + 8y = 0$
10. $x^2 + y^2 - 8x + 4y + 4 = 0$

Sketch the curves in Problems 11–30.

11. $x^2 + y^2 = 25$
12. $x^2 + y^2 = 36$
13. $x^2 + y^2 = 4$
14. $x^2 + y^2 = 9$
15. $4x^2 + 4y^2 = 1$
16. $9x^2 + 9y^2 = 1$
17. $(x-3)^2 + y^2 = 9$
18. $x^2 + (y-2)^2 = 4$
19. $(x-1)^2 + (y-1)^2 = 1$
20. $(x+2)^2 + (y+2)^2 = 4$
21. $(x-3)^2 + (y-2)^2 = 4$
22. $(x+4)^2 + (y-3)^2 = 9$
23. $(x+3)^2 + (y+4)^2 = 25$
24. $(x-4)^2 + (y+2)^2 = 16$
25. $(x-3)^2 + (y-4)^2 = 25$
26. $(x+4)^2 + (y-2)^2 = 4$
27. $\left(x - \frac{1}{2}\right)^2 + y^2 = \frac{1}{4}$
28. $x^2 + \left(y + \frac{1}{2}\right)^2 = \frac{1}{4}$
29. $\left(x - \frac{1}{2}\right)^2 + \left(y + \frac{3}{2}\right)^2 = \frac{25}{4}$
30. $\left(x + \frac{5}{2}\right)^2 + \left(y - \frac{1}{2}\right)^2 = \frac{9}{4}$

B *In Problems 31–40 write the equation of the circle with the given requirements.*

31. Center at $(3, -4)$ and radius 5
32. Center at $(-2, 3)$ and radius 4
33. Center at $(-4, 5)$ and tangent to the x-axis
34. Center at $(7, -5)$ and tangent to the y-axis
35. End points of a diameter at $(1, 6)$ and $(-7, 6)$
36. End points of a diameter at $(-1, -1)$ and $(7, 5)$
37. End points of a diameter at $(-8, 6)$ and $(16, -4)$

38. End points of a diameter at $(3, -2)$ and $(-5, 4)$

39. Center at $(-1, -2)$ containing the point $(-5, 2)$

40. Center at $(3, -1)$ containing the point $(-1, 2)$

Sketch the graphs of the equations in Problems 41–55, noting any restrictions on the variables.

41. $x^2 + y^2 - 4x - 10y + 28 = 0$

42. $x^2 + y^2 + 14x + 2y + 41 = 0$

43. $x^2 + y^2 + 2x - y + 1 = 0$

44. $x^2 + y^2 - 3x + 4y = 0$

45. $9x^2 + 9y^2 + 36y + 28 = 6x$

46. $4x^2 - 8y + 4y^2 + 1 = 4x$

47. $x^2 + y^2 + 16x + 61 = 2y : y \leqslant 1$

48. $x^2 + y^2 + 14x + 41 = 2y : y \geqslant 1$

49. $x^2 + y^2 + 16x + 49 = 2y : y \leqslant 1$

50. $x^2 + y^2 + 14x + 25 = 2y : y \geqslant 1$

51. $4x^2 + 4y^2 = 4x + 16y - 8 : x \geqslant 1$

52. $4x^2 + 4y^2 = 12x + 16y - 9 : x \geqslant 2$

53. $x^2 + y^2 + 32x + 241 = 2y : x \geqslant 16, y \leqslant 1$

54. $x^2 + y^2 + 6y + 4 = 4x : x \geqslant 2, y \geqslant -3$

55. $x^2 + y^2 + 6y = 7 : x \leqslant 2, y \geqslant -3$

C *Sketch the graphs of the systems of inequalities in Problems 56–65.*

56. $\begin{cases} x^2 + y^2 \leqslant 25 \\ y \geqslant x \end{cases}$

57. $\begin{cases} x^2 + y^2 \geqslant 16 \\ 4y \leqslant x \end{cases}$

58. $\begin{cases} x^2 + y^2 \leqslant 36 \\ |x| > 2 \end{cases}$

59. $\begin{cases} 4x^2 + 4y^2 > 25 \\ |y| \leqslant 4 \end{cases}$

60. $\begin{cases} x^2 + y^2 \leqslant 16 \\ x^2 + y^2 + 2y \geqslant 8 \end{cases}$

61. $\begin{cases} x^2 + y^2 \leqslant 16 \\ x^2 + y^2 - 6x \geqslant 16 \end{cases}$

62. $\begin{cases} x^2 + y^2 + 6x \leqslant 6y - 9 \\ x^2 + y^2 + 4x \geqslant 4y - 4 \end{cases}$

63. $\begin{cases} x^2 + y^2 - 12 \leqslant 6x + 4y \\ x^2 + y^2 + 2x \leqslant 11 - 4y \end{cases}$

64. $\begin{cases} x^2 + y^2 + 2y \leqslant 6x - 1 \\ x^2 + y + 8 \leqslant 6x \end{cases}$

65. $\begin{cases} x^2 + y^2 + 4x \leqslant 6y + 3 \\ x + y^2 - 6y + 7 \leqslant 0 \end{cases}$

66. A circle centered at the origin is represented by $x^2 + y^2 = r^2$. Find the equations of the semicircles shown below.

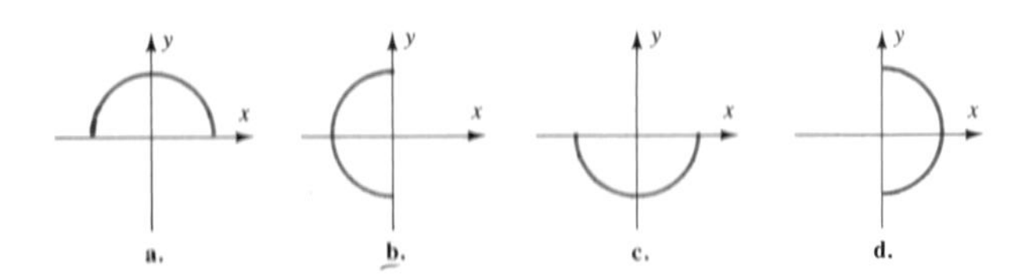

67. Show that the center of the circle circumscribed about a right triangle is the midpoint of the hypotenuse.

3.5 Ellipses and Hyperbolas

The graphs of the general second-degree equation,

$$Ax^2 + Bxy + Cy^2 + Dx + Ey + F = 0$$

have an interesting geometric representation. A *right circular cone* is formed by revolving a line about a nonparallel fixed line, called its *axis*. A plane passing perpendicular to the axis through the cone forms a circle at the intersection. The *conic sections—parabola, circle, ellipse,* and *hyperbola*—are curves formed by plane sections of the cone at various angles, as shown in Figure 3.12.

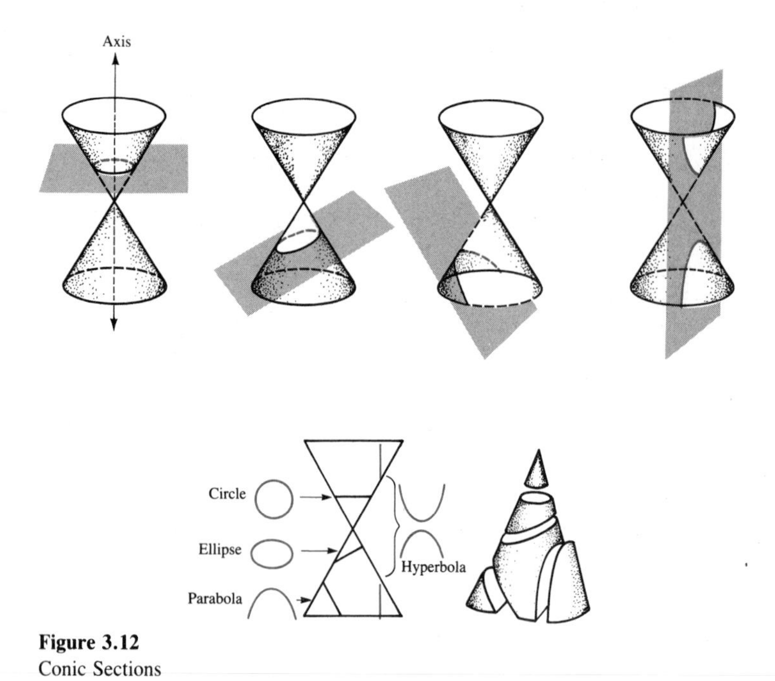

Figure 3.12
Conic Sections

The parabola and the circle have been considered in Sections 3.3 and 3.4. Now, let us define the ellipse. An **ellipse** is the set of points such that the *sum* of the distances from two fixed points, called **foci,** is constant.

In this section, we will consider only a special case of the ellipse, namely that ellipse in which the foci are symmetrically located on a coordinate axis. First, let $F_1(-c, 0)$ and $F_2(c, 0)$ be the foci and let the constant distance be $2a$. Notice that the center of this ellipse is at (0, 0), as shown in Figure 3.13. Let $P(x, y)$ be any point on the ellipse. Since the sum of the distances is the constant $2a$,

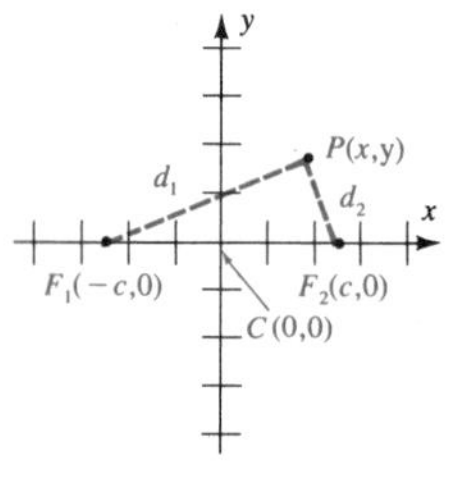

Figure 3.13

$$d_1 + d_2 = 2a$$

$$\sqrt{(x + c)^2 + (y - 0)^2} + \sqrt{(x - c)^2 + (y - 0)^2} = 2a$$

Isolate the radical.
$$\sqrt{(x + c)^2 + y^2} = 2a - \sqrt{(x - c)^2 + y^2}$$

Square both sides.
$$(x + c)^2 + y^2 = 4a^2 - 4a\sqrt{(x - c)^2 + y^2} + (x - c)^2 + y^2$$

Simplify.
$$2cx = 4a^2 - 4a\sqrt{(x - c)^2 + y^2} - 2cx$$

Isolate the radical.
$$4a\sqrt{(x - c)^2 + y^2} = 4a^2 - 4cx$$

Divide by 4 and square.
$$a^2(x - c)^2 + a^2y^2 = a^4 - 2a^2cx + c^2x^2$$

Simplify.
$$a^2x^2 + a^2c^2 + a^2y^2 = a^4 + c^2x^2$$

Let $b^2 = a^2 - c^2$.
$$(a^2 - c^2)x^2 + a^2y^2 = a^2(a^2 - c^2)$$

Divide by a^2b^2.
$$b^2x^2 + a^2y^2 = a^2b^2$$

$$\frac{x^2}{a^2} + \frac{y^2}{b^2} = 1$$

Let $x = 0$ to obtain the y-intercepts $\pm b$; and let $y = 0$ to obtain the x-intercepts $\pm a$.

EQUATION OF AN ELLIPSE

> The equation of an **ellipse** with center at (0, 0) is
>
> $$\frac{x^2}{a^2} + \frac{y^2}{b^2} = 1$$
>
> with x-intercepts $(\pm a, 0)$ and y-intercepts $(0, \pm b)$

Similarly, if $F_1(0, c)$ and $F_2(0, -c)$ are the foci and $a^2 = b^2 - c^2$, the same equation results. Thus, if $a > b$, the foci are on the x-axis, and if $a < b$, the foci are on the y-axis.

Example 1 Sketch: $\frac{x^2}{9} + \frac{y^2}{4} = 1$

Solution The center of the ellipse is (0, 0). The x-intercepts are ± 3, and the y-intercepts are ± 2. The foci can also be found, since

$$c^2 = a^2 - b^2$$
$$c^2 = 9 - 4$$
$$c = \pm\sqrt{5}$$

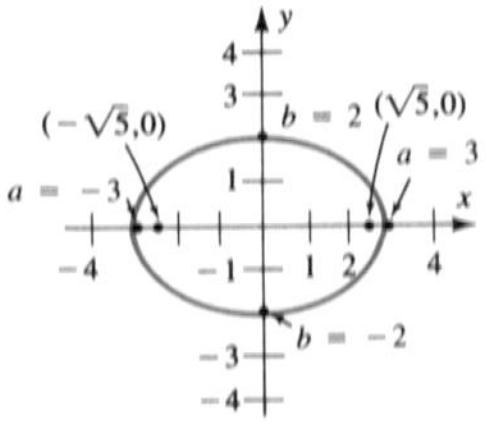

□

Example 2 Sketch: $25x^2 + 16y^2 = 400$

Solution First divide by 400 to get a constant of 1; then simplify to obtain the standard form:

$$\frac{25x^2}{400} + \frac{16y^2}{400} = \frac{400}{400}$$

$$\frac{x^2}{16} + \frac{y^2}{25} = 1$$

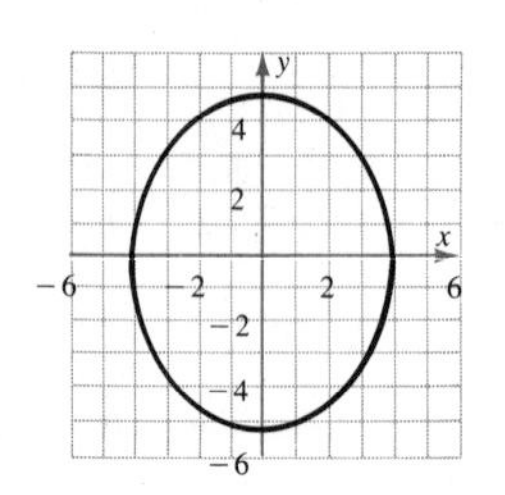

The center of the ellipse is (0, 0). The x-intercepts are ± 4, and the y-intercepts are ± 5. □

A **hyperbola** is the set of points such that the *difference* of the distances to two fixed points, called the **foci,** is a constant. As with the ellipse, we will consider only the special cases of the foci symmetrically arranged on the coordinate axes (see Figure 3.14).

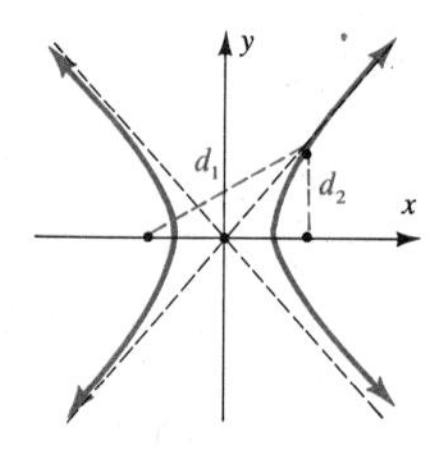

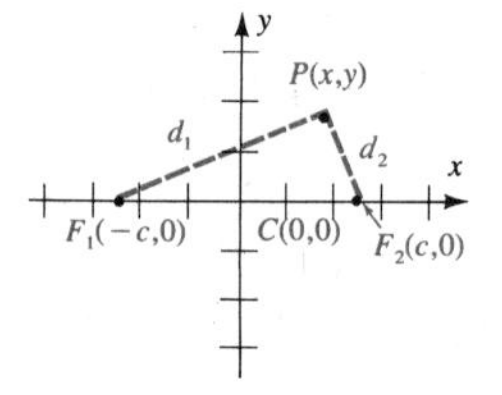

Figure 3.14

Thus, if $(-c, 0)$ and $(c, 0)$ are the foci and $2a$ is the constant difference, then

$$|d_1 - d_2| = 2a$$

$$\left|\sqrt{(x + c)^2 + (y - 0)^2} - \sqrt{(x - c)^2 + (y - 0)^2}\right| = 2a$$

The simplification is similar to that for the ellipse. (You are asked to fill in the details in Problem 50 in Problem Set 3.5.) The result after this simplification is

$$\frac{x^2}{a^2} + \frac{y^2}{a^2 - c^2} = 1$$

which looks like the equation for the ellipse. However, notice that $2a < 2c$ and $a < c$, so $a^2 - c^2 < 0$. Let $-b^2 = a^2 - c^2$, or $b^2 = c^2 - a^2$; then the equation of the hyperbola is

$$\frac{x^2}{a^2} - \frac{y^2}{b^2} = 1$$

If we start with the foci $(0, -c)$ and $(0, c)$ on the y-axis and a constant difference of $2a$, then we obtain the equation of a hyperbola in the form

$$\frac{y^2}{a^2} - \frac{x^2}{b^2} = 1$$

EQUATIONS OF HYPERBOLAS

> The equation of a **hyperbola** with center at (0, 0) is
>
> $$\frac{x^2}{a^2} - \frac{y^2}{b^2} = 1^2$$
>
> with foci $(\pm c, 0)$, ***x*-intercepts** $(\pm a, 0)$, and constant difference $2a$; or
>
> $$\frac{y^2}{a^2} - \frac{x^2}{b^2} = 1^2$$
>
> with foci $(0, \pm c)$, ***y*-intercepts** $(0, \pm a)$, and constant difference $2a$.

An important aid to graphing hyperbolas is the drawing of two lines called the *asymptotes* of the hyperbola. A line is called an asymptote of a curve if the distance between the curve and the line approaches but never reaches zero. To draw these asymptotes, first draw a rectangle with sides parallel to the coordinate axes with intercepts $\pm a$, $\pm b$. The asymptotes are the lines drawn through the corners of the rectangle as illustrated by Examples 3 and 4.

Example 3 Sketch: $\dfrac{x^2}{4} - \dfrac{y^2}{9} = 1$

Solution The center of the hyperbola is (0, 0), $a = 2$, and $b = 3$. Construct the rectangle from which the asymptotes can be drawn and used as guidelines to sketch the curve. □

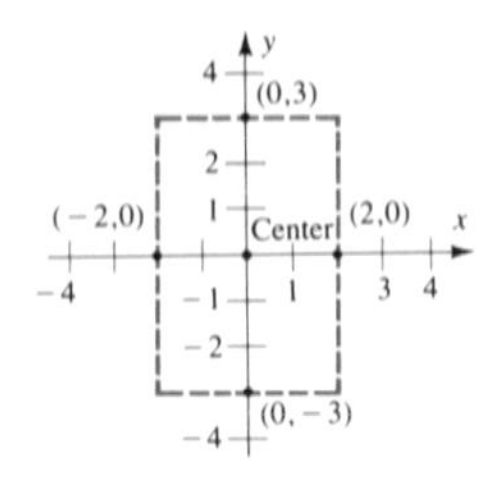

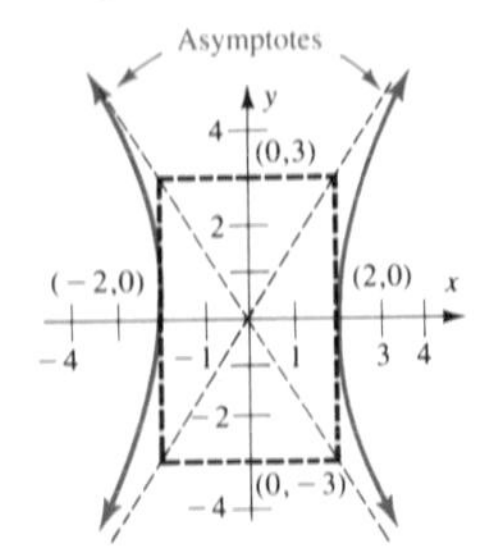

Example 4 Sketch: $\dfrac{y^2}{16} - \dfrac{x^2}{25} = 1$

Solution The hyperbola is again centered at (0, 0), with $a = 4$, and $b = 5$. Construct the rectangle and the asymptotes, and notice that this hyperbola opens upward and downward instead of horizontally, as in Example 3.

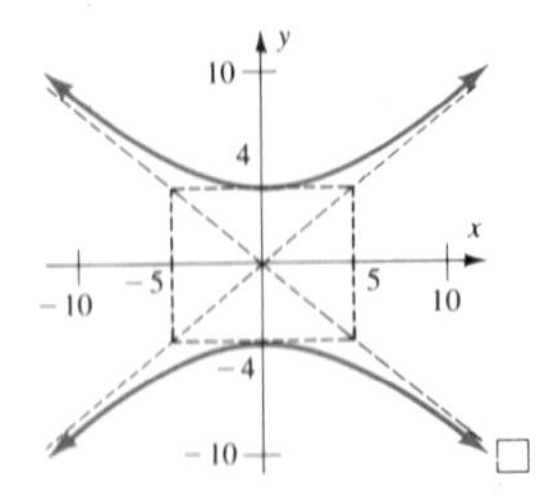

□

The equations of the conic sections have, with the exception of the parabola, been second-degree in both x and y. The equation $Ax^2 + By^2 = C$ may be identified as a circle, an ellipse, a hyperbola, or one of the degenerate cases, as shown in Table 3.3.

Table 3.3 Graphs of $Ax^2 + By^2 = C$

Graph	Equation $Ax^2 + By^2 = C$; A, B, and C Real
Circle	$A = B$; A, B, and C have same sign
Ellipse	A, B, and C have same sign
Hyperbola	A and B have opposite signs; $C \neq 0$
Two Intersecting Straight Lines	A and B have opposite signs; $C = 0$
Point	A and B have same sign; $C = 0$
No Graph	A and B have same sign; C has the opposite sign

Note that the case of two intersecting lines corresponds to the asymptotes of the hyperbola. That is, if

$$\frac{x^2}{a^2} - \frac{y^2}{b^2} = 0 \qquad \text{Factor, difference of squares}$$

$$\left(\frac{x}{a} - \frac{y}{b}\right)\left(\frac{x}{a} + \frac{y}{b}\right) = 0$$

$$\frac{x}{a} - \frac{y}{b} = 0 \quad \text{or} \quad \frac{x}{a} + \frac{y}{b} = 0$$

$$\frac{y}{b} = \frac{x}{a} \qquad\qquad \frac{y}{b} = -\frac{x}{a}$$

$$y = \frac{b}{a}x \qquad\qquad y = -\frac{b}{a}x$$

$$y = \pm\frac{b}{a}x$$

In Examples 5–12 the guides listed in Table 3.3 are used to identify the graph of each equation.

Example 5 $4x^2 + 9y^2 = 36$

4, 9, and 36 are positive; thus, the graph is an **ellipse.** □

Example 6 $4x^2 + 9y = 36$

No y^2 term; thus, the graph is a **parabola.** □

Example 7 $9x^2 - 16y^2 = 144$

9 and -16 are opposite in sign and $144 \neq 0$; thus, the graph is a **hyperbola.** □

Example 8 $9x - 16y = 144$

No x^2 and y^2 terms; thus, the graph is a **line.** □

Example 9 $25x^2 + 25y^2 = 1$

$25 = 25$, and 25, 25, and 1 are positive; thus, the graph is a **circle.** □

Example 10 $25x^2 + 4y^2 = 0$

25 and 4 are positive and $C = 0$; thus, the graph is a **point** $(0, 0)$. □

Example 11 $25x^2 - 4y^2 = 0$

25 and -4 are opposite in sign and $C = 0$; thus, the graph is **two intersecting lines** through the origin. □

Example 12 $25x^2 + 4y^2 = -100$

25 and 4 are positive and -100 is negative; thus, there is **no graph.** □

The conic sections have many interesting and useful properties. The planets, for example, move in elliptical orbits with the sun at one focus, while most comets move in hyperbolic paths. Comets on hyperbolic paths obviously are seen just one time on earth. Halley's Comet is one notable exception, following an elliptical path. Thus, this famous comet returns to be seen on earth periodically, every 76 years. The next appearance of Halley's Comet is due in early 1986.

A navigational system called *LORAN* (LOng RAnge Navigation) is based on the properties of the hyperbola. The difference in the times at which a signal is received at two different points is proportional to the difference between the distances from the two points to the origin of the signal. (Note that this is equivalent to the definition of the hyperbola as a set of points such that the difference of the distances to two fixed points is a constant.) Hence, the origin of the signal is on a hyperbola. If two quantities are inversely proportional, the graph of the relationship is hyperbolic.

Figure 3.15
Reflection on an Ellipse

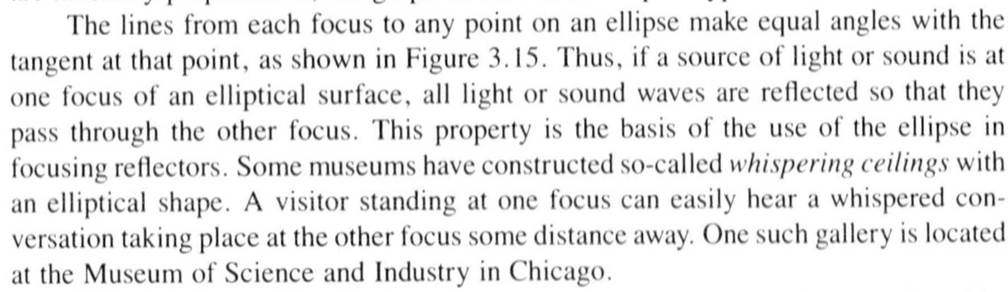

The lines from each focus to any point on an ellipse make equal angles with the tangent at that point, as shown in Figure 3.15. Thus, if a source of light or sound is at one focus of an elliptical surface, all light or sound waves are reflected so that they pass through the other focus. This property is the basis of the use of the ellipse in focusing reflectors. Some museums have constructed so-called *whispering ceilings* with an elliptical shape. A visitor standing at one focus can easily hear a whispered conversation taking place at the other focus some distance away. One such gallery is located at the Museum of Science and Industry in Chicago.

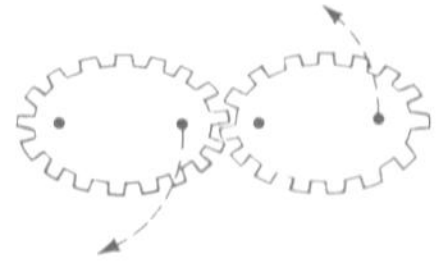

Figure 3.16
Elliptical Gears

Gears are sometimes elliptical in shape to produce a nonuniform motion. The gears in Figure 3.16 will remain in contact if each turns around its left-hand focus as shown, after beginning with all four foci aligned. Steam boilers presumably have their greatest strength if they are elliptical with axes in the ratio 2:1.

These varied applications only suggest the extent of the usefulness of these curves and you may be able to add to this small list yourself.

Problem Set 3.5

A *Name the graph of each equation in Problems 1–20.*

1. $x^2 + y^2 = 25$

2. $x^2 - y^2 = 25$

3. $4x^2 - y^2 = 4$

4. $4x^2 + y^2 = 4$

5. $4x^2 + y^2 = -4$

6. $4x^2 - y^2 = -4$

7. $4x^2 - y = 4$

8. $4x + y^2 = 4$

9. $16x^2 + 25y^2 = 100$

10. $16x^2 + 25y^2 = -100$

11. $49x + 64y = 0$

12. $49x^2 + 64y^2 = 0$

13. $25x^2 - 9y^2 = 36$

14. $25x^2 - 9y^2 = 225$

15. $4x^2 + 9y^2 = 36$

16. $4x^2 + 9y^2 = 0$

17. $x^2 - y^2 = 81$

18. $x^2 + y^2 = -81$

19. $x^2 - y^2 = 0$

20. $x^2 + y^2 = 0$

Sketch the curves in Problems 21–30.

21. $\frac{x^2}{9} + \frac{y^2}{16} = 1$

22. $\frac{x^2}{16} + \frac{y^2}{9} = 1$

23. $\frac{x^2}{9} - \frac{y^2}{16} = 1$

24. $\frac{y^2}{16} - \frac{x^2}{9} = 1$

25. $\frac{x^2}{25} + \frac{y^2}{36} = 1$

26. $\frac{x^2}{49} - \frac{y^2}{16} = 1$

27. $\frac{x^2}{49} + y^2 = 1$

28. $x^2 + \frac{y^2}{49} = 1$

29. $\frac{y^2}{144} - \frac{x^2}{100} = 1$

30. $\frac{y^2}{144} + \frac{x^2}{100} = 1$

B *Sketch the curves in Problems 31–40, if possible.*

31. $9x^2 + y^2 = 9$

32. $x^2 - 9y^2 = 9$

33. $4x^2 - 81y^2 = 324$

34. $81y^2 - 4x^2 = 324$

35. $9x^2 + 4x^2 = -36$

36. $9x^2 - 4y^2 = -36$

37. $4x^2 + 25y^2 = 0$

38. $4x^2 - 25y = 0$

39. $4x^2 - 25y^2 - 100 = 0$

40. $4x^2 + 25y^2 + 100 = 0$

Sketch the graphs of the equations in Problems 41–45, noting the restrictions on the variables.

41. $\frac{x^2}{4} + \frac{y^2}{9} = 1 : |x| \leqslant 2$

42. $\frac{x^2}{4} - \frac{y^2}{9} = 1 : |x| \leqslant 4$

43. $\frac{y^2}{9} - \frac{x^2}{16} = 1 : |y| \geqslant 4$

44. $\frac{x^2}{9} + \frac{y^2}{16} = 1 : |y| \geqslant 2$

45. $\frac{x^2}{6} - \frac{y}{6} = x - \frac{11}{6} : y < 5$

C *Use the distance formula and the definition of the ellipse to find the equation of each ellipse from the information given in Problems 46–49.*

46. Foci at $(4, 0)$ and $(-4, 0)$ and constant sum 10

47. Foci at $(\pm 3, 0)$ and constant sum of distances 10

48. Foci at $(0, \pm 15)$ and sum of distances 34

49. Foci at $(0, 12)$ and $(0, -12)$ and constant sum 26

50. Use the distance formula and the definition of hyperbola to derive the equation

$$\frac{x^2}{a^2} - \frac{y^2}{b^2} = 1$$

The following is needed for Problems 51–56.
The orbit of a planet can be described by the equation of an ellipse,

$$\frac{x^2}{a^2} + \frac{y^2}{b^2} = 1$$

with the sun at one focus. The orbit is commonly identified by its **major axis** *$2a$ and its* **eccentricity** *e:*

$$e = \frac{c}{a} = \sqrt{1 - \frac{b^2}{a^2}} \qquad a \geqslant b > 0$$

where c is the distance between the center and a focus.

51. The major axis of the earth's orbit is approximately 186,000,000 miles and its eccentricity is $\frac{1}{62}$. Find the greatest distance, called the **apogee,** and the smallest distance, called the **perigee,** of the sun from the earth.

52. The perigee of the planet Mercury with the sun is approximately 28 million miles, and the eccentricity of the orbit is $\frac{1}{5}$. Find the major axis of Mercury's orbit. (See Problem 51.)

53. The moon's orbit is elliptical with the earth at one focus. If the major axis of the orbit is approximately 378,000 miles and the apogee is approximately 199,000 miles, then what is the eccentricity of the moon's orbit? (See Problem 51.)

54. Show that the eccentricity of an ellipse is *always* nonnegative and less than 1.

55. Show that if $e = 0$, the ellipse is a circle.

56. Show that if the ellipse is a circle, then $e = 0$.

57. Show that for any point (x, y) on the hyperbola

$$\frac{x^2}{a^2} - \frac{y^2}{b^2} = 1 \qquad \text{or} \qquad b^2x^2 - a^2y^2 = a^2b^2$$

the product of the distances to the asymptotes is a constant.

*3.6 Translation and Conics

Recall that the graph of $y = ax^2$ was discovered to be a parabola with vertex at the origin, opening vertically. The graph of $y - k = a(x - h)^2$ is then *congruent* (identical in size and shape) to $y = ax^2$, but the vertex is at (h, k). That is, the vertex is moved

*Optional

h units horizontally and k units vertically from $(0, 0)$ to (h, k). In fact, since the curves are congruent, we may think of every point on $y = ax^2$ moved h units horizontally and k units vertically to form $y - k = a(x - h)^2$. Similarly, $x = ay^2$ and $x - h = a(y - k)^2$, or $x^2 + y^2 = r^2$ and $(x - h)^2 + (y - k)^2 = r^2$ behave in the same way. A curve that has been shifted to a new position in this way is said to be *translated*.

TRANSLATION

If every x and y in the equation of a graph are replaced by $x - h$ and $y - k$, then the new equation is the equation of a congruent graph **translated** h units horizontally and k units vertically.

For example, $x^2 + y^2 = 25$ is the equation of a circle with radius 5 and center at $(0, 0)$, and $(x - 1)^2 + (y + 2)^2 = 25$ is the equation of a circle with radius 5 and center at $(1, -2)$.

This principle also may be applied to graphs of ellipses and hyperbolas with centers at the origin. As we discussed in Section 3.5, the equation

$$\frac{x^2}{a^2} + \frac{y^2}{b^2} = 1$$

is that of an ellipse with center at $(0, 0)$, horizontal axis $2a$, and vertical axis $2b$.

STANDARD FORM OF AN ELLIPSE

$$\frac{(x - h)^2}{a^2} + \frac{(y - k)^2}{b^2} = 1$$

is the equation of an ellipse with center at (h, k), horizontal axis $2a$, and vertical axis $2b$.

Once the center is located, the graph is sketched in essentially the same way as if the point (h, k) were the origin.

Example 1 Sketch: $\dfrac{(x - 1)^2}{25} + \dfrac{(y + 2)^2}{16} = 1$

Solution By inspection: $(h, k) = (1, -2)$, $a = 5$, and $b = 4$.

1. The center is at $(1, -2)$.

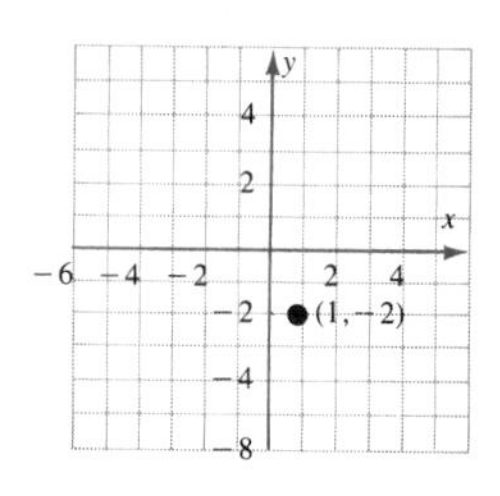

2. The ends of the horizontal axis are 5 units horizontally to either side of $(1, -2)$, and the ends of the vertical axis are 4 units above and below $(1, -2)$.

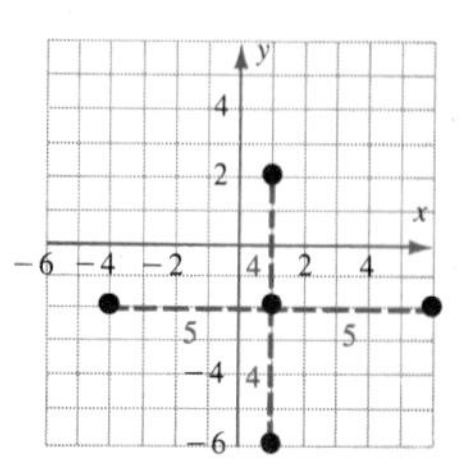

3. Using these key points, sketch the ellipse.

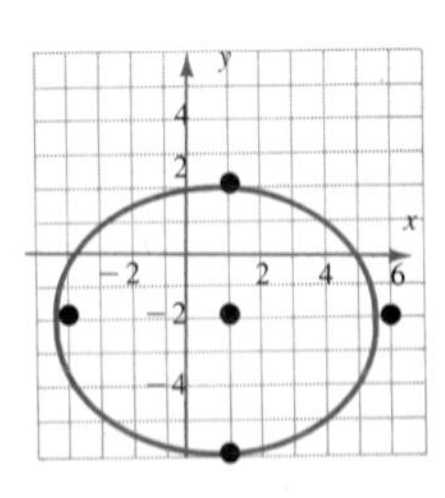

□

Example 2 Sketch: $9(x + 3)^2 + 4(y - 2)^2 = 36$

Solution First, divide both sides by 36 to obtain the constant 1:

$$\frac{9(x+3)^2}{36} + \frac{4(y-2)^2}{36} = \frac{36}{36}$$

$$\frac{(x+3)^2}{4} + \frac{(y-2)^2}{9} = 1$$

Now sketch with $(h, k) = (-3, 2)$, $a = 2$, and $b = 3$.

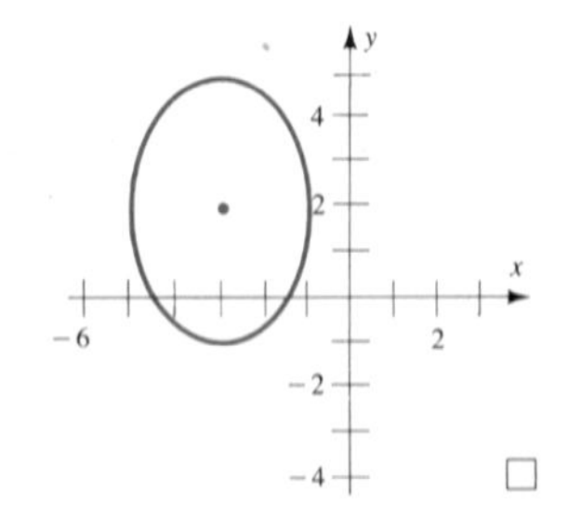

□

An equation of an ellipse may be written in standard form by using the same technique employed for the circle. The square is completed on both variables, but unlike the circle, the coefficients of x^2 and y^2 are not equal in this case.

Example 3 Rewrite $4x^2 + y^2 + 8x - 4y - 8 = 0$ in standard form, identify the center, and find a and b.

Solution

$$4x^2 + y^2 + 8x - 4y - 8 = 8$$

$$4x^2 + 8x \quad + y^2 - 4y = 8$$ Isolate the variables and group.

$$4(x^2 + 2x \quad) + (y^2 - 4y \quad) = 8$$ Factor constant term from each group.

$$4(x^2 + 2x + 1) + (y^2 - 4y + 4) = 8 + 4 + 4$$ Complete the square on x and on y.

$$4(x+1)^2 + (y-2)^2 = 16$$ Factor and simplify.

$$\frac{(x+1)^2}{4} + \frac{(y-2)^2}{16} = 1$$ Divide by 16.

The center is at $(-1, 2)$, $a = 2$, and $b = 4$. □

Example 4 Rewrite $9x^2 + 4y^2 - 90x + 8y + 193 = 0$ in standard form.

Solution

$$9x^2 + 4y^2 - 90x + 8y + 193 = 0$$

$$9x^2 - 90x \qquad + 4y^2 + 8y \qquad = -193 \qquad \text{Isolate variables.}$$

$$9(x^2 - 10x \qquad) + 4(y^2 + 2y \qquad) = -193 \qquad \text{Factor.}$$

$$9(x^2 - 10x + 25) + 4(y^2 + 2y + 1) = -193 + 225 + 4 \qquad \text{Complete the square.}$$

$$9(x - 5)^2 + 4(y + 1)^2 = 36 \qquad \text{Factor and simplify.}$$

$$\mathbf{\frac{(x - 5)^2}{4} + \frac{(y + 1)^2}{9} = 1} \qquad \text{Divide by 36.}$$

□

Example 5 Find the equation of the ellipse with center at $(-6, 4)$ and tangent to both the coordinate axes. Make a sketch and note that $a = -6$ and $b = 4$. Thus, the equation is

$$\frac{[x - (-6\ ]^2}{6^2} + \frac{(y - 4)^2}{4^2} = 1$$

$$\mathbf{\frac{(x + 6)^2}{36} + \frac{(y - 4)^2}{16} = 1}$$

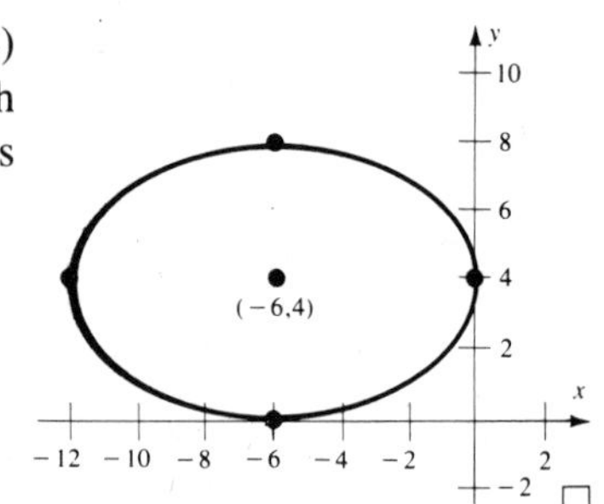

□

Similarly, we have a standard form for the equations of hyperbolas with center at the point (h, k).

STANDARD FORM OF A HYPERBOLA

> The equations of a hyperbola with center at (h, k) are
>
> $$\mathbf{\frac{(x - h)^2}{a^2} - \frac{(y - k)^2}{b^2} = 1} \qquad \mathbf{\frac{(y - k)^2}{a^2} - \frac{(x - h)^2}{b^2} \quad 1}$$

Example 6 Sketch: $\dfrac{(x - 5)^2}{4} - \dfrac{(y + 3)^2}{49} = 1$

Solution By inspection, $(h, k) = (5, -3)$, $a = 2$, and $b = 7$.

1. The center is at $(5, -3)$.
2. Construct the rectangle from which the asymptotes can be drawn and used as guidelines to sketch the curve.

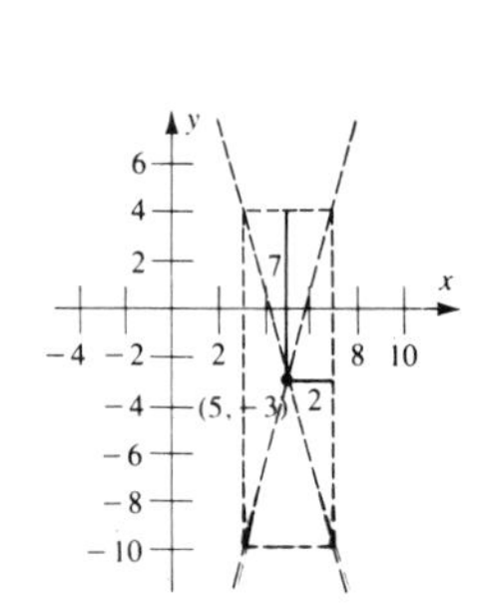

3. Sketch the hyperbola opening horizontally.

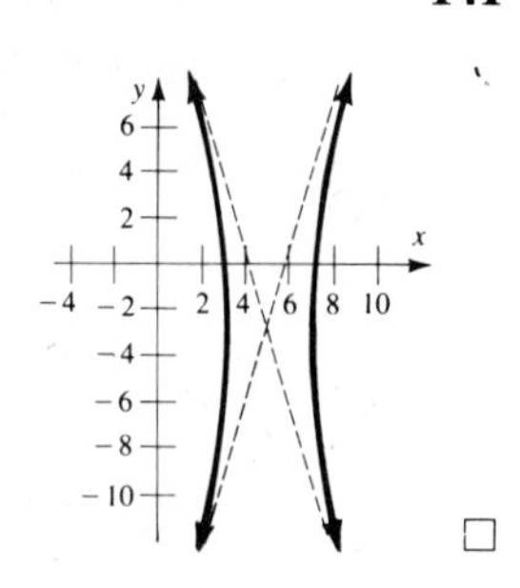

□

Example 7 Sketch: $(y + 2)^2 - \frac{x^2}{9} = 1$

Solution Rewrite as

$$\frac{(y+2)^2}{1} - \frac{(x-0)^2}{9} = 1$$

The hyperbola opens vertically, and is centered at $(0, -2)$ with $a = 1$ and $b = 3$.

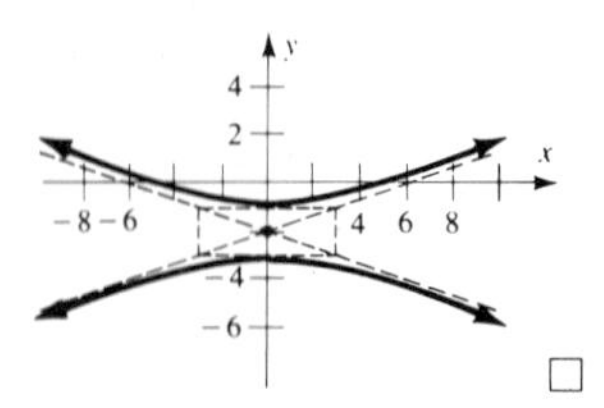

□

As with the ellipse, the equation of the hyperbola may have to be rewritten in standard form. The procedure is similar to that for the ellipse. Note, however, that since the hyperbola is a difference you must be careful with the sign when you factor and when you complete the square.

Example 8 Rewrite $x^2 - y^2 + 2x + 2y - 4 = 0$ in standard form and identify the center and lengths of the axes.

Solution

$$x^2 - y^2 + 2x + 2y - 4 = 0$$

$$x^2 + 2x - y^2 + 2y = 4$$ Isolate the variable terms.

$$(x^2 + 2x \quad) - (y^2 - 2y \quad) = 4$$ Group the terms. *Watch the signs.*

$$(x^2 + 2x + 1) - (y^2 - 2y + 1) = 4 + 1 - 1$$ Complete the squares. *Watch the signs.*

$$(x + 1)^2 - (y - 1)^2 = 4$$ Factor.

$$\mathbf{\frac{(x+1)^2}{4} - \frac{(y-1)^2}{4} = 1}$$ Divide by 4.

The hyperbola opens horizontally. **The center is at $(-1, 1)$; $a = 2$; $b = 2$.** □

Example 9 Sketch: $x^2 - 4y^2 + 6x + 16y - 3 = 0$

Solution

$$x^2 + 6x \quad - 4y^2 + 16y = 3$$
$$(x^2 + 6x \quad) - 4(y^2 - 4y \quad) = 3$$
$$(x^2 + 6x + 9) - 4(y^2 - 4y + 4) = 3 + 9 - 16$$
$$(x + 3)^2 - 4(y - 2)^2 = -4$$
$$\frac{(x + 3)^2}{-4} - \frac{4(y - 2)^2}{-4} = \frac{-4}{-4} \quad \text{Divide by } -4$$
$$\mathbf{\frac{(y - 2)^2}{1} - \frac{(x + 3)^2}{4} = 1}$$

The hyperbola has center at $(-3, 2)$, opens vertically, and has $a = 1$ and $b = 2$. □

Problem Set 3.6

A *Sketch the curves in Problems 1–10.*

1. $\dfrac{(x - 2)^2}{9} + \dfrac{(y - 3)^2}{16} = 1$
2. $\dfrac{(x - 4)^2}{16} + \dfrac{(y - 1)^2}{9} = 1$
3. $\dfrac{(x + 4)^2}{9} - \dfrac{(y - 3)^2}{16} = 1$
4. $\dfrac{(y - 3)^2}{16} - \dfrac{(x + 4)^2}{9} = 1$
5. $\dfrac{(x - 5)^2}{25} + \dfrac{(y + 6)^2}{36} = 1$
6. $\dfrac{(x + 4)^2}{49} - \dfrac{(y - 7)^2}{16} = 1$
7. $\dfrac{(x + 7)^2}{49} + (y - 3)^2 = 1$
8. $(x + 5)^2 - \dfrac{(y - 2)^2}{49} = 1$
9. $\dfrac{(y - 8)^2}{144} + \dfrac{(x - 10)^2}{100} = 1$
10. $\dfrac{(y - 10)^2}{144} - \dfrac{(x - 8)^2}{100} = 1$

B *Rewrite the equations in Problems 11–30 in standard form. Identify the curve and the center as well as a and b for each.*

11. $9(x + 5)^2 + 16(y - 3)^2 = 144$
12. $25(x - 4)^2 + 4(y + 7)^2 = 100$
13. $81(y - 4)^2 - 36(x + 6)^2 = 324$
14. $36(x - 5)^2 - 100(y - 2)^2 = 900$
15. $(x - 3)^2 - (y + 5)^2 = 49$
16. $(y - 6)^2 - (x - 8)^2 = -64$
17. $25x^2 + 4y^2 - 16y - 84 = 0$
18. $4x^2 + 49y^2 - 36x - 115 = 0$
19. $4x^2 - 9y^2 - 8x - 32 = 0$
20. $4y^2 - 9x^2 + 18x - 45 = 0$
21. $y^2 - 4x^2 + 2y - 8x + 1 = 0$
22. $4y^2 - x^2 - 16y + 4x - 8 = 0$
23. $x^2 - y^2 - 10x - 14y - 33 = 0$
24. $x^2 - y^2 - 4x + 10y + 15 = 0$
25. $4x^2 - y^2 + 24x + 4y + 16 = 0$
26. $x^2 - 9y^2 - 4x - 18y - 14 = 0$
27. $4x^2 + 9y^2 - 16x - 18y - 19 = 0$
28. $9x^2 + 4y^2 + 36x + 8y + 4 = 0$
29. $4x^2 + 81y^2 + 24x - 162y - 207 = 0$
30. $25x^2 + 9y^2 + 50x - 36y - 164 = 0$

C **31.** Consider two fixed points $(-m, 0)$ and $(m, 0)$, and the set of points such that the product of the slopes of the lines that join it to the fixed points is $-n^2$. Show that this set of points is an ellipse.

32. Consider two equal elliptical gears in contact when the four foci are aligned (see the figure in the margin). Show that if each gear turns about its left-hand focus, the gears stay in contact.

Problems 33–37 refer to $Ax^2 + Cy^2 + Dx + Ey + F = 0$ where A, C, D, E, and F are real numbers.

33. Show that the graph is an ellipse if $A > C > 0$, and

$$\frac{D^2}{4A} + \frac{E^2}{4C} > F$$

34. Show that the graph is two intersecting lines if $A > 0 > C$, and

$$\frac{D^2}{4A} + \frac{E^2}{4C} \neq F$$

35. Show that the graph is a single point if $A > C > 0$, and

$$\frac{D^2}{4A} + \frac{E^2}{4C} = F$$

36. Show that the graph is a hyperbola if $A > 0 > C$, and

$$\frac{D^2}{4A} + \frac{E^2}{4C} \neq F$$

37. Show that the equation has no graph if $A > C > 0$, and

$$\frac{D^2}{4A} + \frac{E^2}{4C} < F$$

3.7 Summary and Review

This chapter discussed the tools of the coordinate system: plotting points, distance, midpoint, and slope. These tools are used to develop equations for lines, certain parabolas, circles, and special cases of the ellipse and hyperbola. In the next two chapters, we will encounter other curves and new tools for use in the coordinate plane.

CHAPTER OBJECTIVES

After studying this chapter, you should be able to:

1. Plot points in the coordinate plane
2. Find distance, midpoint, and slope, given two points
3. Use slope to establish parallelism and perpendicularity
4. Find the equation of a line when given two points, the slope and y-intercept, the slope and any point, or equivalent information

5. Recognize the standard, slope–intercept, and point–slope forms of the equation of a straight line and sketch the graph of the given line or associated half-plane, with restrictions
6. Sketch the graph of a quadratic equation as a parabola and find the vertex, the direction in which it opens, and other points on the curve
7. Sketch the graph of a quadratic inequality or system of inequalities involving quadratics
8. Sketch the graph of a circle, find the center and radius of a circle, and sketch the graph of an inequality or system of inequalities involving circles
9. Find the equations of certain parabolas or circles, given sufficient data
10. Sketch the graph of an ellipse or hyperbola

Problem Set 3.7

Fill in the word or words necessary to make the statements in Problems 1–30 complete and correct.

CONCEPT PROBLEMS

[3.1] **1.** A coordinate plane is used to plot ______ pairs.

[3.1] **2.** The axes divide the coordinate plane into four regions called ______ .

[3.1] **3.** The ______ is the set of all replacements for the first component of the ordered pair.

[3.1] **4.** Every point on the graph of an equation has coordinates that ______ the equation, and every ordered pair satisfying the equation has ______ that lie on the graph.

[3.1] **5.** The difference $x_2 - x_1$ is called the ______ change or Δx.

[3.1] **6.** The ______ between two points $P_1(x_1, y_1)$ and $P_2(x_2, y_2)$ is $\sqrt{\Delta x^2 + \Delta y^2}$, where $\Delta x =$ ______ and $\Delta y =$ ______ .

[3.1] **7.** The midpoint between $P_1(x_1, y_1)$ and $P_2(x_2, y_2)$ is (______ , ______).

[3.1] **8.** The slope of the line containing $P_1(x_1, y_1)$ and $P_2(x_2, y_2)$ is $m =$ ______ .

[3.1] **9.** Two lines are parallel if and only if the slopes of the lines are ______ .

[3.1] **10.** The product of the slopes of two ______ lines is negative one.

[3.2] **11.** $y - k = m(x - h)$ is the equation of the line through ______ with ______ m.

[3.2] **12.** An x-intercept is a point where a graph crosses the ______ .

[3.2] **13.** The ______ form of the equation of a line is $y = mx + b$.

[3.2] **14.** The linear equation $Ax + By + C = 0$ is said to be ______ form.

[3.2] **15.** A line divides the coordinate plane into three parts: the line itself and two ______ .

[3.2] **16.** The graph of a system of inequalities is the ______ of the solution sets of the members of the system.

[3.3] **17.** The graph of $y = ax^2 + bx + c$, $a \neq 0$, is a(n) ______ .

[3.3] **18.** The graph of $y - k = a(x - h)^2$ has ________ at (h, k) and opens ________ if $a < 0$ and ________ if $a > 0$.

[3.3] **19.** k is the ________ value of y in $y - k = a(x - h)^2$ when $a < 0$.

[3.3] **20.** The graph of ________ has vertex at (h, k) and opens to the right if ________ and to the left if ________ .

[3.3] **21.** The points (x, y) and $(x, -y)$ are symmetric with respect to the ________ .

[3.3] **22.** The points (x, y) and ________ are symmetric with respect to the line $y = x$.

[3.4] **23.** The graph of ________ is a circle with center at (h, k) and radius r.

[3.5] **24.** The conic sections (parabola, circle, ellipse, and hyperbola) are curves formed by the intersection of a(n) ________ and a(n) ________ .

[3.5] **25.** An ellipse is the set of points such that the ________ of the distances from two fixed points is a constant.

Problems 26 and 27 refer to the equation for an ellipse:

$$\frac{x^2}{a^2} + \frac{y^2}{b^2} = 1$$

[3.5] **26.** This is the equation of an ellipse with ________ at (0, 0).

[3.5] **27.** This ellipse has x-intercepts ________ and ________ .

Problems 28–30 refer to the equations for hyperbolas:

$$\frac{x^2}{a^2} - \frac{y^2}{b^2} = 1 \qquad \text{or} \qquad \frac{y^2}{a^2} - \frac{x^2}{b^2} = 1$$

[3.5] **28.** A hyperbola is the set of points such that the ________ of the distances to two fixed points is ________ .

[3.5] **29.** The first equation is a hyperbola with ________ -intercepts at ________ and ________ .

[3.5] **30.** A hyperbola is sketched by first drawing a rectangle and its ________ .

REVIEW PROBLEMS

[3.1] *Find the distance between the given points in Problems 31–40.*

31. (0, 0), (8, 6)

32. (0, 8), (6, 0)

33. (7, 2), (−5, −3)

34. (9, −2), (−6, 6)

35. (1, −3), (−2, −7)

36. (3, −8), (−2, 4)

37. (−2, −1), (2, 2)

38. (−6, −3), (9, 5)

39. (−9, 1), (7, −3)

40. (8, −5), (−4, 0)

[3.1] *Find the midpoint between the given points in Problems 41–50.*

41. (3, 6), (5, 2)

42. (7, 4), (1, 8)

43. (4, −1), (2, 5)

44. (2, −5), (6, 3)

45. (−7, 5), (3, −1)

46. (8, −6), (−4, 2)

47. (−2, −1), (2, 2)

48. (−6, −3), (9, 5)

49. (−9, 1), (7, −3)

50. (8, −5), (−4, 0)

[3.1] *Find the slope of the line containing the given points in Problems 51–60.*

51. $(2, 1), (5, 4)$

52. $(7, 6), (3, 4)$

53. $(-5, 9), (4, 3)$

54. $(8, -2), (2, 2)$

55. $(7, -9), (1, -4)$

56. $(-3, -12), (9, 3)$

57. $(-2, -1), (2, 2)$

58. $(-6, -3), (9, 5)$

59. $(-9, 1), (7, -3)$

60. $(8, -5), (-4, 0)$

[3.2] *Find the equation of the line from the information given in Problems 61–70, and write its equation in standard form.*

61. y-intercept -5; slope $-\frac{2}{3}$

62. x-intercept 4; slope $\frac{3}{4}$

63. Slope $\frac{3}{5}$; through $(-2, 7)$

64. Slope $-\frac{7}{2}$; through $(2, -5)$

65. Vertical line through $(6, -4)$

66. Horizontal line through $(7, -3)$

67. Through $(2, 1)$ and $(5, -1)$

68. Through $(2, -1)$ and $(-5, 1)$

69. Through $(-2, 1)$ and $(-5, -1)$

70. Through $(-2, 1)$ and $(-5, 1)$

[3.2] *In Problems 71–80 graph the solutions.*

71. $y = -2x + 5$

72. $y = 3x - 4$

73. $y = \frac{1}{4}x - 2$

74. $y = -\frac{1}{2}x + 4$

75. $y = \frac{1}{2}x - 5 : -2 \leq x \leq 4$

76. $y - 2 = \frac{2}{5}(x + 3) : |x - 1| \leq 3$

77. $3x + 2y > 6$

78. $2x - 5y + 10 \leq 0$

79. $\begin{cases} y \leq -\frac{3}{4}x + 3 \\ x \geq 0 \\ y \geq 0 \end{cases}$

80. $\begin{cases} y - 7 < \frac{3}{4}(x - 4) \\ y \geq -\frac{1}{4}x + 4 \\ 4 < x < 8 \end{cases}$

[3.3] *In Problems 81–90 graph the solutions.*

81. $y - 8 = -\frac{1}{2}(x + 1)^2$

82. $x + 5 = 2(y - 3)^2$

83. $y = x^2 - 10x \pm 23$

84. $y > x^2 - 6x + 10$

85. $3x + 2y^2 + 12y + 9 = 0$

86. $4x = 4y^2 + 4y + 3$

87. $\begin{cases} x < -(y - 4)^2 \\ x \geq -4 \end{cases}$

88. $\begin{cases} y > \frac{1}{2}(x - 3)^2 \\ y \leq x + 1 \end{cases}$

89. $\begin{cases} x^2 - 4x + 4 + y \leq -3 \\ x + y \geq -3 \end{cases}$

90. $\begin{cases} y < x^2 - 6x + 10 \\ 4y > x^2 - 6x + 13 \end{cases}$

[3.4, 3.5] *In Problems 91–100 sketch the solutions.*

91. $(x + 1)^2 + (y - 2)^2 = 25$

92. $(x - 2)^2 + (y + 1)^2 \leq 9$

93. $x^2 + y^2 - 2x + 6y = 3$

94. $x^2 + y^2 \leq 6x + 8y$

95. $\frac{x^2}{49} + \frac{y^2}{25} = 1$

96. $\frac{x^2}{4} + \frac{y^2}{16} = 1$

97. $\frac{y^2}{9} - \frac{x^2}{1} = 1$

98. $\frac{y^2}{16} - \frac{x^2}{9} = 1$

99. $\begin{cases} x^2 - 4y - 8 < 4x - y^2 \\ x + y \geq 2 \end{cases}$

100. $\begin{cases} (x-4)^2 + (y-2)^2 \geq 25 \\ (x-2)^2 + (y-6)^2 \leq 16 \end{cases}$

*[3.6] *Identify and sketch each curve in Problems 101–120.*

101. $2x - y - 8 = 0$

102. $2x + y - 10 = 0$

103. $4x^2 - 16y = 0$

104. $\dfrac{(x-3)^2}{4} - \dfrac{(y+2)}{6} = 1$

105. $\dfrac{(x-3)^2}{9} - \dfrac{(y+2)^2}{25} = 1$

106. $\dfrac{x-3}{9} + \dfrac{y-2}{25} = 1$

107. $(x+3)^2 + (y-2)^2 = 0$

108. $9(x+3)^2 + 4(y-2) = 0$

109. $9(x+3)^2 - 4(y-2)^2 = 0$

110. $x^2 + 8(y-2)^2 = 16$

111. $x^2 + 64(y+4)^2 = 16$

112. $x^2 + y^2 - 3y = 0$

113. $4x^2 - 3y^2 - 24y - 112 = 0$

114. $y^2 - 4x + 2y + 21 = 0$

115. $x^2 - 4y^2 - 6x - 8y - 11 = 0$

116. $x^2 + 4x + 12y + 64 = 0$

117. $y^2 - 6y - 4x + 5 = 0$

118. $100x^2 - 7y^2 + 98y - 368 = 0$

119. $x^2 + y^2 + 2x - 4y - 20 = 0$

120. $4x^2 + 12x + 4y^2 + 4y + 1 = 0$

---

*Optional

CHAPTER 4

Functions

Gottfried Wilhelm von Leibniz (1646–1716)

The concept of a function is suggested by all the processes of nature where we observe natural phenomena varying according to distance or to time. Nearly all the "known" functions have presented themselves in the attempt to solve geometrical, mechanical, or physical problems.

J. T. Mertz
A History of European Thought in the Nineteenth Century,
Edinburgh and London, 1903, p. 696

Historical Note

The word *function* was used as early as 1694 by Gottfried von Leibniz to denote any quantity connected with a curve. Leibniz was one of the most universal geniuses of the 17th century, and as a teenager he came up with many of the great ideas in mathematics. However, because of his youth, his ideas were not fully accepted at the time. He was refused a doctorate at the University of Leibzig because of his youth, even though he had completed all the requirements. Among other things, Leibniz invented the calculus, exhibited an early calculating machine that he invented, and distinguished himself in law, philosophy, and linguistics. His ideas on functions were generalized by other mathematicians, including P. G. Lejeune-Dirichlet (1805–1859).

Around 1815, functions were being considered that were no longer "nice," and it was Dirichlet who, in 1837, suggested a very broad definition of a function—in fact, the one that leads to the definition used in this book. Dirichlet proposed a very "badly behaved" function: When x is rational, let $y = c$, and when x is irrational, let $y = d \neq c$. This function, often known as *Dirichlet's function,* is so pathological that it is often used to test hypotheses about functions.

*Optional

Chapter Overview

The notion of a function is one of the most important and unifying ideas in mathematics. The first three chapters of this book presented a review of much of your previous mathematics, and now you are ready to study functions. In this book we use the idea of a function to tie together the topics that are introduced in later chapters.

4.1 Functions

The central theme of this course is the idea of a function. A function is essentially a relationship between two sets of numbers, frequently defined by an equation. This equation may be viewed as a rule that describes how one quantity is determined by another quantity.

The temperature T at a certain location may depend on the time t. Then the temperature is said to be a function of time. The area of a circle A is given by $A = \pi r^2$, where r is the radius of the circle. The area is determined by the radius, so it may be said that the area is a function of the radius. The determined variable is called the *dependent variable*, and the other variable is the *independent variable*. The independent variable is always the first component of an ordered pair. The **domain** of a function is the set of values of the independent variable for which it is defined. *In this book, the domain is the set of all real numbers that make sense for the independent variable, unless otherwise specified.* When we use the variables x and y in this book, we will always mean that x is the independent variable and y is the dependent variable. The set of all corresponding values of the dependent variable is called the **range.**

FUNCTION

A **function** is a rule that assigns to each element of the domain a single element in its range.

The definition in the box says that to each x in the domain there corresponds exactly one y in the range. These pairs of corresponding values may be viewed as ordered pairs (x, y) and the definition could be reworded: *A function is a set of ordered pairs (x, y) in which no two different ordered pairs have the same first element x.*

It is customary to give functions letter names, such as f, g, f_1, or F. If y is the value of the function f corresponding to x, it is written $y = f(x)$ and is read "y is equal to the value of the function f at x," or, more briefly, "y equals f at x," or "y equals f of x." The symbol $f(x)$ does *not* mean multiplication; rather, it is a single symbol representing the second component of the ordered pair (x, y).

Functions may be described by sets, by tables, by equations, or by graphs, as well as in words. The examples that follow should help clarify this notion and its notation.

Functions Can Be Represented as Sets

Example 1 $f = \{(0, 0), (1, 1), (2, 4), (3, 9), (4, 16)\}$
Notation: $f(0) = 0, f(1) = 1, f(2) = 4, f(3) = 9$, and $f(4) = 16$ □

Example 2 $g = \{(0, -3), (1, -6), (2, -9), (3, -12), (4, -16)\}$
Notation: $g(0) = -3, g(1) = -6, g(2) = -9, g(3) = -12$, and $g(4) = -16$ □

Example 3 $h = \{(0, 0), (1, 1), (1, 4), (3, 9), (4, 16)\}$ *Not* a function

A member of the domain (namely, 1) is associated with *more* than a single second component.

Do not use $h(x)$ notation because h is *not* a function. □

Functions Can Be Represented by Tables

Example 4 The function T is represented by the relationship shown in the following table:

T

Time (24 Hour Clock)	Temperature (Celsius)	Notation
6	8°	$T(6) = 8$
10	14°	$T(10) = 14$
12	17°	$T(12) = 17$
14	18°	$T(14) = 18$
18	14°	$T(18) = 14$

□

Functions Can Be Represented by Equations

Example 5 $F(x) = x^2$

Notation:

$F(3) = 9$ since $3^2 = 9$

Replace x by 3

$F(4) = 16$ since $4^2 = 16$

Replace x by 4

Each value of x is associated with one value for y, so this function could be written $F = \{(x, F(x)) | F(x) = x^2\}$. This would be too formal for our purposes. Just remember that when you write $F(x) = x^2$, you are specifying a function called F and a set of ordered pairs whose first components are represented by x and second components by $F(x)$ or by x^2 [since $F(x) = x^2$].

$$F(w) = w^2$$
$$F(2x + 3) = (2x + 3)^2$$
$$= 4x^2 + 12x + 9$$

□

Example 6 The equation $x = y^2$ does *not* define y as a function of x, because you can find a value of x that is associated with *more* than a single number. For example, if $x = 4$, then

$$4 = y^2$$
$$y = \pm 2$$
□

Example 7 $A(r) = \pi r^2$

Notation: $A(2) = \pi(2)^2 = 4\pi$, $A(3) = 9\pi$, $A(5) = 25\pi$ □

Example 8 $H(x) = 3x^2 + 2x + 1$

Notation: $H(2) = 3 \cdot 2^2 + 2 \cdot 2 + 1 = 17$ $H(-4) = 3(-4)^2 + 2(-4) + 1 = 48 - 8 + 1 = 41$ □

Functions Can Be Represented by Graphs

Example 9 The function L is represented by the line shown.

Notation: Since $(-1, -4)$, $(0, -2)$, $(1, 0)$, $(2, 2)$, and $(3, 4)$ are ordered pairs that represent points on line L, $L(-1) = -4$, $L(0) = -2$, $L(1) = 0$, $L(2) = 2$, and $L(3) = 4$.

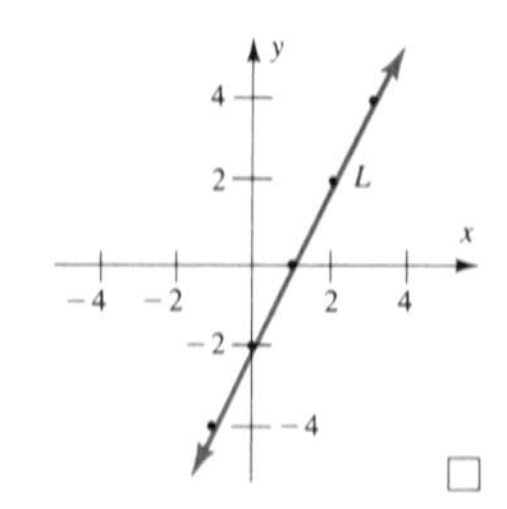

□

Example 10 The function P is represented by the graph shown. Here P is a function, since for each value of x there is one and only one value of y.

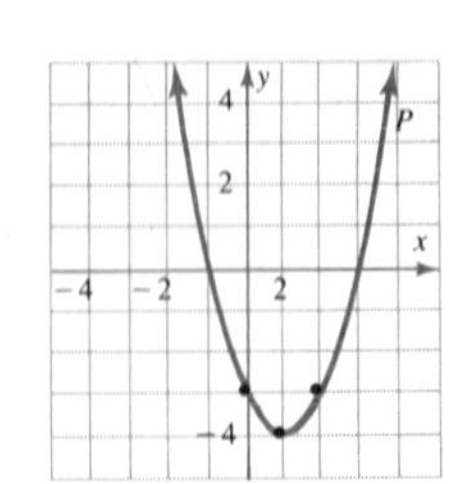

□

Example 11 Consider the set Q of ordered pairs, graphed below.

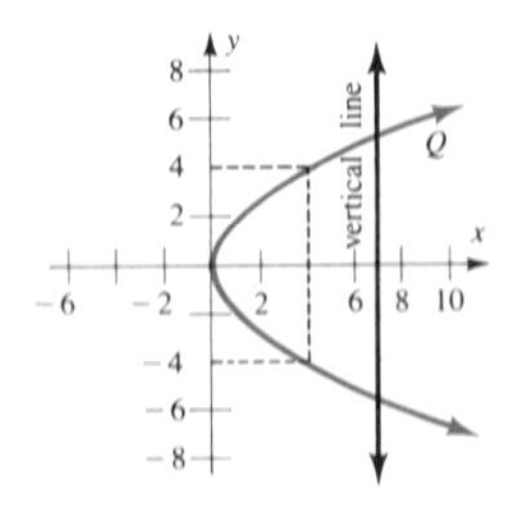

When given a graph, you can use a **vertical line test** to determine whether the graph is a function: If a vertical line can be drawn so that it crosses the graph at more than one point, then the graph is not a function.

There is at least one value of x that is associated with more than a single number; thus, Q does not represent a function. □

There are, of course, other ways to represent functions. Not all sets of ordered pairs are functions; nevertheless, the notion of a function is an important unifying concept in mathematics. Functional notation is particularly useful. You should learn its use immediately:

$f(x)$ — x is a member of the domain.

$f(x)$ — $f(x)$ is a member of the range

That is, $f(x)$ is the *number* associated with x, and f is the *function* representing the rule itself.

For Examples 12–18, let $f(x) = x^2 + 1$, *and* $g(x) = (x + 1)^2$.

Example 12 Find $f(1)$ and $g(1)$.

Solution

$$f(1) = 1^2 + 1 = 2 \qquad g(1) = (1 + 1)^2 = 4$$ □

Example 13 Find $f(-3)$ and $g(-3)$.

Solution

$$f(-3) = (-3)^2 + 1 = 9 + 1 = 10$$

$$g(-3) = (-3 + 1)^2 = (-2)^2 = 4$$ □

Example 14 Find $f(w)$ and $g(w)$.

Solution

$$f(w) = w^2 + 1$$

$$g(w) = (w + 1)^2 = w^2 + 2w + 1$$ □

Example 15 Find $3f(w + 1)$ and $-2g(w + 1)$.

Solution

$$\begin{aligned} 3f(w + 1) &= 3[(w + 1)^2 + 1] \\ &= 3[(w^2 + 2w + 1) + 1] \\ &= 3(w^2 + 2w + 2) \\ &= 3w^2 + 6w + 6 \end{aligned}$$

$$\begin{aligned} -2g(w + 1) &= -2(w + 1 + 1)^2 \\ &= -2(w + 2)^2 \\ &= -2(w^2 + 4w + 4) \\ &= -2w^2 - 8w - 8 \end{aligned}$$ □

Example 16 Find $f(w + h)$ and $g(w + h)$.

Solution

$$\begin{aligned} f(w + h) &= (w + h)^2 + 1 \\ &= w^2 + 2wh + h^2 + 1 \end{aligned}$$

$$\begin{aligned} g(w + h) &= (w + h + 1)^2 \\ &= w^2 + wh + w + wh + h^2 + h + w + h + 1 \\ &= w^2 + 2wh + h^2 + 2w + 2h + 1 \end{aligned}$$ □

Example 17 Find $f(w + h) - f(w)$ and $g(w + h) - g(w)$. Use the results of Examples 14 and 16.

Solution

$$\begin{aligned} f(w + h) - f(w) &= (w^2 + 2wh + h^2 + 1) - (w^2 + 1) \\ &= 2wh + h^2 \end{aligned}$$

$$g(w + h) - g(w) = (w^2 + 2wh + h^2 + 2w + 2h + 1) - (w^2 + 2w + 1)$$
$$= 2wh + h^2 + 2h$$ □

Example 18 Find $\dfrac{f(w + h) - f(w)}{h}$ and $\dfrac{g(w + h) - g(w)}{h}$. Use the results of Example 17.

Solution

$$\frac{f(w + h) - f(w)}{h} = \frac{2wh + h^2}{h} = 2w + h \qquad \frac{g(w + h) - g(w)}{h} = \frac{2wh + h^2 + 2h}{h} = 2w + h + 2$$ □

Problem Set 4.1

A *In Problems 1–20 let* $f(x) = 2x + 1$ *and* $g(x) = 2x^2 - 1$. *Find the requested values.*

1. **a.** $f(0)$ **b.** $f(2)$ **c.** $f(-3)$ **d.** $f(\sqrt{5})$ **e.** $f(\pi)$

2. **a.** $f(1)$ **b.** $g(1)$ **c.** $f(\sqrt{3})$ **d.** $g(\sqrt{3})$ **e.** $g(1 + \sqrt{3})$

3. **a.** $f(w)$ **b.** $g(w)$ **c.** $g(t)$ **d.** $g(v)$ **e.** $f(m)$

4. **a.** $f(t)$ **b.** $f(p)$ **c.** $f(t + 1)$ **d.** $g(t + 1)$ **e.** $f(t^2)$

5. **a.** $f(1 + \sqrt{2})$ **b.** $g(1 + \sqrt{2})$ **c.** $g(t + 3)$ **d.** $f(t^2 + 2t + 1)$ **e.** $g(m - 1)$

6. **a.** $f(x + 2)$ **b.** $g(x + 2)$ **c.** $f(t + h)$ **d.** $g(t + h)$ **e.** $f(x + h)$

7. $2f(x + 2)$

8. $4g(x + 2)$

9. $-3f(t^2 + 2t + 1)$

10. $f(2t^2 + t)$

11. $3f(x + 2) - 4g(x + 2)$

12. $f(t^2 + 2t + 1) - g(t + 3)$

13. $f(2t^2 + t) - 3f(t - 1)$

14. $f(2t^2 + t) - g(t - 4)$

15. $\dfrac{f(t + 3) - f(t)}{3}$

16. $\dfrac{g(t + 2) - g(t)}{2}$

17. $\dfrac{f(t + h) - f(t)}{h}$

18. $\dfrac{g(t + h) - g(t)}{h}$

19. $\dfrac{f(x + h) - f(x)}{h}$

20. $\dfrac{g(x + h) - g(x)}{h}$

Determine whether the sets given in Problems 21–32 are functions.

21. $\{(6, 3), (9, 4), (7, -1), (5, 4)\}$

22. $\{(3, 6), (4, 9), (-1, 7), (4, 5)\}$

23. $\{6, 9, 7, 5\}$

24. $\{10, 20, 30, 40\}$

25. $y = 5x + 2$

26. $y \leq 5x + 2$

27. $y = -1$ if x is a rational number, and $y = 1$ if x is an irrational number.

28. $y = -1$ if x is a positive integer, and $y = 1$ if x is a negative integer.

29. $\{(x, y) \mid y = \text{Closing price of Xerox stock on July 1 of year } x\}$

30. $\{(x, y) \mid x = \text{Closing price of Xerox stock on July 1 of year } y\}$

31. (x, y) is a point on a circle with center $(2, 3)$ and radius 4.

32. (x, y) is a point on a line passing through $(2, 3)$ and $(4, 5)$.

B

33. $F(t) = \dfrac{t + 1}{t - 1}$; find $F(0)$

34. $G(m) = (2m - 3)(m + 1)$; find $G(2)$

35. $H(t) = \dfrac{(t + 4)(t - 3)}{t + 4}$; find $H(1)$

36. $I(m) = \dfrac{m^2 + 5m + 6}{m + 3}$; find $I(11)$

37. $N(x) = \dfrac{4x^2 - 17x - 15}{4x + 3}$; find $N(9)$

38. $R(x) = \dfrac{x^3 - 1}{x^2 + x + 1}$; find $R(17)$

39. $S(y) = \dfrac{y^3 + 1}{y^2 - y + 1}$; find $S(14)$

40. $T(x) = \dfrac{(3x + 1)(4x - 1)}{4x - 1}$; find $T(\frac{1}{3})$

41. $V(p) = \sqrt{p^2 + 3p + 7}$; find $V(3)$

42. $W(p) = -\sqrt{p^2 + 4p + 16}$; find $W(-4)$

In Problems 43–50 find: $\dfrac{f(x + h) - f(x)}{h}$ *for the given function f.*

43. $f(x) = 5x^2$

44. $f(x) = 3x^2$

45. $f(x) = x^2 + 5$

46. $f(x) = 3x^2 + 2x$

47. $f(x) = x^2 - x + 2$

48. $f(x) = 2x^2 + 3x - 4$

49. $f(x) = \dfrac{1}{x}$

50. $f(x) = \dfrac{x + 1}{x - 1}$

Economics *For Problems 51–60 use the table below, which reflects the purchasing power of the dollar from 1944 to 1984. Let x represent the year; let the domain be the years 1944, 1954, 1964, 1974, and 1984; and let*

$r(x)$ = *Price of a pound of round steak*
$s(x)$ = *Price of a 5 pound bag of sugar*
$b(x)$ = *Price of a loaf of bread*
$c(x)$ = *Price of a pound of coffee*
$e(x)$ = *Price of a dozen eggs*
$m(x)$ = *Price of a half gallon of milk*
$g(x)$ = *Price of a gallon of gasoline*

Year	Round Steak (1 lb)	Sugar (5 lb)	Bread (Loaf)	Coffee (1 lb)	Eggs (1 doz)	Milk ($\frac{1}{2}$ gal)	Gasoline (1 gal)
1944	$ .45	$ .34	$ .09	$ .30	$ .64	$ .29	$ .21
1954	.92	.52	.17	1.10	.60	.45	.29
1964	1.07	.59	.21	.82	.57	.48	.30
1974	1.78	2.08	.36	1.31	.84	.78	.53
1984	2.15	1.49	1.29	2.69	1.15	1.08	1.52

51. Find each value:
a. $r(1964)$ **b.** $m(1954)$ **c.** $g(1944)$ **d.** $c(1974)$ **e.** $s(1984)$

52. Find each value:
a. $s(1964)$ **b.** $b(1954)$ **c.** $c(1944)$ **d.** $e(1974)$ **e.** $m(1984)$

53. Find each value:
a. $s(1984) - s(1944)$ **b.** $b(1984) - b(1944)$

54. Find each value:
a. $r(1984) - r(1944)$ **b.** $m(1984) - m(1944)$

55. **a.** Find the change in the pricé of eggs from 1944 to 1984.
b. Write the change in the price of eggs using function notation.

56. **a.** Find the change in the price of coffee from 1944 to 1984.
b. Write the change in the price of coffee using function notation.

57. Find (to the nearest cent): $\dfrac{g(1944 + 40) - g(1944)}{40}$

58. Find (to the nearest cent): $\dfrac{r(1944 + 40) - r(1944)}{40}$

59. State in words whether you can attach any meaning to the number found in Problem 57.

60. State in words whether you can attach any meaning to the number found in Problem 58.

C *Let d be a function that represents the distance an object falls (neglecting air resistance) in t seconds. Find the distance the object falls for the intervals of time given in Problems 61–66 if* $d(t) = 16t^2$.

61. From $t = 2$ to $t = 6$

62. From $t = 2$ to $t = 4$

63. From $t = 2$ to $t = 3$

64. From $t = 2$ to $t = 2 + h$

65. From $t = x$ to $t = x + h$

66. Give a physical interpretation for: $\dfrac{d(x + h) - d(x)}{h}$

4.2 Special Types of Functions

In the last two chapters, you studied linear and quadratic equations. Recall that the graphs of the first-degree equation

$$Ax + By + C = 0$$

are lines, and the graphs of the second-degree equations ($a \neq 0$)

$$y = ax^2 + bx + c \quad \text{and} \quad x = ay^2 + by + c$$

are parabolas. Not all lines and parabolas are functions, as shown in Figure 4.1.

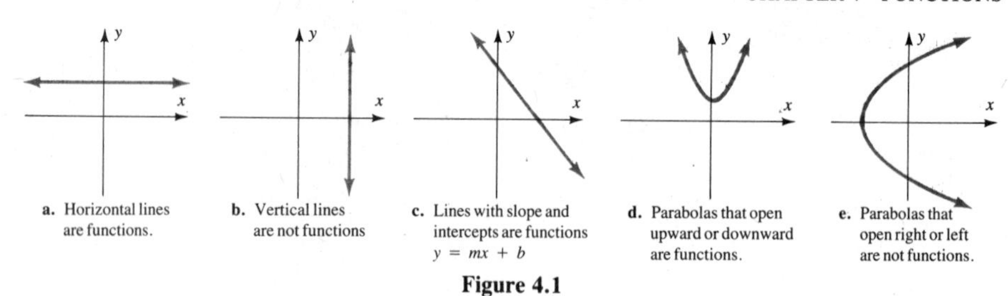

a. Horizontal lines are functions.
b. Vertical lines are not functions
c. Lines with slope and intercepts are functions $y = mx + b$
d. Parabolas that open upward or downward are functions.
e. Parabolas that open right or left are not functions.

Figure 4.1

The first special types of functions to be considered are those that are *increasing*, *decreasing*, or *constant*.

INCREASING, DECREASING, AND CONSTANT FUNCTIONS

Let S be a subset of the domain of a function f. Then:

f is **increasing** on S if $f(x_1) < f(x_2)$ whenever $x_1 < x_2$ in S
f is **decreasing** on S if $f(x_1) > f(x_2)$ whenever $x_1 < x_2$ in S
f is **constant** on S if $f(x_1) = f(x_2)$ for every x_1 and x_2 in S

Example 1 Consider the function g graphed in the figure. Note that g is:

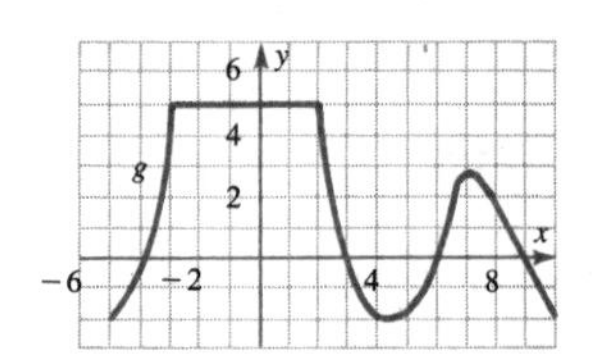

Increasing on the interval $-5 \leq x \leq -3$
Constant on the interval $-3 \leq x \leq 2$
Decreasing on the interval $2 \leq x \leq 4$

Increasing on the interval $4 \leq x \leq 7$
Decreasing on the interval $7 \leq x \leq 10$ □

Any horizontal line also serves as an example of a constant function (see Figure 4.1a). All nonvertical lines are functions, and these lines can be written in slope–intercept form:

$$y = mx + b$$

Since y and $f(x)$ both represent the second component of a particular ordered pair,

$$f(x) = mx + b$$

This is called a **linear function** (see Figure 4.1c).

For parabolas:

$y = ax^2 + bx + c$ is a function
Each x value yields exactly one y value. See Figure 4.1d.

$x = ay^2 + by + c$ is *not* a function
If $x = 0$, then $ay^2 + by + c = 0$ has two solutions whenever $b^2 - 4ac > 0$. See Figure 4.1e.

Thus, $f(x) = ax^2 + bx + c \quad a \neq 0$ defines a **quadratic function.**

It is possible that a function is defined by different equations for different parts of the domain. In this case, each part of the domain is graphed separately.

Example 2 Graph: $f(x) = \begin{cases} x - 3 & \text{if } x \geq 3 \\ 3 - x & \text{if } x < 3 \end{cases}$

Solution First graph $y = x - 3$, as shown in figure a. The solid part of the line is the part of the graph for $x \geq 3$. Next, graph $y = 3 - x$ for $x < 3$, as shown in figure b. The result is the part shown as solid lines in figure c.

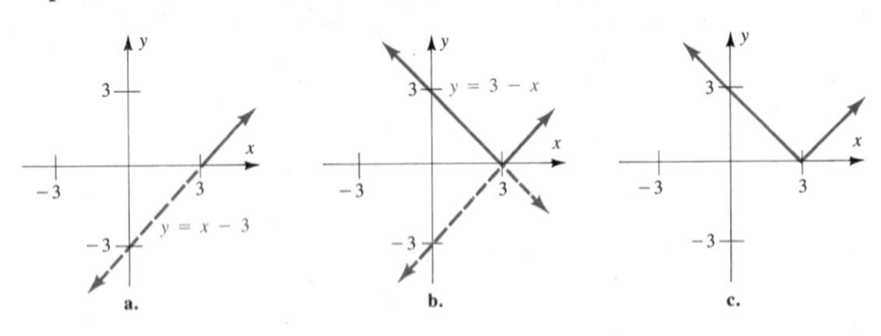

□

Example 3 Graph: $g(x) = \begin{cases} -3 & \text{if } x \leq -3 \\ x & \text{if } -3 < x < 3 \\ 3 & \text{if } x \geq 3 \end{cases}$

Solution Graph each part separately and then combine as shown.

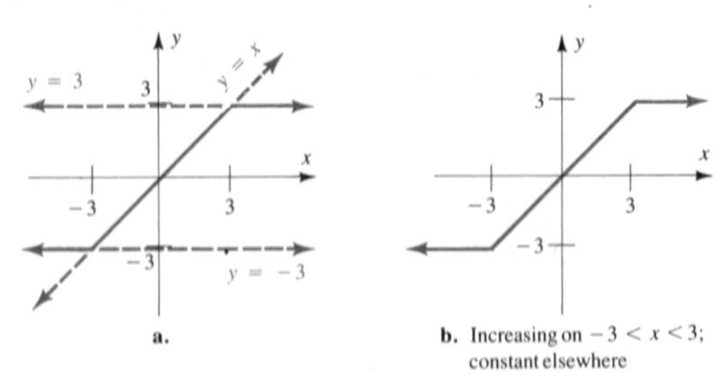

b. Increasing on $-3 < x < 3$; constant elsewhere

□

Absolute Value Function

The *absolute value function* is very similar to the two preceding examples. In Section 1.2, absolute value was defined as

$$|x| = \begin{cases} x & \text{if } x \geq 0 \\ -x & \text{if } x < 0 \end{cases}$$

The **absolute value function** is defined by

$$f(x) = |x|$$

For example, if

$$f(x) = |x - 3|$$

first apply the definition of absolute value to write

$$f(x) = \begin{cases} x - 3 & \text{if } x - 3 \geq 0 \\ -(x - 3) & \text{if } x - 3 < 0 \end{cases}$$

Note that this is the function that was graphed in Example 2.

Example 4 Graph: $f(x) = |x - 3| + 2$

Solution First, apply the definition of absolute value:

$$f(x) = \begin{cases} (x-3)+2 & \text{if} \quad x-3 \geqslant 0 \\ -(x-3)+2 & \text{if} \quad x-3<0 \end{cases}$$

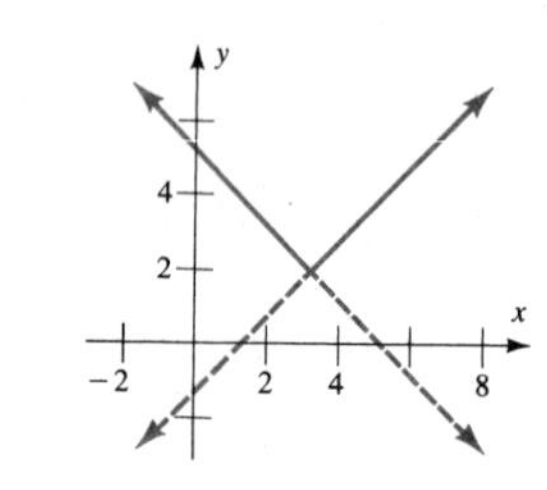

Simplify:

$$f(x) = \begin{cases} x-1 & \text{if} \quad x \geqslant 3 \\ 5-x & \text{if} \quad x<3 \end{cases}$$

To graph f, graph the linear function

$$y = x - 1 \qquad \text{for} \quad x \geqslant 3$$

Next, graph the linear function

$$y = 5 - x \qquad \text{for} \quad x < 3$$

The result is shown by the solid lines (in color) in the illustration. □

Example 5 Graph: $g(x) = \frac{1}{2}|x + 4|$

Solution

$$g(x) = \begin{cases} \frac{1}{2}(x+4) & \text{if} \quad x+4 \geqslant 0 \\ -\frac{1}{2}(x+4) & \text{if} \quad x+4<0 \end{cases}$$

Simplify:

$$g(x) = \begin{cases} \frac{1}{2}x+2 & \text{if} \quad x \geqslant -4 \\ -\frac{1}{2}x-2 & \text{if} \quad x<-4 \end{cases}$$

Graph the components separately, as shown by the solid lines (in color). □

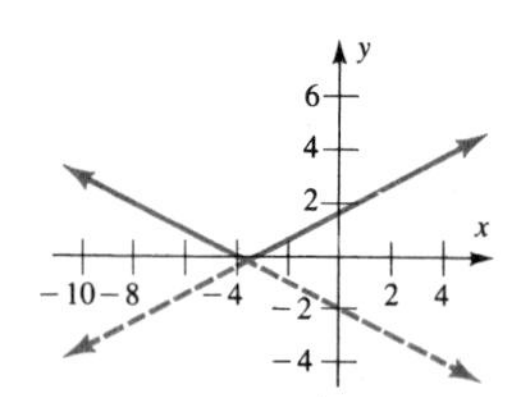

*Greatest Integer Function

An old TV quiz show required that several contestants guess the price of an item, and the contestant coming the *closest without exceeding the true value* won the item. This, basically, is the idea of another function, the *greatest integer function*, or *step function*. This rule provides a means by which to assign an integral value to the function. Consider the following:

$$G(x) = \begin{cases} 0 & \text{if} \quad 0 \leqslant x < 1 \\ 1 & \text{if} \quad 1 \leqslant x < 2 \\ 2 & \text{if} \quad 2 \leqslant x < 3 \\ 3 & \text{if} \quad 3 \leqslant x < 4 \\ 4 & \text{if} \quad 4 \leqslant x < 5 \end{cases} \qquad \begin{array}{lll} G(0) = 0 & \text{since} & 0 \leqslant 0 < 1 \\ G(\frac{1}{3}) = 0 & \text{since} & 0 \leqslant \frac{1}{3} < 1 \\ G(\frac{5}{2}) = 2 & \text{since} & 2 \leqslant \frac{5}{2} < 3 \end{array}$$

$\vdots$

*Optional

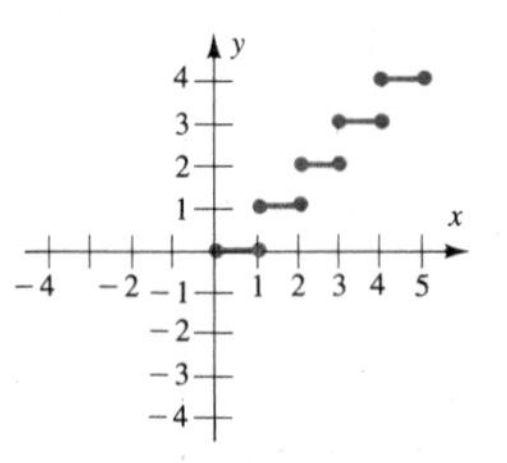

Figure 4.2

Note that the domain for this function is $\{x|0 \leq x < 5\}$ and the range is $\{0, 1, 2, 3, 4\}$. Also, notice that each part of the graph in Figure 4.2 is simply a constant function and that the result looks like the steps of a stairway. As can be seen, the function is easy enough to understand, but if the domain were very large it would become cumbersome to write, so the following notation is used.

GREATEST INTEGER FUNCTION

$$f(x) = [\![x]\!]$$

is called the **greatest integer function** and is defined by

$$[\![x]\!] = n \qquad \text{if} \qquad n \leq x < n + 1$$

where n is an integer.

The definition says that $f(x)$ is equal to the *largest* integer that does not exceed the value of x.

Example 6 Let $f(x) = [\![x]\!]$ and find $f(5.3)$, $f(-5.3)$, $f(\pi)$, and $f\left(\frac{3}{4}\right)$.

Solution

$$f(5.3) = [\![5.3]\!] = 5$$
$$f(-5.3) = [\![-5.3]\!] = -6$$
$$f(\pi) = [\![\pi]\!] = 3$$
$$f\left(\tfrac{3}{4}\right) = [\![\tfrac{3}{4}]\!] = 0$$

□

Example 7 Graph $f(x) = x + [\![x]\!]$ for $-3 \leq x < 3$.

Solution Use the definition of $[\![x]\!]$:

$$f(x) = \begin{cases} x - 3 & \text{if} \quad -3 \leq x < -2 \\ x - 2 & \text{if} \quad -2 \leq x < -1 \\ x - 1 & \text{if} \quad -1 \leq x < 0 \\ x & \text{if} \quad 0 \leq x < 1 \\ x + 1 & \text{if} \quad 1 \leq x < 2 \\ x + 2 & \text{if} \quad 2 \leq x < 3 \end{cases}$$

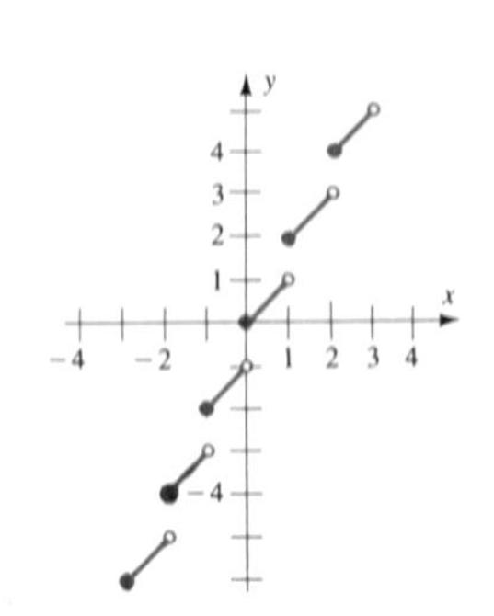

Each of these linear functions is graphed for their respective domains, as shown by the solid line segments.

□

There are many real-life examples of the greatest integer function: the postage on a first-class letter weighing x ounces, the charges for traveling x miles in a taxi, and the toll for a telephone call lasting x minutes.

Example 8 Write a function that expresses the charges for a telephone call that costs \$2.00 for the first 3 minutes and 50¢ for each additional minute or fraction thereof. Graph the function.

Solution

$$P(x) = \begin{cases} 2 & \text{if } 0 < x \leq 3 \\ 2 - .5[\![3 - x]\!] & \text{if } x > 3 \end{cases}$$

The graph is shown below.

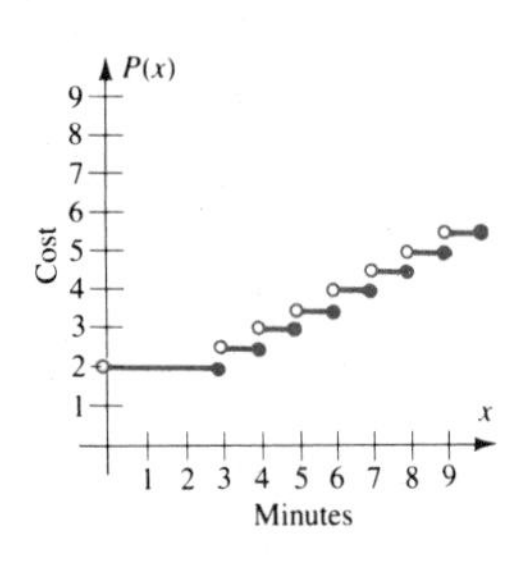

$$\begin{aligned} P(1) &= 2 \quad \text{since} \quad 0 < 1 \leq 3 \\ P(2) &= 2 \quad \text{since} \quad 0 < 2 \leq 3 \\ P(3) &= 2 \quad \text{since} \quad 0 < 3 \leq 3 \\ P(3.1) &= 2 - .5[\![3 - 3.1]\!] \quad \text{since} \quad 3.1 > 3 \\ &= 2 - .5[\![-.1]\!] \\ &= 2 - .5(-1) \\ &= 2.5 \\ P(4) &= 2 - .5[\![3 - 4]\!] \quad \text{since} \quad 4 > 3 \\ &= 2 + .5 \\ &= 2.5 \end{aligned}$$

□

Problem Set 4.2

A *In Problems 1–10 find:*

a. $f(1)$ **b.** $f(5.3)$ **c.** $f(\pi)$ **d.** $f(-\frac{1}{2})$ **e.** $f(-5.3)$

1. $f(x) = \begin{cases} -1 & \text{if } x < 0 \\ 0 & \text{if } x = 0 \\ 1 & \text{if } x > 0 \end{cases}$

2. $f(x) = \begin{cases} 1 & \text{if } x < \pi \\ 2 & \text{if } x = \pi \\ 3 & \text{if } x > \pi \end{cases}$

3. $f(x) = \begin{cases} -x & \text{if } x < -3 \\ 2x & \text{if } -3 \leq x \leq 3 \\ x^2 & \text{if } x > 3 \end{cases}$

4. $f(x) = \begin{cases} 10x & \text{if } x \leq 0 \\ x^3 & \text{if } 0 < x < 3 \\ 0 & \text{if } x \geq 3 \end{cases}$

5. $f(x) = [\![x]\!]$

6. $f(x) = |x|$

7. $f(x) = |x| + 1$

8. $f(x) = |x + 1|$

9. $f(x) = |x| + x$

10. $f(x) = [\![x]\!] + x$

Graph the functions given in Problems 11–40.

11. $f(x) = |x|$

12. $f(x) = [\![x]\!]$

13. $f(x) = \begin{cases} x - 2 & \text{if } x \geq 2 \\ 2 - x & \text{if } x < 2 \end{cases}$

14. $f(x) = \begin{cases} x + 3 & \text{if } x \geq -3 \\ -x - 3 & \text{if } x < -3 \end{cases}$

15. $g(x) = \begin{cases} -2 & \text{if } x \leq -2 \\ 0 & \text{if } -2 < x < 2 \\ 2 & \text{if } x \geq 2 \end{cases}$

16. $g(x) = \begin{cases} 1 & \text{if } x \leq -3 \\ \frac{1}{2}x + \frac{5}{2} & \text{if } -3 < x < 3 \\ 4 & \text{if } x \geq 3 \end{cases}$

17. $g(x) = \begin{cases} -3 & \text{if } x < 0 \\ 2 & \text{if } x = 0 \\ 1 & \text{if } x > 0 \end{cases}$

18. $r(x) = \begin{cases} -4 & \text{if } x < 1 \\ 3 & \text{if } x = 1 \\ -2 & \text{if } x > 1 \end{cases}$

19. $f(x) = \begin{cases} x - 1 & \text{if } x \geq 2 \\ -x + 3 & \text{if } x < 2 \end{cases}$

20. $g(x) = \begin{cases} x - 1 & \text{if } x \geq -1 \\ -x - 3 & \text{if } x < -1 \end{cases}$

21. $f(x) = |x| + 2$

22. $g(x) = 3 - |x|$

23. $h(x) = |x + 3|$

24. $k(x) = |x - 2|$

25. $f(x) = |2x| - 3$

26. $G(x) = |3x| + 2$

27. $m(x) = |x|^2$

28. $n(x) = |x^2|$

29. $f(x) = |x - 2| + 1$

30. $g(x) = |x + 1| - 2$

B

31. $b(x) = [\![x]\!] + 2$

32. $d(x) = [\![x + 2]\!]$

33. $f(x) = 2[\![x]\!] + 1$

34. $g(x) = [\![x]\!] - 2$

35. $h(x) = [\![x]\!] - x$

36. $j(x) = [\![x]\!] + |x|$

C

37. $r(x) = |x - 2| + |x|$

38. $s(x) = |x + 1| - |x|$

39. $F(x) = |x| - |x - 1|$

40. $G(x) = |x| - |x + 2|$

41. The charges for a certain telephone call are 75¢ for the first 3 minutes and 25¢ for each additional minute or fraction thereof. Write a function that gives the cost of a call lasting x minutes.

42. The telephone company charges $1.00 for the first 3 minutes for a certain call and 35¢ for each additional minute or fraction thereof. Write a function that gives the cost of a call lasting x minutes.

43. If a letter costs 20¢ for the first ounce and 17¢ for each additional ounce or fraction thereof, write a function that gives the cost of mailing a letter weighing x ounces.

44. If a letter costs 10¢ for the first 2 ounces and 6¢ for each additional ounce or fraction thereof, write a function that gives the cost of mailing a letter weighing x ounces.

45. If a taxi charges 80¢ plus 10¢ per each $\frac{1}{10}$ of a mile or fraction thereof, write a function that gives the cost of a taxi ride of x miles.

46. If the charges for a taxi are $1.50 plus 25¢ per each $\frac{1}{2}$ mile or fraction thereof, write a function that gives the cost of a taxi ride of x miles.

4.3 Algebra of Functions

Suppose functions f and g are defined by $f(x)$ and $g(x)$; then the sum and difference of f and g can be defined.

FUNCTION ADDITION AND SUBTRACTION

The **sum** of the functions f and g is defined by

$$(f + g)(x) = f(x) + g(x)$$

The **difference** of the functions f and g is defined by

$$(f - g)(x) = f(x) - g(x)$$

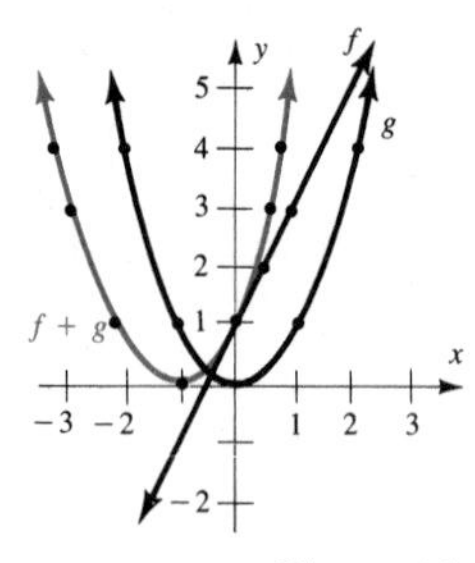

Figure 4.3

Let $f(x) = 2x + 1$ and $g(x) = x^2$. These functions can be graphed as shown in Figure 4.3. Next, consider the functions $(f + g)(x)$ and $(f - g)(x)$:

$$\begin{aligned}(f+g)(x) &= f(x) + g(x) \\ &= 2x + 1 + x^2 \\ &= x^2 + 2x + 1\end{aligned} \qquad \begin{aligned}(f-g)(x) &= 2x + 1 - x^2 \\ &= -x^2 + 2x + 1\end{aligned}$$

The graph of $(f + g)(x)$ is shown in color in Figure 4.3. It can be sketched by graphing

$$\begin{aligned}(f+g)(x) &= x^2 + 2x + 1 \\ &= (x + 1)^2\end{aligned}$$

Complete the square in order to graph; in this example it is already a perfect square so simply complete the factoring step.

or it can be sketched by adding the ordinates:

$$\begin{aligned}(f+g)(1) &= f(1) + g(1) \\ &= 3 + 1 \\ &= 4\end{aligned}$$

This can be done with points on the graph in Figure 4.3.

$$\begin{aligned}(f+g)(2) &= f(2) + g(2) \\ &= 5 + 4 \\ &= 9\end{aligned}$$

Can you graph $(f - g)(x)$?

In these examples, the domain of both f and g is the set of real numbers, so the domain of $f + g$ and $f - g$ is also the set of real numbers. In general, the domain of both $f + g$ and $f - g$ is the intersection of the domains of f and g.

Example 1 Find $f + g$, $f - g$, $(f + g)(-1)$, $(f + g)(5)$, $(f - g)(-1)$, and $(f - g)(5)$, where $f(x) = x^2$ and $g(x) = x + 4$.

Solution

$$\begin{aligned}(f+g)(x) &= f(x) + g(x) \\ &= x^2 + x + 4\end{aligned} \qquad \begin{aligned}(f-g)(x) &= f(x) - g(x) \\ &= x^2 - (x + 4) \\ &= x^2 - x - 4\end{aligned}$$

$$\begin{aligned}(f+g)(-1) &= (-1)^2 + (-1) + 4 \\ &= 4\end{aligned} \qquad \begin{aligned}(f-g)(-1) &= (-1)^2 - (-1) - 4 \\ &= -2\end{aligned}$$

$$\begin{aligned}(f+g)(5) &= 5^2 + 5 + 4 \\ &= 34\end{aligned} \qquad \begin{aligned}(f-g)(5) &= 5^2 - 5 - 4 \\ &= 16\end{aligned}$$

□

If functions f and g are defined by $f(x)$ and $g(x)$, then the product and quotient of f and g can be defined.

FUNCTION MULTIPLICATION AND DIVISION

The **product** of the functions f and g is defined by

$$(fg)(x) = f(x) \cdot g(x)$$

The **quotient** of the functions f and g is defined by

$$(f/g)(x) = \frac{f(x)}{g(x)}$$

Let $f(x) = 2x + 1$ and $g(x) = x^2$. We find $(fg)(x)$ and $(f/g)(x)$ as follows:

$$(fg)(x) = f(x) \cdot g(x) = (2x+1)(x^2) = 2x^3 + x^2$$

$$(f/g)(x) = \frac{f(x)}{g(x)} = \frac{2x+1}{x^2} \quad (x \neq 0)$$

The domain of fg is the intersection of the domains of f and g. The domain of f/g is the intersection of the domains of f and g from which values that cause $g(x) = 0$ are excluded.

Example 2 Find fg, f/g, $(fg)(-1)$, $(fg)(5)$, $(f/g)(-1)$, and $(f/g)(5)$, where $f(x) = x^2$ and $g(x) = x + 4$.

Solution

$$(fg)(x) = f(x) \cdot g(x) = x^2(x+4) = x^3 + 4x^2$$

$$(f/g)(x) = \frac{f(x)}{g(x)} = \frac{x^2}{x+4} \quad (x \neq -4)$$

$$(fg)(-1) = (-1)^3 + 4(-1)^2 = 3$$

$$(f/g)(-1) = \frac{(-1)^2}{-1+4} = \frac{1}{3}$$

$$(fg)(5) = 5^3 + 4(5)^2 = 225$$

$$(f/g)(5) = \frac{5^2}{5+4} = \frac{25}{9}$$

□

Example 3 Find fg and f/g if

$$f = \{(0, 0), (-1, 1), (2, 4), (3, 9), (5, 25)\}$$
$$g = \{(0, -5), (-1, 0), (2, -3), (4, -2), (5, -1)\}$$

Solution The intersection of the domains is $\{0, -1, 2, 5\}$.

$$(fg)(0) = f(0) \cdot g(0) = 0(-5) = 0$$
$$(fg)(-1) = f(-1) \cdot g(-1) = 1(0) = 0$$
$$(fg)(2) = f(2) \cdot g(2) = 4(-3) = -12$$
$$(fg)(5) = f(5) \cdot g(5) = 25(-1) = -25$$

Thus, $\boldsymbol{fg = \{(0, 0), (-1, 0), (2, -12), (5, -25)\}}$.

For f/g we must also exclude from the domain any values of x for which g is 0; thus, -1 is also excluded and the domain of f/g is $\{0, 2, 5\}$.

$$(f/g)(0) = \frac{f(0)}{g(0)} = \frac{0}{-5} = 0$$

$$(f/g)(2) = \frac{f(2)}{g(2)} = \frac{4}{-3} = -\frac{4}{3}$$

$$(f/g)(5) = \frac{f(5}{g(5)} = \frac{25}{-1} = -25$$

Thus, $f/g = \{(0, 0), (2, -\frac{4}{3}), (5, -25)\}$, and $(f/g)(-1)$ is not defined. □

Composition of Functions

In addition to the obvious ways of combining functions by adding, subtracting, multiplying, and dividing them, there is another way functions can be combined, which is called *composition of functions*. As an example of composition, suppose the cost of manufacturing x calculators is $50x + 200$. This means the cost of each calculator is

$$c(x) = \frac{50x + 200}{x} \qquad x > 0$$

so the cost depends on the number of calculators produced. Suppose also that a store sells these calculators by marking the price up 20%. That is, for an item costing $\$k$, the price is

$$\begin{aligned} p(k) &= k + .20k \\ &= 1.20k \end{aligned}$$

Ultimately, the price depends on the number produced by finding

$$\begin{aligned} p[c(x)] &= p\left[\frac{50x + 200}{x}\right] \\ &= 1.20\left(\frac{50x + 200}{x}\right) \\ &= 60 + \frac{240}{x} \end{aligned}$$

This function is called the *composition of* c by p and is denoted by $p \circ c$.

COMPOSITION OF FUNCTIONS

The **composite function of f by g,** denoted by $g \circ f$, is the function defined by

$$(g \circ f)(x) = g[f(x)]$$

The domain of $g \circ f$ is the subset of the domain of f containing those values for which $g \circ f$ is defined.

Example 4 If $f(x) = x^2$ and $g(x) = x + 4$, find the following:

a. $g \circ f$ **b.** $f \circ g$ **c.** $(g \circ f)(5)$ **d.** $(f \circ g)(5)$

Solution **a.**

$$\begin{aligned} (g \circ f)(x) &= g[f(x)] \\ &= g[x^2] \\ &= x^2 + 4 \end{aligned}$$

b.

$$\begin{aligned} (f \circ g)(x) &= f[g(x)] \\ &= f[x + 4] \\ &= (x + 4)^2 \\ &= x^2 + 8x + 16 \end{aligned}$$

c. $(g \circ f)(5) = 5^2 + 4$
$= 29$

d. $(f \circ g)(5) = 5^2 + 8(5) + 16$
$= 81$ □

Example 5 Find $f \circ g$ and $g \circ f$ if

$$f = \{(0,0), (-1,1), (-2,4), (-3,9), (5,25)\}$$
$$g = \{(0,-5), (-1,0), (2,-3), (4,-2), (5,-1)\}$$

Solution

g	f	$(f \circ g)$
$0 \to -5 \to$	Not defined	
$-1 \to 0 \to$	$0 \to 0$	$(-1,0)$
$2 \to -3 \to$	$-3 \to 9$	$(2,9)$
$4 \to -2 \to$	$-2 \to 4$	$(4,4)$
$5 \to -1 \to$	$-1 \to 1$	$(5,1)$

f	g	$(g \circ f)$
$0 \to 0 \to$	$0 \to -5$	$(0,-5)$
$-1 \to 1 \to$	Not defined	
$-2 \to 4 \to$	$4 \to -2$	$(-2,-2)$
$-3 \to 9 \to$	Not defined	
$5 \to 25 \to$	Not defined	

$f \circ g = \{(-1, 0), (2, 9), (4, 4), (5, 1)\}$ $\quad g \circ f = \{(0, -5), (-2, -2)\}$ □

Notice in Examples 4 and 5 that $f \circ g \neq g \circ f$, so you must be careful about the order in which you carry out composition.

Problem Set 4.3

A *In Problems 1–5 let* $f(x) = 2x - 3$ *and* $g(x) = x^2 + 1$. *Find:*

1. $(f + g)(5)$ **2.** $(f - g)(3)$ **3.** $(fg)(2)$
4. $(f/g)(4)$ **5.** $(f \circ g)(2)$

In Problems 6–10 let $f(x) = \dfrac{x - 2}{x + 1}$ *and* $g(x) = x^2 - x - 2$. *Find:*

6. $(f + g)(2)$ **7.** $(f - g)(5)$ **8.** $(fg)(102)$
9. $(f/g)(99)$ **10.** $(f \circ g)(1)$

In Problems 11–15 let $f(x) = \dfrac{2x^2 - x - 3}{x - 2}$ *and* $g(x) = x^2 - x - 2$. *Find:*

11. $(f + g)(-1)$ **12.** $(f - g)(2)$ **13.** $(fg)(9)$
14. $(f/g)(102)$ **15.** $(f \circ g)(0)$

In Problems 16–20 let $f = \{(0, 1), (1, 4), (2, 7), (3, 10)\}$ *and* $g = \{(0, -3), (1, -1), (2, 1), (3, 3)\}$. *Find:*

16. $(f + g)(1)$ **17.** $(f - g)(3)$ **18.** $(fg)(2)$
19. $(f/g)(0)$ **20.** $(f \circ g)(2)$

In Problems 21–25 let $f = \{(5, 3), (6, 2), (7, 9), (8, 12)\}$ *and* $g = \{(5, 8), (6, 5), (7, 4), (8, 3)\}$. *Find:*

21. $(f + g)(6)$ **22.** $(f - g)(7)$ **23.** $(fg)(5)$

24. $(f/g)(8)$ **25.** $(f \circ g)(6)$

In Problems 26–30 let $f = \{(5, 9), (10, 29), (15, 39), (20, 49)\}$ *and* $g = \{(5, 4), (10, 5), (15, 6), (20, 9)\}$. *Find:*

26. $(f + g)(10)$ **27.** $(f - g)(5)$ **28.** $(fg)(15)$

29. $(f/g)(20)$ **30.** $(f \circ g)(10)$

In Problems 31–34 let $f(x) = 2x - 3$ *and* $g(x) = x^2 + 1$. *Find:*

31. $(f + g)(x)$ **32.** $(f - g)(x)$

33. $(fg)(x)$ **34.** $(f/g)(x)$

In Problems 35–38 let $f(x) = \dfrac{x - 2}{x + 1}$ *and* $g(x) = x^2 - x - 2$. *Find:*

35. $(f + g)(x)$ **36.** $(f - g)(x)$

37. $(fg)(x)$ **38.** $(f/g)(x)$

In Problems 39–42 let $f(x) = \dfrac{2x^2 - x - 3}{x - 2}$ *and* $g(x) = x^2 - x - 2$. *Find:*

39. $(f + g)(x)$ **40.** $(f - g)(x)$

41. $(fg)(x)$ **42.** $(f/g)(x)$

In Problems 43–46 let $f(x) = 4x + 1$ *and* $g(x) = x^3 + 3$. *Find:*

43. $(f + g)(x)$ **44.** $(f - g)(x)$

45. $(fg)(x)$ **46.** $(f/g)(x)$

In Problems 47–50 let $f(x) = x^3 - 1$ *and* $g(x) = x - 1$. *Find:*

47. $(f + g)(x)$ **48.** $(f - g)(x)$

49. $(fg)(x)$ **50.** $(f/g)(x)$

B *In Problems 51–60 find* $f \circ g$ *and* $g \circ f$.

51. $f = \{(0, 1), (1, 3), (2, 0), (3, 2)\}$; $g = \{(0, 3), (1, 2), (2, 1), (3, 0)\}$

52. $f = \{(0, 2), (1, 0), (2, 3), (3, 1)\}$; $g = \{(0, 1), (1, 3), (2, 0), (3, 2)\}$

53. $f(x) = 2x - 3$; $g(x) = \dfrac{x + 3}{2}$ **54.** $f(x) = 3x + 1$; $g(x) = \dfrac{x - 1}{3}$

55. $f(x) = \frac{1}{2}x + 1$; $g(x) = 2x - 2$ **56.** $f(x) = 2 - \frac{1}{3}x$; $g(x) = 6 - 3x$

57. $f(x) = 2x - 3$; $g(x) = x^2 + 1$ **58.** $f(x) = \dfrac{x - 2}{x + 1}$; $g(x) = x^2 - x - 2$

59. $f(x) = x^2$; $g(x) = x^2 - x - 2$ **60.** $f(x) = 4x + 1$; $g(x) = x^3 + 3$

C **61.** *Physics* Suppose that the volume of a certain cone is given by the function

$$V(h) = \frac{\pi h^3}{12}$$

where h is the height. Furthermore, suppose that the height is expressed as a function of time by $h(t) = 2t$.

a. Find the volume for $t = 2$.

b. Express the volume as a function of time by finding $V \circ h$.

c. If the domain of V is $\{h | 0 < h \leq 6\}$, find the domain of h; that is, what are the permissible values for t?

62. *Physics* The surface area of a spherical balloon is given by

$$S(r) = 4\pi r^2$$

Suppose that the radius r is expressed as a function of time as $r(t) = 3t$.

a. Find the surface area for $t = 2$.

b. Express the surface area as a function of time by finding $S \circ r$.

c. If the domain of S is $\{r | 0 < r < 8\}$, find the domain of r; that is, what are the permissible values for t?

63. If $f(x) = x^2$, then $f(1/x) = (1/x)^2 = 1/x^2 = 1/f(x)$. Give an example of a function for which $f(1/x) \neq 1/f(x)$.

64. If $f(x) = x$, then $f(x^2) = x^2 = [f(x)]^2$. Give an example of a function for which $f(x^2) \neq [f(x)]^2$.

65. If $f(x) = x^2$, then $(f \circ f)(x) = x^4 = f(x) \cdot f(x)$. Give an example of a function for which $(f \circ f)(x) \neq f(x) \cdot f(x)$.

66. If $f(x) = 1 + 1/x$, find each value:

a. $(f \circ f)(x)$ **b.** $(f \circ f \circ f)(x)$ **c.** $(f \circ f \circ f \circ f)(x)$

d. Without doing any additional algebra, guess the value of $(f \circ f \circ f \circ f \circ f)(x)$ by noticing a pattern in parts a–c.

67. Let $f(x) = \sqrt{x}$. Choose *any* positive x. Find a numerical value for $(f \circ f)(x)$, $(f \circ f \circ f)(x)$, and $(f \circ f \circ f \circ f)(x)$. If this procedure is repeated a large number of times,

$$(f \circ f \circ f \circ \cdots \circ f)(x)$$

can you predict the outcome for *any* x?

4.4 Inverse Functions

In mathematics, the ideas of "opposite operations" and "inverse properties" are very important. The basic notion of an opposite operation or an inverse property is to "undo" a previously performed operation. For example, pick a number and call it x; then:

	x	Think: I pick 5.
Add 7:	$x + 7$	Think: Now I have $5 + 7 = 12$.

The opposite operation returns you to x:

Subtract 7:	$x + 7 - 7 = x$	Think: $12 - 7 = 5$, my original number.

We now want to apply this idea to functions. Pick a number and call it x:

x Think: I'll pick 3 this time.

Now evaluate some given function f for the number you picked; suppose we let $f(x) = 2x + 9$:

$f(x) = 2x + 9$ Think: $f(3) = 2(3) + 9 = 15$

The *inverse function,* call it f^{-1} if it exists, is a function so that $f^{-1}(15)$ is 3, the original number. In symbols, $f(3) = 15$, so

$f^{-1}[f(x)] = x$ Think: $f^{-1}[f(3)] = 3$

Of course, for f^{-1} to be an inverse function, it must "undo" the effect of f for *each and every* member of the domain. This may be impossible if f is a function such that two x values give the same y value. For example, if $f(x) = x^2$, then $f(2) = 4$ and $f(-2) = 4$. So we need a function f^{-1} so that

$f^{-1}(4) = 2$ and *also* $f^{-1}(4) = -2$

This would require the inverse to yield *two* values, namely 2 and -2; this *violates* the definition of a function. So it is necessary to limit the given function f to a class of functions, called *one-to-one functions,* for which no two elements of the domain lead to the same element in the range.

ONE-TO-ONE FUNCTION

A function f is said to be **one-to-one** if

$f(x_1) = f(x_2)$ implies $x_1 = x_2$

for elements x_1 and x_2 in the domain of f.

Each element in the domain of f^{-1} is the range element in the corresponding ordered pair for f, and each element in the domain of f is the range element in the corresponding pair for f^{-1}. Thus, for every x in the domain of f,

$$(f^{-1} \circ f)(x) = x$$

and for every x in the domain of f^{-1},

$$(f \circ f^{-1})(x) = x$$

In fact, the following relationship between functions, their inverses, and composition is used as a definition of inverse function.

INVERSE FUNCTIONS

If f is a one-to-one function with domain D and range R, and g is a function with domain R and range D such that

$(g \circ f)(x) = x$ for every x in D
$(f \circ g)(x) = x$ for every x in R

then f and g are called **inverses.** The function g is the **inverse function** of f and is denoted by f^{-1}.

Example 1 Show that $f(x) = \frac{1}{2}x + 1$ and $g(x) = 2x - 2$ are inverses.

Solution

$$(f \circ g)(x) = \tfrac{1}{2}(2x - 2) + 1 \quad \text{and} \quad (g \circ f)(x) = 2\left(\tfrac{1}{2}x + 1\right) - 2$$
$$= x - 1 + 1 \qquad\qquad = x + 2 - 2$$
$$= x \qquad\qquad = x$$

Thus, $(f \circ g)(x) = x = (g \circ f)(x)$ and f and g are inverses. □

Example 2 Show that $f(x) = x^3 + 3$ and $g(x) = (x - 3)^{1/3}$ are inverses.

Solution

$$(f \circ g)(x) = [(x - 3)^{1/3}]^3 + 3 \quad \text{and} \quad (g \circ f)(x) = [(x^3 + 3) - 3]^{1/3}$$
$$= x - 3 + 3 \qquad\qquad = [x^3]^{1/3}$$
$$= x \qquad\qquad = x$$

Thus, f and g are inverse functions. □

This definition can be used to show that two functions are inverses, but it *cannot* be used to *find* the inverse of a given function. To find the inverse it is helpful to visualize a function as a set of ordered pairs. Suppose we pick a number, say 3, and evaluate a function f at 3 to find $f(3) = 15$. Then, $(3, 15)$ is an element of f. Now the inverse function f^{-1} requires that 15 is changed back into 3; that is, $f^{-1}(15) = 3$ so that $(15, 3)$ is an element of f^{-1}. Thus,

If function f has element (x, y), then inverse function f^{-1} must have element (y, x).

INVERSE FUNCTION

> If the function f is the set of ordered pairs of the form (a, b), and if f is one-to-one, then f^{-1} is the **inverse function** of f and is the set of ordered pairs of the form (b, a).

Example 3 Find the inverse of $f = \{(0, 2), (1, 0), (2, 3), (3, 1)\}$.

Solution Interchange first and second components to obtain

$$f^{-1} = \{(2, 0), (0, 1), (3, 2), (1, 3)\} \quad \text{or} \quad \{(\mathbf{0}, \mathbf{1}), (\mathbf{1}, \mathbf{3}), (\mathbf{2}, \mathbf{0}), (\mathbf{3}, \mathbf{2})\}$$ □

It should be evident from the definition and Example 3 that the domain and range of f^{-1} are simply the range and domain, respectively, of f. Given $y = f(x)$, which is a one-to-one function, the inverse f^{-1} is defined by $x = f(y)$. However, in order to rewrite $x = f(y)$ in the form $y = f^{-1}(x)$, you must solve for y. For example, given $y = 2x - 1$, its inverse is $x = 2y - 1$ (interchanging x and y) or $y = \frac{1}{2}x + \frac{1}{2}$ (solving for y). That is, $x = f(y)$ is equivalent to $y = f^{-1}(x)$. This is used to find the inverse of given functions, as illustrated in the following example.

Example 4 Find f^{-1} if $f(x) = 2 - \frac{1}{3}x$.

Solution

$$y = 2 - \tfrac{1}{3}x$$

$$x = 2 - \tfrac{1}{3}y \qquad \text{Interchange } x \text{ and } y.$$

$$\tfrac{1}{3}y = 2 - x \qquad \text{Then solve for } y.$$
$$y = 6 - 3x$$

or

$$\mathbf{f^{-1}(x) = 6 - 3x}$$ □

Let us summarize the steps illustrated by Example 4.

PROCEDURE FOR FINDING AN EQUATION FOR AN INVERSE FUNCTION

1. Let $y = f(x)$ be a given *one-to-one* function.
2. Interchange x and y in the given equation.
3. Solve for y. The resulting function defined by the equation $y = f^{-1}(x)$ is the inverse of f.

The domain of f and the range of f^{-1} must be equal as well as the domain of f^{-1} and the range of f.

Example 5 Find g^{-1} if $g(x) = x^3 + 3$.

Solution

$$y = x^3 + 3$$
$$x = y^3 + 3 \qquad \text{Interchange } x \text{ and } y.$$
$$y^3 = x - 3 \qquad \text{Solve for } y.$$
$$y = (x - 3)^{1/3}$$
$$g^{-1}(x) = (x - 3)^{1/3}$$ □

Example 6 Find t^{-1} if t is defined by $t(x) = x^2$.

Solution $t(x) = x^2$ is not a one-to-one function, so it has no inverse function. □

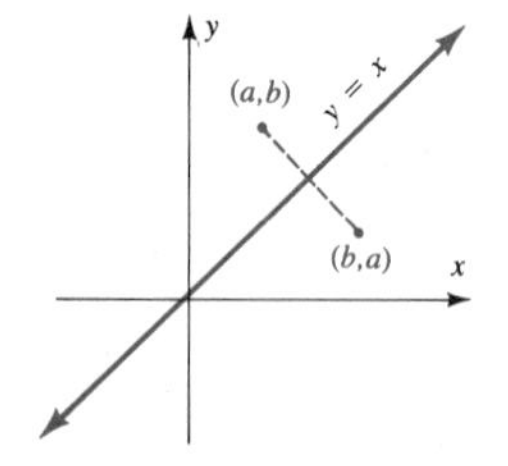

Figure 4.4

The graphs of f and f^{-1} have an interesting relationship. Notice that the points (a, b) and (b, a) are symmetric with respect to the line $y = x$. (See Figure 4.4.) That is, the points are on opposite sides of and equidistant from the line. The points are called *reflections in the line*. Since for every ordered pair (a, b) in f, the ordered pair (b, a) is in f^{-1}, the graphs of $y = f(x)$ and $y = f^{-1}(x)$ are reflections in the line $y = x$.

Example 7 Show that $y = 2 - \frac{1}{3}x$ and $y = 6 - 3x$ (inverses found in Example 4) are symmetric with respect to the line $y = x$ by graphing each.

Solution The graphs are shown in the figure. □

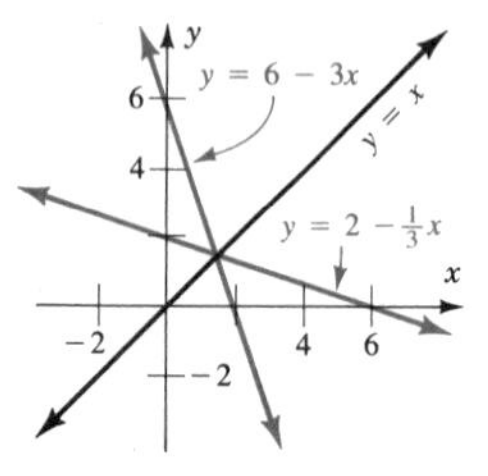

Example 8 Given the graph of $y = f(x)$ shown at the right, sketch the graph of its inverse.

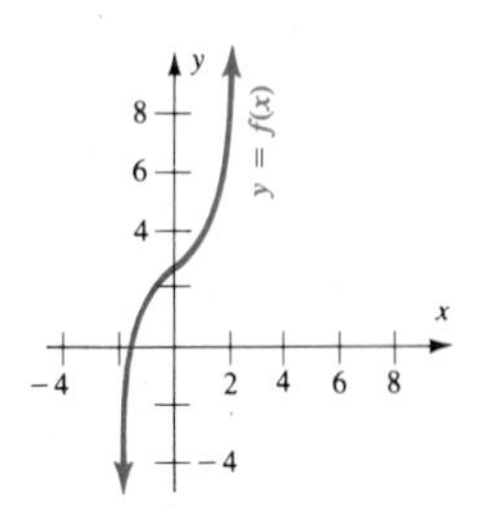

Solution First, sketch the line $y = x$. Then sketch the reflection of $y = f(x)$ in the line to obtain the graph of $y = f^{-1}(x)$, as shown in the figure.

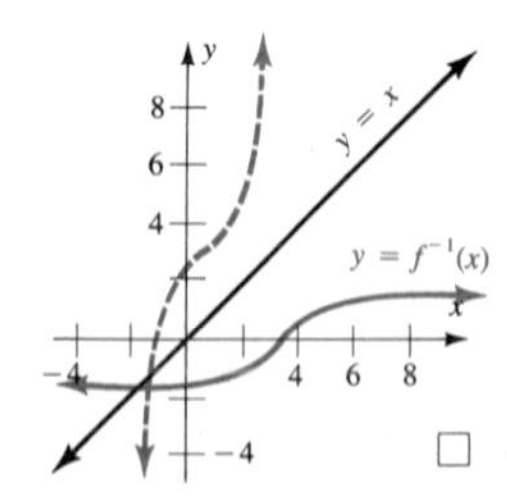

□

Problem Set 4.4

A *Determine whether the functions in Problems 1–18 are one-to-one.*

1. $f = \{(0, 4), (1, 5), (2, 6), (3, 7)\}$

2. $g = \{(1, 4), (2, 3), (3, 2), (4, 1)\}$

3. $h = \{(-2, 4), (-1, 1), (0, 0), (1, 1), (2, 4)\}$

4. $j = \{(-2, 8), (-1, -1), (0, 0), (1, 1), (2, 8)\}$

5. $f(x) = 5x - 3$

6. $g(x) = 3x + 7$

7. $F(x) = x^2 + 1$

8. $G(x) = 2x^2$

9. $h(x) = \dfrac{1}{x}$

10. $m(x) = \dfrac{1}{x^3 - 1}$

11. $j(x) = \sqrt{x}$

12. $K(x) = 3\sqrt{x}$

13. $n(x) = \dfrac{x^2 + 1}{x}$

14. $t(x) = x^3 - 2x^2 + x + 5$

15. $f(x) = \sqrt{25 - x^2}$

16. $g(x) = -\sqrt{36 - x^2}$

17. $f(x) = |1 - x|$

18. $g(x) = |49 - x^2|$

For each of Problems 19–30 determine whether the functions are inverses by using composition of functions.

19. $f(x) = 5x;\ \ g(x) = \frac{1}{5}x$

20. $f(x) = -3x;\ \ g(x) = \frac{1}{3}x$

21. $f(x) = 3x - 2;\ \ g(x) = \dfrac{x + 2}{3}$

22. $f(x) = 2x - 3;\ \ g(x) = \dfrac{x + 3}{2}$

23. $f(x) = \frac{1}{4}x + 1;\ \ g(x) = 4(x - 1)$

24. $f(x) = 3(x + 1);\ \ g(x) = \frac{1}{3}x - 3$

25. $f(x) = x - 1;\ \ g(x) = 1 - x$

26. $f(x) = \dfrac{1}{x};\ \ g(x) = \dfrac{1}{x}$

27. $f(x) = \frac{1}{x-3}$; $g(x) = \frac{1+3x}{x}$

28. $f(x) = \frac{1}{2x+1}$; $g(x) = \frac{1-x}{2x}$

29. $f(x) = \frac{3x}{2x-1}$; $g(x) = \frac{x}{2x-3}$

30. $f(x) = \frac{x}{x+2}$; $g(x) = \frac{2x}{x-1}$

Each equation in Problems 31–42 defines a one-to-one function. Find an equation for each inverse and solve explicitly for y.

31. $y = x + 2$

32. $y = x - 4$

33. $f(x) = 5x - 2$

34. $g(x) = 7x + 3$

35. $3x + 5y = 15$

36. $5x + 3y = 15$

37. $F(x) = \frac{2}{3}x + \frac{7}{3}$

38. $G(x) = \frac{3}{5}x - \frac{2}{5}$

39. $y = x^3$

40. $y = 1 - x^3$

41. $y = x^3 + 1$

42. $y = x^5 - 32$

B *For each of Problems 43–56 graph the given function and its inverse on the same set of coordinate axes.*

43. $f(x) = x + 2$; $f^{-1}(x) = x - 2$

44. $g(x) = x - 4$; $g^{-1}(x) = x + 4$

45. $h(x) = 5x - 2$; $h^{-1}(x) = \frac{x+2}{5}$

46. $j(x) = 7x + 3$; $j^{-1}(x) = \frac{x-3}{7}$

47. $f(x) = x^2$ domain $x \geq 0$
$f^{-1}(x) = \sqrt{x}$ domain $x \geq 0$

48. $g(x) = x^2$ domain $x \leq 0$
$g^{-1}(x) = -\sqrt{x}$ domain $x \geq 0$

49. $f(x) = x^2 - 2$ domain $x \geq 0$
$f^{-1}(x) = \sqrt{x+2}$ domain $x \geq -2$

50. $g(x) = |x - 1|$ domain $x \geq 1$
$g^{-1}(x) = |x + 1|$ domain $x \geq 0$

51. $f(x) = \frac{1}{4}x - 2$

52. $g(x) = \frac{1}{3}x + 1$

53. $h(x) = \frac{1}{3}x - \frac{5}{3}$

54. $k(x) = -\frac{1}{4}x + \frac{3}{4}$

55. $m(x) = 4x - 2$

56. $n(x) = 3x + 6$

57. Consider the figure at the right.
  a. What are the coordinates of A and B?
  b. What is the slope of the line passing through A and B?

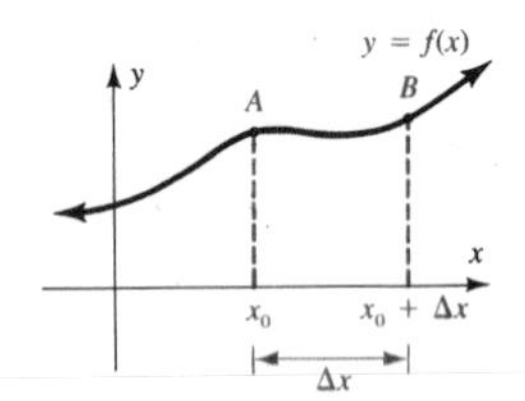

58. Consider the figure at the right.
  a. What are the coordinates of A and B?
  b. What is the slope of the line passing through A and B?

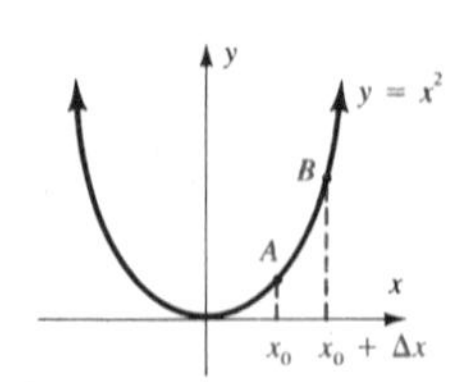

59. Answer Problem 58b, where $x_0 = 1$ and Δx is:
  a. $\Delta x = 2$ **b.** $\Delta x = 1$ **c.** $\Delta x = .5$ **d.** $\Delta x = .25$
  e. $\Delta x = .1$ **f.** $\Delta x = .01$ **g.** $\Delta x = .001$

60. Consider the figure at the right.

a. What are the coordinates of A and B?

b. What is the slope of the line passing through A and B?

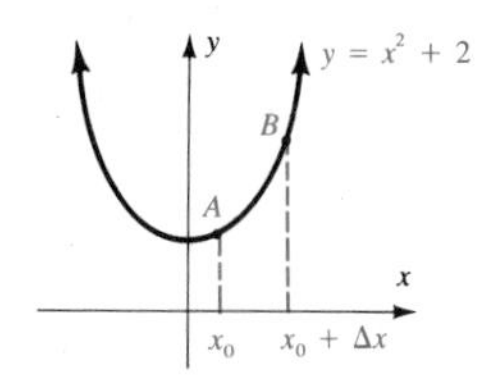

61. Answer Problem 60b, where $x_0 = 1$ and Δx is:

a. $\Delta x = 1$ **b.** $\Delta x = .5$ **c.** $\Delta x = .25$ **d.** $\Delta x = .1$
e. $\Delta x = .01$ **f.** $\Delta x = .001$ **g.** $\Delta x = .0001$

*4.5 Variation

Sometimes functional relationships are indicated by using some special terminology common to the sciences. For example, the function d defined by $d(t) = 16t^2$ gives the distance d that an object falls in a given time t, which can be described by this statement: "The distance a body falls from rest *varies directly as* the square of the time it falls (disregarding air resistance)." Several types of variation are defined.

VARIATION

If variables x and y are related so that (for any constant k):

I. $y = kx$, we say **y varies directly as x,** or *y is directly proportional to x*

II. $y = kx^2$, we say *y varies directly as the square of x*, or *y is directly proportional to the square of x*

III. $y = \frac{k}{x}$, we say **y varies inversely as x,** or *y is inversely proportional to x*

IV. $y = \frac{k}{x^2}$, we say *y varies inversely as the square of x*, or *y is inversely proportional to the square of x*

Example 1 The surface area of a sphere varies directly as the square of the radius. If the surface area is 16π cm^2 when the radius is 2 cm, what is the surface area when the radius is 10 cm?

Solution *Step 1* Use the definition of variation to write an equation.
If S is the surface area and r is the radius, then

$$S = kr^2$$

Step 2 Use the given ordered pair to determine the value of k given $(r, S) = (2, 16\pi)$:

$$16\pi = k(2)^2$$
$$4\pi = k$$

*Optional

Step 3 The formula is therefore $S = 4\pi r^2$. Now S can be found for any given value of r; in particular, if $r = 10$ cm, then

$$S = 4\pi(10 \text{ cm})^2$$
$$\mathbf{S = 400\pi \text{ cm}^2}$$ □

Example 2 The intensity of light is inversely proportional to the square of the distance from its source. If a certain light source projects 7.5 ft-candles at 10 ft, what is the intensity of this lamp at 5 ft?

Solution *Step 1* $I = \dfrac{k}{d^2}$ I is intensity and d is distance (in feet)

Step 2 $7.5 = \dfrac{k}{10^2}$

$750 = k$

Step 3 The formula is therefore $I = \dfrac{750}{d^2}$; if $d = 5$, then

$$I = \frac{750}{5^2}$$
$$= 30$$

The intensity at 5 ft is 30 ft-candles. □

Other types of variation involve more than one variable.

VARIATION

If the variables x, y, and z are related so that (for any constant k):

V. $z = kxy$, we say **z varies jointly as x and y**

VI. z can be written as a *combination* of the other variation types, we may have, for example,

$z = k\dfrac{x^2}{y^3}$ *z varies directly as the square of x and inversely as the cube of y*

Example 3 The stiffness of a rectangular beam varies jointly as the breadth and the cube of its depth. Write an equation to represent this relationship.

Solution $\mathbf{S = kwd^3}$ S is stiffness of beam, w is breadth, and d is depth □

Example 4 The *f*-stop of a camera varies directly as the focal length of the lens and inversely as the diameter of the lens opening. Write an equation to represent this relationship.

Solution $\mathbf{N = k\left(\dfrac{f}{d}\right)}$ N is *f*-stop, f is focal length, and d is diameter of lens opening □

Problem Set 4.5

A *Write each statement in Problems 1–24 as an appropriate equation, using k as the constant of variation. Be sure to identify your variables if they are not given.*

1. x varies directly as y.

2. m varies inversely as n.

3. s is directly proportional to the square of t.

4. v is directly proportional to the square of t.

5. A varies jointly as ℓ and w.

6. V varies jointly as ℓ, w, and h.

7. V is directly proportional to the cube of r.

8. S varies jointly as the square of r and θ.

9. C varies directly as t and inversely as r.

10. x varies directly as the cube of y and inversely as the square of z.

11. The volume of a sphere varies directly as the cube of its radius.

12. The distance that a spring stretches is directly proportional to the force applied.

13. The current in an electrical circuit is inversely proportional to the resistance.

14. *Electrical engineering* The current in a wire varies directly as the electromotive force and inversely as the resistance.

15. *Business* The simple interest earned in a given time varies jointly as the principal and the interest rate.

16. *Physics* The pressure exerted by a liquid at a given point is directly proportional to the depth of that point below the surface of the liquid.

17. *Earth science* The force with which the earth attracts an object above the earth's surface varies inversely with the square of the distance from the center of the earth.

18. *Music* The frequency of vibration of air in an open pipe organ is inversely proportional to the length of the pipe.

19. *Chemistry* The volume of a gas varies directly as the temperature and inversely as the pressure.

20. *Physics* The strength of a rectangular beam varies directly as its width and the square of its depth.

21. *Physics* The kinetic energy varies jointly as the mass and the square of the velocity.

22. *Physics* The stiffness of a beam varies jointly as the breadth and depth and inversely as the square of the length.

23. *Physics* The amount of heat put out by an electrical appliance varies jointly as time and resistance and the square of the current.

24. *Physics* The centripetal force of a body moving in a circular path at a constant speed varies inversely as the radius of the path.

B

25. If w varies directly as t, find w when $t = 15$ if it is known that $w = 7$ when $t = 5$.

26. If t varies inversely as s, find t when $s = 13$ if it is known that $t = 2$ when $s = 5$.

27. If p is directly proportional to the square of q, find p when $q = 2$ if it is known that $p = 5$ when $q = 4$.

28. If x is directly proportional to the square of z, find x when $z = 5$ if it is known that $x = 10$ when $z = 2$.

29. If A varies jointly as b and h, find A when $b = 10$ and $h = 6$ if it is known that $A = 28$ when $b = 7$ and $h = 4$.

30. If V varies jointly as ℓ, w, and h, find V when $\ell = 7$, $w = 2$, and $h = 2$ if it is known that $V = 30$ when $\ell = 5$, $w = 3$, and $h = 2$.

31. The volume of a sphere varies directly as the cube of its radius. Find the volume when the radius is 6 cm if it is known that the volume of the sphere is $36\pi\ \text{cm}^3$ when the radius is 3 cm.

32. *Physics* Determine how far a force of 15 kg will stretch a spring if the distance that a spring stretches is directly proportional to the force applied. A preliminary experiment shows that a force of 20 kg will stretch the spring 12 cm.

33. *Electrical engineering* Determine the current in an electrical circuit when the resistance is 5 ohms if the current is inversely proportional to the resistance. It is known that a resistance of 4 ohms in this circuit will produce a current of 25 amps.

34. *Business* Simple interest earned in a given time varies jointly as the principal and the interest rate. How much will $1500 at 8% earn in the same period that $1000 earns $350 at 7%?

35. *Physics* The pressure exerted by a liquid at a given point is directly proportional to the depth of that point below the surface of the liquid. What is the pressure on a submarine at 300 ft if the water pressure on that submarine is 400 pounds per square foot at a depth of 100 ft?

36. *Physics* The frequency of vibration of air in an open pipe organ is inversely proportional to the length of the pipe. How long should a pipe be to cause the air to vibrate 16 times per second, if an 8 ft length vibrates at 64 times per second?

37. *Chemistry* If the volume of a gas varies directly as the temperature and inversely as the pressure, find the volume at 360° Kelvin and a pressure of 30 kilograms per square centimeter. A preliminary experiment has shown that this gas occupies a volume of 5 liters at 320° Kelvin and a pressure of 16 kg/cm^2.

C

38. If $y = f(x)$ varies directly as x, show that: $\dfrac{f(x_1)}{f(x_2)} = \dfrac{x_1}{x_2}$

39. If $y = f(x)$ and $\dfrac{f(x_1)}{f(x_2)} = \dfrac{x_2}{x_1}$, show that $f(x)$ varies inversely as x.

4.6 Summary and Review

The notion of a function is one of the most important and fundamental concepts in mathematics. Much of the groundwork for the study of calculus is contained in this chapter. For example, the basic ideas of calculus concern the notions of the limit of a *function,* the derivative of a *function,* and the integral of a *function.* All require a thorough understanding of this chapter. For example, the definition of derivative involves the calculation of

$$\frac{f(x+h)-f(x)}{h}$$

for a given function f.

CHAPTER OBJECTIVES

After studying this chapter, you should be able to:

1. Use functional notation
2. Graph linear functions
3. Graph quadratic functions
4. Recognize and graph constant, linear, and absolute value functions
*5. Recognize, graph, and use the greatest integer function to write equations for certain applied problems
6. Add, subtract, multiply, and divide functions
7. Find the composition of functions
8. Find the inverse of a one-to-one function
*9. Write equations from statements of variation
*10. Solve variation problems when initial values are given

Problem Set 4.6

CONCEPT PROBLEMS

Fill in the word or words to make the statements in Problems 1–17 complete and correct.

[4.1] **1.** A function f of a real number x is ______________________________
____.

[4.1] **2.** Using functional notation, f represents ________ whereas $f(x)$ represents ________ and x is ________ .

[4.2] **3.** A constant function f is given by ________ .

[4.2] **4.** A linear function g is given by ________ .

[4.2] **5.** A quadratic function h is given by ________ .

[4.2] **6.** An absolute value function F is given by ________ .

*[4.2] **7.** If $f(x) = [\![x]\!]$, where $[\![x]\!] = n$ if ________ is called the ________.

[4.3] **8.** $(f+g)(x)$ means ________ .

*Optional

[4.3] **9.** $(f - g)(x)$ means ______ .

[4.3] **10.** $(fg)(x)$ means ______ .

[4.3] **11.** $(f/g)(x)$ means ______ .

[4.3] **12.** $(f \circ g)(x)$ means ______ .

[4.4] **13.** If f and g are inverse functions, then ______ and ______ .

[4.4] **14.** If f is a one-to-one function, then f^{-1} is obtained by ______ x and y and then solving for y.

*[4.5] **15.** $y = kx$ means that y varies ______________________ .

*[4.5] **16.** $y = k/x$ means that y varies ______________________ .

*[4.5] **17.** $p = k\frac{x^3}{y^2}$ means that ______________________ .

REVIEW PROBLEMS

[4.1] *In Problems 18–39 let* $f(x) = 5x - 3$ *and* $g(x) = 2x^2 - 9$. *Find the requested values.*

18. $f(3)$
19. $g(3)$
20. $f(10)$
21. $g(4)$
22. $f(t)$
23. $g(t)$
24. $g(t + 1)$
25. $f(t + 1)$
26. $f(x + h)$
27. $g(x + h)$
28. $g(3 - h)$
29. $f(5 - h)$
30. $3f(x)$
31. $f(3x)$
32. $2f(w) + 5$
33. $2f(w + 5)$
34. $f(x - h) - f(x)$
35. $g(x - h) - g(x)$
36. $\dfrac{g(t + h) - g(t)}{2}$
37. $\dfrac{f(t + h) - f(t)}{5}$
38. $\dfrac{f(x + h) - f(x)}{h}$
39. $\dfrac{g(x + h) - g(x)}{h}$

[4.2] *Graph the functions given in Problems 40–47.*

40. $f(x) = 4 - 3x$

41. $f(x) = -\frac{2}{3}x + \frac{5}{3}$

42. $f(x) = \begin{cases} x - 1 & \text{if } x \geq 3 \\ 5 - x & \text{if } x < 3 \end{cases}$

43. $f(x) = \begin{cases} 3x - 1 & \text{if } 0 < x < 1 \\ -5x + 7 & \text{if } 1 \leq x \leq 2 \\ -3 & \text{if } 2 < x < 5 \end{cases}$

44. $y - 3 = 2(x + 4)^2$

45. $f(x) = 3x^2 - 6x + 1$

46. $f(x) = |x - 3| + 2$

***47.** $f(x) = [\![x + 1]\!] - 2$

*[4.2] **48.** *Business* The handling charge on mail orders at a boutique is 5% of the amount if the order is less than \$30; otherwise it is \$1.50. Write a function that gives the mailing charge on an order of x dollars.

[4.3] *In Problems 49–60 let* $m(y) = 2y^2 + 3$ *and* $n(z) = 3z - 4$. *Find the requested values.*

49. $(m + n)(3)$
50. $(m - n)(1)$
51. $(mn)(0)$
52. $(m/n)(2)$
53. $(m \circ n)(3)$
54. $(n \circ m)(3)$
55. $(m + n)(x)$
56. $(m - n)(x)$
57. $(mn)(x)$

58. $(m/n)(x)$ **59.** $(m \circ n)(x)$ **60.** $(n \circ m)(x)$

[4.3] *In Problems 61–76 let* $F(x) = \dfrac{x+4}{2}$ *and* $G(x) = 5x - 4$. *Find the requested values.*

61. $(G + F)(2)$ **62.** $(G - F)(2)$ **63.** $(GF)(2)$

64. $(F/G)(2)$ **65.** $(G/F)(2)$ **66.** $(F \circ G)(2)$

67. $(G \circ F)(2)$ **68.** $(G + F)(x)$ **69.** $(G - F)(x)$

70. $(GF)(x)$ **71.** $(G/F)(x)$ **72.** $(F/G)(x)$

73. $(F \circ G)(x)$ **74.** $(G \circ G)(x)$ **75.** $\dfrac{G(x+h) - G(x)}{h}$

76. $\dfrac{F(x+h) - F(x)}{h}$

[4.4] *Determine whether the pairs of functions in Problems 77–86 are inverse functions.*

77. $f(x) = 2 - x;\ \ g(x) = x - 2$ **78.** $F(x) = 1 - x;\ \ G(x) = 1 - x$

79. $F(x) = 3x - 2;\ \ G(x) = \frac{2}{3}x + \frac{3}{2}$ **80.** $f(x) = 4x + 2;\ \ g(x) = \frac{1}{4}x - \frac{1}{2}$

81. $s(x) = x^2;\ \ t(x) = \sqrt{x}$ **82.** $s(x) = 3x^2;\ \ t(x) = \sqrt{\frac{1}{3}x}$

83. $S(x) = \dfrac{1}{x-1};\ \ T(x) = \dfrac{x+1}{x}$ **84.** $S(x) = \dfrac{2}{x};\ \ T(x) = \dfrac{x}{2}$

85. $f(x) = \dfrac{1}{x-3};\ \ g(x) = \dfrac{x+3}{x}$ **86.** $h(x) = \dfrac{x+2}{x};\ \ k(x) = \dfrac{2}{x-1}$

[4.4] *Find the inverse of each of the functions given in Problems 87–92.*

87. $f(x) = 2x - 5$ **88.** $f(x) = 3x + 2$ **89.** $g(x) = \dfrac{3}{x+1}$

90. $g(x) = \dfrac{7}{x+3}$ **91.** $h(x) = 3 - \frac{1}{2}x$ **92.** $h(x) = 2 - \frac{1}{3}x$

*[4.5] *Write each of the statements in Problems 93–98 as an equation.*

93. x varies directly as the square of y.

94. z varies inversely as w.

95. x varies directly as y and inversely as z.

96. P is directly proportional to the cube of t.

97. T is inversely proportional to the square of s.

98. B varies directly as the square of x and inversely as the cube of y.

*[4.5] *Solve each proportion in Problems 99–104.*

99. If x varies directly as the square of y, find x when $y = 4$ if it is known that $x = 500$ when $y = 10$.

---

*Optional

100. If z varies inversely as w, find z when $w = 10$ if it is known that $w = 5$ when $z = 20$.

101. If x varies directly as y and inversely as z, find x when y is 108 and z is 5 if it is known that $x = 1$ when $y = 54$ and $z = 6$.

102. If P is directly proportional to the cube of t, find P when $t = 5$ if it is known that $P = 27$ when $t = 3$.

103. Suppose T is inversely proportional to the square of s and an experiment shows that $T = 5$ when $s = 10$. Find T when $s = 15$.

104. *Physics* The pressure varies directly as the force applied and inversely as the area to which it is applied. If the pressure is 2000 kg/cm^2 when a force of 4 kg is applied to an area of .002 cm^2, what is the pressure when a force of 6 kg is applied to an area of .01 cm^2?

CHAPTER 5

Polynomial and Rational Functions

Leonhard Euler
(1707–1783)

The business of concrete mathematics is to discover the equations which express the mathematical laws of the phenomenon under consideration; and these equations are the starting point of the calculus, which must obtain from them certain quantities by means of others.

Auguste Comte
Positive Philosophy,
Book 1, Chapter 2

Historical Note

Leonhard Euler was the most prolific mathematician of all time. His works, if assembled, would fill 60 to 80 volumes. He was the first, in 1734, to use the notation $f(x)$ for functions. His tremendous powers of concentration were truly amazing. He often worked with several small children underfoot. When interrupted, he patiently answered questions or settled disputes, and then was able to continue his train of thought, sometimes even rocking a baby while at work on a difficult problem. Euler's father was a pastor whose avocation was mathematics, which had a profound influence on his son. Euler's early religious training gave him a simple, unquestioning faith that enabled him to accept his blindness in later years with courage. Euler's blindness for the last 12 years of his life did not impair his mathematical research. In fact, Euler is quoted as saying that after becoming blind, he was able to concentrate on his mathematics even better.

Chapter Overview

This chapter opens the door to higher mathematics by introducing you to the solution and graphing of polynomial equations of degree greater than 2. The key idea in this process is an understanding of synthetic division, which is introduced in Section 5.1. Toward the end of this chapter, we discuss the graphs of rational functions, which will prepare you for the types of functions you will see in calculus.

5.1 Synthetic Division and the Factor Theorem

The first part of this chapter is concerned with polynomial functions.

POLYNOMIAL FUNCTION

A function P is a **polynomial function** in x if

$$P(x) = a_n x^n + a_{n-1}x^{n-1} + a_{n-2}x^{n-2} + \cdots + a_1 x + a_0$$

where n is an integer greater than or equal to zero and the coefficients $a_0, a_1, a_2, \ldots, a_{n-1}, a_n$ are real numbers.

Notice that if $n = 2$ (degree 2), then $P(x)$ is a *quadratic function;* if $n = 1$ (degree 1), then $P(x)$ is a *linear function;* and if $n = 0$ (degree 0), then $P(x)$ is a *constant function*. This chapter focuses on polynomial functions with degree greater than 2 since constant, linear, and quadratic functions have been discussed previously.

The key to solving and graphing polynomial functions is a process called *synthetic division*. Consider $2x^4 - x^3 - 16x^2 - 3x + 18$ divided by $x - 1$:

$$\begin{array}{r l}
 & 2x^3 + x^2 - 15x - 18 \\
x - 1 \,\big) & 2x^4 - x^3 - 16x^2 - 3x + 18 \\
 & \underline{2x^4 - 2x^3} \\
 & \quad x^3 - 16x^2 \\
 & \quad \underline{x^3 - x^2} \\
 & \quad\quad -15x^2 - 3x \\
 & \quad\quad \underline{-15x^2 + 15x} \\
 & \quad\quad\quad -18x + 18 \\
 & \quad\quad\quad \underline{-18x + 18} \\
 & \quad\quad\quad\quad 0
\end{array}$$

If the division is rewritten without variables, as shown below, the repetition occurring in the division becomes obvious. First, the numbers in color show that the quotient is repeated within the array. Second, several lines of numbers are repetitions of the numbers directly above them.

```
            2    1 − 15 − 18
     ______________________
− 1 )2 − 1 − 16  −3   18
     2 − 2
     ─────
         1 − 16
         1  −1
     ──────────
           − 15  −3
           − 15  15
           ─────────
                − 18   18
                − 18   18
                ─────────
                        0
```

If the repetition is eliminated, the division appears as

```
            2    1 −  15 − 18
     ________________________
− 1 )2 − 1 − 16   −3    18
       − 2
     ─────
         1
            − 1
     ──────────
          − 15
                  15
          ──────────
                − 18
                        18
                ──────────
                         0
```

There are *still* repetitions, as shown in color above. If these numbers are deleted and the result compressed, all the essential elements of the division are retained. Next the numbers shown in color at the top are written at the bottom:

```
            2    1 − 15 − 18
− 1 )2 − 1 − 16 − 3   18
       − 2 − 1   15   18
     ───────────────────
```

−1	2	−1	−16	−3	18
		−2	−1	15	18
	2	1	−15	−18	0

Observe that the operations are performed in a zigzag fashion in this compact form:

−1	2	−1	−16	−3	18
	Bring down ↓	↓ Subtract	↓ Subtract	↓ Subtract	↓ Subtract
		−2	−1	15	18
	2	1	−15	−18	0

Multiply (−1)(2) Multiply (−1)(1) Multiply (−1)(−15) Multiply (−1)(−18)

Finally, if the **subtraction is replaced by addition** and **−1 is replaced by 1** (this is the idea of adding the opposite), the bottom line, which represents the quotient, is unchanged:

1⌋	2	−1	−16	−3	18
	Bring down ↓	Add 2	Add 1	Add −15	Add −18
	2	1	−15	−18	0

Multiply (1)(2); Multiply (1)(1); Multiply (1)(−15); Multiply (1)(−18)

This scheme is now in the format called **synthetic division:**

1⌋	2	−1	−16	−3	18
		2	1	−15	−18
	2	1	−15	−18	0

This synthetic division shows the polynomial $2x^4 - x^3 - 16x^2 - 3x + 18$ divided by $x - 1$, with a quotient of $2x^3 + x^2 - 15x - 18$ and a remainder of 0. Synthetic division is used for dividing a polynomial by a *linear factor* $x - r$. That is, the divisor must be first degree *and* must have a leading coefficient of *one*.

Example 1 Divide $6x^3 - 19x^2 + x + 6$ by $x - 3$.

Solution The divisor is of the form $x - r$, where $r = 3$, so to apply synthetic division, use the coefficients of the dividend and use 3 as the divisor:

3⌋	6	−19	1	6
		18	−3	−6
	6	−1	−2	0

3(6); 3(−1); 3(−2)

Thus, $6x^3 - 19x^2 + x + 6 = (x - 3)(6x^2 - 1x - 2)$, and $x - 3$ is a factor since the remainder is 0. □

Example 2 Divide $x^3 - 3x^2 + 4x - 1$ by $x + 2$.

Solution The divisor is of the form $x - r$, where $r = -2$ [Notice: $x + 2 = x - (-2)$.]

−2⌋	1	−3	4	−1
		−2	10	−28
	1	−5	14	−29

The quotient with remainder is: $x^2 - 5x + 14 - \dfrac{29}{x + 2}$ □

A polynomial is a sum of terms, usually arranged in descending powers, or degrees, of one variable. If $P(x)$ is an nth-degree polynomial function of a single variable x, it can be written as

$$P(x) = a_nx^n + a_{n-1}x^{n-1} + \cdots + a_2x^2 + a_1x + a_0 \qquad a_n \neq 0$$

Dividing $P(x)$ by $x - r$ will produce a polynomial quotient $Q(x)$ and a constant remainder R, which can be written

$$\frac{P(x)}{x-r} = Q(x) + \frac{R}{x-r} \qquad \text{or} \qquad P(x) = (x-r)Q(x) + R$$

If the division "comes out even," the remainder is 0 and $x - r$ is a factor of $P(x)$. That is,

$$P(x) = (x-r)Q(x)$$

Therefore, in the solution of a polynomial equation $P(x) = 0$,

$$x - r = 0 \qquad \text{or} \qquad Q(x) = 0$$

Thus, r is a root and $Q(x) = 0$ is called the **depressed equation,** which is solved separately.

To summarize:

FACTOR THEOREM

> For a polynomial $P(x)$, $P(x) = (x-r)Q(x) + R$, and if $R = 0$, then $x - r$ is a factor of $P(x)$ and r is a root of $P(x) = 0$.

Example 3 Is $\frac{1}{2}$ a root of $2x^3 + x^2 - 13x + 6 = 0$?

Solution

$$\begin{array}{r|rrrr} \frac{1}{2} & 2 & 1 & -13 & 6 \\ & & 1 & 1 & -6 \\ \hline & 2 & 2 & -12 & 0 \end{array}$$

Yes, $\frac{1}{2}$ is a root since $R = 0$. Now, we can use this result to find the other roots:

$$\begin{aligned} 2x^3 + x^2 - 13x + 6 &= \left(x - \tfrac{1}{2}\right)(2x^2 + 2x - 12) \\ &= \left(x - \tfrac{1}{2}\right)(2)(x^2 + x - 6) \\ &= (2x-1)(x+3)(x-2) \end{aligned}$$

The complete solution is $\left\{\frac{1}{2}, -3, 2\right\}$. □

In Example 3 above, $x = \frac{1}{2}$ was found to be a solution. Then the depressed equation $2x^2 + 2x - 12 = 0$ was solved to find the other roots of the original equation. If the depressed equation is not of degree 2 or less, it may have to be attacked synthetically.

Example 4 Divide $2x^4 - 9x^2 + 23x - 7$ by $x + 3$, and write the quotient and remainder.

Solution Since there is no x^3 term, a coefficient of 0 is used in that position:

$$\begin{array}{r|rrrrr} -3 & 2 & 0 & -9 & 23 & -7 \\ & & -6 & 18 & -27 & 12 \\ \hline & 2 & -6 & 9 & -4 & 5 \end{array}$$

$$\mathbf{2x^3 - 6x^2 + 9x - 4 + \frac{5}{x+3}}$$

□

Example 5 Find k so that $x - 2$ is a factor of $2x^3 - 3x^2 + 4kx - 2$.

Solution

$$\begin{array}{r|rrrr} 2 & 2 & -3 & 4k & -2 \\ & & 4 & 2 & 8k+4 \\ \hline & 2 & 1 & 4k+2 & 8k+2 \end{array}$$

The remainder $8k + 2$ must be zero if $x - 2$ is a factor, so

$$\begin{aligned} 8k + 2 &= 0 \\ 8k &= -2 \\ \mathbf{k} &= \mathbf{-\frac{1}{4}} \end{aligned}$$

□

Example 6 Reduce: $\dfrac{2x^3 - 9x^2 + 7x + 6}{(x+3)(x-2)}$

Solution If the rational expression reduces, then $x + 3$ or $x - 2$, or both, must be factors of the numerator.

$$\begin{array}{r|rrrr} -3 & 2 & -9 & 7 & 6 \\ & & -6 & 45 & -156 \\ \hline & 2 & -15 & 52 & -150 \end{array}$$

$x + 3$ is *not* a factor.

$$\begin{array}{r|rrrr} 2 & 2 & -9 & 7 & 6 \\ & & 4 & -10 & -6 \\ \hline & 2 & -5 & -3 & 0 \end{array}$$

$x - 2$ is a factor.

$$\begin{aligned} \frac{2x^3 - 9x^2 + 7x + 6}{(x+3)(x-2)} &= \frac{(x-2)(2x^2 - 5x - 3)}{(x+3)(x-2)} \\ &= \frac{(x-2)(2x+1)(x-3)}{(x+3)(x-2)} \\ &= \mathbf{\frac{(2x+1)(x-3)}{x+3}} \end{aligned}$$

Factor quadratic as usual—you do not need to use synthetic division.

□

Problem Set 5.1

A *Use synthetic division to find the quotient in Problems 1–10. The remainder in each case is 0.*

1. $\dfrac{3x^3 - 2x^2 + 4x - 75}{x - 3}$

2. $\dfrac{3x^3 + 2x^2 - 4x + 8}{x + 2}$

3. $\dfrac{x^4 - 6x^3 + x^2 - 8}{x + 1}$

4. $\dfrac{x^4 - 3x^3 + x + 6}{x - 2}$

5. $\dfrac{2x^4 - 15x^2 + 8x - 3}{x + 3}$

6. $\dfrac{2x^4 - 15x - 2}{x - 2}$

7. $\dfrac{4x^5 - 3x^4 - 5x^3 + 4}{x - 1}$

8. $\dfrac{4x^3 - 3x^2 - 5x + 2}{x + 1}$

9. $\dfrac{3x^3 - 2x^2 + 4x - 24}{x - 2}$

10. $\dfrac{3x^3 + 2x^2 + 4x + 24}{x + 2}$

Use synthetic division to find the quotient of the first polynomial divided by the second polynomial in Problems 11–22.

11. $x^3 + 2x^2 - x - 2;\ x - 1$
12. $x^3 + 2x^2 - x - 2;\ x + 1$
13. $x^3 + 2x^2 - x - 2;\ x + 2$
14. $x^3 - 2x^2 - x + 2;\ x + 1$
15. $x^3 - 2x^2 - x + 2;\ x - 1$
16. $x^3 - 2x^2 - x + 2;\ x - 2$
17. $x^3 - 4x^2 - 11x + 30;\ x - 2$
18. $x^3 - 4x^2 - 11x + 30;\ x + 3$
19. $x^3 - 4x^2 - 11x + 30;\ x - 5$
20. $x^3 - 8x^2 + x + 42;\ x + 2$
21. $x^3 - 8x^2 + x + 42;\ x - 3$
22. $x^3 - 8x^2 + x + 42;\ x - 7$

Use synthetic division to find the quotient and remainder in Problems 23–30.

23. $\dfrac{2x^3 - 3x^2 + 4x - 10}{x - 2}$

24. $\dfrac{3x^3 - 7x^2 - 5x + 2}{x - 3}$

25. $\dfrac{x^4 - 5x^3 + 2x^2 - x + 3}{x - 2}$

26. $\dfrac{x^4 - 3x^3 - 4x^2 + 2x - 5}{x - 4}$

27. $\dfrac{2x^4 + 5x^3 + 2x^2 + 5x + 2}{x + 2}$

28. $\dfrac{5x^4 + 10x^3 - 20x^2 - 12x - 2}{x + 3}$

29. $\dfrac{x^4 - 20x^2 - 10x - 50}{x - 5}$

30. $\dfrac{x^5 - 3x^4 + 2x^2 - 5}{x + 2}$

B *Use synthetic division to find the quotient of the first polynomial divided by the second polynomial in Problems 31–50.*

31. $6x^3 + 17x^2 - 5x - 6;\ x + 3$
32. $6x^3 + 7x^2 - 30x + 9;\ x + 3$
33. $6x^3 + 17x^2 - 5x - 6;\ x + \frac{1}{2}$
34. $6x^3 + 7x^2 - 30x + 9;\ x - \frac{1}{3}$
35. $6x^3 + 17x^2 - 5x - 6;\ x - \frac{2}{3}$
36. $6x^3 + 7x^2 - 30x + 9;\ x - \frac{3}{2}$
37. $4x^4 - 4x^3 - 3x^2 + 2x + 1;\ x - 1$
38. $4x^4 - 4x^3 - 3x^2 + 4x - 1;\ x - 1$
39. $4x^4 - 4x^3 - 3x^2 + 2x + 1;\ x + \frac{1}{2}$
40. $4x^4 - 4x^3 - 3x^2 + 4x - 1;\ x - \frac{1}{2}$
41. $6x^4 + x^3 - 55x^2 - 9x + 9;\ x + 3$
42. $6x^4 + 7x^3 - 27x^2 - 28x + 12;\ x + 2$
43. $6x^4 + x^3 - 55x^2 - 9x + 9;\ x - 3$
44. $6x^4 + 7x^3 - 27x^2 - 28x + 12;\ x - 2$
45. $6x^4 + x^3 - 55x^2 - 9x + 9;\ x + \frac{1}{2}$
46. $6x^4 + 7x^3 - 27x^2 - 28x + 12;\ x - \frac{1}{3}$
47. $6x^4 + x^3 - 55x^2 - 9x + 9;\ x - \frac{1}{3}$

48. $6x^4 + 7x^3 - 27x^2 - 28x + 12;\ \ x + \frac{3}{2}$

49. $6x^4 + x^3 - 7x^2 - x + 1;\ \ x - \frac{1}{3}$

50. $6x^4 + x^3 - 7x^2 - x + 1;\ \ x + \frac{1}{2}$

Reduce the rational expressions in Problems 51–60.

51. $\dfrac{2x^3 - 3x^2 - 8x - 3}{(x - 3)(x + 2)}$

52. $\dfrac{2x^3 - 3x^2 - 8x - 3}{(x - 2)(x + 1)}$

53. $\dfrac{2x^3 - 3x^2 - 8x - 3}{(2x - 3)(x + 1)}$

54. $\dfrac{2x^4 + 5x^3 - 16x^2 - 45x - 18}{(x - 1)(x + 2)}$

55. $\dfrac{2x^4 + 5x^3 - 16x^2 - 45x - 18}{(x - 3)(x + 4)}$

56. $\dfrac{2x^4 + 5x^3 - 16x^2 - 45x - 18}{(x + 1)(x + 3)}$

57. $\dfrac{3x^3 - 2x^2 - 27x + 18}{(x - 3)(x + 2)}$

58. $\dfrac{3x^3 - 2x^2 - 27x + 18}{(x - 3)(3x - 2)}$

59. $\dfrac{6x^4 - x^3 - 25x^2 + 4x + 4}{(x - 3)(x + 2)}$

60. $\dfrac{6x^4 - x^3 - 25x^2 + 4x + 4}{(x - 2)(3x + 1)}$

C **61.** Find the value of k for which $x + 1$ is a factor of $4x^3 - 4x^2 + kx + 4$.

62. Find the value of h for which $x - 2$ is a factor of $3x^3 - 5x^2 + hx + 4$.

63. What is m in $3x^3 + mx^2 - 7x + 6$ if $x + 3$ is a factor?

64. What is n in $4x^3 - 2nx^2 - 8x + 6$ if $x - 3$ is a factor?

65. If $x - 4$ is a factor of $6x^3 + 13x^2 + 2kx - 40$, then what is the value of k?

66. If $x + 1$ is a factor of $3x^4 - x^3 + hx^2 + x + 2$, then what is the value of h?

67. Find m so that $x + 1$ is a factor of $5x^3 + m^2x^2 + 2mx - 3$.

68. Find n so that $x - 4$ is a factor of $x^3 - n^2x^2 - 8nx - 16$.

69. Find k and m so that $x^3 - mx^2 + 2x - 8k$ is divisible by $(x - 1)(x + 2)$.

70. Find two values of k in $x^6 + 2kx^5 - kx^4 - 5k^2x^3 - 11x^2 + k^4x + 3k^2$ if $x - k$ is a factor.

5.2 Graphing Polynomial Functions

As we mentioned in the last section, dividing a polynomial $P(x)$ by the binomial $x - r$ produces a polynomial quotient $Q(x)$ and a constant remainder R. This can be written as

$$P(x) = (x - r)Q(x) + R$$

Let $x = r$. Then

$$P(r) = (r - r)Q(r) + R$$
$$P(r) = 0 \cdot Q(r) + R$$
$$P(r) = R$$

It appears that the remainder R is the value of the polynomial $P(x)$ at $x = r$. This result is called the **Remainder Theorem.**

REMAINDER THEOREM

> If $P(x)$ is a polynomial and $P(x) = (x - r)Q(x) + R$, then $P(r) = R$.

Example 1 If $P(x) = 2x^3 - x^2 - 18x + 9$, find $P(2)$.

Solution

$$\begin{array}{r|rrrr} 2 & 2 & -1 & -18 & 9 \\ & & 4 & 6 & -24 \\ \hline & 2 & 3 & -12 & \mathbf{-15 = R = P(2)} \end{array}$$

□

Example 2 If $P(x) = x^3 - x^2 + x + 1$, find $P(1)$.

Solution

$$\begin{array}{r|rrrr} 1 & 1 & -1 & 1 & 1 \\ & & 1 & 0 & 1 \\ \hline & 1 & 0 & 1 & \mathbf{2 = R = P(1)} \end{array}$$

□

The Remainder Theorem can be used to tabulate values for a function, in order to sketch its graph. Synthetic division is used to obtain the results with a minimum of calculation. When synthetic division is used to evaluate $P(r)$, as in the examples, it is called **synthetic substitution.** Suppose $y = P(x)$ is a polynomial equation and you wish to sketch its graph. If $y = P(x) = (x - r)Q(x) + R$, then $P(r) = R$ and $(x, y) = [x, P(x)] = (r, R)$.

Example 3 Sketch the graph of $y = 2x^4 - 5x^3 + 5x - 2$.

Solution First, evaluate the equation at each integer from -3 to $+3$ to get some initial points for the sketch. A shortened form of synthetic substitution is used, since each value is substituted into the original polynomial and the depressed equation is rarely of interest.

	2	−5	0	5	−2	
−3	2	−11	33	−94	280	(−3, 280)
−2	2	−9	18	−31	60	(−2, 60)
−1	2	−7	7	−2	0	(−1, 0)
0					−2	(0, −2)
1	2	−3	−3	2	0	(1, 0)
2	2	−1	−2	1	0	(2, 0)
3	2	1	3	14	40	(3, 40)

Plotting these points (see the figure), does not tell a great deal about the behavior of the graph. Additional points are needed, so values halfway between the values plotted are found. Notice that decimals, rather than fractions, are used since a calculator is very convenient and efficient for these calculations.

	2	−5	0	5	−2	
−1.5	2	−8	12	−13	17.5	(−1.5, 17.5)
−.5	2	−6	3	3.5	−3.75	(−.5, −3.75)
.5	2	−4	−2	4	0	(.5, 0)
1.5	2	−2	−3	.5	−1.25	(1.5, −1.25)
2.5	2	0	0	5	10.5	(2.5, 10.5)
.75	2	−3.5	2.6	3.03	.27	(.75, .27) Approximate values

Since graphs of polynomial functions are "smooth" or continuous, the additional points seem to indicate the behavior well enough to be able to sketch the curve, as shown in the figure, with greater assurance.

□

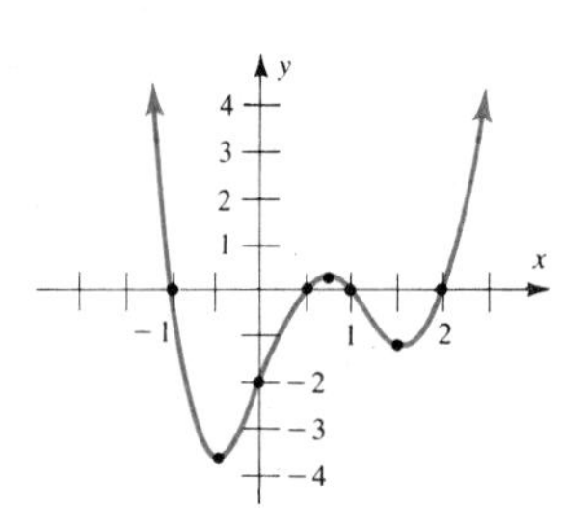

Example 4 Sketch the graph of $y = 2x^3 - 3x^2 - 12x + 17$.

Solution Again, start with integer values:

	2	−3	−12	17	
−3	2	−9	15	−28	(−3, −28)
−2	2	−7	2	13	(−2, 13)
−1	2	−5	−7	24	(−1, 24)
0				17	(0, 17)
1	2	−1	−13	4	(1, 4)
2	2	1	−10	−3	(2, −3)
3	2	3	−3	8	(3, 8)
4	2	5	8	49	(4, 49)

Then plot the points. The scales on the axes will often have to be adjusted to accommodate the values obtained.

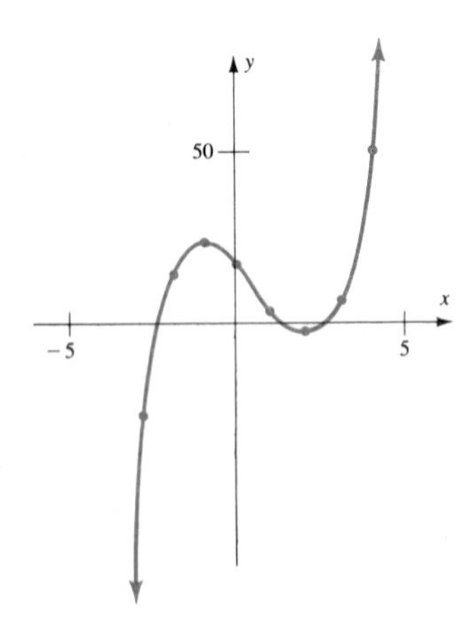

Connect the points to create a smooth continuous curve. Of course, our curve is an approximation. Locating additional points would give a better approximation, but it would still be an approximation.

□

The graphs of polynomial functions have *turning points* where the functions change from increasing to decreasing or from decreasing to increasing. A polynomial of degree n has at most $n - 1$ turning points. In calculus you will find how to locate the exact position of these turning points. For now, however, you will need to rely on synthetic division and plotting points to sketch graphs. If a polynomial function can be factored into linear factors, the number of plotted points can be reduced as illustrated by Example 5.

Example 5 Sketch the graph of $y = (x + 1)(x - 2)(3x - 2)$.

Solution Begin by locating the *critical values* for the polynomial function. Recall that these are x values for which $y = 0$:

$$(x + 1)(x - 2)(3x - 2) = 0 \quad \leftarrow y = 0$$

$$x = -1, 2, \frac{2}{3}$$

These critical values divide the x-axis into four regions:

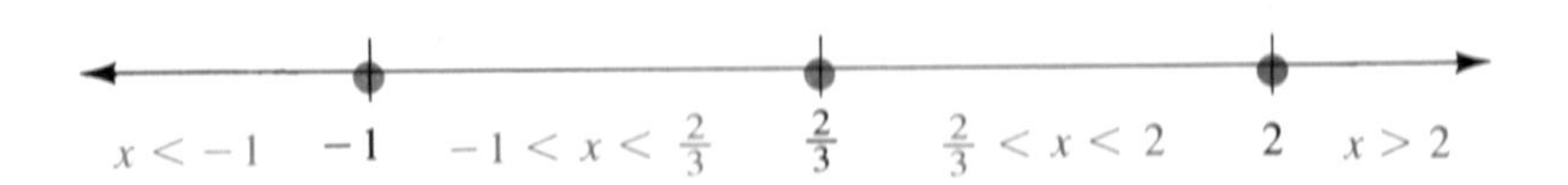

Since $y = 0$ for $x = -1$, $x = \frac{2}{3}$, and $x = 2$, it follows that y must be either positive or negative in each of the four regions. That is, you need to determine whether

y is positive or negative in *each* of the four regions. To do this, select an x value in *each* region and substitute that value into the function, as shown below:

Factors

$x + 1$:	$-$	$+$	$+$	$+$
$x - 2$:	$-$	$-$	$-$	$+$
$3x - 2$:	$-$	$-$	$+$	$+$
	NEG -1	POS $\frac{2}{3}$	NEG 2	POS

Test value, say $x = -10$

For the interval $x < -1$, select *any* number less than -1; we select $x = -10$. Then

$x + 1$: $-10 + 1$ is negative; write $-$
$x - 2$: $-10 - 2$ is negative; write $-$
$3x - 2$: $3(-10) - 2$ is negative; write $-$

(NEG)(NEG)(NEG) = NEG — Write NEG for the interval below number line.

Repeat these steps for *each* interval. Now, wherever you have labeled the number line as NEG, the graph is below the axis; and wherever you have labeled it POS, the graph is above the axis.

Plot some points using synthetic division to draw the graph, as shown in the figure. Synthetic division requires the coefficients in polynomial (not factored) form. Thus,

$$y = (x + 1)(x - 2)(3x - 2)$$
$$= 3x^3 - 5x^2 - 4x + 4$$

	3	-5	-4	4	
-2	3	-11	18	-32	Need a point to the left of the critical value -1.
$-.3$	3	-5.9	-2.2	4.7	Need a point between -1 and 0; plot $(-.3, 4.7)$.
0				4	Plot (0, 4).
1	3	-2	-6	-2	Plot $(1, -2)$.
1.5	3	$-.5$	-4.8	-3.1	Need a point between 1 and 2; plot $(1.5, -3.1)$.
3	3	4	8	28	

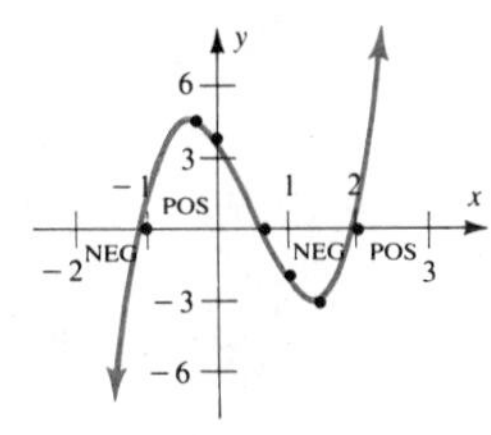

Graph of $y = (x + 1)(x - 2)(3x - 2)$
or $y = 3x^3 - 5x^2 - 4x + 4$ □

Problem Set 5.2

A *In Problems 1–10 find the values specified for the given functions by synthetic substitution.*

1. $f(x) = x^3 - 2x^2 + x - 5$; $f(1), f(2), f(-1), f(-3), f(10)$
2. $g(x) = x^3 + 4x^2 - 3x + 2$; $g(1), g(2), g(-1), g(-3), g(5)$
3. $h(y) = 2y^4 - 7y^3 + 5y^2 + 3y + 1$; $h(-1), h(-2), h\left(\frac{1}{2}\right)$
4. $k(z) = 6z^4 + 4z^3 - 5z^2 + 4z - 2$; $k(-2), k(1), k\left(\frac{1}{3}\right)$
5. $m(x) = 6x^4 - 5x^3 + 7x + 4$; $m(2), m\left(-\frac{2}{3}\right), m\left(-\frac{1}{2}\right)$
6. $n(x) = 8x^5 - 2x^4 + 10x^3 - 5x - 4$; $n(1), n(-1), n\left(-\frac{3}{4}\right)$
7. $f(x) = (x - 2)(x + 3)(2x - 5)$; $f(2), f(-3), f\left(\frac{5}{2}\right), f(1), f(-2)$
8. $g(x) = (x - 1)(x - 4)(2x + 1)$; $g(1), g(4), g\left(-\frac{1}{2}\right), g(-2), g(-1)$
9. $h(x) = (x + 1)(x + 2)(3x - 1)$; $h(-1), h(1), h(-2), h(2), h\left(\frac{1}{3}\right)$
10. $k(x) = x(x - 3)(x + 3)$; $k(0), k(1), k(2), k(3), k(-3)$

B *Sketch the graph of each polynomial function in Problems 11–30.*

11. $f(x) = x^3 - 3x^2 + 10$
12. $g(x) = x^3 + 3x^2 + 11$
13. $y = 3x^4 - x^3 - 14x^2 + 4x + 8$
14. $y = 5x^4 + 3x^3 - 22x^2 - 12x + 8$
15. $y = x^4 - x^3 - 3x^2 + 2x + 4$
16. $y = x^4 - 2x^2 - 4x + 3$
17. $g(x) = 2x^3 - 3x^2 - 12x + 3$
18. $h(x) = 2x^3 + 3x^2 - 12x + 48$
19. $y = x^4 - 7x^2 - 2x + 2$
20. $y = x^4 - 14x^3 + 58x^2 - 46x - 9$
21. $y = (x - 1)(x + 1)(x + 3)$
22. $y = (x - 1)(x - 4)(x + 3)$
23. $y = (x + 1)(x + 3)(2x - 5)$
24. $y = (x - 1)(x - 4)(2x + 1)$
25. $y = (x + 1)(x + 2)(3x - 1)$
26. $y = x(x - 3)(x + 3)$
27. $y = 3x^2(x - 3)(x + 1)$
28. $y = 5x^2(x - 4)(x + 2)$
29. $y = x^2(x^2 - 1)$
30. $y = x^2(x^2 - 4)$

C *Sketch the graph of each polynomial function in Problems 31–34.*

31. $y = x^5 + x^4 - 5x^3 - 5x^2 + 4x + 4$
32. $y = x^5 + 2x^4 - 5x^3 - 10x^2 + 4x + 8$
33. $y = x^5 - 15x^3 + 10x^2 + 60x - 12$
34. $y = x^5 - 3x^4 - 6x^3 + 10x^2 + 21x - 18$

5.3 Rational Root Theorem

The roots, or solutions, of polynomial equations are also sometimes called **zeros of the polynomial.** If $P(x)$ is a polynomial, then r is called a *zero* of P if $P(r) = 0$. As points were plotted to sketch a polynomial function, it was evident that if a polynomial is positive at a and negative at b, then somewhere between a and b it must have at least one real zero.

LOCATION THEOREM

> If $P(x)$ is a polynomial, and $P(a)$ and $P(b)$ are opposite in sign, then there is at least one real zero on the interval between a and b.

A theorem that can be used as a guide to find the zeros of a polynomial is the **Rational Root Theorem.** This theorem allows the construction of a list of all possible rational roots *before* work begins.

RATIONAL ROOT THEOREM

> If $P(x) = a_nx^n + \cdots + a_2x^2 + a_1x + a_0$ has integer coefficients and $P(x)$ has a rational zero m/n (in lowest terms), then m is a factor of a_0 and n is a factor of a_n.

This theorem states that all rational roots of $P(x) = 0$ are of the form m/n, where the constant term is divisible by m and the leading coefficient is divisible by n. To use this theorem, take all factors of a_0, and they will be the integer possibilities. Then divide these integers by the factors of a_n to obtain a full list of the rational possibilities.

Example 1 List all possible rational roots for $6x^3 - 7x^2 + 11x + 35 = 0$.

Solution

$a_0 = 35$ with factors	$a_n = 6$ with factors
$\pm 1, \pm 5, \pm 7, \pm 35$	$\pm 1, \pm 2, \pm 3, \pm 6$

All possibilities are the integer factors of a_0 divided in turn by each factor of a_n

$$\pm 1, \quad \pm 5, \quad \pm 7, \quad \pm 35$$

$$\pm\frac{1}{2}, \quad \pm\frac{5}{2}, \quad \pm\frac{7}{2}, \quad \pm\frac{35}{2}$$

$$\pm\frac{1}{3}, \quad \pm\frac{5}{3}, \quad \pm\frac{7}{3}, \quad \pm\frac{35}{3}$$

$$\pm\frac{1}{6}, \quad \pm\frac{5}{6}, \quad \pm\frac{7}{6}, \quad \pm\frac{35}{6}$$

□

The Location Theorem shows where to look for rational zeros. The Rational Root Theorem lists the possibilities for rational zeros. A final theorem sets upper and lower bounds on the possibilities. An **upper bound** of a set of numbers is a number that is greater than or equal to every number in the set; a **lower bound** is a number that is less than or equal to every number in the set.

UPPER AND LOWER BOUND THEOREM

If all the sums are the *same sign* in the synthetic division of $P(x)$ by $x - a$, then a is an **upper bound** for the roots of $P(x) = 0$ when $a \geq 0$.

If the sums *alternate in sign* in the synthetic division of $P(x)$ by $x - b$, then b is a **lower bound** for the roots of $P(x) = 0$ when $b \leq 0$.

Example 2 Determine upper and lower bounds for the equation of Example 1.

Solution

	6	−7	11	35	
0	6	−7	11	35	
1	6	−1	10	45	
5	6	23	126	665	Upper bound
	+	+	+	+	Because $x = 5$ is an upper bound, all listed possibilities that are greater than 5 need not be considered further.
−1	6	−13	24	11	
−5	6	−37	196	−945	Lower bound
	+	−	+	−	

We have found that 5 is an upper bound and −5 is a lower bound, but there are an infinite number of upper and lower bounds. That is, since 5 is an upper bound, then 6 must also be an upper bound. In fact, every number larger than 5 is an upper bound. In calculus you will consider the *least upper bound* and *greatest lower bound.* □

Example 3 Solve: $2x^5 - 23x^3 + x^2 + 61x + 39 = 0$

Solution All possible rational roots are

$$\pm 1, \quad \pm 3, \quad \pm 13, \quad \pm 39, \quad \pm\frac{1}{2}, \quad \pm\frac{3}{2}, \quad \pm\frac{13}{2}, \quad \pm\frac{39}{2}$$

No other rational roots are possible, so begin by trying these.

	2	0	−23	1	61	39		
1	2	2	−21	−20	41	80		Do not forget to insert the zero coefficient for the missing x^4 term.
−1	2	−2	−21	22	39	0	$x = -1$	$R = 0$, so use the depressed equation.
−1	2	−4	−17	39	0		$x = -1$	Try −1 again and again until it does not work, or you may miss a root that occurs more than once.
−1	2	−6	−11	50				
3	2	2	−11	6				
−3	2	−10	13	0			$x = -3$	

$$2x^2 - 10x + 13 = 0$$

Use the quadratic formula.

$$x = \frac{10 \pm 2i}{4}$$

$$x = \frac{5 \pm i}{2}$$

$$\left\{ -1, -1, -3, \frac{5+i}{2}, \frac{5-i}{2} \right\}$$

□

In Example 3, −1 is a **multiple root** of the equation. This means that the polynomial has two factors of $(x + 1)$.

$$\begin{aligned} 2x^5 - 23x^3 + x^2 + 61x + 39 &= (x+1)(x+1)(x+3)(2x^2 - 10x + 13) \\ &= (x+1)^2(x+3)(2x^2 - 10x + 13) \end{aligned}$$

In this case, the polynomial has a zero of **multiplicity** 2.

Example 4 Solve: $8x^3 - 12x^2 - 66x + 35 = 0$

Solution Possibilities:

$$\pm 1, \quad \pm 5, \quad \pm 7, \quad \pm 35$$

$$\pm\frac{1}{2}, \quad \pm\frac{5}{2}, \quad \pm\frac{7}{2}, \quad \pm\frac{35}{2}$$

$$\pm\frac{1}{4}, \quad \pm\frac{5}{4}, \quad \pm\frac{7}{4}, \quad \pm\frac{35}{4}$$

$$\pm\frac{1}{8}, \quad \pm\frac{5}{8}, \quad \pm\frac{7}{8}, \quad \pm\frac{35}{8}$$

Try the integer possibilities first. Arrange the possibilities in numerical order so that the Location Theorem can be more easily applied.

	8	−12	−66	35	
5	8	28	74	405	Upper bound
1	8	−4	−70	−35	*Root* between 5 and 1
0				35	*Root* between 1 and 0
−1	8	−20	−46	81	*Root* between −1 and −5
−5	8	−52	194	−935	Lower bound

The Location Theorem tells you there are three intervals containing roots: between 5 and 1, between 1 and 0, and between −1 and −5. The interval between 5 and 1 has only five possibilities: $\frac{5}{2}, \frac{7}{2}, \frac{5}{4}, \frac{7}{4}$, and $\frac{35}{8}$. Check them one at a time:

	8	−12	−66	35	
$\frac{5}{2}$	8	8	−46	−80	
$\frac{7}{2}$	8	16	−10	0	$x = \frac{7}{2}$

$$\begin{aligned} 8x^2 + 16x - 10 &= 0 \\ 4x^2 + 8x - 5 &= 0 \\ (2x - 1)(2x + 5) &= 0 \\ x = \frac{1}{2} \quad \text{or} \quad x &= -\frac{5}{2} \end{aligned}$$

The depressed equation is quadratic, so is now easy to factor directly.

$$\left\{\mathbf{\frac{7}{2}, \frac{1}{2}, -\frac{5}{2}}\right\}$$

□

Example 5 Solve: $x^4 + 2x^3 - 4x^2 - 2x + 3 > 0$

Solution First factor the polynomial $f(x) = x^4 + 2x^3 - 4x^2 - 2x + 3$. The possibilities are ± 1 and ± 3 from the Rational Root Theorem.

	1	2	−4	−2	3	
3	1	5	11	31	96	Upper bound
1	1	3	−1	−3	0	$x = 1$
1	1	4	3	0		$x = 1$

$x^2 + 4x + 3 = 0$ This is the depressed equation.
$(x + 1)(x + 3) = 0$
$x = -1$ or $x = -3$

Now write the inequality in factored form:

$$(x - 1)^2(x + 1)(x + 3) > 0$$
$$(x - 1)(x - 1)(x + 1)(x + 3) > 0$$

The critical values are 1, -1, and -3, as shown in the figure:

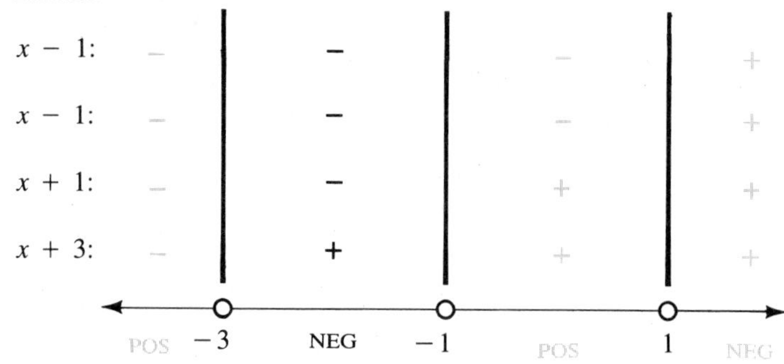

The solution is the part shown in color: $\boldsymbol{x < -3}$ **or** $\boldsymbol{-1 < x < 1}$ **or** $\boldsymbol{x > 1}$. This answer could also be stated as $\boldsymbol{x < -3}$ **or** $\boldsymbol{x > -1, x \neq 1}$. □

Problem Set 5.3

Solve the polynomial equations and inequalities in Problems 1–30.

A

1. $6x^3 + 5x^2 - 2x - 1 = 0$
2. $10x^3 - 7x^2 - 4x + 1 = 0$
3. $x^4 + x^3 - 7x^2 - 13x - 6 = 0$
4. $x^4 - 2x^3 - 8x^2 + 18x - 9 = 0$
5. $x^4 + 2x^3 - 13x^2 - 14x + 24 = 0$
6. $x^4 + 2x^3 - 16x^2 - 2x + 15 = 0$
7. $x^4 - 4x^3 + 4x - 1 = 0$
8. $x^4 - 4x^2 - 4x - 1 = 0$
9. $x^4 - 4x^3 + 6x^2 - 4x + 1 = 0$
10. $x^4 - 4x^3 - 2x^2 + 4x + 1 = 0$
11. $v^4 + 2v^3 - 13v^2 - 14v + 24 = 0$
12. $w^4 - 2w^3 - 13w^2 + 14w + 24 = 0$
13. $8a^3 - 10a^2 - 17a + 10 < 0$
14. $10b^3 - b^2 - 17b - 6 > 0$
15. $6x^3 - 19x^2 + 4x - 3 = 0$
16. $6y^3 - 17y^2 - 5y + 6 = 0$
17. $2p^3 - 9p^2 + p + 12 \leq 0$
18. $3q^3 - 19q^2 + 4q + 12 \geq 0$
19. $x^5 + 3x^4 + 4x = 5x^3 + 15x^2 - 12$
20. $x^5 + 30x^2 + 9x = 3x^4 + 10x^3 + 27$

B

21. $3x^4 + 14x^3 + 14x^2 = 8(x + 1)$
22. $x^4 - 12x^3 - 13 = 6(3 - 2x - 5x^2)$
23. $2x^4 - x^3 + 2x = 1$
24. $2x^3 + 4x + 3 = 3x^2$
25. $6x^3 - 35x^2 - 7x + 6 > 0$
26. $6x^3 + 11x^2 - 57x - 20 \leq 0$
27. $12x^3 + 4x^2 - 17x + 6 = 0$
28. $12x^3 + 8x^2 - 13x + 3 = 0$
29. $10x^3 + 11x^2 < 2x + 3$
30. $12x^3 - 10x^2 \leq 4x - 2$

Show that the equations in Problems 31–34 have no rational roots.

31. $x^3 - 2x^2 + 3x - 4 = 0$

32. $2x^3 + 5x^2 - 3x + 1 = 0$

33. $x^4 + 4x^3 - x^2 - 2x + 3 = 0$

34. $3x^4 - x^3 + 4x^2 + 2x - 2 = 0$

35. The dimensions of a rectangular box are consecutive integers, and its volume is 504 ft^3. Find the dimensions of the box.

36. A slice 1 inch thick is cut from a cube, leaving a volume of 180 cubic inches. What was the length of a side of the original cube?

37. The volume of a cube is doubled when the length is increased 3 in., width is increased 6 in., and height is decreased 2 in. What is the length of an original side?

38. If the length and width are increased by 4 cm each and the height is reduced by 2 cm, the volume of a cube is doubled. What is the length of an original side?

39. The hypotenuse of a right triangle is 9 cm longer than one of its legs. If the area of the figure is 60 cm^2, then what is the length of each side if they are integers?

40. A side of a right triangle is 1 in. shorter than the hypotenuse. The area of the triangle is 30 in.^2. What is the length of each side if they are integers?

C *Solve the polynomial equations and inequalities in Problems 41–46.*

41. $18x^5 + 3x^4 + 32x^3 + 5x^2 = 8x + 2$

42. $60x^4 + 88x^3 + 60 = 213x^2 + 121x$

43. $6 - 16x^4 < 5x(8x^2 + 9x + 2)$

44. $2[3(x^4 + 1) - 19x^2] < 5x(x^2 + 1)$

45. $12x^5 + 4x^4 - 83x^3 + 47x^2 + 20x - 12 = 0$

46. $24x^5 - 22x^4 + 25x^3 - 20x^2 + x + 2 = 0$

5.4 Fundamental Theorem of Algebra

The coefficients of a polynomial yield much valuable information. For example, the Rational Root Theorem uses the coefficients to provide a list of possible rational roots. However, more information can be obtained from the coefficients. The *number* of roots can be found by looking at the changes of signs in the coefficients using a result called **Descartes' Rule of Signs.**

DESCARTES' RULE OF SIGNS

If $P(x)$ is a polynomial arranged in descending powers of x with real coefficients, then:

1. The number of **positive** real roots of $P(x) = 0$ is equal to the number of variations in sign in $P(x)$ or is less than that number by an even counting number.
2. The number of **negative** real roots of $P(x) = 0$ is equal to the number of variations in sign in $P(-x)$ or is less than that number by an even counting number.

Example 1 Use Descartes' Rule of Signs to investigate the nature of the roots of

$$x^5 + 3x^4 - 2x^3 - x^2 + 4x - 5 = 0$$

Solution

$$P(x) = x^5 + 3x^4 - 2x^3 - x^2 + 4x - 5$$

+ + − − + − First, note signs of the coefficients.

1 2 3 Then count the number of variations of signs.

There are three variations, so there are **either three positive real roots or one positive real root.**

$$P(-x) = (-x)^5 + 3(-x)^4 - 2(-x)^3 - (-x)^2 + 4(-x) - 5$$ First, find $P(-x)$.

$$= -x^5 + 3x^4 + 2x^3 - x^2 - 4x - 5$$

− + + − − − Count variations.

1 2

There are two variations, so there are **either two negative real roots or no negative real roots.**

□

Example 2 Investigate the nature of the roots of $2x^2 + 8x + 5 = -x^4$.

Solution

$$P(x) = x^4 + 0x^3 + 2x^2 + 8x + 5$$ First, reorder the terms.

+ + + + A zero coefficient is not counted.

0 positive real roots

$$P(-x) = x^4 + 0x^3 + 2x^2 - 8x + 5$$

+ + − +

2 or 0 negative real roots

□

Note in Example 2 that zero coefficients are ignored when counting variations in sign. Further, with no positive roots and at most two negative real roots, there are at most two real roots. One may suspect that a fourth-degree equation has four solutions. The other solutions must then be complex.

In 1799 a 22-year-old graduate student named Karl Gauss proved in his doctoral thesis that every polynomial equation has at least one solution. This, of course, is an assumption that you have made throughout your study of algebra—from the time you solved first-degree equations in beginning algebra until now. It is an idea so basic to algebra that it is called the **Fundamental Theorem of Algebra.**

FUNDAMENTAL THEOREM OF ALGEBRA

> Every polynomial equation in a single variable with real or complex coefficients has at least one root.

If the equation has one solution, a depressed equation of one degree less may be obtained. That new equation, according to the Fundamental Theorem, has a root. This root may now be used to obtain an equation of lower degree. The result of this process is suggested by the following corollary:

COROLLARY

> Every polynomial of degree n in a single variable can be decomposed into n linear factors.

The Factor Theorem then tells us that each linear factor yields a solution. Thus, the corollary could be stated as follows:

COROLLARY

> Every polynomial equation of degree n in a single variable has n roots.

Of course, the roots need not be distinct or real. Consider the following examples.

Example 3 Show that $x^9 - x^5 + x^4 + x^2 + 1 = 0$ has at least six nonreal complex solutions.

Solution

$$P(x) = x^9 - x^5 + x^4 + x^2 + 1$$

$$+ \quad - \quad + \quad + \quad +$$

2 or 0 positive real roots

$$P(-x) = -x^9 + x^5 + x^4 + x^2 + 1$$

$$- \quad + \quad + \quad + \quad +$$

1 negative real root

The polynomial equation has one negative real root and at most two positive real roots. However, it has nine solutions since it is degree 9; thus, there are at least six nonreal complex solutions. □

Example 4 Show that $x^6 - 2x^3 + x^2 - 2x + 2 = 0$ has at least two nonreal complex roots.

Solution

$$P(x) = x^6 - 2x^3 + x^2 - 2x + 2$$

$$+ \quad - \quad + \quad - \quad +$$

4, 2, or 0 positive real roots

$$P(-x) = x^6 + 2x^3 + x^2 + 2x + 2$$

0 negative real roots

The polynomial equation has six roots, and at most four of these are real (positive). Thus, at least two roots are complex and nonreal. □

If synthetic division is used on the equation in Example 4, two positive real roots are quickly found:

	1	0	0	−2	1	−2	2
1	1	1	1	−1	0	−2	0
1	1	2	3	2	2	0	
	+	+	+	+	+		

Notice that $x = 1$ is a root of *multiplicity* 2. When counting the number of roots using Descartes' Rule, multiple roots are *not* counted as a single root. A root of multiplicity 2 counts as two roots and roots of multiplicity n count as n roots.

Note further that the depressed equation produced by the synthetic division has all positive coefficients; thus, all the positive roots have been found. Hence, the polynomial equation $x^6 - 2x^3 + x^2 - 2x + 2 = 0$ has exactly two positive real roots and four roots that are not real.

Many of the polynomial equations solved in this chapter have had integer coefficients but irrational or nonreal solutions. If you know one such solution, then the next theorem tells you how to find another.

CONJUGATE PAIR THEOREM

If $P(x) = 0$ is a polynomial equation with rational coefficients, then:

1. When $m + \sqrt{n}$ is a root, $m - \sqrt{n}$ is also a root, where m and n are rational and $\sqrt{n}$ is irrational.
2. When $a + bi$ is a root, $a - bi$ is also a root, where a and b are real numbers.

If you know one irrational or complex root of a polynomial equation, then you know that its conjugate is a root. Further, when these values are used to obtain a depressed equation, the remaining roots may be found more easily.

Example 5 Solve $x^4 - 4x - 1 = 4x^3$, given that $2 + \sqrt{5}$ is one root.

Solution Rearrange terms to get $x^4 - 4x^3 - 4x - 1 = 0$. Now, using the known root, we get

$2+\sqrt{5}$	1	−4	0	−4	−1
		$2+\sqrt{5}$	1	$2+\sqrt{5}$	1
	1	$-2+\sqrt{5}$	1	$-2+\sqrt{5}$	0

And $2 - \sqrt{5}$ must be a root also:

$2-\sqrt{5}$	1	$-2+\sqrt{5}$	1	$-2+\sqrt{5}$
		$2-\sqrt{5}$	0	$2-\sqrt{5}$
	1	0	1	0

Using the depressed equation, the remaining roots can now be found:

$$x^2 + 1 = 0$$
$$x = \pm i$$
$$\{2 \pm \sqrt{5}, \pm i\}$$

□

Example 6 Solve $x^4 + 2x^3 = 4x + 4$, given that $-1 - i$ is one root.

Solution Rearrange terms to get $x^4 + 2x^3 - 4x - 4 = 0$.

$-1-i$	1	2	0	-4	-4
		$-1-i$	-2	$2+2i$	4
	1	$1-i$	-2	$-2+2i$	0

$-1+i$	1	$1-i$	-2	$-2+2i$
		$-1+i$	0	$2-2i$
	1	0	-2	0

$$x^2 - 2 = 0$$
$$x = \pm\sqrt{2}$$
$$\{\mathbf{-1 \pm i, \pm\sqrt{2}}\}$$

□

Example 7 Solve $2x^6 + x^5 + 2x^3 - 6x^2 + x - 4 = 0$, given that i is a multiple root.

Solution If i is a multiple root, then $-i$ is too. So use these values to discover a depressed equation that can be solved.

	2	1	0	2	-6	1	-4
i	2	$1+2i$	$-2+i$	$1-2i$	$-4+i$	$-4i$	0
$-i$	2	1	-2	1	-4	0	
i	2	$1+2i$	$-4+i$	$-4i$	0		
$-i$	2	1	-4	0			

$$2x^2 + x - 4 = 0$$
$$x = \frac{-1 \pm \sqrt{1+32}}{4} = \frac{-1 \pm \sqrt{33}}{4}$$
$$\left\{\pm i, \frac{-1 \pm \sqrt{33}}{4}\right\}$$

□

Problem Set 5.4

A *Investigate the nature of the real roots of the equations in Problems 1–6, using Descartes' Rule of Signs.*

1. $3x^3 + 7x - 1 = 0$

2. $x^3 + 2x + 2 = 0$

3. $x^3 - 9x^2 + 22x - 24 = 0$

4. $x^3 - 5x^2 - 18x + 72 = 0$

5. $2x^2 + 6x - 3x^3 - 4 = 0$

6. $5x^3 - 2x^4 + x^2 - 7 = 0$

In Problems 7–12 let $f(x) = x^4 - 6x^3 + 15x^2 - 2x - 10$ to find the requested value.

7. $f(i)$ **8.** $f(-i)$ **9.** $f(\sqrt{2})$

10. $f(-\sqrt{3})$ **11.** $f(2 + i)$ **12.** $f(1-\sqrt{3})$

In Problems 13–18 let $g(x) = x^4 - 8x^3 + 21x^2 - 14x - 10$ to find the requested value.

13. $g(-i)$ **14.** $g(i)$ **15.** $g(\sqrt{3})$

16. $g(-\sqrt{2})$ **17.** $g(3 - i)$ **18.** $g(1+\sqrt{2})$

19. Is $1 + \sqrt{2}$ a root of $x^3 - 2x^2 - x + 1 = 0$? If it is, name another root.

20. Is $1 + i$ a root of $x^3 - 4x^2 + 6x - 4 = 0$? If it is, name another root.

21. Is $1 - 2i$ a root of $x^3 - x^2 + 3x + 5 = 0$? If it is, name another root.

B *Solve the equations in Problems 22–25 if the given value is a root.*

22. $x^3 - 2x^2 + 4x - 8 = 0;\ 2i$

23. $x^4 + 13x^2 + 36 = 0;\ -3i$

24. $x^4 - 6x^2 + 25 = 0;\ 2 + i$

25. $x^4 - 4x^3 + 3x^2 + 8x - 10 = 0;\ 2 + i$

26. Show that $x^n + 1 = 0$ has no real roots if n is even and one real root if n is odd.

27. Show that $x^n - 1 = 0$ has exactly one real root if n is odd and two if n is even.

28. Use Descartes' Rule and the Location Theorem to show that the polynomial equation $x^3 + x^2 - 6x + 1 = 0$ has exactly two positive real roots.

29. Use Descartes' Rule and the Location Theorem to show that the polynomial equation $x^4 + x^3 - 8x^2 - 3x + 2 = 0$ has exactly two negative real roots.

30. Show that $x^5 - 10x + 2 = 0$ has two positive real roots, one negative real root, and two roots that are not real.

31. Show that $x^6 + 2x - 1$ has one positive real root, one negative real root, and four roots that are not real.

C *Solve the equations in Problems 32–37 if the given value is a root.*

32. $2x^4 - 5x^3 + 9x^2 - 15x + 9 = 0;\ i\sqrt{3}$

33. $2x^4 - x^3 - 13x^2 + 5x + 15 = 0;\ -\sqrt{5}$

34. $3x^5 + 10x^4 - 8x^3 + 12x^2 - 11x + 2 = 0;\ -2 + \sqrt{5}$

35. $2x^5 + 9x^4 - 3x^2 - 8x - 42 = 0;\ i\sqrt{2}$

36. $x^5 - 11x^4 + 24x^3 + 16x^2 - 17x + 3 = 0;\ 2 + \sqrt{3}$

37. $x^5 - 2x^4 + 6x^3 + 24x^2 + 5x + 26 = 0;\ 2 + 3i$

38. Solve $x^6 + x^5 - 3x^4 - 4x^3 + 4x + 4 = 0$ if $-\sqrt{2}$ is a multiple root.

39. Solve $x^4 - 4x^3 + 14x^2 - 20x + 25 = 0$ if $1 + 2i$ is a multiple root.

40. Show that $x^2 - 2x + 1 + 2i = 0$ has a root of $2 - i$ but *not* $2 + i$. Does this contradict the Conjugate Pair Theorem? Why, or why not?

41. Show that $x^2 - 3x - 2 - 3\sqrt{2} = 0$ has a root of $3 + \sqrt{2}$ but *not* of its conjugate. Does this contradict the Conjugate Pair Theorem? Why, or why not?

42. Show that $x^3 + ax^2 + bx - 1 = 0$ has at least one positive real root.

43. Show that $x^3 + ax^2 + bx + 1 = 0$ has at least one negative real root.

44. Show that if a polynomial equation has all positive real roots, then the signs of the coefficients must alternate, positive with negative.

45. Show that if an nth-degree polynomial equation with real coefficients has n real roots, then the number of positive real roots is equal to the number of variations in sign (not less).

5.5 Graphing Rational Functions

In order to evaluate and graph polynomial functions, we used synthetic division and the division algorithm. That is, we considered $P(x)/D(x)$ for polynomials $P(x)$ and $D(x)$, where $D(x) \neq 0$. Now consider this quotient from another viewpoint. An expression of the form $P(x)/D(x)$ is called a *rational function.*

RATIONAL FUNCTION

A **rational function** f is the quotient of polynomial functions $P(x)$ and $D(x)$; that is,

$$f(x) = \frac{P(x)}{D(x)} \qquad \text{where } D(x) \neq 0$$

Rational equations were solved in Chapter 2, and now we want to graph rational functions. The domain of a rational function is the set of all real x such that the denominator is not zero. When writing a rational function, we will assume this domain. That is, if

$$f(x) = \frac{x + 3}{x - 2}$$

it is not necessary to write $x \neq 2$ since this is implied in the definition of a rational function. However, if there are common factors in the numerator and denominator, as in

$$\begin{aligned} g(x) &= \frac{2x^2 + 5x - 3}{x + 3} \\ &= \frac{(2x - 1)(x + 3)}{x + 3} \\ &= 2x - 1 \qquad x \neq -3 \end{aligned}$$

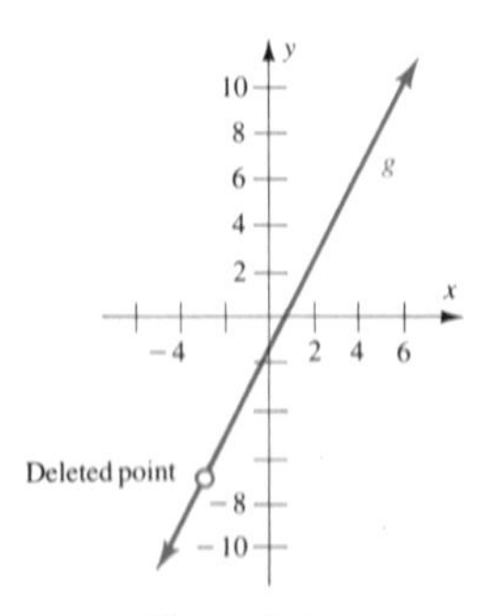

Figure 5.1
Graph of
$g(x) = \dfrac{2x^2 + 5x - 3}{x + 3}$

it *is* necessary to write $x \neq -3$ since this is not implied when the simplified form $g(x) = 2x - 1$ is written. This means that the graph g is the same as that of the linear function $y = 2x - 1$ with the point at $x = -3$ deleted from the domain, as shown in Figure 5.1.

Example 1 Graph: $f(x) = \dfrac{x^3 + 4x^2 + 7x + 6}{x + 2}$

Solution Simplify if possible:

These factors were obtained by synthetic division:

$$f(x) = \frac{(x + 2)(x^2 + 2x + 3)}{x + 2}$$
$$= x^2 + 2x + 3 \qquad x \neq -2$$

$$\begin{array}{r|rrrr} -2 & 1 & 4 & 7 & 6 \\ & & -2 & -4 & -6 \\ \hline & 1 & 2 & 3 & 0 \end{array}$$

The graph of f is the same as for the quadratic function $y = x^2 + 2x + 3$ with the point at $x = -2$ deleted from the domain. To sketch this parabola, complete the square:

$$y - 3 = x^2 + 2x$$
$$y - 3 + 1 = x^2 + 2x + 1$$
$$y - 2 = (x + 1)^2$$

The vertex is at $(-1, 2)$ and the parabola opens upward. It is drawn with the point where $x = -2$ deleted, as shown in the figure.

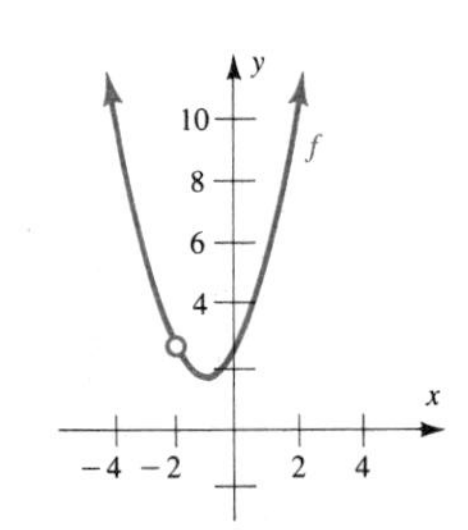

Graph of $f(x) = \dfrac{x^3 + 4x^2 + 7x + 6}{x + 2}$ □

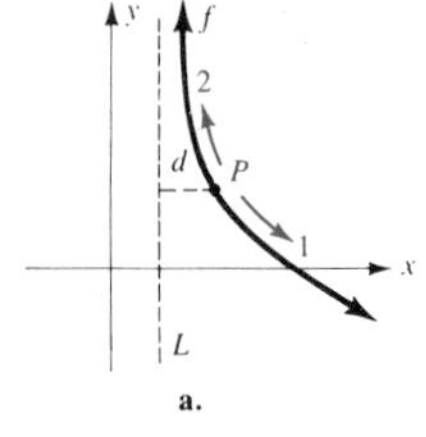

a.

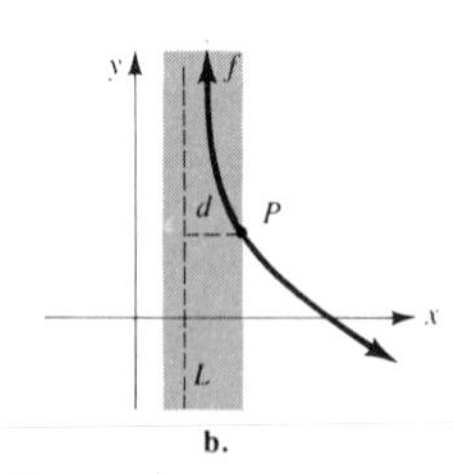

b.

Figure 5.2
An Asymptote to a Curve

In order to understand the graph of a rational function that does not reduce to a polynomial function with a deleted point, it is necessary to understand the notion of an *asymptote* introduced in Section 3.5. Recall that an **asymptote** is a line having the property that the distance from a point P on the curve to the line approaches zero as the distance from P to the origin increases without bound and P is *on a suitable part of the curve*. This last phrase (in italics) is best illustrated by considering Figure 5.2, where L is an asymptote for the function f. Consider P and d, the distance from P to the line L, as shown in Figure 5.2a. Now the distance from P to the origin can increase in two ways, depending on whether P moves along the curve in direction 1 or direction 2. In direction 1 the distance d increases without bound, but in direction 2 the distance d approaches zero. Thus, if you consider the portion of the curve in the shaded region of Figure 5.2b, you see that the conditions of the definition of an asymptote apply. Even though some rational functions may not have asymptotes, there are three types of asymptotes that occur frequently enough to merit consideration. These are *vertical, horizontal,* and *slant asymptotes*. An example of each is shown in Figure 5.3.

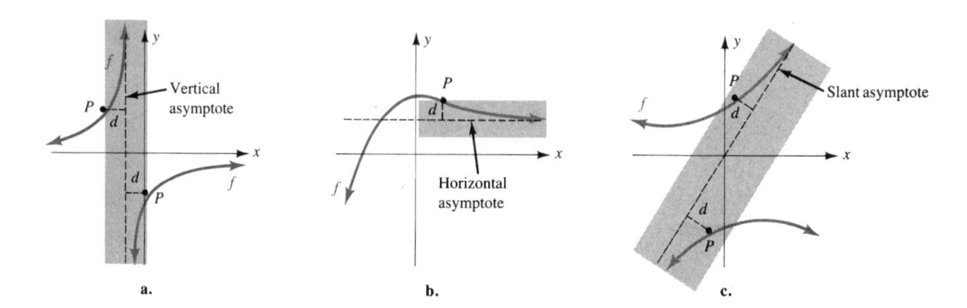

Figure 5.3
Examples of Asymptotes

ASYMPTOTES

If $f(x) = \dfrac{P(x)}{D(x)}$, where $P(x)$ and $D(x)$ have no common factors, then:

1. The line $x = r$ is a **vertical** asymptote if $D(r) = 0$.
2. The line $y = 0$ is a **horizontal** asymptote if the degree of $P(x)$ is less than the degree of $D(x)$.

   The line $y = \dfrac{a_n}{b_n}$ is a **horizontal** asymptote if the degree of $P(x)$ is the same as the degree of $D(x)$ and

   $$P(x) = a_n x^n + \cdots + a_0 \quad \text{and} \quad D(x) = b_n x^n + \cdots + b_0$$
3. The line $y = mx + b$ is a **slant** asymptote if the degree of $P(x)$ is 1 more than the degree of $D(x)$ and

   $$f(x) = \frac{P(x)}{D(x)} = mx + b + \frac{R}{D(x)}$$

Example 2 will serve as a model for many types of rational functions.

Example 2 Sketch: $y = \dfrac{1}{x}$

Solution $y = \dfrac{1}{x}$ is symmetric with respect to the origin and has a vertical asymptote at $x = 0$ as well as a horizontal asymptote at $y = 0$. The graph is obtained by plotting points.

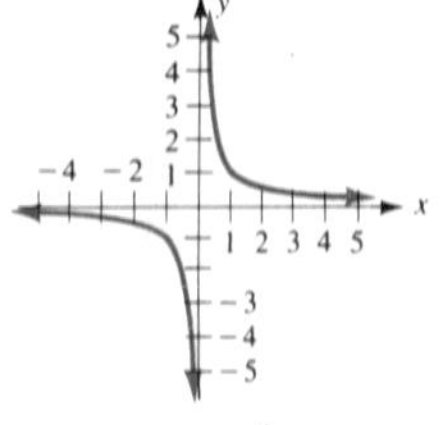

Graph of $y = \dfrac{1}{x}$ □

Variations of the curve drawn in Example 2 are shown in Figure 5.4.

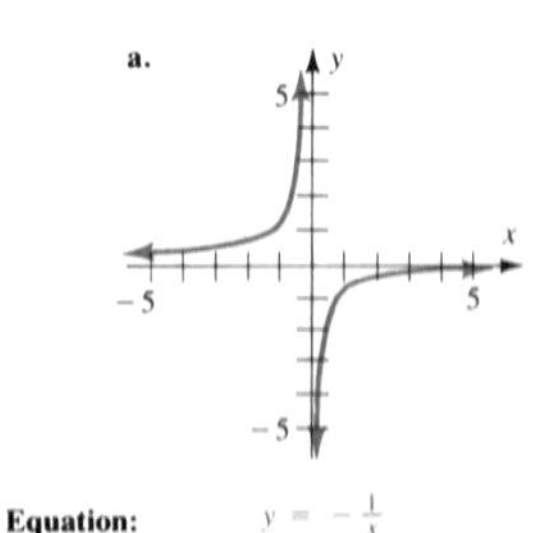

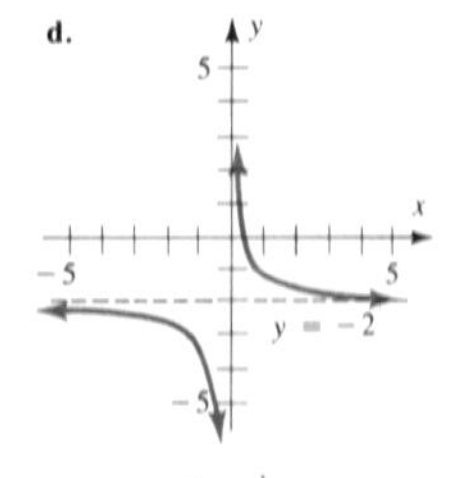

	a.	b.	c.	d.
Equation:	$y = -\frac{1}{x}$	$y = \frac{2}{x}$	$y = \frac{1}{x-3}$	$y + 2 = \frac{1}{x}$
Compare with Example 2:	Reflected about the x-axis	Each point will be twice as far from the x-axis	Vertical asymptote is translated 3 units to the right	Horizontal asymptote is translated 2 units down

Figure 5.4
Graphs of Rational Functions

Example 3 Sketch: $y = \dfrac{2x^2 - 3x + 5}{x^2 - x - 2}$

Solution Reduce, if possible; then find the asymptotes:

$$x^2 - x - 2 = (x - 2)(x + 1)$$ So the vertical asymptotes are $x = 2$ and $x = -1$.

The degrees of the numerator and denominator are the same, so a horizontal asymptote is

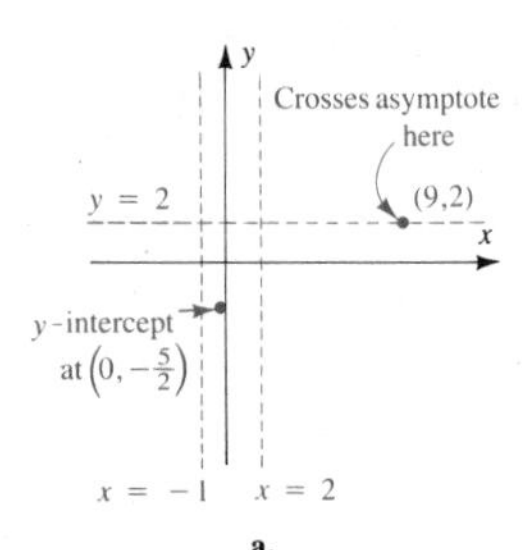

a.

$$y = \frac{2}{1} = 2$$ Horizontal asymptote

Draw the asymptotes and plot some points (see figure a). Some convenient points to plot are the x-intercepts (set $y = 0$ and solve for x); the y-intercepts (set $x = 0$ and solve for y); and the points where the graph passes through an asymptote. For $y = 0$, we have

$$0 = \frac{2x^2 - 3x + 5}{x^2 - x - 2}$$ Multiply both sides by $x^2 - x - 2$

$$0 = 2x^2 - 3x + 5$$ No real roots since $b^2 - 4ac = 9 - 4 \cdot 2 \cdot 5 < 0$

For $x = 0$, we have

$$y = \frac{2 \cdot 0^2 - 3 \cdot 0 + 5}{0^2 - 0 - 2}$$

$$= -\frac{5}{2}$$ The y-intercept is $(0, -\frac{5}{2})$.

The graph of a function will not pass through a vertical asymptote. It can, however, pass through a horizontal or slant asymptote. For this example, $y = 2$ is the equation of the horizontal asymptote. To find the point(s) of intersection, if any, substitute $y = 2$ into the original equation of the function. That is,

$$2 = \frac{2x^2 - 3x + 5}{x^2 - x - 2}$$

$$2x^2 - 2x - 4 = 2x^2 - 3x + 5 \qquad x \neq 2, -1$$

$$-2x - 4 = -3x + 5$$

$$x = 9$$

The curve passes through the point (9, 2). That is, (9, 2) is on the curve and also on the horizontal asymptote.

To draw the graph, plot some additional points as shown in figure b.

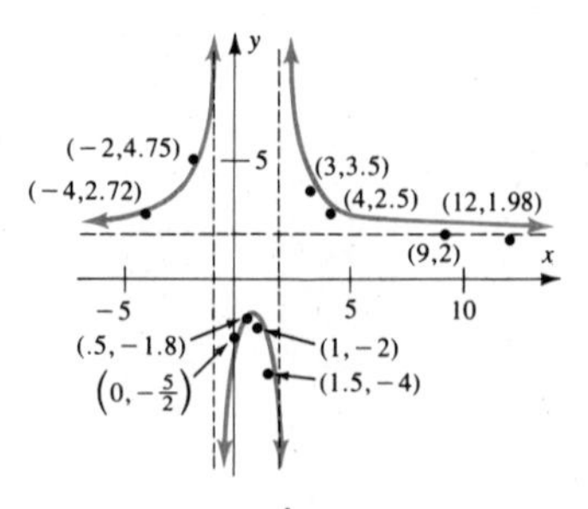

b.

Graph of $y = \dfrac{2x^2 - 3x + 5}{x^2 - x - 2}$ □

The next example illustrates a slant asymptote.

Example 4 Let f be a function defined by the equation

$$y = \frac{2x^2 - 5x + 1}{x - 3}$$

Sketch the graph.

Solution There are no horizontal asymptotes; the vertical asymptote is $x = 3$ (by inspection). There is a slant asymptote since the degree of the numerator is 1 more than the degree of the denominator:

	2	−5	1
3	2	1	4
	↑	↑	↑
	m	b	Remainder; $f(x) = 2x + 1 + \dfrac{4}{x - 3}$

The line $y = mx + b$ is a slant asymptote, so for this example, $y = 2x + 1$ is the slant asymptote.

The y-intercept is $\left(0, -\frac{1}{3}\right)$ and is found by

$$y = \frac{2 \cdot 0^2 - 5 \cdot 0 + 1}{0 - 3} = -\frac{1}{3}$$

The x-intercepts are (2.3, 0) and (.2, 0) and are found by

$$0 = \frac{2x^2 - 5x + 1}{x - 3}$$

$$2x^2 - 5x + 1 = 0 \qquad \text{Multiply both sides by } x - 3,\ x \neq 3$$

$$x = \frac{5 \pm \sqrt{17}}{4}$$

$$\approx 2.3, .2$$

For the intersection of the slant asymptote $y = 2x + 1$ and the graph, substitute $2x + 1$ for y in the original equation of the function:

$$\begin{aligned} 2x + 1 &= \frac{2x^2 - 5x + 1}{x - 3} \\ (2x + 1)(x - 3) &= 2x^2 - 5x + 1 \\ 2x^2 - 5x - 3 &= 2x^2 - 5x + 1 \\ -3 &= 1 \end{aligned}$$

This is a false equation, so there is no solution; the graph does not pass through the asymptote.

Plot some additional points to find the graph as shown in the figure.

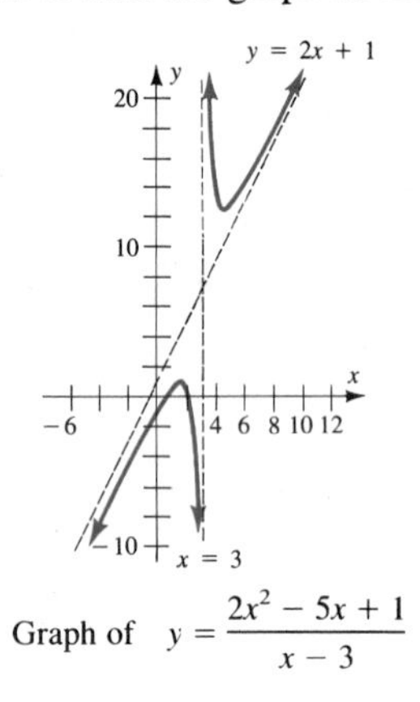

Graph of $y = \dfrac{2x^2 - 5x + 1}{x - 3}$

Problem Set 5.5

A *Graph the function defined by each equation given in Problems 1–33.*

1. $y = \dfrac{x^2 - x - 12}{x + 3}$

2. $y = \dfrac{x^2 + x - 2}{x - 1}$

3. $y = \dfrac{x^2 - x - 6}{x + 2}$

4. $y = \dfrac{2x^2 - 13x + 15}{x - 5}$

5. $y = \dfrac{6x^2 - 5x - 4}{2x + 1}$

6. $y = \dfrac{15x^2 + 13x - 6}{3x - 1}$

B **7.** $y = \dfrac{1}{x}$

8. $y = \dfrac{-1}{x}$

9. $y = \dfrac{1}{x} + 2$

10. $y = -\dfrac{1}{x} + 1$

11. $y = \dfrac{4}{x^2}$

12. $y = \dfrac{-2}{x^2}$

13. $y = \dfrac{1}{x + 2}$

14. $y = \dfrac{1}{x - 3}$

15. $y = \dfrac{-1}{x + 3}$

16. $y = \dfrac{1}{x - 4}$

17. $y = \dfrac{2x^3 - 3x^2 - 32x - 15}{x^2 - 2x - 15}$

18. $y = \dfrac{3x^3 + 5x^2 - 26x + 8}{x^2 + 2x - 8}$

19. $y = \dfrac{x^3 + 6x^2 + 10x + 4}{x + 2}$

20. $y = \dfrac{x^3 + 9x^2 + 15x - 9}{x + 3}$

21. $y = \dfrac{x^3 + 12x^2 + 40x + 40}{x + 2}$

22. $y = \dfrac{x^3 + 5x^2 + 6x}{x + 3}$

23. $y = \dfrac{2x^2 + 2}{x^2}$

24. $y = \dfrac{2x^2 - 1}{x^2}$

25. $y = \dfrac{x^2}{x - 4}$

26. $y = \dfrac{4x}{x^2 - 2}$

27. $y = \dfrac{-x^2}{x - 1}$

28. $y = \dfrac{x^3}{(x - 1)^2}$

29. $y = \dfrac{x^3}{x^2 - 1}$

30. $y = \dfrac{x^4}{x^2 - 1}$

C 31. $y = \dfrac{x^2}{x^3 - x^2 - 20x}$

32. $y = \dfrac{x^2}{20x - x^2 - x^3}$

33. $y = \dfrac{x^2 + 3x - 2}{x^2 + 2x - 8}$

*5.6 Partial Fractions

In algebra, rational expressions are added by finding common denominators; for example:

$$\frac{5}{x-2} + \frac{3}{x+1} = \frac{5(x+1) + 3(x-2)}{(x-2)(x+1)}$$

$$= \frac{8x - 1}{(x-2)(x+1)}$$

In calculus, however, it is sometimes necessary to break apart the expression

$$\frac{8x - 1}{(x-2)(x+1)}$$

into two fractions with denominators that are linear. The technique for doing this is called the **method of partial fractions.**

The rational expression

$$f(x) = \frac{P(x)}{D(x)}$$

can be **decomposed** into partial fractions if there are no common factors and if the degree of $P(x)$ is less than the degree of $D(x)$. If the degree of $P(x)$ is greater than or equal to the degree of $D(x)$, then use either long division or synthetic division to obtain a polynomial plus a proper fraction. For example,

*Optional

$$\frac{x^4 + 2x^3 - 4x^2 + x - 3}{x^2 - x - 2} = x^2 + 3x + 1 + \underbrace{\frac{8x - 1}{x^2 - x - 2}}$$

This was found by long division.

Proper fraction
This is the part that is decomposed into partial fractions.

Now look at the proper fraction. There is a theorem that says this proper fraction can be written as a sum,

$$F_1 + F_2 + \cdots + F_j$$

where *each* F_i is of the form

$$\frac{A}{(x - r)^n} \quad \text{or} \quad \frac{Ax + B}{(x^2 + st + t)^n}$$

This tells us that any polynomial with real coefficients can be expressed as a product of real linear and quadratic factors. We begin by focusing on the first form.

PARTIAL FRACTION DECOMPOSITION; LINEAR FACTORS

Let $f(x) = P(x)/D(x)$, where $P(x)$ and $D(x)$ have no common factors and the degree of $P(x)$ is less than the degree of $D(x)$. Also suppose that $D(x) = (x - r)^n$. Then $f(x)$ can be decomposed into partial fractions:

$$\frac{A_1}{x - r} + \frac{A_2}{(x - r)^2} + \cdots + \frac{A_n}{(x - r)^n}$$

Example 1 Decompose $\dfrac{8x - 1}{x^2 - x - 2}$ into partial fractions.

Solution

$$\frac{8x - 1}{x^2 - x - 2} = \frac{8x - 1}{(x - 2)(x + 1)}$$

First, factor the denominator, if possible, and make sure there are no common factors.

$$= F_1 + F_2$$

Break up the fraction into parts, each with a linear factor.

$$= \frac{A}{x - 2} + \frac{B}{x + 1}$$

The task is to find A and B.

$$= \frac{A(x + 1) + B(x - 2)}{(x - 2)(x + 1)}$$

Obtain a common denominator on the right.

Now, multiply both sides of this equation by the least common denominator, which is $(x - 2)(x + 1)$ for this example:

$$8x - 1 = A(x + 1) + B(x - 2)$$

Substitute, one at a time, the values that cause each of the factors in the least common denominator to be zero.

Let $x = -1$:

$$\begin{aligned} 8x - 1 &= A(x + 1) + B(x - 2) \\ 8(-1) - 1 &= A(-1 + 1) + B(-1 - 2) \\ -9 &= 0 + B(-3) \\ -9 &= -3B \\ 3 &= B \end{aligned}$$

Let $x = 2$:

$$\begin{aligned} 8x - 1 &= A(x + 1) + B(x - 2) \\ 8(2) - 1 &= A(2 + 1) + B(2 - 2) \\ 15 &= 3A \\ 5 &= A \end{aligned}$$

If $A = 5$ and $B = 3$, then

$$\frac{\mathbf{8x - 1}}{\mathbf{(x - 2)(x + 1)}} = \frac{\mathbf{5}}{\mathbf{x - 2}} + \frac{\mathbf{3}}{\mathbf{x + 1}}$$

□

Example 2 illustrates the process if there is a repeated linear factor.

Example 2 Decompose $\dfrac{x^2 - 6x + 3}{(x - 2)^3}$ by using the method of partial fractions.

Solution

$$\frac{x^2 - 6x + 3}{(x - 2)^3} = \frac{A}{x - 2} + \frac{B}{(x - 2)^2} + \frac{C}{(x - 2)^3}$$

Multiply both sides by $(x - 2)^3$:

$$x^2 - 6x + 3 = A(x - 2)^2 + B(x - 2) + C$$

Let $x = 2$.

$$\begin{aligned} (2)^2 - 6(2) + 3 &= A(2 - 2)^2 + B(2 - 2) + C \\ 4 - 12 + 3 &= 0 + 0 + C \\ -5 &= C \end{aligned}$$

Notice that with repeated factors we cannot find all the numerators as we did in Example 1. Now substitute $C = -5$ into the original equation and simplify by combining terms on the right side:

$$\begin{aligned} x^2 - 6x + 3 &= A(x - 2)^2 + B(x - 2) - 5 \\ &= A(x^2 - 4x + 4) + B(x - 2) - 5 \\ &= Ax^2 - 4Ax + Bx + 4A - 2B - 5 \\ &= Ax^2 + (-4A + B)x + (4A - 2B - 5) \end{aligned}$$

If the polynomials on the left and right sides of the equality are equal, then the coefficients of the like terms must be equal. That is,

$$x^2 - 6x + 3 = Ax^2 + (-4A + B)x + (4A - 2B - 5)$$

$$A = 1$$

$$-4A + B = -6$$

$$4A - 2B - 5 = 3$$

If $A = 1$, then

$$\begin{aligned} -4A + B &= -6 \\ -4(1) + B &= -6 \\ B &= -2 \end{aligned}$$

Check: If $A = 1$ and $B = -2$, then

$$\begin{aligned} 4A - 2B - 5 &= 4(1) - 2(-2) - 5 \\ &= 4 + 4 - 5 \\ &= 3 \end{aligned}$$

Thus,

$$\mathbf{\frac{x^2 - 6x + 3}{(x-2)^3} = \frac{1}{x-2} + \frac{-2}{(x-2)^2} + \frac{-5}{(x-2)^3}}$$

□

We will now consider quadratic factors.

PARTIAL FRACTION DECOMPOSITION; QUADRATIC FACTORS

Let $f(x) = P(x)/D(x)$, where $P(x)$ and $D(x)$ have no common factors and the degree of $P(x)$ is less than the degree of $D(x)$. Also, if $D(x) = (x^2 + sx + t)^m$, then $f(x)$ can be decomposed into partial fractions:

$$\frac{A_1x + B_1}{x^2 + sx + t} + \frac{A_2x + B_2}{(x^2 + sx + t)^2} + \cdots + \frac{A_mx + B_m}{(x^2 + sx + t)^m}$$

Example 3 Decompose: $f(x) = \dfrac{2x^3 + 3x^2 + 3x + 2}{(x^2 + 1)^2}$

Solution

$$\frac{2x^3 + 3x^2 + 3x + 2}{(x^2 + 1)^2} = \frac{Ax + B}{x^2 + 1} + \frac{Cx + D}{(x^2 + 1)^2}$$

Multiply by $(x^2 + 1)^2$:

$$2x^3 + 3x^2 + 3x + 2 = (Ax + B)(x^2 + 1) + Cx + D$$

This time, $x^2 + 1 \neq 0$ in the set of real numbers, so multiply out the right side:

$$\begin{aligned} 2x^3 + 3x^2 + 3x + 2 &= Ax^3 + Bx^2 + Ax + B + Cx + D \\ &= Ax^3 + Bx^2 + (A + C)x + (B + D) \end{aligned}$$

Equate the coefficients of the similar terms on the left and right:

$$A = 2$$
$$B = 3$$
$$A + C = 3 \quad \leftarrow \text{If } A = 2\text{, then } 2 + C = 3 \text{ and } C = 1.$$
$$B + D = 2 \quad \leftarrow \text{If } B = 3\text{, then } 3 + D = 2 \text{ and } D = -1.$$

Thus,

$$\frac{2x^3 + 3x^2 + 3x + 2}{(x^2 + 1)^2} = \frac{2x + 3}{x^2 + 1} + \frac{x - 1}{(x^2 + 1)^2}$$ □

Problem Set 5.6

A *Decompose each fraction in Problems 1–36 by using the method of partial fractions.*

1. $\frac{x^2 + 2x + 5}{x^3}$ **2.** $\frac{3x^2 - 2x + 1}{x^3}$ **3.** $\frac{2x^2 - 5x + 4}{x^3}$

4. $\frac{1}{(x + 2)(x + 3)}$ **5.** $\frac{1}{(x + 4)(x + 5)}$ **6.** $\frac{1}{(x + 3)(x + 2)}$

7. $\frac{7x - 10}{(x - 2)(x - 1)}$ **8.** $\frac{11x - 1}{(x - 1)(x + 1)}$ **9.** $\frac{7x + 2}{(x + 2)(x - 4)}$

10. $\frac{2x + 10}{x^2 + 7x + 12}$ **11.** $\frac{2x - 14}{x^2 + x - 6}$ **12.** $\frac{7x - 7}{2x^2 - 5x - 3}$

B **13.** $\frac{4(x - 1)}{x^2 - 4}$ **14.** $\frac{34 - 5x}{48 - 14x + x^2}$ **15.** $\frac{x - 7}{20 - 9x + x^2}$

16. $\frac{5x^2 - 5x - 4}{x^3 - x}$ **17.** $\frac{4x^2 - 7x - 3}{x^3 - x}$ **18.** $\frac{2x^2 - 18x - 12}{x^3 - 4x}$

19. $\frac{2x - 1}{(x - 2)^2}$ **20.** $\frac{4x - 22}{(x - 5)^2}$ **21.** $\frac{x^2 + 5x + 1}{x(x + 1)^2}$

22. $\frac{5x^2 - 2x + 2}{x(x - 1)^2}$ **23.** $\frac{2x^2 + 7x + 2}{(x + 1)^3}$ **24.** $\frac{7x - 3x^2}{(x - 2)^3}$

25. $\frac{x}{x^2 + 4x - 5}$ **26.** $\frac{x}{x^2 - 2x - 3}$ **27.** $\frac{7x - 1}{x^2 - x - 2}$

28. $\frac{10x^2 - 11x - 6}{x^3 - x^2 - 2x}$ **29.** $\frac{-17x - 6}{x^3 + x^2 - 6x}$ **30.** $\frac{12 + 9x - 6x^2}{x^3 - 5x^2 + 4x}$

C **31.** $\frac{5x^2 - 6x + 7}{(x - 1)(x^2 + 1)}$ **32.** $\frac{x^2}{(x + 1)(x^2 + 1)}$ **33.** $\frac{x^3}{(x - 1)^2}$

34. $\frac{x^3}{(x + 1)^2}$ **35.** $\frac{2x^3 - 3x^2 + 6x - 1}{1 - x^4}$ **36.** $\frac{2x^3 - 7x^2 + 8x - 7}{x^2 - 4x + 4}$

5.7 Summary and Review

If you can divide a polynomial $P(x)$ by a linear expression, you can check for factors. If you find factors of $P(x)$, you know the zeros of $y = P(x)$, the roots or solutions of $P(x) = 0$, the critical values of $P(x) < 0$ or $P(x) > 0$, and therefore can solve polynomial equations and inequalities. But polynomial division is difficult, bulky, and time-consuming. Synthetic division, then, is a welcome innovation since it is simple, brief, and economical. Being able to perform many divisions easily and quickly opens up a whole new realm of equation solving.

The *Factor Theorem* and *Remainder Theorem* form the logical basis, but the *Location, Rational Root, Upper and Lower Bounds,* and *Conjugate Pair Theorems,* as well as the *Fundamental Theorem* with its corollaries, provide tools to make synthetic division an effective problem-solving device.

Graphing rational functions is an important topic in calculus and provides an introduction to the essential calculus topics of limits and continuity.

CHAPTER OBJECTIVES

When you complete your study of this chapter, you should be able to:

1. Use synthetic division to
   a. divide $P(x)$ by $x - r$
   b. find the remainder R
   c. identify the depressed equation
   d. factor certain polynomials into linear factors
2. Graph polynomial functions using synthetic division
3. Graph polynomial functions in factored form
4. Solve certain polynomial equations and inequalities
5. List all possible rational roots for a given polynomial equation
6. Locate intervals containing real zeros for a polynomial function
7. Use multiplicity and conjugates to aid in equation solving
8. Assist equation solving by application of the Fundamental Theorem
9. Use Descartes' Rule of Signs to help solve equations
10. Graph rational functions

*11. Decompose a rational expression into partial fractions

Problem Set 5.7

CONCEPT PROBLEMS

Fill in the word or words necessary to make the statements in Problems 1–20 complete and correct.

Consider the polynomial $P(x) = a_nx^n + a_{n-1}x^{n-1} + \cdots + a_1x + a_0$, *where* $P(x) = Q(x)(x - r) + R$ *in Problems 1–14.*

[5.1] **1.** The number n is a(n) ______ . If $n = 2$, then $P(x)$ is a(n) ______ ; if $n = 1$, then $P(x)$ is a(n) ______ ; and if $n = 0$, then $P(x)$ is a(n) ______ .

[5.1] **2.** If $R = 0$, then the resulting polynomial equation is called the ______ .

---

*Optional

[5.2] **3.** If $R = 0$, then $x - r$ is a(n) ______ of $P(x)$.

[5.2] **4.** If $x = r$, then ______ .

[5.3] **5.** If $P(a)$ and $P(b)$ are opposite in sign, then there is ______ .

[5.3] **6.** If $P(x) = 0$ has a rational root, p/q, then p is a factor of ______ and q is a factor of ______ .

[5.3] **7.** If ______________ , then a is an upper bound for the roots of $P(x) = 0$ when ______________ .

[5.3] **8.** If ______________ , then b is a lower bound for the roots of $P(x) = 0$ when ______________ .

[5.3] **9.** If $P(x)$ has three factors of $x - r$, then r is ______________ .

[5.4] **10.** If $P(x)$ has real coefficients, the number of positive roots of $P(x) = 0$ is ______________ .

[5.4] **11.** If $P(x)$ has real coefficients, the number of negative roots of $P(x) = 0$ is ______________ .

[5.4] **12.** $P(x) = 0$ has ______ roots.

[5.4] **13.** If $P(x)$ has rational coefficients and $x - 2 + \sqrt{5}$ is a factor, then ______ is also a factor.

[5.4] **14.** If $P(x) = 0$ has ______ coefficients and $5 - 3i$ is a root, then $5 + 3i$ is also a root.

[5.5] **15.** A rational function f is defined as ______________ .

[5.5] **16.** If $g(x) = \dfrac{a_m x^m + \cdots + a_0}{b_n x^n + \cdots + b_0}$ is a reduced fraction, then $y = 0$ is a horizontal asymptote if ______________ .

[5.5] **17.** If g is defined as shown in Problem 16, the line $x = 0$ is a vertical asymptote if ______ .

[5.5] **18.** If g is defined as shown in Problem 16, the line $y = a_m/b_n$ is a(n) ______ asymptote if ______________ .

[5.5] **19.** If g is defined as shown in Problem 16, the line $y = mx + b$ is a(n) ______ asymptote if ______ and $g(x) = mx + b + \dfrac{R}{D(x)}$.

*[5.6] **20.** Any polynomial with real coefficients can be expressed as a product of real ______ and ______ factors.

REVIEW PROBLEMS

[5.1] *In Problems 21–30 use synthetic division to find the quotient and the remainder when the first polynomial is divided by the second polynomial.*

21. $x^3 - 2x^2 - 3x - 20; \quad x - 4$

22. $x^4 + 2x^3 - x^2 + x - 15; \quad x + 3$

23. $x^3 - x - 6; \quad x - 2$

24. $x^3 - 7x + 6; \quad x + 3$

25. $4x^3 - 4x^2 + 7x - 3; \quad x - \frac{1}{2}$

26. $9x^3 + 8x - 8; \quad x - \frac{2}{3}$

*Optional

27. $3x^3 + 7x^2 - 2;\ \ x + 2$

28. $4x^3 - 6x^2 - 6x + 5;\ \ x - 2$

29. $5x^3 - 18x^2 + 13x - 10;\ \ x - 3$

30. $3x^4 - 4x^3 - 7x^2 + 2x - 4;\ \ x - 2$

[5.2] *Let $f(x) = 2x^5 - x^4 - 15x^3 - 5x^2 + 13x + 6$. Evaluate the function in Problems 31–36.*

31. $f(0)$ **32.** $f(1)$ **33.** $f(2)$ **34.** $f(3)$ **35.** $f(-1)$ **36.** $f\left(-\frac{1}{2}\right)$

[5.4] *Let $g(x) = x^4 - 2x^2 + 4x - 8$. Evaluate the function in Problems 37–42.*

37. $g(0)$ **38.** $g(\sqrt{2})$ **39.** $g(-\sqrt{2})$ **40.** $g(i)$ **41.** $g(2i)$ **42.** $g(-$

[5.2] *Sketch the graph of each polynomial function in Problems 43–48.*

43. $f(x) = x^3 + 2x^2 + 3x$

44. $f(x) = x^4 - 10x^2 + 4x + 8$

45. $f(x) = 2x^5 - x^4 - 15x^3 - 5x^2 + 13x + 6$

46. $f(x) = (x - 4)(x + 3)(x - 1)$

47. $f(x) = x(x - 3)(x + 2)$

48. $f(x) = x^2(x^2 - 9)$

[5.3] *Solve the polynomial equations in Problems 49–60.*

49. $x^3 - 2x^2 - x + 2 = 0$

50. $x^3 + x^2 - 9x - 9 = 0$

51. $x^3 + 4x^2 - 11x - 30 = 0$

52. $x^3 + x^2 - 14x - 24 = 0$

53. $6x^3 - x^2 = 4(8x - 5)$

54. $x^2(6x - 19) = 9(x - 4)$

55. $x^4 + 7x^3 + 18x^2 + 20x + 8 = 0$

56. $8x^4 - 12x^3 - 10x^2 + 3x + 2 = 0$

57. $x^4 - 3x^3 - 9x^2 + 25x - 6 = 0$

58. $x^4 - 3x^3 - 15x^2 + 45x - 28 = 0$

59. $x^4 + 5x^2 - 36 = 0$

60. $x^4 - 3x^3 + 3x^2 + 37x - 78 = 0$

[5.3] *Solve the polynomial inequalities in Problems 61–74.*

61. $(x + 2)(3x - 2)(x + 4) \geqslant 0$

62. $(x - 3)(5x - 3)(2x + 1) \leqslant 0$

63. $x(x + 5)(x - 7) < 0$

64. $x^2(x - 3)(x + 4) > 0$

65. $x(x - 2)^2(x + 1) < 0$

66. $x(x + 3)^2(x - 5) \geqslant 0$

67. $x^3 - 2x^2 - x + 2 \geqslant 0$

68. $x^3 + x^2 - 9x - 9 \leqslant 0$

69. $x^3 - 19x - 30 \leqslant 0$

70. $x^3 + x^2 - 14x - 24 \geqslant 0$

71. $6x^3 - x^2 - 32x + 20 \geqslant 0$

72. $15x^3 + 38x^2 - 23x - 6 > 0$

73. $x^4 - x^3 < 13x^2 - 25x + 12$

74. $2x^4 + x^3 < 14x^2 + 4x - 24$

[5.3] **75.** A rectangular box can be tripled in volume if the same amount is added to each dimension. If the box is originally 3 in. × 5 in. × 7 in., how much is added to each dimension?

[5.4] **76.** Show that the equation $x^9 - x^7 + x^6 + x^2 + 1 = 0$ has at least six nonreal roots.

[5.4] **77.** Solve $x^4 + 6x^3 = x^2 + 12x + 2$ if it is known that $2\sqrt{2} - 3$ is a root.

[5.4] **78.** Find the value of k for which $2 + i$ is a root of $x^4 - 3x^3 - x^2 + 13x + k = 0$.

[5.5] *Graph the functions in Problems 79–88.*

79. $y = \frac{3}{x}$

80. $y = \frac{-3}{x^2}$

81. $y = \frac{2}{x - 2}$

82. $y = \frac{3x^2 - 14x - 5}{x - 5}$

83. $y = \frac{x^3 - 3x^2 + x - 3}{x - 3}$

84. $y = \frac{3x^2 + 2x - 5}{x - 1}$

85. $y = \frac{2x^2 - 3x + 1}{x^2 - 3x + 2}$

86. $y = \frac{x^3 - x - 6}{x - 2}$

87. $y = \frac{2x^2 - x + 1}{x - 1}$

88. $y = \frac{x}{x^2 - 1}$

*[5.6] *Decompose each rational function given in Problems 89–100 into a sum of partial fractions.*

89. $\frac{3x^2 + 5x - 2}{x^3}$

90. $\frac{2x^2 + 7x - 15}{x + 5}$

91. $\frac{2x + 14}{x^2 - x - 6}$

92. $\frac{2x - 18}{x^2 - 2x - 3}$

93. $\frac{2x^2 - 5x + 3}{x^2 - 5x + 6}$

94. $\frac{2x^2 - 3x - 11}{x^2 - 2x - 15}$

95. $\frac{3x + 2}{(x - 1)^2}$

96. $\frac{5x - 13}{(x - 2)^2}$

97. $\frac{5x^2 + x + 4}{x^2 + 2x}$

98. $\frac{2x^2 - x + 6}{x^2 - 3x}$

99. $\frac{2x^3 + x^2 - 1}{x^2 - x}$

100. $\frac{3x^3 - 7x^2 + 3x - 4}{x^2 - 2x}$

*Optional

CHAPTER 6

Exponential and Logarithmic Functions

John Napier
(1550–1617)

The invention of logarithms came upon the world as a bolt from the blue. No previous work led up to it, foreshadowed it or heralded its arrival. It stands isolated, breaking in upon human thought abruptly without borrowing from the work of other intellects or following known lines of mathematical thought.

Lord Moulton
Napier Tercentenary Memorial Volume,
London, 1915, p. 1.

Historical Note

The Scotsman John Napier was the inventor of logarithms. He devised them to ease some of the burden of calculation in astronomy, engineering, and related sciences. Napier originally used the term *artificial number* but later adopted *logarithm,* a word from the Greek, meaning "ratio number."

He spent most of his life at an impressive ancestral estate, Merchiston Castle, near Edinburgh. Much of his energy was squandered on the religious and political debates of the times. He believed that his reputation would be secured in posterity by a book in which he endeavored to prove that the Pope was an Antichrist and that the world was going to end between 1688 and 1700. The book ran 21 editions! He also wrote about war machines and "devices for sayling under water," together with plans and drawings. Some of his machines were notably similar to modern weapons like the submarine and the tank. Many thought him mentally unbalanced and suspected that he dealt in the black arts.

Napier used mathematics and science as a diversion from his serious argumentation on the topical political and religious issues. It is curious that his hobby has had such a lasting and profound influence, while the work that he thought was important has long since been discounted.

Chapter Overview

The exponential and logarithmic functions introduced in this chapter offer a real contrast to the algebraic functions of prior work. First, the variable is the *exponent*, not the base as previously. As a result, the functions have unusual properties that require special attention but provide interesting results, including a rich variety of applications. A new number, *e*, is encountered arising naturally from the exponential function. Due to the computation necessary in real-world applications a calculator becomes a valuable device in this chapter.

*Optional

6.1 Exponential Functions

Growth and decay are important concepts in today's complicated technology and in the study of the impact of this technology on society. Population growth, radioactive decay, growth of investment from interest, profit falling in the face of rising overhead, and many other applications in the social sciences and behavioral sciences, biology, and management are described in terms of *exponential functions*.

EXPONENTIAL FUNCTION

The function f is an **exponential function** if

$$f(x) = b^x \qquad \text{where } b \text{ is a positive constant other than } 1$$

The number b is called the **base** and x is the **exponent.**

So far we have worked only with rational number exponents, so the domain of the function defined above is the set of rational numbers. If the function is graphed, the graph will consist of a series of distinct points. Admittedly, these points would be very close together, but it would be better if the gaps could be filled by making some sense of numbers such as $2^{\sqrt{3}}$ or 5^{π}. The behavior of this function for irrational numbers should be consistent with its behavior for the rational numbers. To ensure this, values at irrational x must have the following property:

For any real number x and rational numbers h and k:

1. If $b \neq 1$, then b^x is a unique positive real number.
2. If $b > 1$ and $h < x < k$, then $b^h < b^x < b^k$.
3. If $0 < b < 1$ and $h < x < k$, then $b^k < b^x < b^h$.

This property will ensure that an irrational power has a real value and that that value is what is expected. Graphs will be smooth with no gaps or jumps. Graphs of exponential functions may now be made using rational values of x with the assurance that a smooth curve through those points will accurately graph all the real values.

Example 1 Estimate $2^{\sqrt{2}}$.

Solution Since $1.4 < \sqrt{2} < 1.5$, then $2^{1.4} < 2^{\sqrt{2}} < 2^{1.5}$. Also,

$$2^{1.4} = 2^{7/5} \approx 2.6 \qquad \text{and} \qquad 2^{1.5} = 2^{3/2} \approx 2.8$$

Thus, $2.6 < 2^{\sqrt{2}} < 2.8$ and **$2^{\sqrt{2}}$ is approximately 2.7.**

Actually, $2^{\sqrt{2}} \approx 2.66514414$, which you can check on a calculator with a y^x key, as shown below.

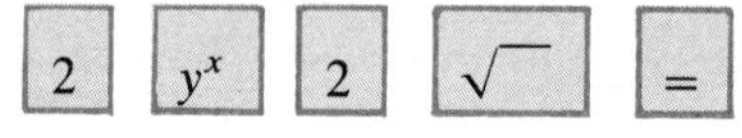

□

Example 2 Sketch the graph of $f(x) = 2^x$.

Solution

x	2^x
−3	.125
−2	.25
−1	.5
0	1
1	2
2	4
3	8
4	16

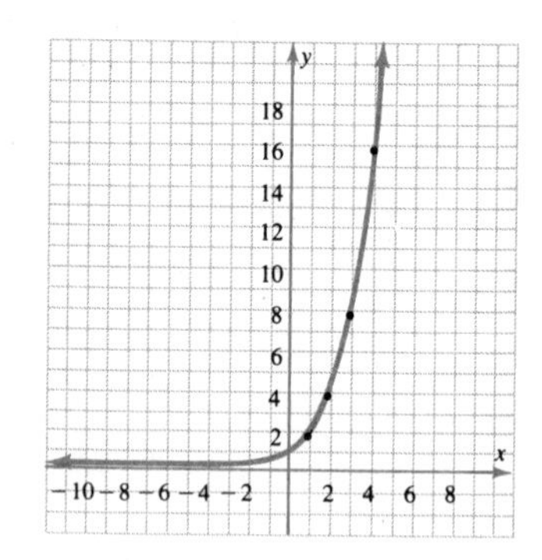

□

Example 3 Sketch the graph of $f(x) = 2^{-x}$.

Solution

x	2^{-x}
−4	16
−3	8
−2	4
−1	2
0	1
1	.5
2	.25
3	.125

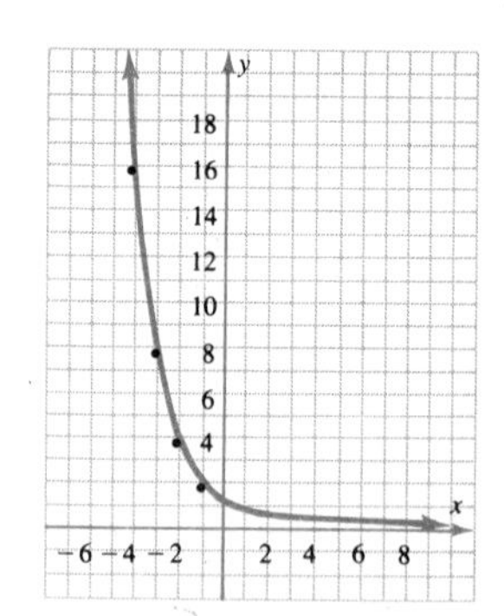

□

Notice the similarity of the graphs of $y = 2^x$ and $y = 2^{-x}$. They are symmetric with respect to the y-axis, as shown in Figure 6.1. This behavior is true in general for $y = b^x$ and $y = b^{-x}$. Additionally, $y = b^x$ is an increasing function for $b > 1$ and a decreasing function for $0 < b < 1$.

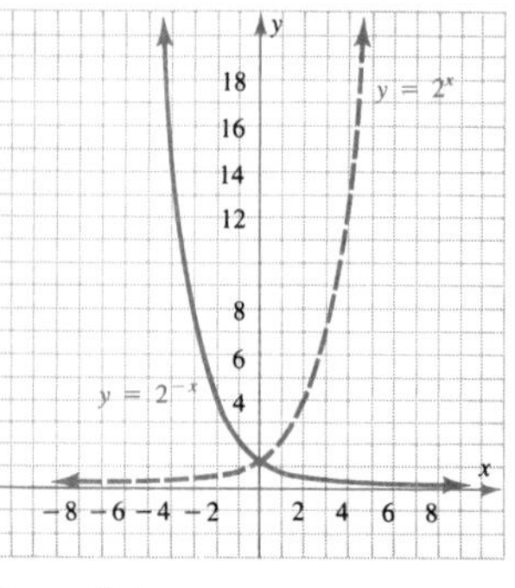

Figure 6.1

Example 4 Sketch the graph of $y = 2^{-x^2}$.

Solution After compiling a table of values, notice that all values except for $x = 0$ are less than 1. Thus, a different scale would be preferable and additional values for $0 < x < 1$ would be helpful to obtain a more accurate sketch.

x	−3	−2	−1	0	1	2	3	.25	.5	.75	1.5
$-x^2$	−9	−4	−1	0	−1	−4	−9				
y	.002	.06	.5	1.0	.5	.06	.002	.96	.84	.68	.21

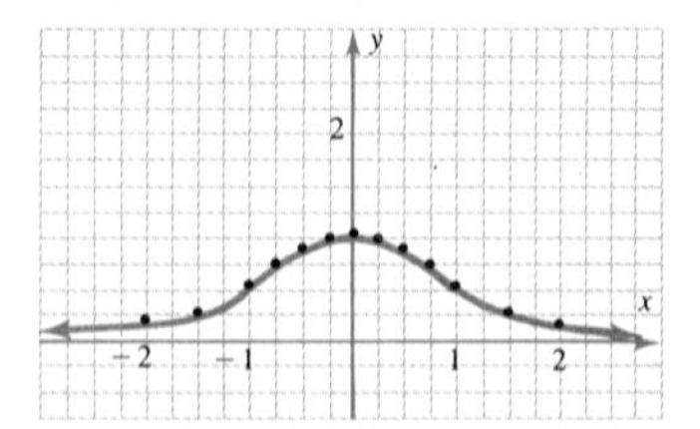

□

Notice in the graphs in the examples that the x-axis is an asymptote like that seen for the hyperbola in Chapter 3 and for rational functions in Chapter 5. As x gets very large or very small, the values of the function become closer and closer to zero, but never reach zero. The number 0 is said to be *the limit of the value of the function*.

Many equations involving exponentials can be solved using a property that results from the fact that the exponential function is one-to-one. This property is very useful when combined with the laws of exponents extended to the real numbers. For two equal powers, the property asserts that if the bases are equal, the exponents must be equal; or if the exponents are equal, the bases will be equal.

PROPERTY OF EXPONENTIALS

For nonzero real x and y, and positive $a, b \neq 1$:

1. If $a^x = a^y$, then $x = y$.
2. If $a^x = b^x$, then $a = b$, if x is odd.
3. If $a^x = b^x$, then $|a| = |b|$, if x is even.

Example 5 Solve for the variable: $5^x = 125$

Solution Write as powers of the same base; then the exponents must be equal:

$$5^x = 125$$
$$5^x = 5^3$$

Thus, $\mathbf{x = 3}$. □

Example 6 Solve for the variable: $36^y = 216$

Solution

$$36^y = 216$$
$$(6^2)^y = 6^3$$
$$6^{2y} = 6^3$$

Thus, $2y = 3$

$\mathbf{y = \frac{3}{2}}$ □

Example 7 Solve for the variable: $8^{z-1} = \frac{1}{4}$

Solution

$$8^{z-1} = \frac{1}{4}$$
$$(2^3)^{z-1} = 2^{-2}$$
$$2^{3z-3} = 2^{-2}$$

Thus, $3z - 3 = -2$

$3z = 1$

$\mathbf{z = \frac{1}{3}}$ □

Example 8 Solve for the variable: $b^3 = 64$

Solution Write with the same exponent; then the bases must be equal:

$$b^3 = 64$$
$$b^3 = 4^3$$

Thus, $\mathbf{b = 4.}$ □

Problem Set 6.1

A *For each of Problems 1–10, sketch the graphs of each pair of functions on a set of coordinate axes.*

1. $y = 2^x;\ y = 3^x$

2. $y = 2^x;\ y = 4^x$

3. $y = 2^x;\ y = 2^{-x}$

4. $y = 3^x;\ y = 3^{-x}$

5. $y = 3^x;\ y = 3^{2x}$

6. $y = 2^x;\ y = 2^{3x}$

7. $y = 3^x;\ y = 3^{x^2}$

8. $y = 2^x;\ y = 2^{x^2}$

9. $y = 2^x;\ y = 2^{|x|}$

10. $y = 3^x;\ y = 3^{-|x|}$

Solve the exponential equations in Problems 11–40.

11. $2^x = 8$

12. $3^x = 27$

13. $5^x = 625$

14. $3^x = 243$

15. $9^x = 3$

16. $8^x = 2$

17. $343 = 7^x$

18. $625 = 5^x$

19. $4 = 8^x$

20. $27^x = 9$

21. $81^x = 27$

22. $125^x = 25$

23. $b^3 = 343$

24. $b^5 = 32$

25. $a^3 = 1331$

26. $3125 = a^5$

27. $25^x = \frac{1}{5}$

28. $27^x = \frac{1}{3}$

29. $8^{2x-2} = \frac{1}{16}$

30. $9^{3x-2} = \frac{1}{27}$

31. $\left(\frac{2}{3}\right)^x = \frac{27}{8}$

32. $\left(\frac{10}{7}\right)^x = .49$

33. $\left(\frac{4}{9}\right)^x = 1.5$

34. $\left(\frac{25}{64}\right)^x = 1.6$

35. $9^{y+1} = 243^{1-2y}$

36. $125^{2-3y} = 625^{y-3}$

37. $\left(\frac{4}{5}\right)^{z+3} = \frac{25}{16}$

38. $\left(\frac{7}{2}\right)^{3-z} = \frac{8}{343}$

39. $(x^2 - 4x + 4)^3 = (27x^3)^2$

40. $64x^6 = (x^2 + 6x + 9)^3$

B *In Problems 41–50 sketch the graphs of each pair of relations on a single set of axes.*

41. $y = 2^x;\ y = 2^{\sqrt{x}}$

42. $y = 3^x;\ y = 3^{-\sqrt{x}}$

43. $y = 3^x;\ y = 3^{1-x}$

44. $y = 2^x;\ y = 2^{2-x}$

45. $y = 2^x;\ y = 2^{1/x}$

46. $y = 3^x;\ y = 3^{-1/x}$

47. $y = 3^{x-1};\ y = 3^{x+1}$

48. $y = 2^{3-x};\ y = 2^{3+x}$

49. $y = 2^{3x-2};\ y = 2^{2x-3}$

50. $y = 3^{2x-1};\ y = 3^{x-2}$

In Problems 51–60 estimate the value of the number as in Example 1; then check with a calculator.

51. $3^{\sqrt{2}}$ **52.** $5^{\sqrt{3}}$ **53.** 4^{π} **54.** 3^{π}

55. $\pi^{\sqrt{10}}$ **56.** $\sqrt{10}^{\pi}$ **57.** $\sqrt{2}^{\sqrt{3}}$ **58.** $\sqrt{3}^{\sqrt{2}}$

59. $(1 + \sqrt{5})^{\sqrt{7}}$ **60.** $(1 + \sqrt{5})^{1 + \sqrt{2}}$

C *Problems 61 and 62 refer to the following:*

Archaeology *Carbon-14 dating is a standard method of determining the age of artifacts. Radioactive substances decompose or decay. The decay rate is measured by* ***half-life,*** *which is the time required for one-half of a substance to decompose. There is a constant percentage of the total carbon present in all living matter that is in the radioactive form, denoted by ^{14}C. Carbon-14 has a half-life of approximately 5600 years. When an organic substance dies, the ^{14}C begins to decay and the amount A remaining after t years is a measure of its age. If A_0 is the original amount of carbon present, then*

$$A = A_0\left(\frac{1}{2}\right)^{t/h}$$

where h is the half-life.

61. Let $A_0 = 100$ and $h = 5600$, and sketch the graph of the decay of ^{14}C by plotting $A = f(t)$. (Take $t = 5600, 11{,}200, \ldots$, and adjust scales accordingly.)

62. The half-life of radium is approximately 1600 years. Sketch the graph for radium in the same manner as Problem 61, where $A_0 = 100$.

Problems 63 and 64 refer to the following:

Space science *A satellite has a radioisotope power supply. The power output P (in watts) is proportional to the amount of the isotope remaining after t days and is*

$$P = P_0\left(\frac{1}{2}\right)^{t/h}$$

where P_0 is the original output and h is the half-life of the isotope.

63. If the satellite has 50 watts power initially from an isotope with a half-life of 173 days, sketch the graph of $P = f(t)$.

64. The satellite in Problem 63 needs a minimum of 6.25 watts to operate a certain mission. How many days after orbit is this mission possible?

6.2 The Natural Base

Exponential functions have a great many applications. A variety of examples are considered here, with emphasis on *growth* and *decay*. Radioactive substances, for example, decompose or decay at varying rates. The decay is measured by the **half-life** of the substance, which is the time required for one-half of the material to decompose. The amount A remaining after time t is given by the exponential equation

$$A = A_0 \left(\tfrac{1}{2}\right)^{t/h}$$

where A_0 is the original amount and h is the half-life (given in the same units as t). Notice that since $0 < \frac{1}{2} < 1$, this is a decreasing function.

Example 1 The half-life of radium-226, denoted by ^{226}Ra, is 1622 years. How much of 110 g of ^{226}Ra remains after 10,000 years?

Solution Here, $A_0 = 110$, $t = 10{,}000$, and $h = 1622$.

$$A = A_0 \left(\tfrac{1}{2}\right)^{t/h}$$

$$A = 110 \left(\tfrac{1}{2}\right)^{10{,}000/1622}$$

$$\approx 110(.0139342)$$

$$\approx 1.53276$$

Approximately 1.5 g of the 110 g remains. □

Note the use of ≈. At nearly every point where a calculator (or table) is used, an approximation is obtained. Calculators are certainly accurate, but only to the limits of their displays. They may, in fact, be incorrect in the last one or two places shown. For a calculator with parentheses, the following scheme may be used to find the value in Example 1:

Using a model without parentheses, the quantity 10,000/1622 will probably have to be stored, as shown in the following alternate scheme:

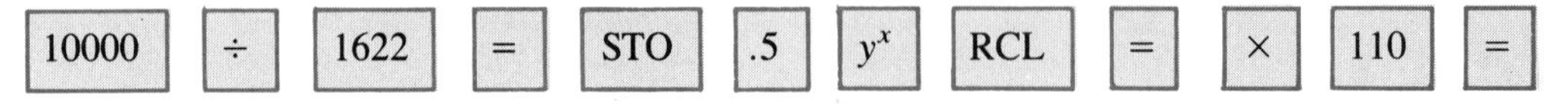

Now consider an example of growth rather than decay, which is an increasing function rather than a decreasing function. Population growth is sometimes gauged by the frequency with which it doubles in size. If the population of a city doubles every n years, then the population P after t years may be approximated by

$$P = P_0 2^{t/n}$$

where P_0 is the initial population.

Example 2 If the population of Arlington, Texas, continued to grow at its 1970–1980 rate, Arlington would double in size approximately every 12 years. If the population of Arlington was 90,229 in 1970, estimate its 1985 population to the nearest thousand.

Solution Here, $P_0 = 90{,}229$, $n = 12$, and $t = 15$.

$$P = P_0 2^{t/n}$$

$$P = 90{,}229(2^{15/12})$$

$$\approx 214{,}601.94$$

$$\approx 215{,}000$$

The 1985 population is approximately 215,000. □

A very practical application of the exponential function is the growth of money at compound interest. In computing compound interest, the principal on which interest is computed is increased by the addition of interest at the end of each interest period. Whenever interest is not actually paid at the end of the period, but instead added to the principal, it is said to be **compounded.** If a principal of P dollars is invested at annual rate r compounded n times a year for t years, then the amount A accumulated is given by

$$A = P\left(1 + \frac{r}{n}\right)^{nt}$$

Example 3 How much is in an account earning 9% compounded quarterly for 5 years, if the initial balance was $1500?

Solution Here, $P = 1500$, $r = .09$, $n = 4$, and $t = 5$.

$$A = P\left(1 + \frac{r}{n}\right)^{nt}$$

$$A = 1500\left(1 + \frac{.09}{4}\right)^{4(5)}$$

$$= 1500(1.0225)^{20}$$

$$\approx 1500(1.56051)$$

$$\approx 2340.76$$

About $2340.76 is accumulated after 5 years. □

Compound interest gives the opportunity to investigate an important and interesting quantity. To do so, consider the formula

$$A = P\left(1 + \frac{r}{n}\right)^{nt}$$

If you let $P = 1$, $t = 1$, and $r = 1$, then

$$A = \left(1 + \frac{1}{n}\right)^n$$

What happens as n increases? That is, if \$1 is invested at 100% interest for 1 year, what happens as the number of times the interest is compounded increases? Interest is often advertised as compounded monthly, weekly, daily, or even instantaneously (or continuously). To investigate the differences, the amount earned may be computed if compounded yearly, semiannually, quarterly, monthly, weekly, daily, hourly, and so on by letting $n = 1, 2, 4, 52, 365, 8760, \ldots$, as shown in Table 6.1.

Table 6.1

n	$\left(1 + \frac{1}{n}\right)$	$\left(1 + \frac{1}{n}\right)^n$	**Amount in Dollars (to the nearest cent)**
1	2	2	2.00
2	1.5	2.25	2.25
4	1.25	2.44	2.44
12	1.083	2.61	2.61
52	1.019	2.69	2.69
365	1.0027	2.7145	2.71
8,760	1.0001142	2.7181	2.72
525,600	1.000001903	2.71826	2.72
31,536,000	1.0000000321	2.7182818	2.72

The table shows that substantial increases occur until interest is calculated weekly or daily; beyond that, the increases are very small. In fact, no matter how often the interest is compounded, the amount accumulated does not exceed an irrational number that is approximately 2.7182818. This number, called the **natural base,** is denoted by the symbol e. Used as the base for exponential functions, this number has a great many applications since it is associated with continuous growth. If interest were compounded continuously—that is, every instant—you could rewrite the compound interest formula as

$$A = Pe^{rt}$$

Consider the difference in Example 3 if the interest is compounded continuously.

Example 4 How much is in an account earning 9% compounded continuously for 5 years, if the initial balance was \$1500?

Solution Here again, $P = 1500$, $t = 5$, and $r = .09$, but this time we use the formula for continuous compounding:

$$A = Pe^{rt}$$

$$A = 1500e^{5(.09)}$$

$$= 1500e^{.45}$$

$$\approx 1500(1.56831)$$

≈ 2352.47

About $2352.47 is in the account after 5 years. □

While the y^x key on the calculator may be used to compute e^x, most scientific calculators have an e^x key. The following keys show how the result in Example 4 may be calculated:

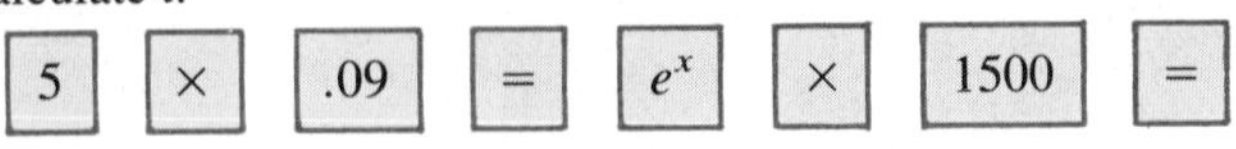

Interest, of course, is generally compounded a finite number of times (discretely) over any time period. However, there are many natural phenomena that grow continuously rather than discretely. Certain kinds of growth or decay are described or approximated by an exponential function of the form

$$y = y_0 e^{kt}$$

where t is the time and k is a constant that depends on the particular phenomenon. If $t = 0$, then the beginning or initial amount is found by substitution:

$$\begin{aligned} y &= y_0 e^{k(0)} \\ &= y_0 e^0 \\ &= y_0 \cdot 1 \\ &= y_0 \end{aligned}$$

Thus, y_0 is the initial amount present—that is, the amount present at time 0.

Example 5 A city has a population of 15,000 in 1985. If the population is approximated by $P = P_0 e^{.04t}$, where t is given in years, what will be the population in the year 2005?

Solution The initial population is 15,000 and the time is 20 years from 1985 to 2005, so $P_0 = 15{,}000$, $t = 20$, and

$$\begin{aligned} P &= 15{,}000e^{.04(20)} \\ &= 15{,}000e^{.8} \\ &\approx 15{,}000(2.22554) \\ &\approx 33{,}383.1 \\ &\approx 33{,}000 \end{aligned}$$

The population will be approximately 33,000 in the year 2005.

□

Recall the comparison of $y = b^x$ and $y = b^{-x}$. If $y = y_0 e^{kt}$, $kt > 0$, represents growth, an increasing function, then $y = y_0 e^{-kt}$ represents decline or decay. A common application is to the radioactive decay or half-life of a substance. Since this is a continuous process, we can substitute the function $y = y_0 e^{-kt}$ for the formula introduced at the beginning of this section.

Example 6 The decay of radioactive argon-39 is described by

$$y = y_0e^{-.173t}$$

where y mg of the argon remains of y_0 mg after t minutes. If you begin with 100 mg of argon-39, then how much is left after 10 minutes?

Solution Here, $y_0 = 100$ and $t = 10$.

$$\begin{aligned} y &= 100e^{-.173(10)} \\ &= 100e^{-1.73} \\ &\approx 100(.177) \\ &= 17.7 \end{aligned}$$

Value obtained from Table II. Better values are available with a calculator.

Thus, 17.7 mg are left after 10 minutes decay. □

Example 7 The atmospheric pressure, P, in pounds per square inch (psi), is approximated by $P = 14.7e^{-.21a}$, where a is the altitude above sea level in miles. What is the pressure:

a. At sea level? **b.** In Denver, the "mile-high city"? **c.** In space 10 miles from earth?

Solution **a.**

$$\begin{aligned} P &= 14.7e^{-.21(0)} \\ &= 14.7e^0 \\ &= 14.7 \end{aligned}$$

Where $a = 0$ at sea level

Approximately 14.7 psi at sea level

b.

$$\begin{aligned} P &= 14.7e^{-.21(1)} \\ &= 14.7e^{-.21} \\ &\approx 14.7(.811) \\ &\approx 11.9 \end{aligned}$$

Where $a = 1$

Approximately 11.9 psi in Denver

c.

$$\begin{aligned} P &= 14.7e^{-.21(10)} \\ &= 14.7e^{-2.10} \\ &\approx 14.7(.122) \\ &\approx 1.80 \end{aligned}$$

Where $a = 10$

Approximately 1.8 psi in space (10 miles) □

Exponential functions can be used to model a wide variety of phenomena. Problem Set 6.2 presents a representative selection.

One last remark on the definition of the natural base e is in order. Recall that as larger values of n were taken, the expression $(1 + 1/n)^n$ became and remained closer and closer to the value of e (see Table 6.1). It is said that the **limit** or **limiting value** is e, and this may be written as shown in the box. The notion and notation of limits will be considered in greater detail in Chapter 8.

DEFINITION OF e

$$\lim_{n\to\infty}\left(1+\frac{1}{n}\right)^n = e$$

where e is called the **natural base** and is approximately 2.71828182845904523536.

Problem Set 6.2

A *The following is necessary for Problems 1–6:*

Physics *The amount A of a radioactive substance remaining after time t is given by*

$$A = A_0\left(\tfrac{1}{2}\right)^{t/h}$$

where A_0 is the original amount and h is the half-life given in the same units as t.

1. The half-life of ^{234}U, uranium-234, is $2.52 \cdot 10^5$ years. How much of a 100 g sample remains after 10,000 years?
2. How much of a 100 g specimen of ^{22}Na, sodium-22, remains after 7 years if its half-life is 2.6 years?
3. Curium-242, ^{242}Cm, has a half-life of 163 days. How much remains of 10 g after 2 years?
4. Neptunium-239, ^{239}Np, has a half-life of 2.34 days. How much remains of 10 g after 1 week?
5. How much of 10 g of ^{198}Pb, lead-198, remains after a day if its half-life is 2.4 hours?
6. How much of 25 g of ^{243}Pu, plutonium-243, remains after a day if its half-life is 4.98 hours?

The following is necessary for Problems 7–10. Give answers to the nearest thousand.

Social science *The population P of a city after t years is given by*

$$P = P_0 2^{t/n}$$

where P_0 is the initial population and n is the number of years necessary for it to double.

7. If the population of Anchorage, Alaska, continued to grow at its 1970–1980 rate, the city would double in size approximately every 5.4 years. Estimate its 1990 population if it was 48,081 in 1970.
8. Aurora, Colorado, would double in size every 8 years if the population continued to grow at its 1970–1975 rate. Estimate its 1985 population if the population was 74,974 in 1970.
9. Every 36 years, Little Rock, Arkansas, would double in population if the population continued to grow at its 1960–1980 rate. Estimate the 1985 population of Little Rock if it was 107,813 in 1960.
10. Springfield, Missouri, had a population of 95,865 in 1960, and it grew from 1960 to 1980 at a rate that would cause it to double every 42.23 years. Estimate Springfield's population in 1990.

B *Problems 11–20 require the following:*

Business *If P represents the principal invested at annual interest rate r compounded n times a year, then A is the amount accumulated in t years:*

$$A = P\left(1 + \frac{r}{n}\right)^{nt}$$

If interest is compounded continuously, the amount A is given by the formula

$$A = Pe^{rt}$$

11. A thousand dollars is left in a bank savings account drawing 7% interest, compounded quarterly for 10 years. What is the balance at the end of that time?

12. A thousand dollars is left in a credit union drawing 7% compounded monthly. What is the balance at the end of 10 years?

13. \$1750 is invested in an account earning $13\frac{1}{2}$% compounded monthly for a 2 year period. What is the account worth at the end of that time?

14. You lend out \$5500 at 10% compounded monthly. If the debt is repaid in 18 months, what is the total owed at the time of repayment?

15. A \$10,000 Treasury Bill earned 16% compounded monthly. If the bill matured in 2 years, what was it worth at maturity?

16. You borrow \$25,000 at $12\frac{1}{4}$% compounded monthly. If you are unable to make any payments the first year, how much do you owe, excluding penalties?

17. A savings institution advertises 7% annual interest, compounded daily. How much more interest would you earn over the bank savings account or credit union in Problems 11 and 12?

18. An $8\frac{1}{2}$% account earns continuous interest. If \$2500 is deposited for 5 years, what is the total accumulated?

19. You lend \$100 at 10% continuous interest. If you are repaid 2 months later, what was owed?

20. If you had a million dollars for just 2 months, how much interest could be earned in an account earning 10% compounded monthly?

The following is necessary for Problems 21–26. Give answers to the nearest 10,000.

Social science *Human population growth can be approximated by*

$$P = P_0e^{rt}$$

where a population of P_0 grows at an annual rate of r to P after t years.

21. The population of the state of Texas grew from 1950 to 1980 at an annual rate of approximately 2%. If the population in 1950 was 7,711,194, what was the population in 1980?

22. Estimate the population of Texas in 1990, using the information in Problem 21.

23. Florida grew in population between 1940 and 1980 at an annual rate of 4.09%. If the population was 1,897,414 in 1940, what was the population in 1980?

24. What is the anticipated population of Florida in the year 2000, according to the data in Problem 23?

25. The population of the city of Los Angeles was 1,970,358 in 1950. It has grown since at an annual rate of 1.36%. Estimate its population in 1980, 1990, and 2000.

26. San Jose, California, has had a phenomenal 6% annual growth since 1950. Estimate its population in 1980, 1990, and 2000, if its population was 95,280 in 1950.

The following is necessary for Problems 27 and 28:

Earth science *The atmospheric pressure P (in psi) is approximated by*

$$P = 14.7e^{-.21a}$$

where a is the altitude above sea level in miles.

27. Mt. McKinley, in Alaska, is the highest point in North America. The elevation is 20,320 feet, or 3.85 miles. What is the pressure at its summit?

28. The lowest land point in the world is the Dead Sea (Israel–Jordan), where the elevation is 1299 feet below sea level ($-.25$ mile). What is the atmospheric pressure at this point?

The following is necessary for Problems 29 and 30:

Physics *The radioactive decay of a substance is expressed by*

$$A = A_0e^{-kt}$$

where the initial amount A_0 decays to an amount A after t years. The positive constant k differs for each substance.

29. Strontium-90 is used in nuclear reactors. What amount of 250 mg of strontium-90 is present after 5 years if $k = .0248$?

30. Radium decays such that $k = .0004$. Find out how much of 1000 mg of radium remains after a century.

C **31.** *Medicine* A healing law for skin wounds states that

$$A = A_0e^{-.1t}$$

where A is the number of square centimeters of unhealed skin after t days when the original area of the wound was A_0. How many days does it take for half of the wound to heal?

32. *Physics* A law of light absorption of a medium for a beam of light passing through is given by

$$I = I_0e^{-rt}$$

where I_0 is the original intensity of the beam in lumens, and I is the intensity after passing through t cm of a medium whose absorption coefficient is r. Find the intensity of a 100 lumen beam after it passes through 2.54 cm of a medium with absorption coefficient .095.

33. *Psychology* A *learning curve* describes the rate at which a person learns certain specific tasks. If N is the number of words per minute typed by a student,

$$N = 80(1 - e^{-.016n})$$

where n is the number of days of instruction. Assuming Joe is an average student, what is his typing rate after 20 days (4 weeks) of instruction?

34. *Psychology* Members of a discussion group tend to be ranked exponentially by the number of times they participate in the discussion. For a group of ten, the number of times P_n, the nth-ranked participant, takes part is given by

$$P_n = P_1 e^{.11(1 - n)}$$

where P_1 is the number of times the first-ranked person participates in the discussion. For each 100 times the top-ranked participant enters the discussion, how many times should the bottom-ranked person be expected to participate?

35. *Psychology* Sketch the *learning curve* in Problem 33. First evaluate the rate for every 2 weeks of instruction ($n = 10, 20, 30, \ldots$).

36. *Psychology* Sketch the *discussion curve* for the ten-person group of Problem 34. (Note that the domain here is the counting numbers less than 11.)

37. *Business* Without benefit of promotion, the yearly sales S of a particular product will decrease at a constant yearly rate. After t years,

$$S = S_0 e^{-kt}$$

where S_0 is the sales in the initial year and k is called the *sales decay constant*. Sketch the graph of the sales decline curve by computing and plotting several points where $S_0 = 1{,}000{,}000$ and $k = .18$.

38. *Physics* The current I (in amperes) in an electrical circuit is

$$I = \frac{E}{R}(1 - e^{-Rt/L})$$

t seconds after the circuit is closed, where E is the voltage, R is resistance (in ohms), and L is inductance (in henrys). Sketch the graph of this function in which $E = 10$, $R = 5$, and $L = 2.1$. Begin by finding the current for each tenth of a second after the switch is flipped.

39. *Physics* The height s of an object after t seconds is given by

$$\text{(A)} \qquad s = v_0 t - 16t^2$$

where v_0 is the initial velocity, if friction (or air resistance) is ignored. This formula is valid only in a vacuum; however, the following more accurately describes the path of the projectile:

$$\text{(B)} \qquad s = \frac{m}{k}\left[\left(v_0 + \frac{32m}{k}\right)(1 - e^{-kt/m}) - 32t\right]$$

where m is the mass and k is a constant for air resistance. Because the graph of equation (A) is a parabola, the graph of (B) cannot be too different. Let $m = 1$, $k = .001$, and $v_0 = 160$, and sketch each. Now compare the two graphs and suggest the reasons for any differences.

40. Estimate the value and make a conjecture about the actual value of the *continued fraction* shown below. [*Hint:* First evaluate $2 + \frac{1}{1}$, then $2 + \dfrac{1}{1 + \frac{1}{2}}$, continuing to take another term each time. A pattern should emerge.]

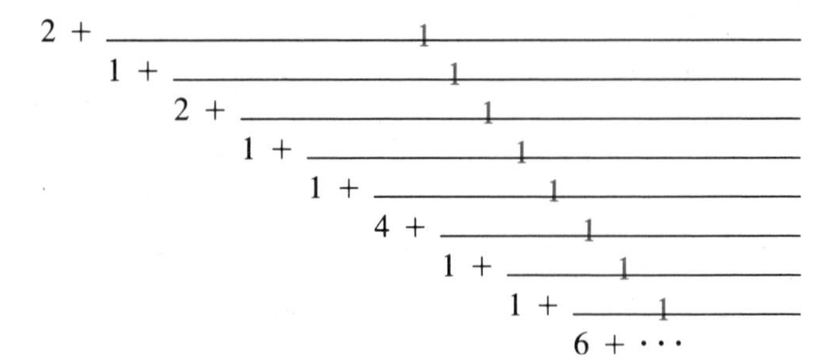

6.3 Logarithmic Functions

The exponential function is seen to be one-to-one and thus has an inverse function; that is, if

$$y = f(x) = b^x \qquad b > 0, b \neq 1$$

then f^{-1} is defined by

$$\mathbf{x = b^y \qquad b > 0, b \neq 1}$$

However, at this point, the means to solve explicitly for $y = f^{-1}(x)$ in terms of x are not available. It can be expressed in words:

y* is the exponent of *b* that yields *x

This is rather wordy. Historically, this new function has been called the **logarithm function** and it is written

$$\mathbf{y = \log_b x}$$

which is read "y is the logarithm (or log) to the base b of x." This means y is the exponent of b that will produce x. The need to write the inverse of the exponential function explicitly in terms of its independent variable leads to the following definition:

LOGARITHMIC FUNCTION

For $b > 0$, $b \neq 1$,

$$x = b^y \qquad \text{is equivalent to} \qquad y = \log_b x$$

The logarithm to base b of x is the exponent y to which b must be raised to equal x.

Note that b is called the *base* for *both* the logarithmic and exponential functions.

The equations $y = \log_b x$ and $x = b^y$ are used interchangeably. An exponential may be rewritten in logarithmic form, and vice versa. If base 10 is used, the logarithms are called **common logarithms** and written $\mathbf{y = \log x}$ with no indication of base. For scientific purposes, the natural base e is used. Such logarithms are called **natural logarithms** and written $\mathbf{y = \ln x}$.

COMMON AND NATURAL LOGARITHMS

$\log x$ means $\log_{10} x$ and $\ln x$ means $\log_e x$

Example 1 Change from exponential form to logarithmic form:

a. $8 = 2^3$ **b.** $\frac{1}{9} = 3^{-2}$ **c.** $64 = 16^{3/2}$ **d.** $1 = e^{\circ}$

Solution

	Exponential	**Logarithmic**
a.	$8 = 2^3$	$\log_2 8 = 3$
b.	$\frac{1}{9} = 3^{-2}$	$\log_3 \frac{1}{9} = -2$
c.	$64 = 16^{3/2}$	$\log_{16} 64 = \frac{3}{2}$
d.	$1 = e^0$	$\ln 1 = 0$

□

Example 2 Change from logarithmic form to exponential form:

a. $\log_{10} 1000 = 3$ **b.** $\log_5 1 = 0$ **c.** $\log_4 8 = \frac{3}{2}$ **d.** $\ln e^2 = 2$

Solution

	Logarithmic	**Exponential**	
a.	$\log_{10} 1000 = 3$	$1000 = 10^3$	Base 10
b.	$\log_5 1 = 0$	$1 = 5^0$	Base 5
c.	$\log_4 8 = \frac{3}{2}$	$8 = 4^{3/2}$	Base 4
d.	$\ln e^2 = 2$	$e^2 = e^2$	Base e

□

Example 3 Find the value of y in $\log_3 (9\sqrt{3}) = y$.

Solution Rewrite in exponential form:

$$3^y = 9\sqrt{3}$$
$$3^y = 3^2 \cdot 3^{.5}$$
$$3^y = 3^{2+.5}$$
$$3^y = 3^{2.5}$$
$$\mathbf{y = 2.5}$$

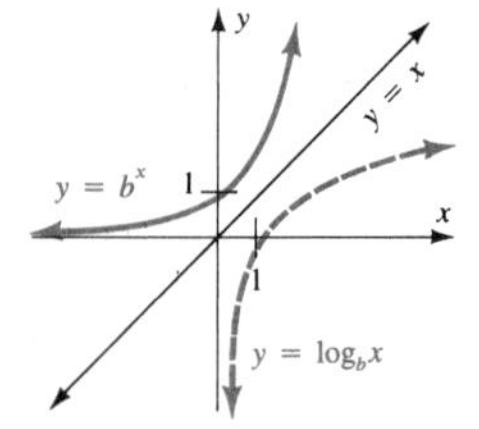

Figure 6.2
Graphs of $y = b^x$ and $y = \log_b x$ where $b > 1$

□

Examining the graph of this new function focuses attention on its more important features. Observe first in Figure 6.2 that since $y = \log_b x$ and $y = b^x$ are inverse functions, they are symmetric with respect to the line $y = x$.

Keep in mind that $y = \log_b x$ is equivalent to $x = b^y$. Using this equivalence will make it easier to compile pairs of values for the graph of $y = \log_b x$. Convenient values may be chosen for y; x can then be computed from $x = b^y$; and points can be plotted to help sketch $y = \log_b x$.

Example 4 Sketch the graph of $y = \log_2 x$.

Solution First, $y = \log_2 x$ is equivalent to $x = 2^y$, so you can compile a table with $y = -2$, -1, 0, 1, 2, 3, and plot the corresponding points.

x	y
$\frac{1}{4}$	-2
$\frac{1}{2}$	-1
1	0
2	1
4	2
8	3

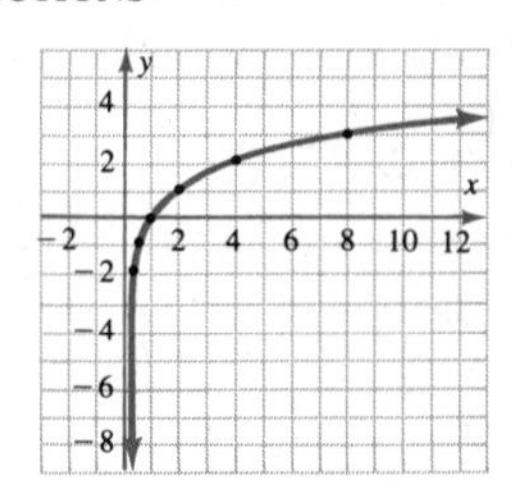

□

Example 5 Sketch the graph of $y = \log_{1/3} x$.

Solution The graph is identical to the graph of $x = \left(\frac{1}{3}\right)^y$. First compile a table from convenient values of y, then plot these points, and finally sketch a smooth curve through the points.

x	y
$\frac{1}{3}$	1
1	0
3	-1
9	-2

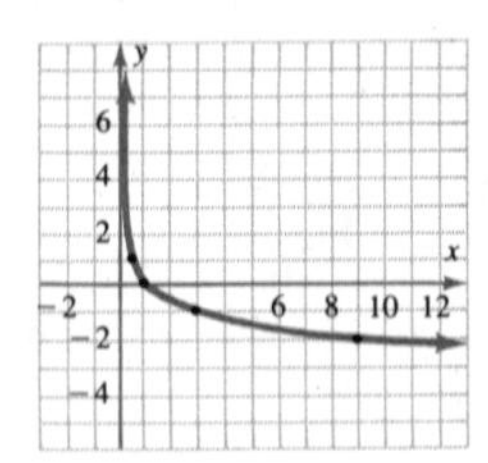

□

From these graphs, note the following important properties:

1. $y = \log_b x$ is increasing if $b > 1$ and decreasing if $0 < b < 1$
2. $y = \log_b x$ is not defined for $x \leqslant 0$
3. $y = \log_b x$ is 0 if and only if $x = 1$; that is, $\log_b 1 = 0$
4. $y = \log_b x$ is 1 if and only if $x = b$; that is, $\log_b b = 1$
5. $y = \log_b x$ is asymptotic to $x = 0$, the y-axis

Since each logarithmic function is equivalent to an exponential, each law of exponents has a counterpart for logarithms. For example, consider the first law, $b^x b^y = b^{x+y}$.

Let $A = b^x$ and $B = b^y$; then $x = \log_b A$ and $y = \log_b B$, and the product $AB = b^x b^y$, or

$$AB = b^{x+y}$$

which can be written in logarithmic form:

$$\log_b AB = x + y$$

Now substitute for x and y:

$$\log_b AB = \log_b A + \log_b B$$

The other laws of exponents yield corresponding properties for logarithms. These properties are summarized in the box.

PROPERTIES OF LOGARITHMS

If A and B are positive, $b > 0$, $b \neq 1$, then:

I. $\log_b AB = \log_b A + \log_b B$
The log of the product of two numbers is the sum of the logs of the numbers.

II. $\log_b \frac{A}{B} = \log_b A - \log_b B$
The log of the quotient of two numbers is the log of the numerator minus the log of the denominator.

III. $\log_b A^n = n \cdot \log_b A$
The log of the nth power of a number is n times the log of the number.

IV. $\log_b 1 = 0$
The log of 1 is 0 for any base.

V. $\log_b \frac{1}{A} = -\log_b A$
The log of the reciprocal of a number is the negative of the log of the number.

VI. **$\log_b m = \log_b n$ is equivalent to $m = n$** if m and n are positive real numbers.
If the logs are equal, then the numbers are equal; and if the numbers are equal, then their logs are equal.

These properties allow you to convert multiplication problems to addition, division to subtraction, and powers and roots to multiplication. For example,

$$\log_b(1745 \cdot 423) = \log_b 1745 + \log_b 423$$

Instead of multiplying these numbers you may add their logarithms. The examples that follow illustrate the use of these properties to expand expressions.

Example 6

$$\begin{aligned} \log_b 21 &= \log_b(3 \cdot 7) \\ &= \log_b 3 + \log_b 7 \end{aligned}$$ □

Example 7

$$\log_b \tfrac{2}{3} = \log_b 2 - \log_b 3$$ □

Example 8

$$\begin{aligned} \log_b 12 &= \log_b(3 \cdot 4) \\ &= \log_b(3 \cdot 2^2) \\ &= \log_b 3 + \log_b 2^2 \\ &= \log_b 3 + 2\log_b 2 \end{aligned}$$ □

Example 9

$$\begin{aligned} \log_b (5\sqrt{33}) &= \log_b [5(3 \cdot 11)^{1/2}] \\ &= \log_b 5 + \log_b (3 \cdot 11)^{1/2} \end{aligned}$$

$$= \log_b 5 + \tfrac{1}{2}\log_b (3 \cdot 11)$$
$$= \log_b 5 + \tfrac{1}{2}(\log_b 3 + \log_b 11)$$ □

Example 10

$$\log_d \frac{1}{k^3} = \log_d k^{-3}$$
$$= -3\log_d k$$ □

Example 11

$$\log_B \frac{a\sqrt[3]{b}}{c^2d} = \log_B \frac{ab^{1/3}}{c^2d}$$
$$= \log_B(ab^{1/3}) - \log_B(c^2d)$$
$$= \log_B a + \log_B b^{1/3} - (\log_B c^2 + \log_B d)$$
$$= \log_B a + \tfrac{1}{3}\log_B b - 2\log_B c - \log_B d$$ □

Historically, these properties made logarithms an important aid to science. Logarithms were devised to ease the complex calculations necessary in science, especially astronomy. Using properties of logarithms and tables of values for logarithms, numerical problems may be greatly simplified. It is paradoxical that logarithms were invented before exponents were widely used. Since our number system is base 10, common logarithms were convenient for numerical calculations.

Finally, reconsider some of the formulas that were used in Problem Set 6.2. A formula may not be in the most useful form. You will frequently have to solve for one of the variables before using the formula. The logarithm function and its properties now make it possible to solve for variables that occur within the exponent. This is illustrated in the following examples.

Example 12 Solve $A = A_0\left(\frac{1}{2}\right)^{t/h}$ for t.

Solution

$$\frac{A}{A_0} = \left(\frac{1}{2}\right)^{t/h}$$ First arrange in exponential form.

$$\frac{t}{h} = \log_{1/2}\frac{A}{A_0}$$ Now rewrite in logarithmic form using the definition of logarithm.

$$t = h\log_{1/2}\frac{A}{A_0}$$ Multiply by h. □

Example 13 Solve $A = Pe^{rt}$ for r.

Solution

$$\frac{A}{P} = e^{rt}$$ Divide by P.

$$rt = \log_e \frac{A}{P}$$ Rewrite in logarithmic form.

$$rt = \ln\frac{A}{P}$$ Base e is the natural logarithm.

$$r = \frac{1}{t}\ln\frac{A}{P}$$ Divide by t. □

Example 14 Solve $I = \frac{E}{R}(1 - e^{-Rt/L})$ for L.

Solution

$$\frac{RI}{E} = 1 - e^{-Rt/L}$$ Multiply by R/E.

$$\frac{RI}{E} - 1 = -e^{-Rt/L}$$ Subtract 1.

$$1 - \frac{RI}{E} = e^{-Rt/L}$$ Multiply by -1.

$$-\frac{Rt}{L} = \ln\left(1 - \frac{RI}{E}\right)$$ Write in logarithmic form.

$$-Rt = L\ln\left(1 - \frac{RI}{E}\right)$$ Multiply by L.

$$L = \frac{-Rt}{\ln\left(1 - \frac{RI}{E}\right)}$$ Divide. □

Problem Set 6.3

A *Rewrite Problems 1–10 in logarithmic form.*

1. $125 = 5^3$ **2.** $243 = 3^5$ **3.** $\frac{1}{9} = 3^{-2}$

4. $4 = \left(\frac{1}{2}\right)^{-2}$ **5.** $8 = 4^{3/2}$ **6.** $9 = 27^{2/3}$

7. $\frac{4}{9} = \left(\frac{3}{2}\right)^{-2}$ **8.** $\frac{1}{3} = 81^{-1/4}$ **9.** $a = b^c$

10. $u = v^w$

Rewrite Problems 11–20 in exponential form.

11. $\log_2 8 = 3$ **12.** $\log_3 81 = 4$ **13.** $\log .01 = -2$

14. $\ln e^2 = 2$ **15.** $\log_4 2 = .5$ **16.** $\log_{27} 3 = \frac{1}{3}$

17. $\log_{1/3} 9 = -2$ **18.** $\log_{1/4} 8 = -\frac{3}{2}$ **19.** $\log_k h = c$

20. $\log_x y = z$

Find the value of x, y, or b in Problems 21–40.

21. $\log_2 8 = x$ **22.** $\log_3 9 = y$ **23.** $\log 100 = x$

24. $\log .0001 = y$ **25.** $\log_{27} 9 = x$ **26.** $\log_4 8 = y$

27. $\log_3 x = 2$ **28.** $\log_5 x = 3$ **29.** $\log_b 16 = 4$

30. $\log_b 125 = 3$ **31.** $\log_7 49 = y$ **32.** $\log_8 (2\sqrt{2}) = y$

33. $\log_y 27 = y$ **34.** $\log_b 4 = b$ **35.** $\log_b 1 = 0$

36. $\log_x x = 1$ **37.** $\ln x = 1$ **38.** $\log x = 2$

39. $\log 100 = y$ **40.** $\ln e^3 = y$

B *For each of Problems 41–50 sketch the graphs of the pairs of functions on the same coordinate plane.*

41. $y = \log_2 x;\ \ y = 2^x$ **42.** $y = \log_3 x;\ \ y = 3^x$

43. $y = \log_2 x;\ \ y = \left(\frac{1}{2}\right)^x$ **44.** $y = \log_3 x;\ \ y = \left(\frac{1}{3}\right)^x$

45. $y = \ln x;\ \ y = e^x$ **46.** $y = \log x;\ \ y = 10^x$

47. $y = \log_2 x;\ \ y = \log_3 x$ **48.** $y = \ln x;\ \ y = \log x$

49. $y = \log x;\ \ y = \log_{.1} x$ **50.** $y = \ln x;\ \ y = \log_{1/e} x$

Use the laws of logarithms to expand the expressions in Problems 51–60.

51. $\log_b (4ac)$ **52.** $\log_b \left(\frac{1}{2} lw\right)$ **53.** $\log(\pi r^2)$

54. $\log(2\pi r)$ **55.** $\log_b(P\sqrt{Q})$ **56.** $\log_b(A^2B^3)$

57. $\log_b \dfrac{rs^2}{\sqrt[3]{t}}$ **58.** $\log_b \dfrac{r^3\sqrt{s}}{t^2}$ **59.** $\ln(Pe^{-rt})$ **60.** $\ln(Ae^{kt})$

Solve for the indicated variable in Problems 61–70.

61. $P = P_0e^{rt}$ for t **62.** $I = I_0e^{-rt}$ for r

63. $P = 14.7e^{-.21a}$ for a **64.** $A = 250e^{-.0248t}$ for t

65. $A = A_0 \left(\frac{1}{2}\right)^{t/h}$ for h **66.** $I = I_0 (10)^{N/10}$ for N

67. $P_n = P_1e^{.11(1 - n)}$ for n **68.** $N = 80(1 - e^{-.016n})$ for n

69. $I = \dfrac{E}{R}(1 - e^{-Rt/L})$ for t **70.** $T = A + (B - A)10^{-kt}$ for k

C **71.** *Find the fallacy in the following development:*

$$3 > 2$$
$$3 \log \tfrac{1}{2} > 2 \log \tfrac{1}{2}$$
$$\log\left(\tfrac{1}{2}\right)^3 > \log\left(\tfrac{1}{2}\right)^2$$
$$\left(\tfrac{1}{2}\right)^3 > \left(\tfrac{1}{2}\right)^2$$
$$\tfrac{1}{8} > \tfrac{1}{4}$$

Problems 72–75 refer to the properties of logarithms listed in the box on page 238.

72. Prove Property II. **73.** Prove Property III.

74. Prove Property IV. **75.** Prove Property V.

6.4 Logarithmic Equations

Most scientific hand-held calculators have both the common and natural logarithm functions. Computing values for these functions is accomplished by entering the given number and pressing the appropriate key.

Example 1 Evaluate log 453.

Solution

log 453 ≈ 2.6560982

Notice that ≈ is used here to indicate that the value is correct to eight places. Your calculator may display fewer (or more) places. In any event, it will probably be sufficient for our applications. □

Example 2 Evaluate ln 7.831.

Solution 7.831 LN

ln 7.831 ≈ 2.0580902 □

Sketching the graphs of more involved functions containing logarithmic functions produces some interesting results. A table of values may be compiled using a calculator. However, it is advisable to consider the domain before compiling values for the graph.

Example 3 Sketch the graph of $y = \log x^2$.

Solution Notice that since $x^2 \geq 0$, all nonzero values of x may be considered. The graph is undefined at $x = 0$, and will have $x = 0$ as a vertical asymptote.

x	y
±1	0
±5	1.398
±10	2
±15	2.352
±20	2.602
±25	2.796
±30	2.954
±35	3.088

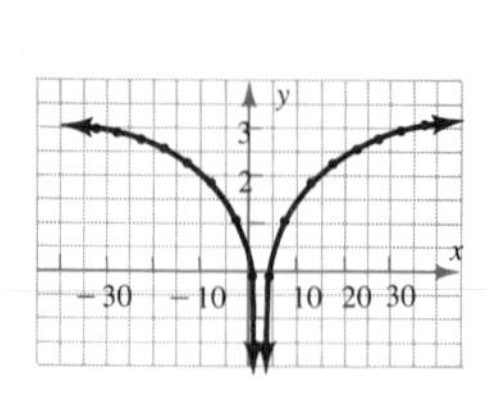

□

Example 4 Sketch the graph of $y = \sqrt{1 - \ln x}$

Solution Notice that x must be positive, and

$$1 - \ln x \geqslant 0$$

$$1 \geqslant \ln x \qquad \textit{Note: } \text{if } 1 = \ln x, \text{ then } x = e$$

$$x \leqslant e$$

So, $0 < x \leqslant e$.

x	y
e	0
2.5	.289
2.0	.554
1.5	.771
1.0	1
.5	1.301
.1	1.817
.01	2.368
0	Undefined

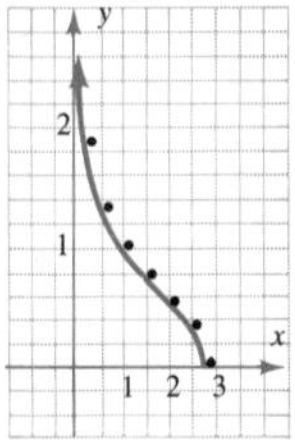

□

Example 5 Evaluate log 3,406,000,000.

Solution

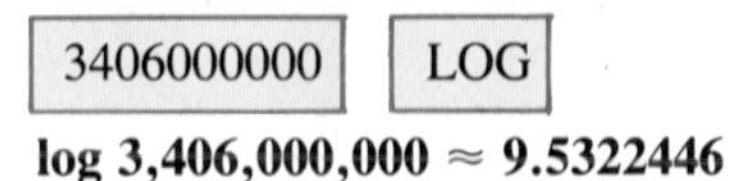

log 3,406,000,000 ≈ 9.5322446 □

It is likely that your calculator would not accept the number in Example 5 in the form given. In general, very large or very small numbers are written in scientific notation (see Section 1.4). Recall that any number may be written as a number between 1 and 10 times a power of 10.

The calculator will accept a number in this form if it has an [EEX], [EE], [EXP], or [SCI] key. Reconsider the number from Example 5.

Example 6 Evaluate log 3,406,000,000 = $\log(3.406 \cdot 10^9)$.

[3.406] [EEX] [9] [LOG]

$\log(3.406 \cdot 10^9) \approx 9.5322446$ □

Example 7 Evaluate ln .00000006572 = $\ln(6.572 \cdot 10^{-8})$.

[6.572] [EEX] [8] [+/−] [LN]

$\ln(6.572 \cdot 10^{-8}) \approx -16.537863$ □

Example 8 Evaluate $\ln(6.023 \cdot 10^{23})$.

[6.023] [EEX] [23] [LN]

$\ln(6.023 \cdot 10^{23}) \approx 54.755043$ □

The number $6.023 \cdot 10^{23}$ found in Example 8 is called *Avogadro's number.* It is the number of molecules in 1 gram-molecule of a substance or the number of atoms in 1 gram-atom of an element.

Example 9 The hydrogen potential, or pH, of a solution is

$$\text{pH} = -\log[\text{H}^+]$$

where $[\text{H}^+]$ is the concentration of hydrogen in aqueous solution in moles per liter. A solution is considered neutral if its pH is 7. When the pH exceeds 7, the solution is alkaline; when it is less than 7, the solution is acid. Grapefruit has a $[H^+]$ of $7.9 \cdot 10^{-4}$. Find its pH to the nearest tenth.

Solution

$$\begin{aligned} pH &= -\log[H^+] \\ &= -\log(7.9 \cdot 10^{-4}) \end{aligned}$$

7.9	EEX	4	+/−	LOG	+/−

$$\begin{aligned} pH &\approx 3.1023729 \\ &\approx 3.1 \end{aligned}$$

The pH of grapefruit is approximately 3.1. □

The properties of logarithms have been used to expand logarithmic expressions. These properties can likewise be used to simplify expressions, and they can be applied to the solution of logarithmic equations.

Example 10 Solve $\log_b 2x + \frac{1}{3}\log_b 27 = \frac{2}{5}\log_b 243 + 2\log_b 2$ for x.

Solution

$$\log_b 2x + \frac{1}{3}\log_b 27 = \frac{2}{5}\log_b 243 + 2\log_b 2$$

Simplify both sides of the equation first.

$$\log_b 2x + \log_b 27^{1/3} = \log_b 243^{2/5} + \log_b 2^2$$

$$\log_b 2x + \log_b 3 = \log_b 9 + \log_b 4$$

$$\log_b 6x = \log_b 36$$

Since $\log_b m = \log_b n$, then $m = n$.

$$6x = 36$$

$$\boldsymbol{x = 6}$$ □

Example 11 Solve $3\log_b 2 - \frac{1}{2}\log_b 12 = 1 - \frac{1}{2}\log_b 3$ for b.

Solution

$$3\log_b 2 - \frac{1}{2}\log_b 12 = 1 - \frac{1}{2}\log_b 3$$

Simplify first.

$$\log_b 2^3 - \log_b \sqrt{12} = \log_b b - \log_b \sqrt{3}$$

Note that $1 = \log_b b$.

$$\log_b \frac{2^3}{\sqrt{12}} = \log_b \frac{b}{\sqrt{3}}$$

Since $\log_b m = \log_b n$, then $m = n$.

$$\frac{2^3}{\sqrt{12}} = \frac{b}{\sqrt{3}}$$

$$\frac{2^3}{2} = b$$

$$\boldsymbol{4 = b}$$ □

Example 12 Solve $\ln(\sqrt{e}\ 3) + \ln x = \ln 2 + \dfrac{1}{2}$ for x.

Solution

$$\ln\frac{\sqrt{e}}{3} + \ln x = \ln 2 + \frac{1}{2}$$

$$\ln\frac{\sqrt{e}}{3} + \ln x = \ln 2 + \frac{1}{2}\ln e \qquad \text{Note that } \ln e = 1.$$

$$\ln\frac{x\sqrt{e}}{3} = \ln 2\sqrt{e} \qquad \text{Simplify.}$$

$$\frac{x\sqrt{e}}{3} = 2\sqrt{e} \qquad \text{Since } \log_b m = \log_b n, \text{ then } m = n.$$

$$\frac{x}{3} = 2$$

$$\boldsymbol{x = 6}$$

Alternate Solution

$$\ln\frac{\sqrt{e}}{3} + \ln x - \ln 2 = \frac{1}{2} \qquad \text{Isolate logarithms.}$$

$$\ln\frac{x\sqrt{e}}{6} = \frac{1}{2} \qquad \text{Simplify.}$$

$$e^{1/2} = \frac{x\sqrt{e}}{6} \qquad \text{Rewrite in exponential form.}$$

$$1 = \frac{x}{6}$$

$$\boldsymbol{6 = x}$$

□

Notice that the solution to Example 10 is independent of the base b. The ordinary calculator has both common and natural logarithm functions, but it is sometimes necessary to evaluate a logarithm with a base other than 10 or e. Common or natural logarithms are sufficient to find the logarithm of a number in any acceptable base.

Recall that

$$\log_b x = y \qquad \text{is equivalent to} \qquad b^y = x$$

and thus,

$$\log b^y = \log x \qquad \text{or} \qquad \ln b^y = \ln x$$

$$y \cdot \log b = \log x \qquad\qquad y \cdot \ln b = \ln x$$

$$y = \frac{\log x}{\log b} \qquad\qquad y = \frac{\ln x}{\ln b}$$

Therefore $\log_b x = \dfrac{\log x}{\log b}$ Therefore $\log_b x = \dfrac{\ln x}{\ln b}$

Can you rework this using $\log_a x$ rather than $\log x$ or $\ln x$?

CHANGE OF BASE PROPERTY

$$\log_b x = \frac{\log x}{\log b} \quad \text{or} \quad \frac{\ln x}{\ln b} \quad \text{or} \quad \frac{\log_a x}{\log_a b}$$

The logarithm of a number in any valid base may be found by dividing the common (or natural) logarithm of the number by the common (or natural) logarithm of the base.

Example 13 Find $\log_3 14$ to the nearest ten-thousandth.

Solution

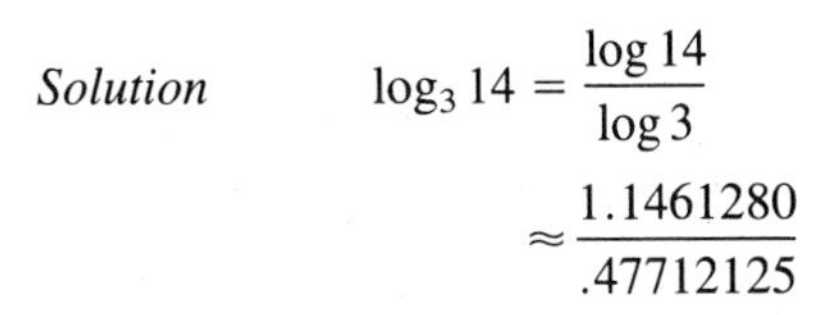

$$\log_3 14 = \frac{\log 14}{\log 3}$$

$$\approx \frac{1.1461280}{.47712125}$$

[14] [LOG] [÷] [3] [LOG] [=]

$$\log_3 14 \approx 2.4021735$$

$$\mathbf{\log_3 14 \approx 2.4022}$$

□

Example 14 Find $\log_\pi \sqrt{2}$ to the nearest ten-thousandth.

Solution

$$\log_\pi \sqrt{2} = \frac{\log \sqrt{2}}{\log \pi}$$

$$\approx \frac{.15051499}{.49714987}$$

$$\approx .30275578$$

$$\mathbf{\log_\pi \sqrt{2} \approx .3028}$$

□

Example 15 Find $\log_5 e$ to the nearest ten-thousandth.

Solution

$$\log_5 e = \frac{\ln e}{\ln 5}$$

$$\approx \frac{1}{1.6094379}$$

$$\approx .62133493$$

$$\mathbf{\log_5 e \approx .6213}$$

□

Example 16 Compute the number of years necessary for a \$1000 investment to grow to \$2500 at 8% interest, compounded quarterly.

Solution Recall the formula

$$A = P\left(1 + \frac{r}{n}\right)^{nt}$$

where P is the principal invested at annual interest rate r compounded n times a year to obtain the amount A in t years. Substitute the given values to obtain

$$2500 = 1000\left(1 + \frac{.08}{4}\right)^{4t}$$

$$2.5 = (1.02)^{4t}$$

$$4t = \log_{1.02} 2.5$$

$$4t = \frac{\log 2.5}{\log 1.02}$$

$$t = \frac{\log 2.5}{4 \log 1.02}$$

2.5	LOG	÷	4	÷	1.02	LOG	=

$$t \approx 11.567792$$

$$t \approx 11.6$$

This result should be rounded to $11\frac{3}{4}$ since compounding is quarterly. (Do you see why it is *not* rounded to the *nearest* quarter?) □

Problem Set 6.4

A *Find the logarithms of the numbers in Problems 1–12.*

1. $\log 45.6$ **2.** $\log 981$ **3.** $\ln .0639$

4. $\ln .00254$ **5.** $\log 19{,}800$ **6.** $\ln 566{,}000$

7. $\ln .3170$ **8.** $\log .1980$ **9.** $\log(3.16 \cdot 10^{-10})$

10. $\ln(5.23 \cdot 10^{-6})$ **11.** $\ln(4.57 \cdot 10^{8})$ **12.** $\log(7.09 \cdot 10^{6})$

For Problems 13–20 sketch the graphs of each function or pair of functions on a coordinate plane.

13. $y = \log x;\ \ y = \log x^3$ **14.** $y = \ln x;\ \ y = \ln x^2$

15. $y = |\ln x|;\ \ y = \ln |x|$ **16.** $y = \log x^2;\ \ y = (\log x)^2$

17. $y = \sqrt{\log x};\ \ y = \log \sqrt{x}$ **18.** $y = \ln \sqrt{x};\ \ y = \ln x^2$

19. $y = \sqrt{1 - \log x}$ **20.** $y = 1 - \sqrt{\ln x}$

The following is needed for Problems 21–30:

Chemistry *The hydrogen potential, or pH, of a solution is*

$$\text{pH} = \log \frac{1}{[H^+]} \quad \text{or} \quad -\log[H^+]$$

where $[H^+]$ *is the concentration of hydrogen ions in aqueous solution given in moles per liter. Find the* pH, *given the* $[H^+]$ *for each substance named, to the nearest tenth.*

21. Vinegar, $1.26 \cdot 10^{-3}$ **22.** Tomatoes, $6.31 \cdot 10^{-5}$

23. Rainwater, $6.31 \cdot 10^{-7}$

24. Soft drink, $1.0 \cdot 10^{-3}$

25. Milk of magnesia, $3.16 \cdot 10^{-11}$

26. Seawater, $3.16 \cdot 10^{-9}$

27. Milk, $3.98 \cdot 10^{-7}$

28. Bread, $3.16 \cdot 10^{-6}$

29. Cherries, $2.51 \cdot 10^{-4}$

30. Lemons, $5.01 \cdot 10^{-3}$

Find the logarithm of each number in Problems 31–40.

31. $\log_\pi e$

32. $\log_e \pi$

33. $\log_3 e$

34. $\log_2 \pi$

35. $\log_{1.03} 3$

36. $\log_{1.05} 2$

37. $\log_{1.44} .000379$

38. $\log_{1.97} 1{,}237{,}000$

39. $\log_{.569} (2.63 \cdot 10^{11})$

40. $\log_{4.58} (3.74 \cdot 10^{-8})$

B *Use the laws of logarithms to solve for x in Problems 41–54.*

41. $\log_b 2 + \frac{1}{2} \log_b 3 + 4 \log_b 5 = \log_b x$

42. $\log_b x = \log_b 5 - 3 \log_b 2 + \frac{1}{4} \log_b 3$

43. $\frac{1}{2} \log_b x = 3 \log_b 5 - \log_b x$

44. $\log_b x - \frac{1}{2} \log_b 2 = \frac{1}{2} \log_b(3x - 4)$

45. $\log 2 = \frac{1}{4} \log 16 - \log x$

46. $\ln e = \ln \frac{\sqrt{2}}{x} - \ln e$

47. $\ln \sqrt{2e} = \ln 1 + \ln x$

48. $\log 10 = \log \sqrt{1000} - \log x$

49. $\frac{1}{2} \log x - \log 100 = 2$

50. $2 = \ln \frac{e^2}{2} + \ln x$

51. $2 \log x + 3 \log 10 = 5$

52. $\log 10^x - 2 = \log 100$

53. $\ln e^3 - \ln x = 1$

54. $\ln 1 + \ln e^x = 2$

The following is needed for Problems 55–58:

Physics *The intensity of sound is measured in decibels. The number of decibels N is given by*

$$N = \log\left(\frac{I}{I_0}\right)^{10} \quad \text{or} \quad 10 \cdot \log \frac{I}{I_0}$$

where I is the power of the sound in watts per square centimeter and I_0 is the power of sound just below the threshold of hearing, $I_0 \approx 10^{-16}$ watt per cm². Find the number of decibels for the power of the given sound.

55. Whisper, 10^{-13} watt per cm²

56. Traffic jam, 10^{-8} watt per cm²

57. Rock concert, $5.23 \cdot 10^{-6}$ watt per cm²

58. Normal conversation, $3.16 \cdot 10^{-10}$ watt per cm²

The following is needed for Problems 59–64:

Business *If P represents the principal invested at r percent annual interest compounded n times a year, then A is the amount accumulated in t years:*

$$A = P\left(1 + \frac{r}{n}\right)^{nt}$$

Find results to the nearest tenth of a percent.

59. How long does it take a sum of money to double if it is earning $9\frac{1}{4}\%$ interest, compounded quarterly?

60. In how many years does a sum of money double while earning $5\frac{1}{2}\%$ interest, compounded quarterly?

61. How long does it take an investment of $1450 to grow to $2000 if it earns 7% interest, compounded monthly?

62. How long does it take to obtain $5400 from an investment of $2100 at 8% interest, compounded monthly?

63. What rate is necessary for an investment to double in 5 years if the investment is compounded quarterly?

64. What rate is necessary for an investment to double in 5 years if the investment is compounded monthly?

The following is needed for Problems 65–70:

Physics *If an object at temperature B is surrounded by air at temperature A, it will gradually cool so that the temperature T, t minutes later, is given by*

$$T = A + (B - A)10^{-kt}$$

This is Newton's Law of Cooling. The constant k depends on the particular object. If we restate the formula, the cooling constant is given by

$$k = \frac{1}{t}\log\frac{B - A}{T - A}$$

65. You draw a tub of hot water ($k = .01$) for a bath. The water is 100°F when drawn and the room is 72°F. If you are called away to the phone, what is the temperature of the water 20 minutes later when you get in?

66. You take a batch of chocolate chip cookies from the oven (250°F) when the room temperature is 74°F. If the cookies cool for 20 minutes and $k = .075$, what is the temperature of the cookies?

67. It is known that the temperature of a given object fell from 120°F to 70°F in an hour when placed in 20°F air. What was the temperature of the object after 30 minutes? [*Hint:* First find the cooling constant for this object.]

68. An object is initially 100°F. In air of 40°F it cools to 45°F in 20 minutes. What is its temperature in 30 minutes?

69. Using the information of Problem 67, how long will it take the object to cool to 40°F?

70. Using the information in Problem 68, how long will it take this object to cool to 75°F?

C **71.** Earth science The *Richter Scale* is a well-known means of measuring the magnitude M of an earthquake:

$$M = \log\frac{A}{A_0}$$

where A is the amplitude of the seismic (shock) waves of a measured earthquake and A_0 is taken as a norm. The size of an earthquake is properly measured by the energy released, E, in ergs. An approximation for this energy is given by

$$\log E = 11.8 + 1.5M$$

What was the energy released by the 1906 San Francisco quake that measured 8.3 on the Richter Scale? Write the answer in scientific notation. (This amount of energy, it is estimated, would be sufficient to provide the entire world's food requirements for a day.)

72. *Physics* A vertically fired projectile will reach a maximum height H as given by

$$H = \frac{1}{K}\left[V_0 - \frac{g}{K}\ln\left(1 + \frac{V_0K}{g}\right)\right]$$

where V_0 is the initial velocity, the gravitational constant g is approximately 32 ft per sec^2, and K is a coefficient of air resistance. Let $K = 2.4$ and sketch the graph of the function $h(V_0) = H$. Use the graph as a basis to discuss the relationship of initial speed on the height attained.

73. Construct a logarithmic scale. On a sheet of paper, draw a line 10 inches long and mark it in tenths and hundredths. Place the edge of a second sheet along this line and use the first as a ruler to mark off logs on the second. At 0, mark 1 since $\log 1 = 0$. Mark 2 at .301, 3 at .477, and so on until you have something similar to the following illustration:

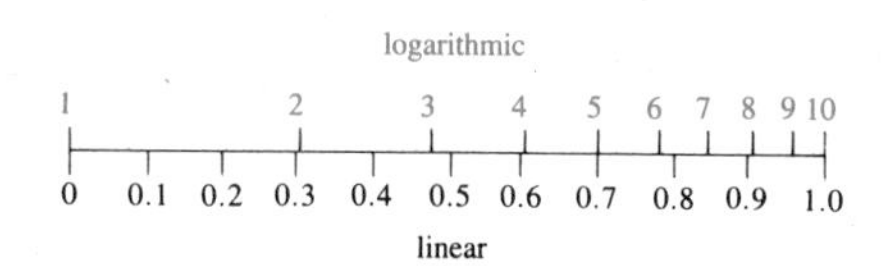

Fill in the scale between 1 and 2 for every tenth, and the others at least at the halves. Using two of these log scales, can you make a simple slide rule? *Explain* how your slide rule multiplies and divides.

74. Construct a coordinate system using logarithmic scales instead of the usual linear scales. The divisions will be irregular instead of evenly spaced. Sketch the graph of $y = x^2$ and explain the result.

75. Using the logarithmic coordinate system, sketch $y = x$ and explain the result.

76. Construct a coordinate system using the logarithmic scale on the vertical axis and the usual linear scale on the horizontal axis. Sketch $y = 2^x$ and explain the result.

77. Construct a coordinate system with a vertical linear scale and a horizontal logarithmic scale. Sketch the graph of $y = \log x$ and explain the result.

*6.5 Additional Applications

There are many applications for the exponential and logarithmic functions. Obviously, all of these problems require a certain amount of nontrivial computation. Without a

*Optional

calculator to perform much of this computation, the work would be very tedious and time-consuming.

The next two examples use the following information: For budgeting, inventory, or tax purposes, businesses must depreciate equipment. An item may not be useful to a company even though it still functions, since it might have become too expensive to maintain. Perhaps it has become obsolete after n years. The value of an item at the end of its useful life is called its scrap value S and is given by

$$S = P(1 - r)^n$$

where P is the purchase price and r is the annual rate of depreciation.

Example 1 A small contractor in light construction depreciates a $40,000 tractor at the rate of 20% a year. What is the scrap value of the equipment in 8 years?

Solution Here, $P = 40{,}000$, $r = .20$, and $n = 8$.

$$\begin{aligned} S &= 40{,}000(1 - .20)^8 \\ &= 40{,}000(.80)^8 \end{aligned}$$

[.8] [y^x] [8] [×] [40000] [=]

$$\begin{aligned} S &\approx 6710.8864 \\ &\approx 6700 \end{aligned}$$

The scrap value is approximately $6700. □

Example 2 An $8000 pickup truck is depreciated to $1500 over 5 years use. What is the rate of depreciation?

Solution Here, $S = 1500$, $P = 8000$, and $n = 5$.

$$\begin{aligned} 1500 &= 8000(1 - r)^5 \\ \frac{1500}{8000} &= (1 - r)^5 \\ \left(\frac{3}{16}\right)^{1/5} &= 1 - r \\ r &= 1 - \left(\frac{3}{16}\right)^{1/5} \\ &= 1 - \left(\frac{3}{16}\right)^{.2} \end{aligned}$$

[3] [÷] [16] [=] [y^x] [.2] [=] [+/−] [+] [1] [=]

$$\begin{aligned} r &\approx .28451546 \\ &\approx .285 \end{aligned}$$

The rate of depreciation is approximately 28.5%. □

More about Growth

Population growth is similar to other types of exponential growth. In t years, the population P_0 will grow to P, as approximated by

$$P = P_0 e^{kt}$$

where k is the growth rate. The constant k may be determined empirically. That is, given population figures from two different years, we can calculate the rate of growth during that period of time. For example, the population figures below show that San Diego has been growing from 1950 to 1980.

City	1950	1960	1970	1980
San Diego, California	334,387	573,224	697,471	870,006

We will use this information in the following examples.

Example 3 Find the annual rate of growth for San Diego from 1970 to 1980.

Solution In this case, $P = 870{,}006$, $P_0 = 697{,}471$, and $t = 10$.

$$870{,}006 = 697{,}471e^{10k}$$

$$\frac{870{,}006}{697{,}471} = e^{10k}$$

$$10k = \ln\frac{870{,}006}{697{,}471}$$

$$k = \frac{1}{10}\ln\frac{870{,}006}{697{,}471}$$

870,006 | ÷ | 697,471 | = | LN | ÷ | 10 | =

$$k \approx .022103917$$

$$\approx .022$$

San Diego grew at an approximate annual rate of 2.2% during the 10 year period 1970–1980. □

Example 4 Predict the population of San Diego in 1990 if it continues to grow at its 1970–1980 rate.

Solution Now we know that $P_0 = 870{,}006$, $k = .022$, and $t = 10$.

$$P = 870{,}006e^{.022(10)}$$

$$P = 870{,}006e^{.22}$$

$$\approx 1{,}084{,}094.2$$

$$\approx 1{,}100{,}000$$

The population would be about 1,100,000 in 1990. □

More about Interest

Compound interest earned on invested principal is an important application of the exponential function. An equally important application of the function involves borrowing money. Essentially, the interest is paid on a diminishing principal instead of being earned on an increasing principal. Such loans are usually repaid in equal monthly installments. If P dollars are borrowed for n months at i monthly interest rate, the monthly payment M on the loan is given by

$$M = \frac{Pi}{1 - (1 + i)^{-n}}$$

Example 5 What is the monthly house payment for a 30 year loan of $60,000 at 15% interest?

Solution Here, $P = 60{,}000$, $i = .15/12 = .0125$, and $n = 30 \cdot 12 = 360$.

$$M = \frac{60{,}000(.0125)}{1 - (1 + .0125)^{-360}}$$

$$= \frac{750}{1 - 1.0125^{-360}}$$

[1.0125] [y^x] [360] [+/−] [=] [+/−] [+] [1] [=]

[1/x] [×] [750] [=]

$$M \approx 758.66641$$

The monthly payment is $758.67. □

Calculating the result in Example 5 is definitely a nontrivial task, which should test your understanding of your calculator. The sequence of calculations may differ slightly from those on your particular model. Nonetheless, notice [+/−] [+] performs subtraction by changing the signs and adding, which [1/x] [×] divides by taking the reciprocal and multiplying.

Is the result in Example 5 what you expected? At $758.67 per month for 30 years, a total of $273,121.20 will be repaid for the loan of $60,000. Of this, $213,121.20 is interest, which is rent on the borrowed money. Examples 6 and 7 compare this with a lower interest rate and a shorter term on the loan.

Example 6 Find the monthly payment and the total paid on $60,000 borrowed for 20 years at 15% interest.

Solution Again, we have $P = 60{,}000$ and $i = .0125$, but this time $n = 20 \cdot 12 = 240$.

$$M = \frac{60{,}000(.0125)}{1 - (1 + .0125)^{-240}}$$

$$= \frac{750}{1 - (1.0125)^{-240}}$$

$$\approx 790.07375$$

$$240M \approx 189{,}616.80$$

The monthly payment is \$790.07 and the total paid is \$189,617. □

Notice that the monthly payment is higher, but there is a substantial savings in total cost over the shorter period.

Example 7 What is the monthly payment and the total amount paid on a \$60,000 loan at 12% for 30 years?

Solution This time, $P = 60{,}000$, $i = .12/12 = .01$, and $n = 360$.

$$M = \frac{60{,}000(.01)}{1 - (1 + .01)^{-360}}$$

$$= \frac{600}{1 - (1.01)^{-360}}$$

$$\approx 617.16755$$

$$360M \approx 222{,}181.20$$

\$222,181 is paid in monthly payments of \$617.17. □

Notice the tremendous difference in the monthly payment when the rate is reduced by 3%. The total cost is lower than it was in Example 5, but not as low as the shorter term at the higher rate (Example 6).

Problem Set 6.5

A *The following is required for Problems 1–6. Give answers to the nearest dollar or nearest tenth of a percent.*

Business *The scrap value S of an item with a useful life of n years is given by*

$$S = P(1 - r)^n$$

where P is the original purchase price and r is the annual rate of depreciation.

1. A delivery truck costs \$12,500 and is used for 6 years. If it is depreciated at a rate of 15%, what is the scrap value?
2. An electric generator costs \$39,000 and has an estimated life of 12 years. What is its scrap value if it is depreciated at 10%?
3. A sand and gravel company purchases a piece of property for \$215,000. It is estimated that the property will have a resale value of \$35,000 after 10 years. What is the rate of depreciation?
4. An apartment building is purchased for \$382,000 and has an effective life of 40 years. If the site is still worth \$50,000 after that period, what is the rate of depreciation?
5. A \$15,000 machine has an expected life of 8 years and a scrap value of \$1000. If the machine is replaced after 5 years, what is its value at the time of replacement?
6. A sales representative depreciates her car over 5 years. If it is purchased for

\$8400 and presumed to have a scrap value of \$950, what is its depreciated value after 2 years?

The following is required for Problems 7–14. Give answers to the nearest thousand people or nearest hundredth of a percent.

Social science *The population P after t years is given by*

$$P = P_0 e^{kt}$$

where P_0 is the initial population and k is the growth rate. Table 6.2 lists the population in eight cities from 1950 to 1980.

Table 6.2
Populations of Eight U.S. Cities

City	1950	1960	1970	1980
Cleveland, Ohio	914,804	876,050	750,879	572,532
Boston, Mass.	801,444	697,197	641,071	562,118
Pittsburgh, Pa.	676,806	604,332	520,089	423,962
Milwaukee, Wis.	637,392	741,324	717,372	632,989
Denver, Colo.	415,786	493,887	514,678	488,765
San Antonio, Tex.	408,442	587,718	654,153	783,296
Memphis, Tenn.	396,000	497,524	623,988	644,838
San Jose, Ca.	95,280	204,196	459,913	625,763

7. Determine the rate of growth for San Antonio, Texas, over the periods 1950–1960, 1960–1970, and 1970–1980.

8. Find the rates of growth for Cleveland, Ohio, over the periods 1950–1960, 1960–1970, and 1970–1980. (Note that the population is decreasing and will have a negative rate of growth.)

9. If Boston, Massachusetts, had continued to decline at its 1950–1970 rate, what would the population have been in 1980? (Note that the decline actually accelerated.)

10. If Denver, Colorado, had continued to grow at its 1960–1970 rate, what would its 1980 population have been? (Note that it actually declined from 1970 to 1980.)

11. San Jose, California, grew by an estimated 164,600 persons from 1970 to 1980. If the city continues to grow at this rate, estimate its population in the year 2000.

12. Pittsburgh, Pennsylvania, decreased by approximately 100,000 persons from 1970 to 1980. If the city were to continue to decline at this rate, estimate its population in 1990.

13. If Memphis, Tennessee, grew uniformly during the decade 1960–1970, estimate its population in 1964.

14. Estimate the 1977 population of Milwaukee, Wisconsin. Assume that its population change was uniform during the 1970–1980 period.

B *The following is required for Problems 15–20:*

Business *If P dollars are borrowed for n months at i monthly interest rate, the monthly payment M on the loan is given by*

$$M = \frac{Pi}{1 - (1 + i)^{-n}}$$

15. A home loan is made for \$45,000 at 9% interest for 30 years. Find the monthly payment.
16. Find the monthly payment for a \$70,000 home loan at 10% for 30 years.
17. The total cost of a new car is \$9673.54. A down payment of \$873.54 is given. The remainder is financed at $19\frac{3}{4}\%$ for 5 years. What is the monthly payment on the new car loan?
18. New furniture is purchased for a total cost of \$2040.47. If a down payment of \$340.47 is made, what is the monthly payment on a loan at $22\frac{1}{2}\%$ for 4 years?
19. An engagement ring is purchased on time. If the \$1100 cost is financed at 17.5% for 6 years, what is the monthly payment?
20. If the ring in Problem 19 is put on a credit card with a 21% charge, what monthly installment would be necessary to pay off the debt in the same time?

Social science *The newspaper article reproduced here is needed for Problems 21–25.*

THE PRESS DEMOCRAT

The Redwood Empire's Leading Newspaper **25** cents

SANTA ROSA, CALIF., SUNDAY, MARCH 28, 1976

World population 4 billion tonight

By EDWARD K. DeLONG

WASHINGTON (UPI) — By midnight tonight, the Earth's population will reach the 4 billion mark, twice the number of people living on the planet just 46 years ago, the Population Reference Bureau said Saturday.

The bureau expressed no joy at the new milestone.

It said global birth rates are too high, placing serious pressures on all aspects of future life and causing "major concern" in the world scientific community, and more than one-third of the present population has yet to reach child-bearing age.

The PRB found cause for optimism, however, in that some governments are stressing birth control to blunt the impact of "explosive growth" and the population growth rate dropped slightly in the past year.

"In 1976, each new dawn brings a formidable increase of approximately 195,000 newborn infants to share the resources of our finite world," it said.

One expert warned that a lack of jobs, rather than too little food, may be the "ultimate threat" facing society as the planet becomes more and more crowded.

It took between two and three million years for the human race to hit the one billion mark in 1850, the PRB said. By 1930, 80 years later, the population stood at 2 billion. A mere 31 years after that, in 1961, it was 3 billion. The growth from 3 to the present 4 billion took just 16 years.

The world could find it has 5 billion people by 1989 — just 13 years from now — if population growth continues at the present rate of 1.8 per cent a year, said Dr. Leon F. Bouvier, vice president of the private, nonprofit PRB.

Bouvier said the newly calculated growth rate is a little lower than the 1.9 per cent estimated last year. Thanks to that slowdown, the passing of the 4 billion milestone came a year later than some demographers had predicted.

"I really think the rate of growth is going to start declining ever so slightly because of declining fertility," Bouvier said. "I think there is some evidence of progress — ever so slow, much too slow."

The new PRB figures show there were 3,982,815,000 people on Earth on Jan. 1. By March 1 the number had grown to 3,994,812,000, the organization said, and by April 1 the total will be 4,000,824,000.

The bureau said its calculations are based on estimates of 328,000 live births per day minus 133,000 deaths.

A growing number of governments are taking steps to slow growth rates, the PRB said.

Singapore appears likely to meet the goal of the two-child family "well before the target date of 1980," it said, and several states in India, which yearly adds the equivalent of the population of Australia, are considering financial incentives to birth control and mandatory sterilization after the birth of two children.

Dr. Paul Ehrlich of Stanford University, one of several population experts contacted by PRB, said he was sad to realize at the age of 44 he had lived through a doubling of Earth's population. He expressed fear the next 44 years could see population growth halted "by a horrifying increase in death rates."

"At this point, hunger does not seem the greatest issue presented by the ever growing number of people," said Dr. Louis M. Hellman, chief of population staff at the Health, Education and Welfare Department.

"Rather, the threat appears to lie in the increasing numbers who can find no work. As these masses of unemployed migrate toward the cities, they create a growing impetus toward political unrest and instability."

21. According to the article, it took 80 years for the world's population to double from 1 billion to 2 billion. What was the rate of growth during that period?

22. The article mentions that the number of people in the world doubled from 2 billion in 1930 to 4 billion in 1976. What was the rate of growth over that period?

23. Using the current rate of growth, estimate the year in which the world population will be 5 billion, 6 billion, 7 billion, and 8 billion.

24. Information subsequent to the article indicates the world population growth is slowing. From 1975 to 1980, the rate was estimated at 1.7%. If this rate continues, in what year will the world figures reach 5 billion, 6 billion, 7 billion, and 8 billion?

25. Using the information in the article, estimate the world population in the year 2001 to the nearest 10 million people.

C **26.** If $a^2 + b^2 = 7ab$, show that $\log(a + b) - \log 3 = \frac{1}{2}(\log a + \log b)$.

27. *Earth science* Use Table 6.3 to answer the following questions:

a. Establish that the use of petroleum has grown exponentially.

b. Estimate the rate of growth of the worldwide use of petroleum.

c. Forecast figures for 1985, 1990, 2000, ...

d. Verify forecast figures for 1985.

e. How do these figures compare to estimated world petroleum reserves? Graph use and remaining reserves on one grid. What conclusions do you reach?

Table 6.3
World Consumption of Petroleum Products

Year	Millions of Barrels	Year	Millions of Barrels
1915	43	1950	340
1920	68	1955	480
1925	110	1960	780
1930	150	1965	1130
1935	170	1970	1670
1940	220	1975	1890
1945	260	1980	2400

6.6 Summary and Review

The functions considered prior to this chapter have been algebraic functions—that is, functions defined by basic algebraic operations on variables and constants. In this chapter we added the exponential and logarithmic functions, with properties quite different from the algebraic functions. These functions also provide a vast increase in the applicability of your mathematics. You have seen applications in business, biology, chemistry, geology, medicine, physics, physiology, psychology, statistics, and other fields. Historically, logarithms were developed as a computational aid, and until the electronic age, they were of tremendous value for that purpose. It is paradoxical that logarithms were developed before exponents came into extensive use. The exponential and logarithmic functions have far greater value in both pure and applied mathematics, while their present use for computation is negligible.

CHAPTER OBJECTIVES

At the conclusion of your study of this chapter, you should be able to:

1. Interchange the logarithmic and exponential form of an expression
2. Sketch the graphs of the exponential and logarithmic functions and understand the graphic relationships among various functions, including $y = b^x$, $y = b^{-x}$, $x = b^y$, and $y = \log_b x$
3. Recognize functions of the form $y = ke^{rx}$ as growth functions for $r > 0$ and decay functions for $r < 0$
4. Apply the basic properties of exponents and logarithms, especially as applied to:
   a. solving exponential and logarithmic equations
   b. evaluating exponential and logarithmic expressions, including scientific notation and tables
5. Make certain applications of these functions given the necessary data, including but not limited to compound interest, population growth, pH, sound intensity, and learning curves

CONCEPT PROBLEMS

Problem Set 6.6

Fill in the word or words that make the statements in Problems 1–20 complete and correct.

[6.1] **1.** The equation $f(x) = b^x$, $b > 0$, $b \neq 1$, defines a(n) ______ function.

[6.1] **2.** For $y = b^x$, b is called the ______ and x is the ______ .

[6.1] **3.** $y = b^x$ is a decreasing function for ______ .

[6.1] **4.** The graph of $y = b^x$ has a(n) ______ at $y = 0$.

[6.1] **5.** For nonzero real x and y and $a > 0$, $a \neq 1$, if $a^x = a^y$, then ______ .

[6.1] **6.** If the exponents of two equal exponentials are equal, then the ______ are equal.

[6.2] **7.** As n increases, the value of $(1 + 1/n)^n$ approaches ______ , which is approximately ______ .

[6.2] **8.** If P is the principal invested at ______ , compounded ______ for ______ , then $A = P(1 + r/n)^{nt}$.

[6.2] **9.** If interest is compounded continuously for t years, then the amount accumulated, A, is given by ______ where P is the original principal and r is the annual rate.

[6.3] **10.** If r is the exponent of s that produces t, then either ______ or ______ .

[6.3] **11.** For $b > 0$, $b \neq 1$, $x = b^y$ is equivalent to $y =$ ______ .

[6.3] **12.** The graph of $y = \log_b x$ is not defined for ______ .

[6.3] **13.** The y-axis is a(n) ______ for the graph of $y = \log_b x$.

[6.3] **14.** The log of the product of two numbers is the ______ of the ______ of the numbers.

[6.3] **15.** The log of ______ is ______ for any base.

[6.3] **16.** If $\log_b x = \log_b y$, then ______ .

[6.3] **17.** If the base is 10, the logarithms are called ______ logarithms.

[6.4] **18.** Population may be approximated by $P = P_0e^{kt}$, where population ______ grows to ______ in ______ years at an annual rate ______ .

[6.4] **19.** The logarithm of a number in any base b is the log (or ln) of the ______ divided by the log (or ln) of ______ .

[6.5] **20.** If ______ dollars are borrowed at ______ monthly rate for ______ months, the monthly payment M is given by $M = \dfrac{Pi}{1 - (1 + i)^{-n}}$.

REVIEW PROBLEMS

[6.1, 6.3] *Sketch the graph of each equation in Problems 21–30.*

21. $y = e^x$ **22.** $y = 2^x$ **23.** $y = \log x$

24. $y = \ln x$ **25.** $y = 3^{|x|}$ **26.** $y = e^{x+1}$

27. $y = -2^{2x}$ **28.** $y = 2^{-x^2}$ **29.** $y = \ln \sqrt{x+1}$

30. $y = 1 + \sqrt{\ln x}$

[6.1] *In Problems 31–39 solve for the variable.*

31. $13^a = 169$ **32.** $6^a = 216$ **33.** $121^a = 1331$

34. $b^{3/4} = 81$ **35.** $343^c = \frac{1}{49}$ **36.** $625^x = \frac{1}{125}$

37. $\left(\frac{100}{49}\right)^x = .7$ **38.** $b^{-2/3} = \frac{100}{121}$ **39.** $\left(\frac{125}{27}\right)^x = .6$

[6.3] *In Problems 40–48 change from exponential to logarithmic form.*

40. $16 = 2^4$ **41.** $3^4 = 81$ **42.** $64 = 2^6$

43. $11^4 = 14{,}641$ **44.** $7^3 = 343$ **45.** $13^3 = 2197$

46. $y = b^x$ **47.** $p = h^k$ **48.** $r = s^t$

[6.3] *In Problems 49–57 change from logarithmic to exponential form.*

49. $3 = \log_3 27$ **50.** $\log_2 16 = 4$ **51.** $5 = \log_5 3125$

52. $\log .01 = -2$ **53.** $\ln 1 = 0$ **54.** $\log 1000 = 3$

55. $y = \log_b x$ **56.** $\log_h r = s$ **57.** $m = \ln n$

[6.3, 6.4, 6.5] *In Problems 58–66 find the value of the variable.*

58. $\log 31.7 = a$ **59.** $b = \log .00683$ **60.** $\log c = 2.4536$

61. $d = \ln 1.05$ **62.** $x = \log_\pi e$ **63.** $-.3245 = \log f$

64. $\log_2 g = 3$ **65.** $\log_h 16 = -.25$ **66.** $\log_\pi 1 = y$

[6.3] *In Problems 67–72 use the laws of logarithms to expand.*

67. $\log ab$ **68.** $\ln ke^{-kt}$ **69.** $\log \dfrac{5(6)^3}{\sqrt{160}}$

70. $\log_b (c\sqrt{d})$ **71.** $\log\left(\frac{I}{I_0}\right)^{10}$ **72.** $\ln\left[A_0\left(\frac{1}{2}\right)^{t/h}\right]$

[6.3] *Solve for the indicated variable in Problems 73–77.*

73. $A = A_0\left(\frac{1}{2}\right)^{\frac{t}{h}}$ for t **74.** $P = P_0e^{rt}$ for r

75. $I = I_0(10)^{\frac{10}{N}}$ for N **76.** $I = \frac{E}{R}(1 - e^L)$ for L

77. $T = A + (B - A)10^{-kt}$ for B

[6.4] *Use the laws of logarithms to solve for x in Problems 78–80.*

78. $\log 2 - \frac{1}{2}\log 9 + 3\log 3 = x$ **79.** $2\ln e = \ln x - \ln 1$

80. $1 = 2\log x - \log 1000$

[6.2] *The following is required for Problems 81–83:*

Physics *The amount A of a radioactive substance remaining after time t is given by*

$$A = A_0\left(\frac{1}{2}\right)^{\frac{t}{h}}$$

where A_0 is the original amount and h is the half-life in the same units as t.

81. The element ^{40}K, potassium-40, has a half-life of approximately $1.26 \cdot 10^9$ years. What amount of an original 100 g sample would remain after a thousand years, to the nearest tenth of a gram?

82. The element ^{85}Kr, krypton-85, has a half-life of approximately 10.6 years. What time, to the nearest tenth of a year, would be necessary for 100 g to decay to 60 g of the element?

83. Find the half-life of the element ^{198}Pb, lead-198, to the nearest tenth, if after 8 hours approximately 9.92 g remain from an original 100 g sample.

[6.2] *The following is required for Problems 84–86:*

Social science *If the population of a city doubles every n years, then the population P after t years may be approximated by*

$$P = P_0 2^{t/n}$$

where P_0 is the initial population.

84. A city of 15,000 is doubling its population every 13.2 years. Approximate its population in 10 years.

85. A city of 15,000 is doubling every 13.2 years. Approximately how many years (to the nearest tenth) will it take this city to grow to 40,000?

86. A city of 15,000 grows to 22,000 in a 10 year period. If this rate of growth continues, approximately how many years (to the nearest tenth) will it take the community to reach a population of 44,000?

[6.2] *The following is required for Problems 87–90:*

Business *If a principal of P dollars is invested at annual rate r compounded n times a year for t years, then the amount A accumulated is given by*

$$A = P\left(1 + \frac{r}{n}\right)^{nt}$$

Calculate the amount after 8 years at $9\frac{1}{4}\%$ on a \$5000 investment if compounded:

87. Quarterly **88.** Monthly **89.** Daily

90. Find the time necessary to double the investment if compounded quarterly.

[6.2] *The following is required for Problems 91–93:*

Business *For continuously compounded interest, the amount A accumulated after t years is*

$$A = Pe^{rt}$$

where P is the principal and r is the annual rate.

91. Calculate the amount after 8 years at $9\frac{1}{4}\%$ on a $5000 investment.

92. Let $P = 1$, $t = 10$, and sketch the graph of the function $A = f(r)$.

93. Find the rate at which the investment doubles in 12 years.

[6.4, 6.5] *The following is required for Problems 94 and 95:*

Chemistry *The* pH *of a solution is* $-\log[H^+]$, *where* $[H^+]$ *is the concentration of hydrogen ions in aqueous solution in moles per liter.*

94. Find the pH of oranges if their $[H^+]$ is $3.16 \cdot 10^{-4}$.

95. Find the $[H^+]$ of ammonia (10% NH_3) with a pH of 11.8.

[6.4] *The following is required for Problems 96 and 97:*

Physics *The intensity of sound is measured in decibels. The number of decibels N is given by*

$$N = \log\left(\frac{I}{I_0}\right)^{10} \quad \text{or} \quad 10 \cdot \log\frac{I}{I_0}$$

where I is the power of the sound in watts per square centimeter and I_0 *is the power of sound just below the threshold of hearing,* $I_0 \approx 10^{-16}$ *watt per* cm^2.

96. Find the number of decibels for a four-engine turbo prop aircraft 100 feet away at takeoff, if the power of the sound is 10^{-4} watt per cm^2.

97. The sound of a four-engine turbo jet aircraft 100 feet away at takeoff is approximately 140 decibels. Find the power of the sound in watts per square centimeter.

[6.4, 6.5] *The following is required for Problems 98 and 99:*

Social science *The population* P_0 *will grow to P in t years, given by*

$$P = P_0e^{rt}$$

where r is the rate of growth. In 1970 the population of Jacksonville, Florida, was 504,265, and the population in 1980 was 541,269.

98. Find the rate of growth between 1970 and 1980.

99. Use this rate to forecast the 1995 population.

[6.5] **100.** *Business* If P dollars are borrowed for n months at i monthly interest rate, the monthly payment M necessary to amortize the loan is

$$M = \frac{Pi}{1 - (1 + i)^{-n}}$$

What is the monthly payment for a 30 year loan at $11\frac{1}{4}\%$ interest on a $65,000 loan?

CHAPTER 7

Matrices and the Determinant Function

Arthur Cayley
(1821–1895)

The new mathematics is sort of supplement to language, affording a means of thought about form and quantity and a means of expression, more exact, compact, and ready than ordinary language.

H. G. Wells,
Mankind in the Making,
London, 1904, pp. 191–192

Historical Note

The Englishman Arthur Cayley was one of the most influential mathematicians of the 19th century and probably second only to Euler as the most prolific writer in the entire history of mathematics. Cayley was a lawyer for 14 years and had published 200–300 papers in mathematics before he accepted the Sadlerian Chair at Cambridge University. He was the first to discuss the topic of matrices, and is generally considered their creator. The matrix has emerged as a powerful tool to manipulate data in the age of the computer.

Cayley's even-tempered personality, tremendous calm, and extraordinary powers of concentration were legend. Cayley worked out his mathematics in his head, often while pacing about, and if interrupted easily resumed his concentration. Once his ideas were clearly formed, he dashed off his work with scarcely an erasure.

Cayley was an accomplished mountain climber, a nature lover, a talented painter (especially in watercolors), and an avid reader of novels. He read thousands of novels during his lifetime, many in French, German, Italian, and Greek as well as English. His wry sense of humor surfaced frequently. When the slant line or "solidus" was first used in 1880 to print fractions in the form 1/2, Cayley wrote the user that it certainly "would give you a strong claim to be president of the Society for the Prevention of Cruelty to Printers."

Chapter Overview

Systems of equations are reviewed in the first section of this chapter, and then the concept of matrices is applied to simplify and extend this work. Later in the chapter, the determinant is introduced as a function of a matrix, providing yet another method of dealing with systems of equations.

7.1 Systems of Equations

Recall that systems of inequalities were solved by graphing. The solution set of a system is the intersection of the solution sets of the individual inequalities. The solution set of a *system of equations* is the set of solutions that satisfies *all* the equations of the system—that is, the *intersection* of the solution sets of the individual equations. Methods of solving a system of linear equations are now reviewed and then applied to nonlinear systems.

Solution by Graphing

Since the graph of each equation in a system of two linear equations in two variables is a line, the solution of the system is the intersection of two lines. In two dimensions, two lines may be related in one of three ways: (a) they intersect at a single point; (b) they are parallel and there is no intersection; or (c) the lines coincide. These possibilities are illustrated by examples below.

Example 1 Solve the systems by graphing:

a. $\begin{cases} 3x - 2y = 15 \\ x + y = 10 \end{cases}$ **b.** $\begin{cases} 2x + 2y = 5 \\ x + y = 10 \end{cases}$ **c.** $\begin{cases} x - \frac{2}{3}y = 5 \\ 3x - 2y = 15 \end{cases}$

Solution The graphs of the systems are shown on page 264, and the solutions are obtained by inspection.

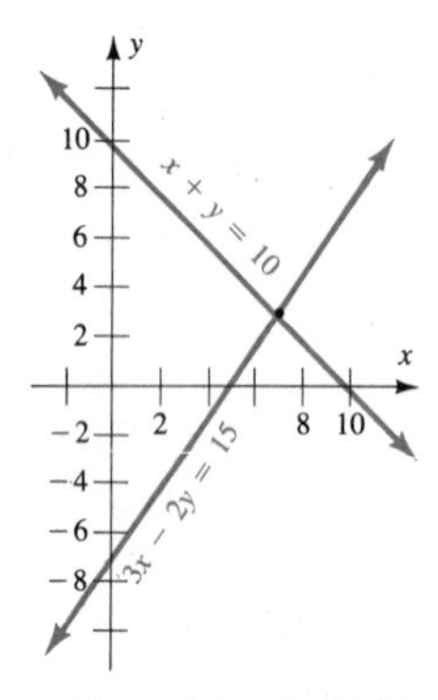

a. The solution is (7, 3) or $x = 7, y = 3$, as shown. The system is called **consistent** when there is a single solution.

Consistent System

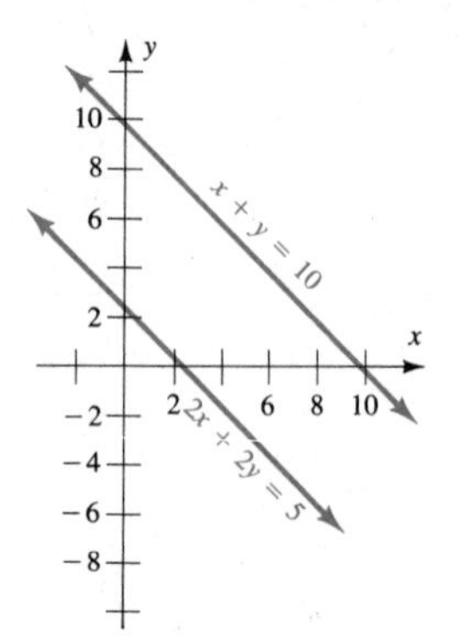

b. The lines are parallel, so there is no point of intersection. Thus, the solution set is empty. In this case, the system is called **inconsistent.**

Inconsistent System

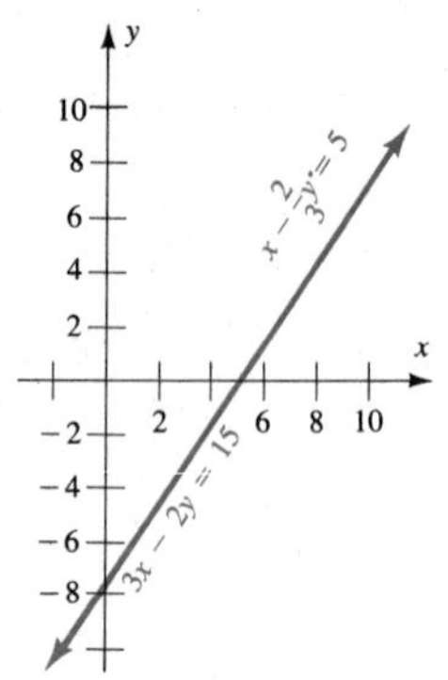

c. The equations represent the same line, and we see that there are infinitely many solutions. Any ordered pair satisfying one equation also satisfies the other. Such a system is called a **dependent** system.

Dependent System □

Solution by the Addition Method

The graphic method is an approximation, since reading the point of intersection depends on the accuracy with which the lines are drawn and the ability to interpret the coordinates. Therefore, other methods of solution are necessary. The **addition method,** also called *linear combination,* requires that you manipulate the equations so that the coefficients of one of the variables are opposites. Then, the equations are added to eliminate the variable.

ADDITION METHOD (LINEAR COMBINATION)

To solve a system of linear equations in two variables:

1. *Multiply* one or both of the equations by constants so that the coefficients of one of the variables are opposites.
2. *Add* corresponding members of the equations to obtain a new equation in a single variable.
3. *Solve* the derived equation for that variable.
4. *Substitute* the value of the one variable into either of the original equations, and solve for the second variable.
5. *State the solution* after *checking* it in the original system.

Example 2

$$\begin{array}{r} \\ 2 \end{array}\begin{cases} 3x - 2y = 15 \\ x + y = 10 \end{cases}$$

1. *Multiply* the second equation by 2; indicated here by the 2 to the left of that equation outside the brace.

$$+\begin{cases} 3x - 2y = 15 \\ 2x + 2y = 20 \end{cases}$$

2. *Add* the two equations, eliminating the y term.

$$5x = 35$$

3. *Solve* for x.

$$x = 7$$

$$(7) + y = 10$$
$$y = 3$$

4. *Substitute* the numerical value of x into *either* of the *original* equations.

Check: $3(7) - 2(3) = 15$
$21 - 6 = 15\surd$
$7 + 3 = 10\surd$

5. *Check and state the solution.*

The solution set is {(7, 3)}. □

Example 3

$$\begin{matrix} \\ -2 \end{matrix}\begin{cases} 2x + 2y = 5 \\ x + y = 10 \end{cases}$$

$$+\begin{cases} 2x + 2y = 5 \\ -2x - 2y = -20 \end{cases}$$

$$0 = -15$$

This result is never true, so the solution set is empty and the equations are **inconsistent.** (See Example 1b.) □

Example 4

$$\begin{matrix} -3 \\ \\ \end{matrix}\begin{cases} x - \frac{2}{3}y = 5 \\ 3x - 2y = 15 \end{cases}$$

$$+\begin{cases} -3x + 2y = -15 \\ 3x - 2y = 15 \end{cases}$$

$$0 = 0$$

This result is always true, so there are infinitely many solutions and the equations are **dependent.** (See Example 1c.) □

Example 5

$$\begin{matrix} 3 \\ 2 \end{matrix}\begin{cases} 3x + 2y = 15 \\ 5x - 3y = 6 \end{cases}$$

1. *Multiply* the first equation by 3 and the second by 2.

$$+\begin{cases} 9x + 6y = 45 \\ 10x - 6y = 12 \end{cases}$$

2. *Add,* eliminating the y term, since the coefficients are opposites.

$$19x = 57$$
$$x = 3$$

3. *Solve* for x.

$$3(3) + 2y = 15$$
$$2y = 6$$
$$y = 3$$
4. *Substitute* to find y.

Check: $3(3) + 2(3) = 15$
$9 + 6 = 15$ ✓
$5(3) - 3(3) = 6$
$15 - 9 = 6$ ✓

5. *Check and state the solution.*

The solution set is $\{(3, 3)\}$. □

Solution by Substitution

A third method of solution is called the **substitution method.** In this method, one of the equations is solved for either variable and the solution is substituted into the other equation.

SUBSTITUTION METHOD

To solve a system of linear equations in two variables:

1. *Solve* one of the equations for one of the variables.
2. *Substitute* the expression that you obtain into the other equation.
3. *Solve* the resulting equation in a single variable for the value of that variable.
4. *Substitute* that value into either of the original equations to determine the value of the other variable.
5. *State the solution* after *checking it* in the original system.

Example 6
$$\begin{cases} 3x - 2y = 15 \\ x + y = 10 \end{cases}$$

Solution Use the second equation to find an expression for y:

$$x + y = 10$$
$$y = -x + 10 \quad \text{1. } \textit{Solve} \text{ for } y.$$
$$3x - 2(-x + 10) = 15 \quad \text{2. } \textit{Substitute} \text{ into the first equation.}$$
$$3x + 2x - 20 = 15$$
$$5x = 35 \quad \text{3. } \textit{Solve} \text{ for } x.$$
$$x = 7$$
$$(7) + y = 10 \quad \text{4. } \textit{Substitute} \text{ to find } y.$$
$$y = 3$$

The solution set is $\{(7, 3)\}$. 5. *State the solution.*

Check: $3(7) - 2(3) = 15$
$21 - 6 = 15$ ✓
$7 + 3 = 10$ ✓ □

Example 7

$$\begin{cases} 3x + 2y = 15 \\ 5x - 3y = 6 \end{cases}$$

Solution Use the second equation:

$$5x - 3y = 6$$
$$3y = 5x - 6$$
$$y = \frac{5x - 6}{3}$$

1. *Solve* for y.

$$3x + 2\left(\frac{5x - 6}{3}\right) = 15$$

2. *Substitute* into the first equation.

$$3x + \frac{10x - 12}{3} = 15$$
$$9x + 10x - 12 = 45$$
$$19x - 12 = 45$$
$$19x = 57$$
$$x = 3$$

3. *Solve* for x.

$$5(3) - 3y = 6$$
$$15 - 3y = 6$$
$$3y = 9$$
$$y = 3$$

4. *Substitute* to find y.

The solution set is $\{(3, 3)\}$

5. *State the solution;* the check is left for the reader.

□

The substitution method may be used in some cases where the addition method does not apply. However, it can be more difficult to use since it is not always easy to solve for one of the variables. Notice in Example 7 that solving for y produces a fraction, whereas the solution of the same system in Example 5 avoids such an expression.

Nonlinear Systems

The methods reviewed above for linear systems may also be applied to systems containing nonlinear equations. Again, graphing is an approximation and only works well when the intersections have integer coordinates. You will see that the addition method may not always apply but that substitution is a universal method.

Example 8 Solve by the addition method:

$$\begin{matrix} \\ -1 \end{matrix}\begin{cases} y = x^2 - 4x + 5 \\ y = 2x - 3 \end{cases}$$

1. Multiply.

$$+\begin{cases} y = x^2 - 4x + 5 \\ -y = \quad -2x + 3 \end{cases}$$
$$0 = x^2 - 6x + 8$$

2. Add.

$$0 = (x-2)(x-4)$$
$$x = 2 \quad \text{or} \quad x = 4$$

3. Solve.

$$y = 2(2) - 3 \qquad y = 2(4) - 3$$
$$y = 1 \qquad y = 5$$

4. Substitute.

The solution set is {(2, 1), (4, 5)}.

5. State the solution. □

Example 9 Solve by the addition method:

$$2\begin{cases} 5x^2 - y^2 = 3 \\ x^2 + 2y^2 = 5 \end{cases}$$

$$\begin{cases} 10x^2 - 2y^2 = 6 \\ x^2 + 2y^2 = 5 \end{cases}$$
$$11x^2 = 11$$
$$x = \pm 1$$

$$(\pm 1)^2 + 2y^2 = 5$$
$$2y^2 = 4$$
$$y^2 = 2$$
$$y = \pm\sqrt{2}$$

The solution set is $\{(1, \sqrt{2}), (1, -\sqrt{2}), (-1, -\sqrt{2}), (-1, \sqrt{2})\}$. □

The system

$$\begin{cases} s^2 - 6st = 1 \\ s^2 - 3st = 2s \end{cases}$$

can be solved by eliminating the st term, but why not eliminate the s^2 terms in the first step? Well, frankly, it will not do any good to eliminate s^2, since

$$+\begin{cases} s^2 - 6st = 1 \\ -s^2 + 3st = -2s \end{cases}$$
$$-3st = 1 - 2s$$

A single equation is obtained, but it still contains both variables. It may not be solved for the value of either variable. Thus, the addition method may not always work for nonlinear systems. As a further example, consider the system

$$\begin{cases} y = x^2 - x - 1 \\ 4x = 7 + 2y - y^2 \end{cases} \quad \text{or} \quad \begin{cases} y - x^2 + x = -1 \\ y^2 - 2y + 4x = 7 \end{cases}$$

for which no combination of multipliers will allow the elimination of one of the variables. For this reason, the substitution method is more general. That is, it may be applied when the addition method is not convenient or will not work.

Example 10 Solve by the substitution method:

$$\begin{cases} x^2 = 2y + 2 \\ x - 3y + 1 = 0 \end{cases}$$

$$x = 3y - 1$$ 1. Solve the second equation for x.

$$(3y - 1)^2 = 2y + 2$$ 2. Substitute into the first equation.

$$9y^2 - 8y - 1 = 0$$

$$(9y + 1)(y - 1) = 0$$ 3. Solve.

$$y = -\frac{1}{9} \quad \text{or} \quad y = 1$$

$$x - 3\left(-\frac{1}{9}\right) + 1 = 0 \qquad x - 3(1) + 1 = 0$$ 4. Substitute.

$$x = -\frac{4}{3} \qquad x = 2$$

The solution set is $\left\{\left(-\frac{4}{3}, -\frac{1}{9}\right), (2, 1)\right\}$**.** 5. State the solution. □

Example 11 Solve by the substitution method:

$$\begin{cases} s^2 - 6st = 1 \\ s^2 - 3st = 2s \end{cases}$$

$$-3st = 2s - s^2$$ 1. Solve the second equation for t.

$$t = \frac{s - 2}{3}, \qquad s \neq 0$$

$$s^2 - 6s\left(\frac{s - 2}{3}\right) = 1$$ 2. Substitute.

$$s^2 - 2s^2 + 4s = 1$$ 3. Solve.

$$s^2 - 4s + 1 = 0$$

$$s = \frac{4 \pm \sqrt{16 - 4(1)(1)}}{2}$$

$$s = 2 \pm \sqrt{3}$$

$$t = \frac{(2 + \sqrt{3}) - 2}{3} \quad \text{or} \quad t = \frac{(2 - \sqrt{3}) - 2}{3}$$ 4. Substitute.

$$t = \frac{\sqrt{3}}{3} \qquad t = -\frac{\sqrt{3}}{3}$$

The solution set is $\left\{\left(2 + \sqrt{3}, \frac{\sqrt{3}}{3}\right), \left(2 - \sqrt{3}, -\frac{\sqrt{3}}{3}\right)\right\}$ 5. State the solution. □

Now reconsider the system that we were unable to solve by the addition method:

$$\begin{cases} y = x^2 - x - 1 \\ 4x = 7 + 2y - y^2 \end{cases}$$

Substitute the first equation into the second one:

$$4x = 7 + 2(x^2 - x - 1) - (x^2 - x - 1)^2$$
$$4x = 7 + 2x^2 - 2x - 2 - x^4 + 2x^3 + x^2 - 2x - 1$$
$$4x = 4 + 3x^2 - 4x - x^4 + 2x^3$$
$$x^4 - 2x^3 - 3x^2 + 8x - 4 = 0$$

This last equation may be solved using the methods described in Chapter 5:

$\frac{p}{q}$: $\pm 1, \pm 2, \pm 4$

	1	−2	−3	8	−4
$x = 1$	1	−1	−4	4	0
$x = 1$	1	0	−4	0	

$$(x - 1)(x - 1)(x^2 - 4) = 0$$
$$(x - 1)^2(x - 2)(x + 2) = 0$$
$$x = 1 \text{ or } x = 2 \text{ or } x = -2$$

$\{(\mathbf{1}, -\mathbf{1}), (\mathbf{2}, \mathbf{1}), (-\mathbf{2}, \mathbf{5})\}$

In summary, some methods developed for linear systems may be applied to nonlinear systems. The addition method, while easier to use than substitution, does not always apply. Both methods eliminate one of the two variables and obtain a single equation in one variable. As long as the derived equation may be solved, the system may be solved.

Problem Set 7.1

A *Solve the systems in Problems 1–10 by graphing.*

1. $\begin{cases} x - y = 2 \\ 2x + 3y = 9 \end{cases}$

2. $\begin{cases} 3x - 4y = 16 \\ -x + 2y = -6 \end{cases}$

3. $\begin{cases} y = 3x + 1 \\ x - 2y = 8 \end{cases}$

4. $\begin{cases} 2x - 3y = 12 \\ -4x + 6y = 18 \end{cases}$

5. $\begin{cases} x - 6 = y \\ 4x + y = 9 \end{cases}$

6. $\begin{cases} 6x + y = -5 \\ x + 3y = 2 \end{cases}$

7. $\begin{cases} 4x - 3y = -1 \\ -2x + 3y = -1 \end{cases}$

8. $\begin{cases} 3x + 2y = 5 \\ 4x - 3y = 1 \end{cases}$

9. $\begin{cases} 2y = 10x - x^2 - 5 \\ y = x + 1 \end{cases}$

10. $\begin{cases} 3y = -2(x^2 + 4x + 1) \\ y = 2x + 6 \end{cases}$

Solve the systems in Problems 11–20 by the addition method, if possible.

11. $\begin{cases} x + y = 16 \\ x - y = 10 \end{cases}$

12. $\begin{cases} x + y = 560 \\ x - y = 490 \end{cases}$

13. $\begin{cases} 6r - 4s = 10 \\ 2s = 3r - 5 \end{cases}$

14. $\begin{cases} 3u + 2v = 5 \\ 4v = -7 - 6u \end{cases}$

15. $\begin{cases} 3a_1 + 4a_2 = -9 \\ 5a_1 + 7a_2 = -14 \end{cases}$

16. $\begin{cases} 5s_1 + 2s_2 = 23 \\ 2s_1 + 7s_2 = 34 \end{cases}$

17. $\begin{cases} \frac{1}{2}y = x - 1 \\ y = 2(1 + 2x - x^2) \end{cases}$

18. $\begin{cases} 2y = x^2 + 6x + 5 \\ y = \frac{1}{2}x + \frac{1}{2} \end{cases}$

19. $\begin{cases} 4y = 3x^2 - 30x + 59 \\ 2y = -3x + 7 \end{cases}$

20. $\begin{cases} 4y = 5x^2 + 30x + 53 \\ 2y = 5x + 19 \end{cases}$

Solve the systems in Problems 21–30 by the substitution method.

21. $\begin{cases} y = 3 - 2x \\ 3x + 2y = -17 \end{cases}$

22. $\begin{cases} 5x - 2y = -19 \\ x = 3y + 4 \end{cases}$

23. $\begin{cases} 2x - 3y = 15 \\ y = \frac{2}{3}x - 5 \end{cases}$

24. $\begin{cases} x + y = 12 \\ .6y = .5(12) \end{cases}$

25. $\begin{cases} 4y + 5x = 2 \\ y = -\frac{5}{4}x + 1 \end{cases}$

26. $\begin{cases} \dfrac{x}{3} - y = 7 \\ x + \dfrac{y}{2} = 7 \end{cases}$

27. $\begin{cases} y + 4x + 2 = 0 \\ y^2 = 4x + 4 \end{cases}$

28. $\begin{cases} x^2 + 4y^2 = 16 \\ x = 4y - 8 \end{cases}$

29. $\begin{cases} x^2 - 12y - 1 = 0 \\ x^2 - 4y^2 - 9 = 0 \end{cases}$

30. $\begin{cases} y^2 = 4 - x \\ x^2 + y^2 = 16 \end{cases}$

B *Solve the systems in Problems 31–52 for all real solutions, using any suitable method.*

31. $\begin{cases} x + y = 7 \\ x - y = -1 \end{cases}$

32. $\begin{cases} x - y = 8 \\ x + y = 2 \end{cases}$

33. $\begin{cases} -x + 2y = 2 \\ 4x - 7y = -5 \end{cases}$

34. $\begin{cases} x - 6y = -3 \\ 2x + 3y = 9 \end{cases}$

35. $\begin{cases} y = 3x + 1 \\ x - 2y = 8 \end{cases}$

36. $\begin{cases} 2x + 3y = 9 \\ x = 6y - 3 \end{cases}$

37. $\begin{cases} x - y = 2 \\ 2x + 3y = 9 \end{cases}$

38. $\begin{cases} 3x - 4y = 16 \\ -x + 2y = -6 \end{cases}$

39. $\begin{cases} 2x - y = 6 \\ 4x + y = 3 \end{cases}$

40. $\begin{cases} 6x + 9y = -4 \\ 9x + 3y = 1 \end{cases}$

41. $\begin{cases} 100x - y = 0 \\ 50x + y = 300 \end{cases}$

42. $\begin{cases} x = \frac{3}{4}y - 2 \\ 3y - 4x = 5 \end{cases}$

43. $\begin{cases} q + d = 147 \\ .25q + .10d = 24.15 \end{cases}$

44. $\begin{cases} x + y = 10 \\ .4x + .9y = .5(10) \end{cases}$

45. $\begin{cases} 2x - y = 2 \\ y^2 = 4x + 4 \end{cases}$

46. $\begin{cases} x^2 + y^2 = 25 \\ x - y + 1 = 0 \end{cases}$

47. $\begin{cases} 2x^2 + 3y^2 = 59 \\ x^2 + 2y^2 = 34 \end{cases}$

48. $\begin{cases} 9x^2 - 4y^2 = 36 \\ x^2 + y^2 = 16 \end{cases}$

49. $\begin{cases} x^2 + 2xy = 8 \\ 3x^2 - 4y^2 = 8 \end{cases}$

50. $\begin{cases} x^2 + xy = 6 \\ y^2 - x^2 = 13 \end{cases}$

51. $\begin{cases} x^2 - xy + y^2 = 21 \\ xy - y^2 = 15 \end{cases}$

52. $\begin{cases} x^2 - xy + y^2 = 3 \\ x^2 + y^2 = 6 \end{cases}$

C *Use a system of equations to solve Problems 53–60.*

53. Find two numbers whose difference is five and whose product is fourteen.

54. Find two numbers whose sum is thirty and the difference of whose squares is three hundred.

55. The product of two numbers added to the larger of the numbers is six. Three times the smaller added to the difference of the numbers is five. Find the numbers.

56. The smaller of two numbers added to twice their product is two. Three times the smaller plus the larger is three. What are the numbers?

57. *Chemistry* A chemist has two solutions of an acid. One is a 40% solution, and the other is a 90% solution. How many liters of each should be mixed to get 10 liters of a 50% solution?

58. *Metallurgy* A craftsman has two alloys of silver. One is 30% silver, and the other is 60% silver. How many grams of each must be mixed to get 16 g of a 37.5% alloy?

59. The area of a right triangle is 60 mm^2, and its hypotenuse is 17 mm. Find the lengths of the legs of the triangle.

60. The area of a rectangle is 60 cm^2, and its diagonal is 13 cm. Find the length and width of the rectangle.

7.2 Matrices

A quick survey of the literature in the biological, natural, and social sciences, in engineering, and in economics will show that scholars are turning to the use of matrices in increasing numbers to solve very complex problems. The computer plays a large part in this movement to matrices. By formulating many problems in the language of matrix algebra, the computer can be applied; without the computer, these problems would be too difficult or too time-consuming to solve.

A **matrix** is a rectangular array of entries. The entries are arranged horizontally in **rows** and vertically in **columns.** We will consider only real number entries. A matrix is said to be of size or **dimension $m \times n$,** where m is the number of rows and n is the number of columns of entries. If $m = n$, the matrix is called a **square matrix.** An $m \times 1$ matrix is called a **column matrix,** since it consists of a single column. Likewise, a $1 \times n$ matrix is called a **row matrix.** A matrix is usually indicated by a capital letter with the dimension sometimes shown as a subscript.

Example 1

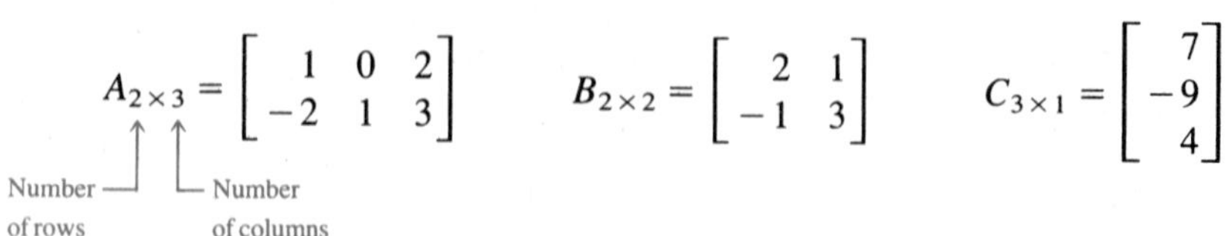

$$A_{2\times 3} = \begin{bmatrix} 1 & 0 & 2 \\ -2 & 1 & 3 \end{bmatrix} \qquad B_{2\times 2} = \begin{bmatrix} 2 & 1 \\ -1 & 3 \end{bmatrix} \qquad C_{3\times 1} = \begin{bmatrix} 7 \\ -9 \\ 4 \end{bmatrix}$$

□

In general, we write

$$A_{m \times n} = [a_{ij}]$$

to indicate the $m \times n$ matrix

$$A = \begin{bmatrix} a_{11} & a_{12} & a_{13} & \cdots & a_{1n} \\ a_{21} & a_{22} & a_{23} & \cdots & a_{2n} \\ \vdots & & & & \vdots \\ a_{m1} & a_{m2} & a_{m3} & \cdots & a_{mn} \end{bmatrix}$$

Notice that a_{23} is the entry in the 2nd row and 3rd column. The first subscript indicates the row, and the second subscript indicates the column.

where the subscripts of the entries indicate the rows and columns of their positions.

EQUALITY OF MATRICES

$$A_{m\times n} = B_{m\times n}$$

if and only if $a_{ij} = b_{ij}$ for each i, j.

This means that two matrices are equal if and only if they are of the same dimension and all corresponding entries are equal.

If two matrices have the same dimension, they are said to be **conformable for addition.** The matrices are added by adding corresponding entries of the given matrices. This is stated formally in the following definition:

MATRIX ADDITION

If $A_{m\times n} = [a_{ij}]$ and $B_{m\times n} = [b_{ij}]$, then

$$A + B = [a_{ij} + b_{ij}]_{m\times n}$$

Use the following matrices to find the indicated sums in Examples 2–5, if possible:

$$A = [4 \quad 2 \quad 6] \qquad B = [1 \quad 7 \quad 2] \qquad C = \begin{bmatrix} 3 & -2 & 1 \\ 2 & 1 & -3 \end{bmatrix}$$

$$D = \begin{bmatrix} 2 & 1 & 0 \\ 4 & 7 & 3 \\ -2 & 0 & 1 \end{bmatrix} \qquad E = \begin{bmatrix} 6 & 1 & 2 \\ 3 & -10 & 4 \\ 1 & 3 & -2 \end{bmatrix}$$

Example 2 $A + B = [4 + 1 \quad 2 + 7 \quad 6 + 2]$

$= [\mathbf{5} \quad \mathbf{9} \quad \mathbf{8}]$ □

Example 3 $A + C$ **Not conformable:** 1×3 and 2×3. □

Example 4 $C + E$ **Not conformable:** 2×3 and 3×3. □

Example 5

$$D + E = \begin{bmatrix} 2+6 & 1+1 & 0+2 \\ 4+3 & 7+(-10) & 3+4 \\ -2+1 & 0+3 & 1+(-2) \end{bmatrix} = \begin{bmatrix} \mathbf{8} & \mathbf{2} & \mathbf{2} \\ \mathbf{7} & \mathbf{-3} & \mathbf{7} \\ \mathbf{-1} & \mathbf{3} & \mathbf{-1} \end{bmatrix}$$ □

One of the attractions of matrices is the ease with which data from tables can be written as a matrix. Consider Table 7.1, which contains sales figures (in thousands of dollars) for a company with offices in San Francisco, Los Angeles, and San Diego.

Table 7.1

	1984			**1985**		
	San Francisco	**Los Angeles**	**San Diego**	**San Francisco**	**Los Angeles**	**San Diego**
Wholesale	150	200	350	175	300	400
Retail	100	150	50	110	100	100

The data may be represented by two matrices. Let S_{84} and S_{85} represent the sales for 1984 and 1985, respectively. Then

$$S_{84} = \begin{bmatrix} 150 & 200 & 350 \\ 100 & 150 & 50 \end{bmatrix} \qquad S_{85} = \begin{bmatrix} 175 & 300 & 400 \\ 110 & 100 & 100 \end{bmatrix}$$

Suppose you wish to know the combined sales for the given years. The answer is the sum of the two matrices:

$$S_{84} + S_{85} = \begin{bmatrix} 150+175 & 200+300 & 350+400 \\ 100+110 & 150+100 & 50+100 \end{bmatrix}$$

$$= \begin{bmatrix} 325 & 500 & 750 \\ 210 & 250 & 150 \end{bmatrix}$$

Continuing with this example, suppose you know that the average profit has been 3% of gross sales. To find the profit in 1985, you need to find 3% of each entry in S_{85}.

This operation of multiplying each entry in a matrix by a number is called **scalar multiplication.**

SCALAR MULTIPLICATION

> If $A_{m \times n} = [a_{ij}]$, then for any real number c
>
> $cA = [ca_{ij}]$

To find the profit for the above example in each category, $(.03)S_{85}$ is calculated:

$$(.03)\,S_{85} = \begin{bmatrix} (.03)\,175 & (.03)\,300 & (.03)\,400 \\ (.03)\,110 & (.03)\,100 & (.03)\,100 \end{bmatrix}$$

$$= \begin{bmatrix} 5.25 & 9 & 12 \\ 3.3 & 3 & 3 \end{bmatrix}$$

Example 6 Find 3A + 2B, if: $A = \begin{bmatrix} 2 & 1 & 0 \\ -2 & 0 & 1 \end{bmatrix} \quad B = \begin{bmatrix} 6 & 1 & 2 \\ 1 & 3 & -2 \end{bmatrix}$

Solution

$$3A + 2B = \begin{bmatrix} 3\cdot 2 & 3\cdot 1 & 3\cdot 0 \\ 3(-2) & 3\cdot 0 & 3\cdot 1 \end{bmatrix} + \begin{bmatrix} 2\cdot 6 & 2\cdot 1 & 2\cdot 2 \\ 2\cdot 1 & 2\cdot 3 & 2(-2) \end{bmatrix}$$

$$= \begin{bmatrix} \mathbf{18} & \mathbf{5} & \mathbf{4} \\ \mathbf{-4} & \mathbf{6} & \mathbf{-1} \end{bmatrix}$$

□

Returning to the 1984–1985 sales example, it may be necessary to find the increase or decrease in sales at the various sites. It is obvious that the change in wholesale sales in San Francisco is found by subtracting: 175 − 150 = 25. To subtract two matrices, corresponding entries are subtracted.

MATRIX SUBTRACTION

> If $A_{m \times n} = [a_{ij}]$ and $B_{m \times n} = [b_{ij}]$, then
>
> $A - B = [a_{ij} - b_{ij}]_{m \times n}$

This is certainly reasonable, since if

$$A - B = A + (-1)B$$

then

$$A - B = [a_{ij} + (-1)b_{ij}]$$
$$= [a_{ij} - b_{ij}]$$

Example 7 Find the change in sales between 1984 and 1985 for the company sales given by S_{84} and S_{85}. (Refer to Table 7.1.)

Solution

$$S_{85} - S_{84} = \begin{bmatrix} 175 - 150 & 300 - 200 & 400 - 350 \\ 110 - 100 & 100 - 150 & 100 - 50 \end{bmatrix}$$

$$= \begin{bmatrix} \mathbf{25} & \mathbf{100} & \mathbf{50} \\ \mathbf{10} & \mathbf{-50} & \mathbf{50} \end{bmatrix}$$ The negative entry shows a decrease in sales □

For this same example, suppose that the profits on wholesale and retail sales are different—for instance, 3% profit on wholesale and 5% on retail sales. These two figures could be arranged in a profit matrix, P:

$$P = [.03 \quad .05]$$

Multiply the wholesale sales by .03 and add .05 times the retail sales to find the total profit at each location, and observe the pattern of such an operation on the entries in the respective matrices.

$$[.03 \quad .05] \quad \text{and} \quad \begin{bmatrix} 175 & 300 & 400 \\ 110 & 100 & 100 \end{bmatrix}$$

First, find the profit in San Francisco in 1985:

$$.03(175) + .05(110) = \mathbf{10.75}$$

$$[.03 \quad .05]\begin{bmatrix} 175 & 300 & 400 \\ 110 & 100 & 100 \end{bmatrix}$$

Similarly, in Los Angeles,

$$.03(300) + .05(100) = \mathbf{14}$$

$$[.03 \quad .05]\begin{bmatrix} 175 & 300 & 400 \\ 110 & 100 & 100 \end{bmatrix}$$

and in San Diego,

$$.03(400) + .05(100) = \mathbf{17}$$

$$[.03 \quad .05]\begin{bmatrix} 175 & 300 & 400 \\ 110 & 100 & 100 \end{bmatrix}$$

The individual profits can be displayed in a 1×3 matrix,

$$[\mathbf{10.75} \quad \mathbf{14} \quad \mathbf{17}]$$

which is called the *product* of P and S_{85}. All these steps may be written as follows:

$$PS_{85} = [.03 \quad .05]\begin{bmatrix} 175 & 300 & 400 \\ 110 & 100 & 100 \end{bmatrix}$$

$$= [.03(175) + .05(110) \quad .03(300) + .05(100) \quad .03(400) + .05(100)]$$

$$= [10.75 \quad 14 \quad 17]$$

The row of P was matched up with each column of S_{85}, but what if P had more than one row? If there were a second row, it would be matched with each column of S_{85} to obtain a second row in the product.

MATRIX MULTIPLICATION

If $A_{m\times n} = [a_{ij}]$ and $B_{n\times p} = [b_{ij}]$, then

$$A_{m\times n}B_{n\times p} = \begin{bmatrix} a_{11} & a_{12} & \cdots & a_{1n} \\ a_{21} & a_{22} & \cdots & a_{2n} \\ \vdots & \vdots & & \vdots \\ a_{i1} & a_{i2} & \cdots & a_{in} \\ \vdots & \vdots & & \vdots \\ a_{m1} & a_{m2} & \cdots & a_{mn} \end{bmatrix} \begin{bmatrix} b_{11} & b_{12} & \cdots & b_{1j} & \cdots & b_{1p} \\ b_{21} & b_{22} & \cdots & b_{2j} & \cdots & b_{2p} \\ \vdots & \vdots & & \vdots & & \vdots \\ b_{n1} & b_{n2} & \cdots & b_{nj} & \cdots & b_{np} \end{bmatrix}$$

$$= [c_{ij}]$$

$$= C_{m\times p}$$

where the element c_{ij} of the product $C_{m\times p}$ is given by

$$c_{ij} = a_{i1}b_{1j} + a_{i2}b_{2j} + a_{i3}b_{3j} + \ldots + a_{in}b_{nj}$$

Example 8 Find AB, if: $A = \begin{bmatrix} 1 & 2 \\ 3 & 4 \end{bmatrix}$ $\quad B = \begin{bmatrix} 4 & -1 & 2 & 6 \\ 1 & 3 & -1 & 2 \end{bmatrix}$

Solution

$$AB = \begin{bmatrix} 1 & 2 \\ 3 & 4 \end{bmatrix} \begin{bmatrix} 4 & -1 & 2 & 6 \\ 1 & 3 & -1 & 2 \end{bmatrix}$$

$$= \begin{bmatrix} 1\cdot 4 + 2\cdot 1 & 1(-1) + 2\cdot 3 & 1\cdot 2 + 2(-1) & 1\cdot 6 + 2\cdot 2 \\ 3\cdot 4 + 4\cdot 1 & 3(-1) + 4\cdot 3 & 3\cdot 2 + 4(-1) & 3\cdot 6 + 4\cdot 2 \end{bmatrix}$$

$$= \begin{bmatrix} \mathbf{6} & \mathbf{5} & \mathbf{0} & \mathbf{10} \\ \mathbf{16} & \mathbf{9} & \mathbf{2} & \mathbf{26} \end{bmatrix}$$

□

To be **conformable** for multiplication, the number of columns of the first matrix must be the same as the number of rows in the second. If the matrices are conformable for multiplication, the product is obtained by considering the ith row of A and the jth column of B to obtain the c_{ij} entry of the product.

Use the following matrices to find the indicated products in Examples 9–12, if possible:

$$A = \begin{bmatrix} 1 & 2 \\ 3 & 4 \end{bmatrix} \quad B = \begin{bmatrix} -1 & 2 \\ 1 & 3 \end{bmatrix} \quad C = \begin{bmatrix} 1 & 2 & 3 \\ 4 & 5 & 6 \end{bmatrix} \quad D = \begin{bmatrix} 1 & 0 & 2 & 1 \\ 2 & 1 & 0 & 0 \\ 3 & 2 & 1 & 0 \end{bmatrix}$$

Example 9
$$AB = \begin{bmatrix} 1 & 2 \\ 3 & 4 \end{bmatrix}\begin{bmatrix} -1 & 2 \\ 1 & 3 \end{bmatrix} = \begin{bmatrix} 1(-1) + 2\cdot 1 & 1\cdot 2 + 2\cdot 3 \\ 3(-1) + 4\cdot 1 & 3\cdot 2 + 4\cdot 3 \end{bmatrix} = \begin{bmatrix} \mathbf{1} & \mathbf{8} \\ \mathbf{1} & \mathbf{18} \end{bmatrix} \quad \square$$

Example 10
$$BA = \begin{bmatrix} -1 & 2 \\ 1 & 3 \end{bmatrix}\begin{bmatrix} 1 & 2 \\ 3 & 4 \end{bmatrix} = \begin{bmatrix} (-1)1 + 2\cdot 3 & (-1)2 + 2\cdot 4 \\ 1\cdot 1 + 3\cdot 3 & 1\cdot 2 + 3\cdot 4 \end{bmatrix} = \begin{bmatrix} \mathbf{5} & \mathbf{6} \\ \mathbf{10} & \mathbf{14} \end{bmatrix} \quad \square$$

Example 11
$$CD = \begin{bmatrix} 1 & 2 & 3 \\ 4 & 5 & 6 \end{bmatrix}\begin{bmatrix} 1 & 0 & 2 & 1 \\ 2 & 1 & 0 & 0 \\ 3 & 2 & 1 & 0 \end{bmatrix}$$
$$= \begin{bmatrix} 1 + 4 + 9 & 0 + 2 + 6 & 2 + 0 + 3 & 1 + 0 + 0 \\ 4 + 10 + 18 & 0 + 5 + 12 & 8 + 0 + 6 & 4 + 0 + 0 \end{bmatrix}$$
$$= \begin{bmatrix} \mathbf{14} & \mathbf{8} & \mathbf{5} & \mathbf{1} \\ \mathbf{32} & \mathbf{17} & \mathbf{14} & \mathbf{4} \end{bmatrix} \quad \square$$

Example 12 D and C are not conformable for multiplication, because their dimensions are 3×4 and 2×3. Therefore, DC is not defined. However, notice that CD is defined. $\square$

Remember that matrix multiplication requires the *first* of the two matrices being multiplied to have the same number of columns as the number of rows in the *second*. If not, they are not conformable for multiplication. Notice in Examples 9 and 10 that $AB \neq BA$ and in Examples 11 and 12 that $CD \neq DC$ since DC is not even defined. That is, matrix multiplication is not commutative. You must therefore be careful of the order of multiplication with matrices.

The Inverse of a Matrix

In algebra, an identity for an operation in a given set is the element of the set that does not change another element when operated on by the identity element. For example,

Addition	**Multiplication**
$a + 0 = 0 + a = a$	$a \cdot 1 = 1 \cdot a = a$
0 is the identity for addition in the real numbers.	1 is the identity of multiplication in the real numbers.

The identity for addition of matrices is the **zero matrix.** The identity for multiplication of matrices is a matrix with 1s on the diagonal and 0s elsewhere. For example,

$$\begin{bmatrix} 1 & 2 & 3 \\ 4 & 5 & 6 \\ 7 & 8 & 9 \end{bmatrix} + \begin{bmatrix} 0 & 0 & 0 \\ 0 & 0 & 0 \\ 0 & 0 & 0 \end{bmatrix} = \begin{bmatrix} 1 & 2 & 3 \\ 4 & 5 & 6 \\ 7 & 8 & 9 \end{bmatrix} \qquad \begin{bmatrix} 1 & 2 & 3 \\ 4 & 5 & 6 \\ 7 & 8 & 9 \end{bmatrix}\begin{bmatrix} 1 & 0 & 0 \\ 0 & 1 & 0 \\ 0 & 0 & 1 \end{bmatrix} = \begin{bmatrix} 1 & 2 & 3 \\ 4 & 5 & 6 \\ 7 & 8 & 9 \end{bmatrix}$$

Identity matrix for 3×3 matrices and addition — Identical

Identity matrix for 3×3 matrices and multiplication — Identical

IDENTITY MATRIX

If $I = [a_{ij}]_{n \times n}$ such that

$$\begin{cases} a_{ij} = 1 & \text{if } i = j \\ a_{ij} = 0 & \text{if } i \neq j \end{cases}$$

then I is called the **identity matrix** of dimension n for multiplication, and

$$IA = AI = A \qquad \text{for any conformable matrix } A$$

Recall the idea of inverse from algebra. The inverse of an element for an operation in a given set is the element of the set that, when operated on by the given element, produces the identity for that operation. For example,

Addition	**Multiplication**
$a + (-a) = (-a) + a = 0$	$a \cdot a^{-1} = a^{-1} \cdot a = 1$
The sum of a real number and its opposite (additive inverse) is 0 (identity for addition).	The product of a nonzero real number and its reciprocal (multiplicative inverse) is 1 (identity for multiplication).

The inverse for multiplication of matrices behaves in a similar way. That is, the product of a matrix and its inverse for multiplication, if it exists, is the identity matrix. Even though each matrix A has an additive inverse, $-A$, it is not particularly useful. The inverse for multiplication, A^{-1} if it exists, is extremely useful. We begin with a formal definition of A^{-1}.

INVERSE OF A MATRIX

If A is a square matrix and if there exists a matrix A^{-1} such that

$$A^{-1}A = AA^{-1} = I$$

then A^{-1} is called the **inverse of A** for multiplication.

It may be verified that two given matrices A and B are inverses by showing that $AB = BA = I$.

Example 13 Verify that the inverse of $A = \begin{bmatrix} 2 & 1 \\ 3 & 2 \end{bmatrix}$ is $B = \begin{bmatrix} 2 & -1 \\ -3 & 2 \end{bmatrix}$.

Solution

$$AB = \begin{bmatrix} 2 & 1 \\ 3 & 2 \end{bmatrix}\begin{bmatrix} 2 & -1 \\ -3 & 2 \end{bmatrix} = \begin{bmatrix} 4-3 & -2+2 \\ 6-6 & -3+4 \end{bmatrix} = \begin{bmatrix} 1 & 0 \\ 0 & 1 \end{bmatrix} = I$$

$$BA = \begin{bmatrix} 2 & -1 \\ -3 & 2 \end{bmatrix}\begin{bmatrix} 2 & 1 \\ 3 & 2 \end{bmatrix} = \begin{bmatrix} 4-3 & 2-2 \\ -6+6 & -3+4 \end{bmatrix} = \begin{bmatrix} 1 & 0 \\ 0 & 1 \end{bmatrix} = I$$

Thus, $\boldsymbol{B = A^{-1}}$. □

Example 14 Show that A and B are inverses, where

$$A = \begin{bmatrix} 0 & 1 & 2 \\ -1 & 1 & 2 \\ 1 & -2 & -5 \end{bmatrix} \quad \text{and} \quad B = \begin{bmatrix} 1 & -1 & 0 \\ 3 & 2 & 2 \\ -1 & -1 & -1 \end{bmatrix}$$

Solution

$$AB = \begin{bmatrix} 0 & 1 & 2 \\ -1 & 1 & 2 \\ 1 & -2 & -5 \end{bmatrix}\begin{bmatrix} 1 & -1 & 0 \\ 3 & 2 & 2 \\ -1 & -1 & -1 \end{bmatrix}$$

$$= \begin{bmatrix} 0+3-2 & 0+2-2 & 0+2-2 \\ -1+3-2 & 1+2-2 & 0+2-2 \\ 1-6+5 & -1-4+5 & 0-4+5 \end{bmatrix}$$

$$= \begin{bmatrix} 1 & 0 & 0 \\ 0 & 1 & 0 \\ 0 & 0 & 1 \end{bmatrix}$$

$$= I$$

$$BA = \begin{bmatrix} 1 & -1 & 0 \\ 3 & 2 & 2 \\ -1 & -1 & -1 \end{bmatrix}\begin{bmatrix} 0 & 1 & 2 \\ -1 & 1 & 2 \\ 1 & -2 & -5 \end{bmatrix}$$

$$= \begin{bmatrix} 0+1+0 & 1-1+0 & 2-2+0 \\ 0-2+2 & 3+2-4 & 6+4-10 \\ 0+1-1 & -1-1+2 & -2-2+5 \end{bmatrix}$$

$$= \begin{bmatrix} 1 & 0 & 0 \\ 0 & 1 & 0 \\ 0 & 0 & 1 \end{bmatrix}$$

$$= I$$

Since $AB = I = BA$, $\mathbf{B = A^{-1}}$. □

Problem Set 7.2

A *In Problems 1–10 perform the indicated matrix operations, if possible, where*

$$A = \begin{bmatrix} 1 & 0 & 2 \\ 3 & -1 & 2 \\ 4 & 1 & 0 \end{bmatrix} \quad B = \begin{bmatrix} 1 & 4 & 0 \\ 3 & -1 & 2 \\ -2 & 1 & 5 \end{bmatrix} \quad C = \begin{bmatrix} 8 & 1 & 6 \\ 3 & 5 & 7 \\ 4 & 9 & 2 \end{bmatrix}$$

1. $B + C$ **2.** $3A - 4B$ **3.** $2B - C$ **4.** AB **5.** AC

6. BC **7.** $AB + AC$ **8.** $A(B + C)$ **9.** $A(BC)$ **10.** $(AB)C$

In Problems 11–20 perform the indicated matrix operations, if possible, where

$$A = \begin{bmatrix} 1 & 2 \\ 4 & 0 \\ -1 & 3 \\ 2 & 1 \end{bmatrix} \quad B = \begin{bmatrix} 4 & 2 \\ -1 & 3 \end{bmatrix} \quad C = \begin{bmatrix} 1 & 0 & 0 & 0 \\ 0 & 1 & 0 & 0 \\ 0 & 0 & 1 & 0 \\ 0 & 0 & 0 & 1 \end{bmatrix}$$

$$D = \begin{bmatrix} 4 & 1 & 3 & 6 \\ -1 & 0 & -2 & 3 \end{bmatrix}$$

11. AB **12.** BA **13.** B^2 **14.** CA **15.** BD
16. DB **17.** $(B + C)A$ **18.** $BA + CA$ **19.** C^3 **20.** CD

In Problems 21–30 use multiplication to determine whether the given matrices are inverses.

21. $\begin{bmatrix} 1 & 2 \\ 2 & 3 \end{bmatrix}, \begin{bmatrix} -3 & 2 \\ 2 & -1 \end{bmatrix}$

22. $\begin{bmatrix} 2 & -5 \\ -1 & 2 \end{bmatrix}, \begin{bmatrix} -2 & -5 \\ -1 & -2 \end{bmatrix}$

23. $\begin{bmatrix} 3 & 5 \\ 4 & 7 \end{bmatrix}, \begin{bmatrix} 7 & -5 \\ -4 & 3 \end{bmatrix}$

24. $\begin{bmatrix} 4 & 7 \\ 5 & 9 \end{bmatrix}, \begin{bmatrix} 9 & -7 \\ -5 & 4 \end{bmatrix}$

25. $\begin{bmatrix} 4 & 3 \\ 2 & 2 \end{bmatrix}, \begin{bmatrix} 1 & -\frac{3}{2} \\ -1 & 2 \end{bmatrix}$

26. $\begin{bmatrix} 2 & 3 \\ 2 & 1 \end{bmatrix}, \begin{bmatrix} -\frac{1}{4} & \frac{3}{4} \\ \frac{1}{2} & -\frac{1}{2} \end{bmatrix}$

27. $\begin{bmatrix} 0 & 1 & 0 \\ 1 & -1 & 0 \\ -1 & 2 & 1 \end{bmatrix}, \begin{bmatrix} -1 & -1 & 0 \\ -1 & 0 & 0 \\ 1 & -1 & -1 \end{bmatrix}$

28. $\begin{bmatrix} 1 & 0 & 0 \\ 0 & 1 & 1 \\ 2 & 0 & 1 \end{bmatrix}, \begin{bmatrix} 1 & 0 & 0 \\ 2 & 1 & -1 \\ -2 & 0 & 1 \end{bmatrix}$

29. $\begin{bmatrix} 6 & -1 & -5 \\ -7 & 1 & 5 \\ -10 & 2 & 11 \end{bmatrix}, \begin{bmatrix} -1 & -1 & 0 \\ -27 & -16 & -5 \\ 4 & 2 & 1 \end{bmatrix}$

30. $\begin{bmatrix} 3 & -2 & 4 \\ 2 & 1 & 2 \\ 5 & 3 & 5 \end{bmatrix}, \begin{bmatrix} -1 & 22 & -8 \\ 0 & -5 & 2 \\ 1 & -19 & 7 \end{bmatrix}$

Business *For Problems 31–40 refer to the table below, which contains sales figures (in thousands of dollars) for a Florida company with branches in Miami, Tampa, and Orlando. Let S_{84} and S_{85} represent the sales for 1984 and 1985.*

	1984			1985		
	Miami	Tampa	Orlando	Miami	Tampa	Orlando
Mail Order	335	280	100	350	270	130
Retail Outlet	250	120	50	290	270	90

31. Write S_{84} and S_{85} as 2×3 matrices.

32. Find $S_{84} + S_{85}$, and interpret this sum.

33. Find $S_{85} - S_{84}$, and interpret this difference.

34. Find $\frac{1}{2}[S_{84} + S_{85}]$, and interpret this matrix.

35. Find $(.08)S_{84}$, and give a plausible interpretation of this scalar product.

36. Find $(.10)S_{85}$, and give a plausible interpretation of this scalar product.

Let $P = [.08 \quad .10]$ *and find:*

37. PS_{84} **38.** PS_{85} **39.** $P[S_{84} + S_{85}]$ **40.** $P[S_{85} - S_{84}]$

B *Business* *For Problems 41–50 refer to the table below, which contains sales figures (in thousands of dollars) for a midwest distributor with offices in Chicago, Detroit, Indianapolis, Milwaukee, and Minneapolis. Let* S_{84} *and* S_{85} *represent the sales for 1984 and 1985.*

	1984					**1985**				
	Chicago	**Detroit**	**Indianapolis**	**Milwaukee**	**Minneapolis**	**Chicago**	**Detroit**	**Indianapolis**	**Milwaukee**	**Minneapolis**
Assembled	3370	1510	740	720	430	3000	1200	700	640	370
Kit Form	3620	1850	430	640	520	3550	1670	480	740	430

41. Write S_{84} and S_{85} as 2×5 matrices.

42. Find $S_{84} + S_{85}$, and interpret this sum.

43. Find $S_{85} - S_{84}$, and interpret this difference.

44. Find $\frac{1}{2}[S_{84} + S_{85}]$, and interpret this matrix.

45. Find $(.15)S_{84}$, and give a plausible interpretation of this scalar product.

46. Find $(.20)S_{85}$, and give a plausible interpretation of this scalar product.

Let $P = [.15 \quad .20]$ *and find:*

47. PS_{84} **48.** PS_{85} **49.** $P[S_{84} + S_{85}]$ **50.** $P[S_{85} - S_{84}]$

51. Let

$$A = [2 \quad 4 \quad 7] \quad \text{and} \quad B = [6 \quad -1 \quad 0]$$

Find a matrix X such that $A + X = 2B$.

52. Let

$$C = \begin{bmatrix} 4 \\ 1 \\ 6 \end{bmatrix} \quad \text{and} \quad D = \begin{bmatrix} -2 \\ 1 \\ 3 \end{bmatrix}$$

Find a matrix Y such that $3C - Y = 2D$.

53. Let

$$E = \begin{bmatrix} 2 & 0 & 7 \\ 4 & -1 & 3 \\ 1 & 3 & 0 \end{bmatrix} \quad \text{and} \quad F = \begin{bmatrix} 1 & 3 & 0 \\ 1 & 3 & -2 \\ -5 & 2 & 1 \end{bmatrix}$$

Find a matrix Z such that $EF + Z = F$.

54. On the basis of your answers to Problems 9 and 10, do you think that the associative property for multiplication holds for matrices that are conformable for multiplication? Show other examples.

55. On the basis of your answers to Problems 7, 8, 17, and 18, do you think that the distributive property for multiplication over addition holds for matrices that are conformable? Show other examples.

56. Let $J = [a_{ij}]$ such that

$$\begin{cases} a_{ij} = 1 & \text{if} \quad i + j = n + 1 \\ a_{ij} = 0 & \text{if} \quad i + j \neq n + 1 \end{cases}$$

Show that $J = J^{-1}$ for $n = 3$.

57. Let A be any 3×3 nonzero matrix. Describe AJ and JA using the definition of J in Problem 56 for $n = 3$, and show that $JJA = A$.

58. Let

$$A = \begin{bmatrix} a & -b \\ -a & b \end{bmatrix}$$

Find A^2, A^3, and A^n, where n is a positive integer.

59. Let

$$A = \begin{bmatrix} v & w \\ -w & v \end{bmatrix} \quad \text{and} \quad B = \begin{bmatrix} x & y \\ -y & x \end{bmatrix}$$

Show that A and B are commutative for matrix multiplication.

60. Let

$$A = \begin{bmatrix} x & 1 - x \\ 1 + x & -x \end{bmatrix}$$

Find A^2, A^3, and A^4. What is A^n, where n is a positive integer?

C **61.** Let $P_{1 \times 3}$ be a matrix representing the percentage weighting of tests, homework, and the final exam in a particular class if:

	Tests	Home-work	Final exam
Percent	50%	20%	30%

Let $C_{3 \times 5}$ be a matrix representing the class record:

	Cathy	Chris	Cindy	Connie	Cori
Tests	92	93	85	78	93
Homework	98	76	90	87	88
Final exam	95	94	86	84	91

Use matrix multiplication to form a matrix that represents the course averages of the students. (C could be a $3 \times n$ matrix for a class of n students.)

The following information is needed for Problems 62–64:

Business *A business produces three products, P_1, P_2, and P_3. Matrix A represents the purchase orders of two clients, C_1 and C_2. Matrix B represents the raw materials, R_1, R_2, R_3, and R_4, used to manufacture each product, and matrix C gives the cost of the raw materials.*

$$\begin{array}{c} \\ C_1 \\ C_2 \end{array}\begin{array}{c} \begin{array}{ccc} P_1 & P_2 & P_3 \end{array} \\ \begin{bmatrix} 9 & 12 & 6 \\ 10 & 20 & 0 \end{bmatrix} \end{array} = A \qquad \begin{array}{c} \\ P_1 \\ P_2 \\ P_3 \end{array}\begin{array}{c} \begin{array}{cccc} R_1 & R_2 & R_3 & R_4 \end{array} \\ \begin{bmatrix} 3 & 4 & 0 & 14 \\ 7 & 9 & 3 & 20 \\ 5 & 12 & 7 & 16 \end{bmatrix} \end{array} = B \qquad \begin{array}{c} \\ R_1 \\ R_2 \\ R_3 \\ R_4 \end{array}\begin{array}{c} \text{Cost} \\ \begin{bmatrix} 10 \\ 5 \\ 15 \\ 20 \end{bmatrix} \end{array} = C$$

Find and interpret the following matrices:

62. AB **63.** BC **64.** ABC

The following information is needed for Problems 65–67:

A Pythagorean triple is a set of three positive integers that may be the lengths of the sides of a right triangle. Let $X = [a \quad b \quad c]$ be a Pythagorean triple and let

$$P = \begin{bmatrix} 2 & 1 & 2 \\ 1 & 2 & 2 \\ 2 & 2 & 3 \end{bmatrix}$$

65. Let $X = [3 \quad 4 \quad 5]$. Show that X and XP are Pythagorean triples.

66. Let $X = [8 \quad 15 \quad 17]$. Show that X and XP are Pythagorean triples.

67. Prove that XP is a Pythagorean triple where $X = [a \quad b \quad c]$ is a Pythagorean triple.

7.3 Matrix Solution of Systems

An important technique applied to matrices is the transformation of a matrix into an **equivalent matrix.**

EQUIVALENT MATRICES

ELEMENTARY ROW OPERATIONS

$A_{m\times n}$ and $B_{m\times n}$ are **equivalent matrices,** written

$$A \sim B \qquad \text{or} \qquad B \sim A$$

if A can be changed into B by a finite number of the following **elementary row operations:**

1. Interchange any two rows.
2. Multiply any row by a nonzero real number.
3. Add a multiple of any row to another row.

The third operation requires a special note. When a multiple of one row is added to a second row, the second row is the only one changed. Observe this procedure carefully in the examples.

To illustrate the usefulness of equivalent matrices consider the following system of three equations in three unknowns:

$$\begin{cases} x + 2y + z = 3 \\ 2x - y - 2z = 3 \\ x + y + 2z = 6 \end{cases}$$

Note that the variables are lined up on one side and the constants are on the other side. This system may be represented as the **augmented matrix**

$$\left[\begin{array}{rrr|r} 1 & 2 & 1 & 3 \\ 2 & -1 & -2 & 3 \\ 1 & 1 & 2 & 6 \end{array}\right]$$

That is, the matrix of coefficients,

$$\begin{bmatrix} 1 & 2 & 1 \\ 2 & -1 & -2 \\ 1 & 1 & 2 \end{bmatrix}$$

is augmented with the matrix of constants. Solving this system, it is seen that each step in the solution is equivalent to an elementary row operation on the corresponding matrix.

Let the rows of the augmented matrix be represented by R_1, R_2, and R_3. The operations on the rows are indicated at the right for each step.

$$\begin{cases} x + 2y + z = 3 \\ 2x - y - 2z = 3 \\ x + y + 2z = 6 \end{cases} \qquad \left[\begin{array}{rrr|r} 1 & 2 & 1 & 3 \\ 2 & -1 & -2 & 3 \\ 1 & 1 & 2 & 6 \end{array}\right] \begin{array}{l} R_1 \\ R_2 \\ R_3 \end{array}$$

The order in which the equations occur is arbitrary, so they may be written in any order. Thus, the system is equivalent to

$$\begin{cases} x + 2y + z = 3 \\ x + y + 2z = 6 \\ 2x - y - 2z = 3 \end{cases} \qquad \left[\begin{array}{rrr|r} 1 & 2 & 1 & 3 \\ 1 & 1 & 2 & 6 \\ 2 & -1 & -2 & 3 \end{array}\right] \begin{array}{l} R_1 \\ R_3 \\ R_2 \end{array}$$

Recalling the addition method, the first equation may be multiplied by -1 and added to the second to obtain another equivalent system:

$$\begin{cases} x + 2y + z = 3 \\ -y + z = 3 \\ 2x - y - 2z = 3 \end{cases} \qquad \left[\begin{array}{ccc|c} 1 & 2 & 1 & 3 \\ 0 & -1 & 1 & 3 \\ 2 & -1 & -2 & 3 \end{array}\right] \begin{array}{l} R_1 \\ (-1)R_1 + R_2 \\ R_3 \end{array}$$

Continuing, x may be eliminated from the third equation by adding -2 times the first equation to the third equation:

$$\begin{cases} x + 2y + z = 3 \\ -y + z = 3 \\ -5y - 4z = -3 \end{cases} \qquad \left[\begin{array}{ccc|c} 1 & 2 & 1 & 3 \\ 0 & -1 & 1 & 3 \\ 0 & -5 & -4 & -3 \end{array}\right] \begin{array}{l} R_1 \\ R_2 \\ (-2)R_1 + R_3 \end{array}$$

The y term may be eliminated from the third equation now by adding -5 times the second to the third:

$$\begin{cases} x + 2y + z = 3 \\ -y + z = 3 \\ -9z = -18 \end{cases} \qquad \left[\begin{array}{ccc|c} 1 & 2 & 1 & 3 \\ 0 & -1 & 1 & 3 \\ 0 & 0 & -9 & -18 \end{array}\right] \begin{array}{l} R_1 \\ R_2 \\ (-5)R_2 + R_3 \end{array}$$

Any equation may be multiplied by a nonzero constant, so multiply the second equation by -1 and the third equation by $-\frac{1}{9}$ to obtain the following:

$$\begin{cases} x + 2y + z = 3 \\ y - z = -3 \\ z = 2 \end{cases} \qquad \left[\begin{array}{ccc|c} 1 & 2 & 1 & 3 \\ 0 & 1 & -1 & -3 \\ 0 & 0 & 1 & 2 \end{array}\right] \begin{array}{l} R_1 \\ (-1)R_2 \\ -\frac{1}{9}R_3 \end{array}$$

This gives the solution $z = 2$, which could be substituted into the second equation to find y, and these two substituted into the first to find x. Alternately, you could continue eliminating terms until you obtain

$$\begin{cases} x + 2y = 1 \\ y = -1 \\ z = 2 \end{cases} \qquad \left[\begin{array}{ccc|c} 1 & 2 & 0 & 1 \\ 0 & 1 & 0 & -1 \\ 0 & 0 & 1 & 2 \end{array}\right] \begin{array}{l} R_1 + (-1)R_3 \\ R_2 + R_3 \\ R_3 \end{array}$$

And finally,

$$\begin{cases} x = 3 \\ y = -1 \\ z = 2 \end{cases} \qquad \left[\begin{array}{ccc|c} 1 & 0 & 0 & 3 \\ 0 & 1 & 0 & -1 \\ 0 & 0 & 1 & 2 \end{array}\right] \begin{array}{l} R_1 + (-2)R_2 \\ R_2 \\ R_3 \end{array}$$

from which the solution to the system could be read from the system or from the corresponding matrix.

Notice that the system and the corresponding matrix

$$\begin{cases} x + 2y + z = 3 \\ 2x - y - 2z = 3 \\ x + y + 2z = 6 \end{cases} \qquad \left[\begin{array}{ccc|c} 1 & 2 & 1 & 3 \\ 2 & -1 & -2 & 3 \\ 1 & 1 & 2 & 6 \end{array}\right]$$

were transformed to

$$\begin{cases} \mathbf{x} & & = \mathbf{3} \\ & \mathbf{y} & = \mathbf{-1} \\ & & \mathbf{z} = \mathbf{2} \end{cases} \qquad \left[\begin{array}{ccc|c} 1 & 0 & 0 & \mathbf{3} \\ 0 & 1 & 0 & \mathbf{-1} \\ 0 & 0 & 1 & \mathbf{2} \end{array}\right]$$

by the elementary row operations listed in the box. In matrix form, a new matrix was obtained with 1s on the main diagonal and 0s above and below. To obtain this matrix, each diagonal element from left to right was chosen as a **pivot** and "cleared out" below, obtaining 0s. That is, first working down the main diagonal, 0s were obtained in each column below the pivot on the diagonal. Second, returning right to left, column-by-column, 0s were obtained above the diagonal. Finally, if any diagonal element is not a 1, the row may be multiplied by a rational number to obtain 1. The final form allows the solutions to be read directly.

Now using elementary row operations, let us rewrite the solution to the same system and see how it would appear. The system is first written as an augmented matrix, and then the elementary row operations are performed. The operations are chronicled by the subscripted R's.

$$\begin{cases} x + 2y + z = 3 \\ 2x - y - 2z = 3 \\ x + y + 2z = 6 \end{cases} \qquad \left[\begin{array}{ccc|c} 1 & 2 & 1 & 3 \\ 2 & -1 & -2 & 3 \\ 1 & 1 & 2 & 6 \end{array}\right] \begin{array}{l} R_1 \\ R_2 \\ R_3 \end{array}$$

$$\sim \left[\begin{array}{ccc|c} 1 & 2 & 1 & 3 \\ 1 & 1 & 2 & 6 \\ 2 & -1 & -2 & 3 \end{array}\right] \begin{array}{l} R_1 \\ R_3 \\ R_2 \end{array} \qquad \sim \left[\begin{array}{ccc|c} 1 & 2 & 1 & 3 \\ 0 & -1 & 1 & 3 \\ 2 & -1 & -2 & 3 \end{array}\right] \begin{array}{l} R_1 \\ (-1)R_1 + R_2 \\ R_3 \end{array}$$

$$\sim \left[\begin{array}{ccc|c} 1 & 2 & 1 & 3 \\ 0 & -1 & 1 & 3 \\ 0 & -5 & -4 & -3 \end{array}\right] \begin{array}{l} R_1 \\ R_2 \\ (-2)R_1 + R_3 \end{array} \qquad \sim \left[\begin{array}{ccc|c} 1 & 2 & 1 & 3 \\ 0 & -1 & 1 & 3 \\ 0 & 0 & -9 & -18 \end{array}\right] \begin{array}{l} R_1 \\ R_2 \\ (-5)R_2 + R_3 \end{array}$$

$$\sim \left[\begin{array}{ccc|c} 1 & 2 & 1 & 3 \\ 0 & 1 & -1 & -3 \\ 0 & 0 & 1 & 2 \end{array}\right] \begin{array}{l} R_1 \\ (-1)R_2 \\ -\frac{1}{9}R_3 \end{array} \qquad \sim \left[\begin{array}{ccc|c} 1 & 2 & 0 & 1 \\ 0 & 1 & 0 & -1 \\ 0 & 0 & 1 & 2 \end{array}\right] \begin{array}{l} R_1 + (-1)R_3 \\ R_2 + R_3 \\ R_3 \end{array}$$

$$\sim \left[\begin{array}{ccc|c} 1 & 0 & 0 & 3 \\ 0 & 1 & 0 & -1 \\ 0 & 0 & 1 & 2 \end{array}\right] \begin{array}{l} R_1 + (-2)R_2 \\ R_2 \\ R_3 \end{array} \qquad \begin{cases} x & & = 3 \\ & y & = -1 \\ & & z = 2 \end{cases}$$

The method illustrated is known as the **Gauss–Jordan method,** and the final form of the augmented matrix is called the **row-reduced form.** In the row-reduced form of the augmented matrix $[A|B]$, A is an identity matrix:

$$\left[\begin{array}{cccccc|c} 1 & 0 & 0 & \cdots & 0 & & b_1 \\ 0 & 1 & 0 & \cdots & 0 & & b_2 \\ \vdots & \vdots & \vdots & & \vdots & & \vdots \\ 0 & 0 & 0 & \cdots & 1 & & b_n \end{array}\right]$$

The Gauss–Jordan method is now illustrated with examples. Notice in the examples how the row-reduced form is obtained by pivoting around each diagonal element, clearing out first below and then above.

Example 1

$$\begin{cases} 2x + y = 1 \\ 6x + 4y = -2 \end{cases}$$

Solution

$$\left[\begin{array}{cc|c} 2 & 1 & 1 \\ 6 & 4 & -2 \end{array}\right] \begin{array}{l} R_1 \\ R_2 \end{array} \quad \sim \left[\begin{array}{cc|c} 2 & 1 & 1 \\ 0 & 1 & -5 \end{array}\right] \begin{array}{l} R_1 \\ R_2 - 3R_1 \end{array}$$

$$\sim \left[\begin{array}{cc|c} 2 & 0 & 6 \\ 0 & 1 & -5 \end{array}\right] \begin{array}{l} R_1 - R_2 \\ R_2 \end{array} \quad \sim \left[\begin{array}{cc|c} 1 & 0 & \mathbf{3} \\ 0 & 1 & \mathbf{-5} \end{array}\right] \begin{array}{l} \frac{1}{2}R_1 \\ R_2 \end{array}$$

Thus, $\boldsymbol{x = 3, y = -5}$. □

Example 2

$$\begin{cases} x - y = 3 \\ 2x - y + z = 8 \\ y + 2z = 5 \end{cases}$$

Solution

$$\left[\begin{array}{ccc|c} 1 & -1 & 0 & 3 \\ 2 & -1 & 1 & 8 \\ 0 & 1 & 2 & 5 \end{array}\right] \begin{array}{l} R_1 \\ R_2 \\ R_3 \end{array} \quad \sim \left[\begin{array}{ccc|c} 1 & -1 & 0 & 3 \\ 0 & 1 & 1 & 2 \\ 0 & 1 & 2 & 5 \end{array}\right] \begin{array}{l} R_1 \\ R_2 - 2R_1 \\ R_3 \end{array}$$

$$\sim \left[\begin{array}{ccc|c} 1 & 0 & 1 & 5 \\ 0 & 1 & 1 & 2 \\ 0 & 0 & 1 & 3 \end{array}\right] \begin{array}{l} R_1 + R_2 \\ R_2 \\ R_3 - R_2 \end{array} \quad \sim \left[\begin{array}{ccc|c} 1 & 0 & 0 & \mathbf{2} \\ 0 & 1 & 0 & \mathbf{-1} \\ 0 & 0 & 1 & \mathbf{3} \end{array}\right] \begin{array}{l} R_1 - R_3 \\ R_2 - R_3 \\ R_3 \end{array}$$

Thus, $\boldsymbol{x = 2, y = -1, z = 3}$. □

Example 3 Find the equation of the parabola of the form $y = ax^2 + bx + c$ containing the points $(-1, 3)$, $(1, -1)$, and $(2, 0)$.

Solution Given three points on a parabola of the form $y = ax^2 + bx + c$, the coefficients a, b, and c can be found with a system of equations. Thus, substitute the points into $y = ax^2 + bx + c$ to obtain three equations in a, b, and c.

$$\begin{array}{rl} (-1, 3): & 3 = a - b + c \\ (1, -1): & -1 = a + b + c \\ (2, 0): & 0 = 4a + 2b + c \end{array}$$

$$\left[\begin{array}{ccc|c} 1 & -1 & 1 & 3 \\ 1 & 1 & 1 & -1 \\ 4 & 2 & 1 & 0 \end{array}\right] \begin{array}{l} R_1 \\ R_2 \\ R_3 \end{array} \quad \sim \left[\begin{array}{ccc|c} 1 & -1 & 1 & 3 \\ 0 & 2 & 0 & -4 \\ 0 & 6 & -3 & -12 \end{array}\right] \begin{array}{l} R_1 \\ R_2 - R_1 \\ R_3 - 4R_1 \end{array}$$

$$\sim \left[\begin{array}{ccc|c} 1 & 0 & 1 & 1 \\ 0 & 1 & 0 & -2 \\ 0 & 0 & -3 & 0 \end{array}\right] \begin{array}{l} R_1 + \frac{1}{2}R_2 \\ \frac{1}{2}R_2 \\ R_3 - 3R_2 \end{array} \quad \sim \left[\begin{array}{ccc|c} 1 & 0 & 0 & \mathbf{1} \\ 0 & 1 & 0 & \mathbf{-2} \\ 0 & 0 & 1 & \mathbf{0} \end{array}\right] \begin{array}{l} R_1 + \frac{1}{3}R_3 \\ R_2 \\ -\frac{1}{3}R_3 \end{array}$$

Thus, $\boldsymbol{a = 1, b = -2, c = 0}$, and the parabola is $y = \mathbf{1}x^2 - \mathbf{2}x + \mathbf{0}$, or $\boldsymbol{y = x^2 - 2x}$. □

It should be noted that not all systems are consistent. It may not be possible to achieve a row-reduced form using the Gauss–Jordan method. You will become aware of this situation when, for instance, all the entries in one of the rows become zero at some step in the solution.

Example 4 Solve: $\begin{cases} x - y = 1 \\ y + z = 2 \\ x + z = 3 \end{cases}$

Solution

$$\left[\begin{array}{ccc|c} 1 & -1 & 0 & 1 \\ 0 & 1 & 1 & 2 \\ 1 & 0 & 1 & 3 \end{array}\right] \sim \left[\begin{array}{ccc|c} 1 & -1 & 0 & 1 \\ 0 & 1 & 1 & 2 \\ 0 & 1 & 1 & 2 \end{array}\right] \begin{array}{l} R_1 \\ R_2 \\ R_3 - R_1 \end{array}$$

$$\sim \left[\begin{array}{ccc|c} 1 & -1 & 0 & 1 \\ 0 & 1 & 1 & 2 \\ 0 & 0 & 0 & 0 \end{array}\right] \begin{array}{l} R_1 \\ R_2 \\ R_3 - R_2 \end{array}$$

The row of 0s indicates that there are *an infinite number of solutions*, since it corresponds to the equation

$$0 \cdot x + 0 \cdot y + 0 \cdot z = 0 \quad \text{or} \quad 0 = 0$$

An equivalent system is

$$\begin{cases} x - y = 1 \\ y + z = 2 \end{cases} \quad \text{or} \quad \begin{cases} x = y + 1 \\ z = 2 - y \end{cases}$$

The solution set may be described as the set of all triples $(y + 1, \ y, \ 2 - y)$. Recall from Section 7.1 that if the solution set is infinite, the equations are called **dependent.** □

In Example 4, if a row of 0s had developed with a nonzero entry in the last column, such as 0, 0, 0, 2, it would correspond to the equation $0 \cdot x + 0 \cdot y + 0 \cdot z = 2$. Such an equation is equivalent to $0 = 2$, and there are no solutions to the system. Recall that if the solution set of the system is empty, the equations are called **inconsistent.**

Problem Set 7.3

A *Solve the systems in Problems 1–30 by the Gauss–Jordan method.*

1. $\begin{cases} x + y = 7 \\ x - y = -1 \end{cases}$

2. $\begin{cases} x - y = 8 \\ x + y = 2 \end{cases}$

3. $\begin{cases} -x + 2y = 2 \\ 4x - 7y = -5 \end{cases}$

4. $\begin{cases} x - 6y = -3 \\ 2x + 3y = 9 \end{cases}$

5. $\begin{cases} 4x - 3y = 1 \\ 5x + 2y = 7 \end{cases}$

6. $\begin{cases} 3x + 7y = 5 \\ 4x + 9y = 7 \end{cases}$

7. $\begin{cases} x - y = 2 \\ 2x + 3y = 9 \end{cases}$

8. $\begin{cases} 3x - 4y = 16 \\ -x + 2y = -6 \end{cases}$

9. $\begin{cases} 2x - y = 6 \\ 4x + y = 3 \end{cases}$

10. $\begin{cases} 6x + 9y = -4 \\ 9x + 3y = 1 \end{cases}$

11. $\begin{cases} x + y = 1 \\ x + z = -1 \\ y + z = -4 \end{cases}$

12. $\begin{cases} x + y = 1 \\ x - z = 2 \\ y - z = -3 \end{cases}$

13. $\begin{cases} x - y = 1 \\ x + z = 1 \\ y - z = 1 \end{cases}$

14. $\begin{cases} x + y = 2 \\ x - z = 1 \\ -y + z = 1 \end{cases}$

15. $\begin{cases} x + 5z = 9 \\ y + 2z = 2 \\ 2x + 3z = 4 \end{cases}$

16. $\begin{cases} x + 2z = 13 \\ 2x + y = 8 \\ -2y + 9z = 41 \end{cases}$

17. $\begin{cases} 4x + y = -2 \\ 3x + 2z = -9 \\ 2y + 3z = -5 \end{cases}$

18. $\begin{cases} 5x + z = 9 \\ x - 5z = 7 \\ x + y = z \end{cases}$

19. $\begin{cases} x + y = -2 \\ y + z = 2 \\ x - y - z = -1 \end{cases}$

20. $\begin{cases} x + y = -1 \\ y + z = -1 \\ x + y + z = 1 \end{cases}$

21. $\begin{cases} 2x + y = 1 \\ y - z = 1 \\ 2x + y + z = 1 \end{cases}$

22. $\begin{cases} x - 2y = 0 \\ x + z = 0 \\ x + 2y + 2z = 0 \end{cases}$

23. $\begin{cases} x + 2z = 13 \\ 2x + y = 8 \\ -2y + 9z = 41 \end{cases}$

24. $\begin{cases} x + 2z = 0 \\ 3x - y + 2z = 0 \\ 4x + y = 6 \end{cases}$

25. $\begin{cases} x - z = 1 - y \\ 2x - 3z = 2 \\ y + 2z = 3 \end{cases}$

26. $\begin{cases} 3x + 3z = y + 2 \\ x - z = 5 \\ x + y = z \end{cases}$

27. $\begin{cases} x + y = 3 \\ x + 2y + z = -2 \\ 2x + 3y + 2z = 1 \end{cases}$

28. $\begin{cases} y - 2x = -2 \\ 2x - y + 4z = -14 \\ -3x + 2y + z = 4 \end{cases}$

29. $\begin{cases} 2y + z = 1 \\ y - x = 1 \\ x + 2z = 4 \end{cases}$

30. $\begin{cases} y - x = 8 \\ y + 2z = 1 \\ x - 3z = 3 \end{cases}$

B *Use the Gauss–Jordan method to solve the systems in Problems 31–40.*

31. $\begin{cases} x + 2y + z = 9 \\ x + y - z = 5 \\ 3x - y + 2z = 12 \end{cases}$

32. $\begin{cases} x + 3y - z = -2 \\ x - y + 2z = -4 \\ 2x + y - 3z = 3 \end{cases}$

33. $\begin{cases} 2x + y - z = 1 \\ x + 3y - 2z = 4 \\ 3x - 2y + z = -2 \end{cases}$

34. $\begin{cases} 2x - y + z = -1 \\ x + 2y + z = 1 \\ x - y - 3z = 2 \end{cases}$

35. $\begin{cases} w + z = 5 \\ x + y = 5 \\ y + z = 3 \\ 2w + y = 10 \end{cases}$

36. $\begin{cases} x + z = 0 \\ w + y = x \\ x + y = 4 \\ y + z = 2 \end{cases}$

37. $\begin{cases} 2x + 2y - z + w = -2 \\ 3y - x + 3z = 0 \\ 3x - 2z - w = 9 \\ y + z - 2w = 9 \end{cases}$

38. $\begin{cases} x + y + z + w = 3 \\ x - y - w = 0 \\ y + 2z + w = 0 \\ 2x - z + w = 3 \end{cases}$

39. $\begin{cases} w + x + y + z = 0 \\ x + y + z = 1 \\ w + x + y = 4 \\ w - y + z = 0 \end{cases}$

40. $\begin{cases} w - x + y + z = 0 \\ w + x - y - z = 2 \\ x + y + z = -3 \\ -w + x + y = 1 \end{cases}$

C *In Problems 41–46 find the equation for the parabola of the form $y = ax^2 + bx + c$ containing the given points.*

41. $(0, 5)$, $(5, 0)$, $(2, -3)$

42. $(1, 2)$, $(3, -2)$, $(6, 7)$

43. $(1, -3)$, $(2, -2)$, $(3, 1)$

44. $(1, 2)$, $(2, 1)$, $(3, -2)$

45. $(\frac{1}{2}, \frac{1}{2})$, $(1, 1)$, $(2, 5)$

46. $(\frac{1}{2}, \frac{5}{4})$, $(1, 2)$, $(2, 5)$

Show that the systems in Problems 47 and 48 have no solutions.

47. $\begin{cases} x + 2y - z = 3 \\ z - 3x - 6y = 1 \\ 4y + 2x - z = 2 \end{cases}$

48. $\begin{cases} 3x - y - 2z = 3 \\ y - 4x + z = 5 \\ 2x - y - 3z = 2 \end{cases}$

Solve the systems in Problems 49–52. Each system has infinitely many solutions. List the solution set as an ordered triple in terms of the variable y.

49. $\begin{cases} x + 2y = 1 \\ 4y + 3z = 3 \\ 3z - 2x = 1 \end{cases}$

50. $\begin{cases} x - 2y = 1 \\ 3x + z = 2 \\ 6y + z = -1 \end{cases}$

51. $\begin{cases} x - 2y + 3z = 3 \\ 2x + 3y - z = 5 \\ x + 5y - 4z = 2 \end{cases}$

52. $\begin{cases} x + y + 3z = -1 \\ x + 3y - 3z = 3 \\ 2x + 5y - 3z = 4 \end{cases}$

7.4 Matrix Solution of Systems, Inverses

Using matrix notation, a system of equations can be written in yet another form. For example, consider

$$\begin{cases} x + 2y + 3z = 1 \\ 2x + 5y + 4z = 2 \\ x - 3y + 2z = 3 \end{cases}$$

Let A be the matrix of the coefficients in the system, and let X be a matrix of the variables in the system:

$$A = \begin{bmatrix} 1 & 2 & 3 \\ 2 & 5 & 4 \\ 1 & -3 & 2 \end{bmatrix} \quad \text{and} \quad X = \begin{bmatrix} x \\ y \\ z \end{bmatrix}$$

Notice that the product

$$AX = \begin{bmatrix} 1 & 2 & 3 \\ 2 & 5 & 4 \\ 1 & -3 & 2 \end{bmatrix} \begin{bmatrix} x \\ y \\ z \end{bmatrix} = \begin{bmatrix} x + 2y + 3z \\ 2x + 5y + 4z \\ x - 3y + 2z \end{bmatrix}$$

is a matrix that has entries exactly like the left side of the system. Now, if B is the column matrix of the constants on the right side of the system,

$$B = \begin{bmatrix} 1 \\ 2 \\ 3 \end{bmatrix}$$

then the system may be written as the matrix equation $AX = B$:

$$\begin{bmatrix} 1 & 2 & 3 \\ 2 & 5 & 4 \\ 1 & -3 & 2 \end{bmatrix} \begin{bmatrix} x \\ y \\ z \end{bmatrix} = \begin{bmatrix} 1 \\ 2 \\ 3 \end{bmatrix}$$

If matrix A has an inverse, the solution of the system $AX = B$ can be found by multiplying both sides of the matrix equation on the left by A^{-1}:

$$\begin{aligned} AX &= B \\ A^{-1}(AX) &= A^{-1}B \\ (A^{-1}A)X &= A^{-1}B \\ IX &= A^{-1}B \\ X &= A^{-1}B \end{aligned}$$

The solutions to the system can then be read from $A^{-1}B$. The problem remains to find a method of computing inverses.

The Gauss–Jordan method transforms the augmented matrix $[A|B]$ by elementary row operations to $[I|A^{-1}B]$ since $A^{-1}B$ has been shown to be the solutions of the matrix equation $AX = B$. That is,

$$[A|B] \sim [I|A^{-1}B]$$

The combined effect on the augmented matrix by the elementary row operations is multiplication by A^{-1}, since $A^{-1}A = I$.

Since the transformation of A into I in the Gauss–Jordan method is equivalent to multiplying by A^{-1}, these same elementary row operations should transform I into A^{-1}:

$$\begin{aligned} [A|I] &\sim [A^{-1}A|A^{-1}I] \\ &\sim [I|A^{-1}] \end{aligned}$$

This means the inverse of a matrix, if one exists, may be computed by elementary row operations on the matrix augmented by the identity matrix. If a matrix has an inverse, the matrix is called **nonsingular.**

Example 1 Find the inverse of: $A = \begin{bmatrix} 2 & 1 \\ 3 & 2 \end{bmatrix}$

Solution First write the augmented matrix $[A|I]$:

$$\left[\begin{array}{cc|cc} 2 & 1 & 1 & 0 \\ 3 & 2 & 0 & 1 \end{array}\right] \begin{matrix} R_1 \\ R_2 \end{matrix} \sim \left[\begin{array}{cc|cc} 2 & 1 & 1 & 0 \\ 1 & 1 & -1 & 1 \end{array}\right] \begin{matrix} R_1 \\ R_2 - R_1 \end{matrix}$$

$$\sim \left[\begin{array}{cc|cc} 1 & 0 & 2 & -1 \\ 1 & 1 & -1 & 1 \end{array}\right] \begin{matrix} R_1 - R_2 \\ R_2 \end{matrix}$$

$$\sim \left[\begin{array}{cc|cc} 1 & 0 & 2 & -1 \\ 0 & 1 & -3 & 2 \end{array}\right] \begin{matrix} R_1 \\ R_2 - R_1 \end{matrix}$$

$[A|I]$ is transformed into $[I|A^{-1}]$ where

$$A^{-1} = \begin{bmatrix} 2 & -1 \\ -3 & 2 \end{bmatrix}$$ □

Example 2 Find the inverse of: $B = \begin{bmatrix} 1 & -1 & 0 \\ 2 & -1 & 1 \\ 0 & 1 & 2 \end{bmatrix}$

Solution Write the augmented matrix $[B|I]$; then use elementary row operations to obtain $[I|B^{-1}]$:

$$\left[\begin{array}{rrr|rrr} 1 & -1 & 0 & 1 & 0 & 0 \\ 2 & -1 & 1 & 0 & 1 & 0 \\ 0 & 1 & 2 & 0 & 0 & 1 \end{array}\right] \begin{matrix} R_1 \\ R_2 \\ R_3 \end{matrix}$$

$$\sim \left[\begin{array}{rrr|rrr} 1 & -1 & 0 & 1 & 0 & 0 \\ 0 & 1 & 1 & -2 & 1 & 0 \\ 0 & 1 & 2 & 0 & 0 & 1 \end{array}\right] \begin{matrix} \\ R_2 - 2R_1 \\ \\ \end{matrix}$$

$$\sim \left[\begin{array}{rrr|rrr} 1 & -1 & 0 & 1 & 0 & 0 \\ 0 & 1 & 1 & -2 & 1 & 0 \\ 0 & 0 & 1 & 2 & -1 & 1 \end{array}\right] \begin{matrix} \\ \\ R_3 - R_2 \end{matrix}$$

$$\sim \left[\begin{array}{rrr|rrr} 1 & -1 & 0 & 1 & 0 & 0 \\ 0 & 1 & 0 & -4 & 2 & -1 \\ 0 & 0 & 1 & 2 & -1 & 1 \end{array}\right] \begin{matrix} \\ R_2 - R_3 \\ \\ \end{matrix}$$

$$\sim \left[\begin{array}{rrr|rrr} 1 & 0 & 0 & -3 & 2 & -1 \\ 0 & 1 & 0 & -4 & 2 & -1 \\ 0 & 0 & 1 & 2 & -1 & 1 \end{array}\right] \begin{matrix} R_1 + R_2 \\ \\ \\ \end{matrix}$$

$$B^{-1} = \begin{bmatrix} -3 & 2 & -1 \\ -4 & 2 & -1 \\ 2 & -1 & 1 \end{bmatrix}$$ □

The solution to the equation $AX = B$ is given by $X = A^{-1}B$. Since the method illustrated above will compute the inverse, if it exists, matrix equations may be solved using the inverse. The following examples use systems for which the inverse matrix has been computed in Examples 1 and 2.

Example 3 Solve with a matrix equation: $\begin{cases} 2x + y = 1 \\ 3x + 2y = -1 \end{cases}$

Solution The system may be represented in the form $AX = B$:

$$\begin{bmatrix} 2 & 1 \\ 3 & 2 \end{bmatrix} \begin{bmatrix} x \\ y \end{bmatrix} = \begin{bmatrix} 1 \\ -1 \end{bmatrix}$$

The inverse of A is known from Example 1, so $X = A^{-1}B$:

$$\begin{bmatrix} x \\ y \end{bmatrix} = \begin{bmatrix} 2 & -1 \\ -3 & 2 \end{bmatrix} \begin{bmatrix} 1 \\ -1 \end{bmatrix}$$

$$\begin{bmatrix} x \\ y \end{bmatrix} = \begin{bmatrix} 3 \\ -5 \end{bmatrix}$$

The solution is (3, −5). □

Example 4 Solve with a matrix equation: $\begin{cases} x - y = 3 \\ 2x - y + z = 8 \\ y + 2z = 5 \end{cases}$

Solution

$$\begin{bmatrix} 1 & -1 & 0 \\ 2 & -1 & 1 \\ 0 & 1 & 2 \end{bmatrix}\begin{bmatrix} x \\ y \\ z \end{bmatrix} = \begin{bmatrix} 3 \\ 8 \\ 5 \end{bmatrix}$$

$$\begin{bmatrix} x \\ y \\ z \end{bmatrix} = \begin{bmatrix} -3 & 2 & -1 \\ -4 & 2 & -1 \\ 2 & -1 & 1 \end{bmatrix}\begin{bmatrix} 3 \\ 8 \\ 5 \end{bmatrix}$$

The inverse is known from Example 2.

$$\begin{bmatrix} x \\ y \\ z \end{bmatrix} = \begin{bmatrix} 2 \\ -1 \\ 3 \end{bmatrix}$$

Thus, $(x, y, z) = (2, -1, 3)$. □

The real advantage of this method, as well as other matrix methods, is the use of the computer. The computer may be programmed to find A^{-1} and to multiply times B to solve $X = A^{-1}B$. The inverse method offers a special advantage if many systems must be solved with the same A and many different B's. However, this method does have its limitations. The Gauss–Jordan method has an advantage because it can be used to solve systems of equations when the number of variables does not equal the number of equations, as well as when A^{-1} does not exist.

Problem Set 7.4

A *In Problems 1–20 find the inverse of each matrix, if it exists.*

1. $\begin{bmatrix} 4 & -7 \\ -1 & 2 \end{bmatrix}$ **2.** $\begin{bmatrix} 3 & 4 \\ 2 & 3 \end{bmatrix}$ **3.** $\begin{bmatrix} 1 & -1 \\ 2 & 3 \end{bmatrix}$

4. $\begin{bmatrix} 3 & -4 \\ -1 & 2 \end{bmatrix}$ **5.** $\begin{bmatrix} -3 & 1 \\ 1 & -2 \end{bmatrix}$ **6.** $\begin{bmatrix} 2 & -3 \\ -4 & 6 \end{bmatrix}$

7. $\begin{bmatrix} 8 & 6 \\ -2 & 4 \end{bmatrix}$ **8.** $\begin{bmatrix} 1 & -\frac{3}{2} \\ -1 & 2 \end{bmatrix}$ **9.** $\begin{bmatrix} -\frac{1}{4} & \frac{3}{4} \\ \frac{1}{2} & -\frac{1}{2} \end{bmatrix}$

10. $\begin{bmatrix} 0 & -5 & 1 \\ -1 & -3 & 1 \\ 1 & 2 & -1 \end{bmatrix}$ **11.** $\begin{bmatrix} 1 & -1 & 0 \\ 2 & -1 & 1 \\ 0 & 1 & 2 \end{bmatrix}$ **12.** $\begin{bmatrix} 4 & 1 & 0 \\ 2 & -1 & 4 \\ -3 & 2 & 1 \end{bmatrix}$

13. $\begin{bmatrix} 1 & 0 & 2 \\ 2 & 1 & 0 \\ 0 & -2 & 9 \end{bmatrix}$ **14.** $\begin{bmatrix} 6 & 1 & 20 \\ 1 & -1 & 0 \\ 0 & 1 & 3 \end{bmatrix}$ **15.** $\begin{bmatrix} 1 & 0 & 2 \\ 3 & -1 & 2 \\ 4 & 1 & 0 \end{bmatrix}$

16. $\begin{bmatrix} 1 & 1 & 0 \\ 1 & 2 & 1 \\ 2 & 3 & 2 \end{bmatrix}$ **17.** $\begin{bmatrix} 1 & -1 & 0 \\ 1 & 0 & 1 \\ 0 & 1 & -1 \end{bmatrix}$ **18.** $\begin{bmatrix} 1 & 1 & 0 \\ 1 & 0 & -1 \\ 0 & -1 & 1 \end{bmatrix}$

19. $\begin{bmatrix} 2 & 1 & 1 \\ 1 & 1 & 1 \\ -1 & 0 & 1 \end{bmatrix}$ **20.** $\begin{bmatrix} 1 & -1 & 3 \\ -1 & 1 & -1 \\ 1 & 0 & 2 \end{bmatrix}$

Solve the systems in Problems 21–40 by solving the corresponding matrix equation with an inverse, if possible. Note that the necessary inverses were found in Problems 1–20.

21. $\begin{cases} 4x - 7y = -5 \\ -x + 2y = 2 \end{cases}$ **22.** $\begin{cases} 3x + 4y = 8 \\ 2x + 3y = 7 \end{cases}$ **23.** $\begin{cases} x - y = 2 \\ 2x + 3y = 9 \end{cases}$

24. $\begin{cases} 3x - 4y = 16 \\ -x + 2y = -6 \end{cases}$ **25.** $\begin{cases} y = 3x + 1 \\ x - 2y = 8 \end{cases}$ **26.** $\begin{cases} 2x - 3y = 12 \\ -4x + 6y = 18 \end{cases}$

27. $\begin{cases} 8x + 6y = -1 \\ 4y - 2x = 3 \end{cases}$ **28.** $\begin{cases} x = \frac{3}{2}y \\ 2y = x - 1 \end{cases}$ **29.** $\begin{cases} \frac{3}{4}y - \frac{1}{4}x = 1 \\ \frac{1}{2}x - \frac{1}{2}y = 2 \end{cases}$

30. $\begin{cases} -5y + z = 9 \\ -x - 3y + z = 0 \\ x + 2y - z = 0 \end{cases}$ **31.** $\begin{cases} x - y = -5 \\ 2x + z = y \\ y + 2z = 13 \end{cases}$

32. $\begin{cases} 4x + y = -2 \\ 2x + y + 4z = -12 \\ -3x + 2y + z = 4 \end{cases}$ **33.** $\begin{cases} x + 2z = 13 \\ 2x + y = 8 \\ -2y + 9z = 41 \end{cases}$

34. $\begin{cases} 6x + y + 20z = 0 \\ x - y = 1 \\ y + 3z = -1 \end{cases}$ **35.** $\begin{cases} x + 2z = 0 \\ 3x - y + 2z = 0 \\ 4x + y = 6 \end{cases}$

36. $\begin{cases} x + y = 3 \\ x + 2y + z = -2 \\ 2x + 3y + 2z = 1 \end{cases}$ **37.** $\begin{cases} x - y = 1 \\ x + z = 1 \\ y - z = 0 \end{cases}$

38. $\begin{cases} x + y = 2 \\ x - z = 1 \\ -y + z = 1 \end{cases}$ **39.** $\begin{cases} 2x + y + z = 17 \\ x + y + z = 7 \\ z - x = 2 \end{cases}$ **40.** $\begin{cases} x - y + 3z = 1 \\ -x + y - z = -1 \\ x + 2z = 1 \end{cases}$

B *Solve the systems in Problems 41–50 by using any convenient method.*

41. $\begin{cases} 4x - 3y = 1 \\ 5x + 2y = 7 \end{cases}$ **42.** $\begin{cases} 3x + 7y = 5 \\ 4x + 9y = 7 \end{cases}$

43. $\begin{cases} x + 2y + z = 9 \\ x + y - z = 5 \\ 3x - y + 2z = 12 \end{cases}$ **44.** $\begin{cases} x + 3y - z = 6 \\ x - y + 2z = 1 \\ 2x + y - 3z = 1 \end{cases}$

45. $\begin{cases} x + z = 5 + 2y \\ 2x + 3z = 4 \\ y + 2z = 2 \end{cases}$ **46.** $\begin{cases} 3x + 3z = y + 2 \\ x - z = 5 \\ x + y = z \end{cases}$

47. $\begin{cases} w + z = 5 \\ x + y = 5 \\ y + z = 3 \\ 2w + y = 10 \end{cases}$

48. $\begin{cases} x + 2y = 1 \\ z = 3 \\ w + x + 3y = 2 \\ 2x + 4y = 0 \end{cases}$

49. $\begin{cases} x + y = 1 \\ w - x = 1 \\ x + z = 0 \\ w + z = 1 \end{cases}$

50. $\begin{cases} w + y = -1 \\ x - y = 2 \\ z - w = 2 \\ x + z = 3 \end{cases}$

C 51. *Chemistry* In order to control a certain type of crop disease, it is necessary to use 23 gallons of chemical A and 34 gallons of chemical B. Commercial spray I has 5 gallons of chemical A and 2 gallons of B. Commercial spray II contains 2 gallons of A and 7 gallons of B. How much of each commercial product should be used to guarantee the proportion required?

52. *Metallurgy* In the manufacture of a certain alloy, 33 kg of metal A are alloyed with 56 kg of metal B. The manufacturer buys and mixes an alloy of 3 kg A to 5 kg B with one containing 4 kg A and 7 kg B because they are readily available. How much of the two alloys should be used to produce the desired final product?

53. *Business* A candy manufacturer mixes chocolate, milk, and almonds to produce three kinds of candy, I, II, and III, in the following proportions:

I: 7 kg chocolate, 5 kg milk, and 1 kg almonds
II: 3 kg chocolate, 2 kg milk, and 2 kg almonds
III: 4 kg chocolate, 3 kg milk, and 3 kg almonds

If 67 kg of chocolate, 48 kg of milk, and 32 kg of almonds are available, how much of each kind of candy can be produced?

54. Using the data in Problem 53, how many kilograms of each kind of candy can be produced with 62 kg of chocolate, 44 kg of milk, and 32 kg of almonds?

Prove the assertions in Problems 55–58 about nonsingular matrices. Assume that the matrices are conformable.

55. If A is nonsingular, then $(A^{-1})^{-1} = A$.
56. If A is nonsingular and $AB = A$, then $B = I$.
57. If A is nonsingular and $AB = AC$, then $B = C$.
58. If A and B are nonsingular, then $(AB)^{-1} = B^{-1}A^{-1}$.

7.5 The Determinant Function

We will now concentrate on a special set of matrices—square matrices. Square matrices have the same number of rows as columns and that number is called the **order** of the matrix. With every square matrix $A_{n \times n}$ with real number entries, there is associated a unique real number called its **determinant,** denoted by $\delta(A)$ or $|A|$. Also, $\delta(A_{n \times n})$ is called a determinant of order n. This relationship defines a function. The domain is the set of square matrices, and the range is the set of real numbers. This may be the first function you have encountered that is not a correspondence between two sets of numbers, but rather between a set of nonnumbers—matrices—and a set of numbers—determinants.

DETERMINANT OF ORDER 2

A determinant of order 2 is

$$\delta(A_{2\times 2}) = \begin{vmatrix} a_{11} & a_{12} \\ a_{21} & a_{22} \end{vmatrix} = a_{11}a_{22} - a_{21}a_{12}$$

That is, a second-order determinant is the difference of cross products:

$$\begin{matrix} a_{11} & & a_{12} \\ & \times & \\ a_{21} & & a_{22} \end{matrix} \qquad a_{11}a_{22} - a_{21}a_{12}$$

Example 1 If $A = \begin{bmatrix} 7 & -5 \\ 2 & 3 \end{bmatrix}$, find $\delta(A)$.

Solution

$$\begin{aligned} \delta(A) = \begin{vmatrix} 7 & -5 \\ 2 & 3 \end{vmatrix} &= (7)(3) - (2)(-5) \\ &= 21 - (-10) \\ &= \mathbf{31} \end{aligned}$$

□

Since determinants resemble matrices, be careful to distinguish between them. Matrices are arrays of numbers, and determinants are real numbers defined by an array. Every square matrix has an associated determinant, but determinants of order greater than 2 are not evaluated as easily as determinants of order 2.

DETERMINANT OF ORDER 3

A determinant of order 3 is defined as

$$\delta(A_{3\times 3}) = \begin{vmatrix} a_{11} & a_{12} & a_{13} \\ a_{21} & a_{22} & a_{23} \\ a_{31} & a_{32} & a_{33} \end{vmatrix} = a_{11}a_{22}a_{33} - a_{11}a_{23}a_{32} - a_{12}a_{21}a_{33} + a_{12}a_{23}a_{31} + a_{13}a_{21}a_{32} - a_{13}a_{22}a_{31}$$

This looks pretty involved, but you may factor by grouping the terms in pairs:

$$\begin{aligned} \delta(A_{3\times 3}) &= a_{11}a_{22}a_{33} - a_{11}a_{23}a_{32} - a_{12}a_{21}a_{33} + a_{12}a_{23}a_{31} + a_{13}a_{21}a_{32} - a_{13}a_{22}a_{31} \\ &= a_{11}(a_{22}a_{33} - a_{23}a_{32}) - a_{12}(a_{21}a_{33} - a_{23}a_{31}) + a_{13}(a_{21}a_{32} - a_{22}a_{31}) \end{aligned}$$

Notice negative signs

$$= a_{11}\begin{vmatrix} a_{22} & a_{23} \\ a_{32} & a_{33} \end{vmatrix} - a_{12}\begin{vmatrix} a_{21} & a_{23} \\ a_{31} & a_{33} \end{vmatrix} + a_{13}\begin{vmatrix} a_{21} & a_{22} \\ a_{31} & a_{32} \end{vmatrix}$$

The 3×3 determinant is now expressed in terms of 2×2 determinants.

$$\begin{vmatrix} a_{11} & a_{12} & a_{13} \\ a_{21} & a_{22} & a_{23} \\ a_{31} & a_{32} & a_{33} \end{vmatrix} = a_{11}\begin{vmatrix} a_{22} & a_{23} \\ a_{32} & a_{33} \end{vmatrix} - a_{12}\begin{vmatrix} a_{21} & a_{23} \\ a_{31} & a_{33} \end{vmatrix} + a_{13}\begin{vmatrix} a_{21} & a_{22} \\ a_{31} & a_{32} \end{vmatrix}$$

Notice negative sign

Example 2

$$\begin{vmatrix} 1 & 2 & -1 \\ 0 & 1 & 2 \\ -1 & 0 & -2 \end{vmatrix} = 1\begin{vmatrix} 1 & 2 \\ 0 & -2 \end{vmatrix} - 2\begin{vmatrix} 0 & 2 \\ -1 & -2 \end{vmatrix} + (-1)\begin{vmatrix} 0 & 1 \\ -1 & 0 \end{vmatrix}$$

$$= 1(-2) - 2(2) + (-1)(1)$$

$$= -2 - 4 - 1$$

$$= \mathbf{-7}$$

□

The expansion

$$a_{11}a_{22}a_{33} - a_{11}a_{23}a_{32} - a_{12}a_{21}a_{33} + a_{12}a_{23}a_{31} + a_{13}a_{21}a_{32} - a_{13}a_{22}a_{31}$$

also may be regrouped by commuting the fourth and fifth terms:

$$\begin{aligned} &a_{11}a_{22}a_{33} - a_{11}a_{23}a_{32} - a_{12}a_{21}a_{33} + a_{13}a_{21}a_{32} + a_{12}a_{23}a_{31} - a_{13}a_{22}a_{31} \\ &= a_{11}(a_{22}a_{33} - a_{23}a_{32}) - a_{21}(a_{12}a_{33} - a_{13}a_{32}) + a_{31}(a_{12}a_{23} - a_{13}a_{22}) \end{aligned}$$

This gives the expansion of the determinant about the first column instead of the first row, as before.

$$\begin{vmatrix} \mathbf{a_{11}} & a_{12} & a_{13} \\ \mathbf{a_{21}} & a_{22} & a_{23} \\ \mathbf{a_{31}} & a_{32} & a_{33} \end{vmatrix} = a_{11}\begin{vmatrix} a_{22} & a_{23} \\ a_{32} & a_{33} \end{vmatrix} - a_{21}\begin{vmatrix} a_{12} & a_{13} \\ a_{32} & a_{33} \end{vmatrix} + a_{31}\begin{vmatrix} a_{12} & a_{13} \\ a_{22} & a_{23} \end{vmatrix}$$

Expressing the 3×3 determinant in terms of 2×2 determinants is called **expansion by minors.**

MINOR

> The minor of a_{ij} is the determinant M_{ij} obtained by deleting the ith row and the jth column of $A_{n \times n}$.

$$\begin{vmatrix} \mathbf{a_{11}} & \mathbf{a_{12}} & \mathbf{a_{13}} \\ \mathbf{a_{21}} & a_{22} & a_{23} \\ \mathbf{a_{31}} & a_{32} & a_{33} \end{vmatrix} \qquad \begin{vmatrix} \mathbf{a_{11}} & \mathbf{a_{12}} & \mathbf{a_{13}} \\ a_{21} & \mathbf{a_{22}} & a_{23} \\ a_{31} & \mathbf{a_{32}} & a_{33} \end{vmatrix}$$

$\begin{vmatrix} a_{22} & a_{23} \\ a_{32} & a_{33} \end{vmatrix}$ is the minor of a_{11} $\qquad$ $\begin{vmatrix} a_{21} & a_{23} \\ a_{31} & a_{33} \end{vmatrix}$ is the minor of a_{12}

Given $\delta(A)$ below, find the minor of the indicated element in Examples 3–5.

$$\delta(A) = \begin{vmatrix} 1 & 2 & -4 \\ 0 & 3 & 4 \\ -1 & -3 & -2 \end{vmatrix}$$

Example 3 $1 = a_{11}$; $\begin{vmatrix} 1 & 2 & -4 \\ 0 & 3 & 4 \\ -1 & -3 & -2 \end{vmatrix}$ $M_{11} = \begin{vmatrix} 3 & 4 \\ -3 & -2 \end{vmatrix} = \mathbf{6}$ □

Example 4 $-3 = a_{32}$; $\begin{vmatrix} 1 & 2 & -4 \\ 0 & 3 & 4 \\ -1 & -3 & -2 \end{vmatrix}$ $M_{32} = \begin{vmatrix} 1 & -4 \\ 0 & 4 \end{vmatrix} = \mathbf{4}$ □

Example 5 $4 = a_{23}$; $\begin{vmatrix} 1 & 2 & -4 \\ 0 & 3 & 4 \\ -1 & -3 & -2 \end{vmatrix}$ $M_{23} = \begin{vmatrix} 1 & 2 \\ -1 & -3 \end{vmatrix} = \mathbf{-1}$ □

It is possible to expand a particular determinant about any row or column by the proper use of signs and obtain the same value. As it occurs, the signs affixed to the minors in the expansion alternate in each expansion. The minor with the proper sign affixed is called the **cofactor** of the corresponding element.

COFACTOR

The determinant A_{ij} is the **cofactor** of a_{ij} and

$$A_{ij} = (-1)^{i+j} M_{ij}$$

The cofactor of a_{ij} is M_{ij} if $i + j$ is even and is $-M_{ij}$ if $i + j$ is odd.

In Examples 6–8, find the cofactor of the indicated element of the determinant given below.

$$\delta(A) = \begin{vmatrix} 1 & 2 & -4 \\ 0 & 3 & 4 \\ -1 & -3 & -2 \end{vmatrix}$$

Example 6 2; $A_{12} = (-1)^{1+2} \begin{vmatrix} 0 & 4 \\ -1 & -2 \end{vmatrix} = (-1)(4) = \mathbf{-4}$ The cofactor of 2 is -4. □

Example 7 -1; $A_{31} = (-1)^{3+1} \begin{vmatrix} 2 & -4 \\ 3 & 4 \end{vmatrix} = (+1)(20) = \mathbf{20}$ The cofactor of -1 is 20. □

Example 8 4; $A_{23} = (-1)^{2+3} \begin{vmatrix} 1 & 2 \\ -1 & -3 \end{vmatrix} = (-1)(-1) = \mathbf{1}$ The cofactor of 4 is 1. □

Cofactors are used in evaluating determinants. Some row or column is arbitrarily chosen and the cofactors of each element in that row or column are combined to find the value of that determinant. This process is illustrated by reworking Example 2 using this technique.

Example 9 Expand $\begin{vmatrix} 1 & 2 & -1 \\ 0 & 1 & 2 \\ -1 & 0 & -2 \end{vmatrix}$ about the third column.

Solution $\begin{vmatrix} 1 & 2 & -1 \\ 0 & 1 & 2 \\ -1 & 0 & -2 \end{vmatrix}$

Note position sign of cofactor.

$$= (-1)\left[(+1)\begin{vmatrix} 0 & 1 \\ -1 & 0 \end{vmatrix}\right] + 2\left[(-1)\begin{vmatrix} 1 & 2 \\ -1 & 0 \end{vmatrix}\right] + (-2)\left[(+1)\begin{vmatrix} 1 & 2 \\ 0 & 1 \end{vmatrix}\right]$$

$$= (-1)\begin{vmatrix} 0 & 1 \\ -1 & 0 \end{vmatrix} - 2\begin{vmatrix} 1 & 2 \\ -1 & 0 \end{vmatrix} + (-2)\begin{vmatrix} 1 & 2 \\ 0 & 1 \end{vmatrix}$$

$$= (-1)(1) - 2(2) + (-2)(1)$$

$$= -1 - 4 - 2$$

$$= \mathbf{-7}$$

□

Example 10 Expand $\begin{vmatrix} 1 & 2 & -1 \\ 0 & 1 & 2 \\ -1 & 0 & -2 \end{vmatrix}$ about the second row.

Solution $\begin{vmatrix} 1 & 2 & -1 \\ 0 & 1 & 2 \\ -1 & 0 & -2 \end{vmatrix}$

$$= 0\left[(-1)\begin{vmatrix} 2 & -1 \\ 0 & -2 \end{vmatrix}\right] + 1\left[(+1)\begin{vmatrix} 1 & -1 \\ -1 & -2 \end{vmatrix}\right] + 2\left[(-1)\begin{vmatrix} 1 & 2 \\ -1 & 0 \end{vmatrix}\right]$$

$$= -0\begin{vmatrix} 2 & -1 \\ 0 & -2 \end{vmatrix} + 1\begin{vmatrix} 1 & -1 \\ -1 & -2 \end{vmatrix} - 2\begin{vmatrix} 1 & 2 \\ -1 & 0 \end{vmatrix}$$

$$= 0 + 1(-3) - 2(2)$$

$$= 0 - 3 - 4$$

$$= \mathbf{-7}$$

□

Example 11 Expand $\begin{vmatrix} 1 & 3 & -2 \\ 1 & -4 & 5 \\ 2 & -2 & 1 \end{vmatrix}$ about the first row.

Solution

$$1\begin{vmatrix} -4 & 5 \\ -2 & 1 \end{vmatrix} - 3\begin{vmatrix} 1 & 5 \\ 2 & 1 \end{vmatrix} + (-2)\begin{vmatrix} 1 & -4 \\ 2 & -2 \end{vmatrix} = 1(6) - 3(-9) + (-2)(6)$$

$$= 6 + 27 - 12$$

$$= \mathbf{21}$$

Do you see where this minus sign comes from?

□

Example 12 Expand $\begin{vmatrix} 1 & 3 & -2 \\ 1 & -4 & 5 \\ 2 & -2 & 1 \end{vmatrix}$ about the second column.

Solution

$$-3\begin{vmatrix} 1 & 5 \\ 2 & 1 \end{vmatrix} + (-4)\begin{vmatrix} 1 & -2 \\ 2 & 1 \end{vmatrix} - (-2)\begin{vmatrix} 1 & -2 \\ 1 & 5 \end{vmatrix} = -3(-9) + (-4)(5) - (-2)(7)$$
$$= 27 - 20 + 14$$
$$= \mathbf{21}$$

Negative positions (arrows point to the $-$ signs before 3 and before (-2))

□

Problem Set 7.5

A *Determine whether the statements in Problems 1–10 are true or false.*

1. $\begin{bmatrix} 1 & 4 \\ 2 & 5 \\ 3 & 6 \end{bmatrix}$ is a 2×3 matrix.

2. $\begin{vmatrix} 1 & 4 \\ 2 & 5 \\ 3 & 6 \end{vmatrix}$ is a determinant of order 3.

Problems 3–8 refer to the determinant:

$$D = \begin{vmatrix} 1 & -2 & 0 \\ -3 & 0 & -1 \\ 2 & 3 & -5 \end{vmatrix}$$

3. The minor of -3 is $\begin{vmatrix} 0 & -1 \\ 3 & -5 \end{vmatrix}$

4. $\begin{vmatrix} 1 & 0 \\ -3 & -1 \end{vmatrix}$ is the minor of 3.

5. The cofactor of -5 is $-\begin{vmatrix} 1 & -2 \\ -3 & 0 \end{vmatrix}$

6. $-\begin{vmatrix} -3 & -1 \\ 2 & -5 \end{vmatrix}$ is the cofactor of -2.

7. $D = 2\begin{vmatrix} -2 & 0 \\ 0 & -1 \end{vmatrix} - 3\begin{vmatrix} 1 & 0 \\ -3 & -1 \end{vmatrix} - 5\begin{vmatrix} 1 & -2 \\ -3 & 0 \end{vmatrix}$

8. $D = 3\begin{vmatrix} -2 & 0 \\ 3 & -5 \end{vmatrix} + 0\begin{vmatrix} 1 & 0 \\ 2 & -5 \end{vmatrix} + 1\begin{vmatrix} 1 & -2 \\ 2 & 3 \end{vmatrix}$

9. If the sum of the row and column numbers of a given entry is even, affix $+$ to its minor to obtain its cofactor.

10. If the sum of the number of the row and the number of the column containing an entry is odd, affix $-$ to its minor to obtain its cofactor.

Evaluate the determinants in Problems 11–30.

11. $\begin{vmatrix} -3 & 7 \\ 2 & -5 \end{vmatrix}$ **12.** $\begin{vmatrix} 6 & -5 \\ -5 & 4 \end{vmatrix}$ **13.** $\begin{vmatrix} 11 & 12 \\ -7 & -9 \end{vmatrix}$

14. $\begin{vmatrix} -8 & -10 \\ 6 & 7 \end{vmatrix}$ **15.** $\begin{vmatrix} -9 & 10 \\ 11 & -12 \end{vmatrix}$ **16.** $\begin{vmatrix} 9 & -15 \\ 8 & -14 \end{vmatrix}$

17. $\begin{vmatrix} 1 & 0 & 0 \\ 0 & 2 & -1 \\ 2 & 1 & 0 \end{vmatrix}$ **18.** $\begin{vmatrix} 2 & -1 & 0 \\ 0 & 2 & 1 \\ 0 & 0 & 1 \end{vmatrix}$ **19.** $\begin{vmatrix} 1 & 2 & 0 \\ -3 & 1 & 0 \\ 2 & -1 & 1 \end{vmatrix}$

20. $\begin{vmatrix} 1 & -2 & 3 \\ 2 & 0 & -1 \\ -3 & 0 & 2 \end{vmatrix}$ **21.** $\begin{vmatrix} 2 & -1 & 2 \\ 0 & 0 & 3 \\ -2 & 3 & -1 \end{vmatrix}$ **22.** $\begin{vmatrix} 3 & 0 & -1 \\ 1 & 2 & -3 \\ -1 & 0 & 2 \end{vmatrix}$

23. $\begin{vmatrix} -1 & 0 & 1 \\ 1 & -1 & 0 \\ 0 & 1 & -1 \end{vmatrix}$ **24.** $\begin{vmatrix} 1 & 0 & -1 \\ 0 & 1 & 1 \\ -1 & 1 & 0 \end{vmatrix}$ **25.** $\begin{vmatrix} 1 & -1 & 0 \\ 0 & 1 & -1 \\ -1 & 0 & 1 \end{vmatrix}$

26. $\begin{vmatrix} 0 & -1 & 2 \\ 1 & 2 & 0 \\ 2 & 1 & -2 \end{vmatrix}$ **27.** $\begin{vmatrix} 3 & -1 & 1 \\ 1 & 0 & 3 \\ 0 & 3 & -1 \end{vmatrix}$ **28.** $\begin{vmatrix} 1 & 2 & 3 \\ 0 & -1 & 2 \\ 3 & 0 & -1 \end{vmatrix}$

29. $\begin{vmatrix} 1 & 1 & 1 \\ 1 & 1 & 1 \\ 1 & 2 & 3 \end{vmatrix}$ **30.** $\begin{vmatrix} 1 & 1 & 1 \\ 1 & 1 & 1 \\ 1 & 3 & 2 \end{vmatrix}$

B *Evaluate the determinants in Problems 31–40 and check by expanding about two different rows (or columns) to obtain the same value.*

31. $\begin{vmatrix} 1 & 1 & 1 \\ 3 & 2 & 1 \\ 2 & 5 & 3 \end{vmatrix}$ **32.** $\begin{vmatrix} 5 & -1 & 1 \\ 6 & -2 & 1 \\ 1 & 2 & 3 \end{vmatrix}$ **33.** $\begin{vmatrix} 1 & -1 & 3 \\ 2 & 5 & -1 \\ -3 & 2 & 1 \end{vmatrix}$

34. $\begin{vmatrix} 3 & -2 & 4 \\ 2 & 3 & -2 \\ 5 & 2 & 3 \end{vmatrix}$ **35.** $\begin{vmatrix} 5 & -7 & 4 \\ 6 & 3 & -5 \\ 4 & 6 & 3 \end{vmatrix}$ **36.** $\begin{vmatrix} 2 & 3 & 4 \\ 5 & -1 & 6 \\ 3 & 7 & -4 \end{vmatrix}$

37. $\begin{vmatrix} 9 & 5 & 4 \\ 5 & 6 & 5 \\ 16 & 4 & 6 \end{vmatrix}$ **38.** $\begin{vmatrix} 5 & 2 & 6 \\ 7 & 4 & 10 \\ 8 & 5 & 13 \end{vmatrix}$ **39.** $\begin{vmatrix} 12 & 15 & 8 \\ 5 & 8 & 3 \\ 9 & 14 & 5 \end{vmatrix}$

40. $\begin{vmatrix} 7 & 17 & 14 \\ 5 & 9 & 4 \\ 8 & 20 & 12 \end{vmatrix}$

C *Evaluate the determinants in Problems 41–44.*

41. $\begin{vmatrix} x & 3 & 4 \\ y & 4 & 5 \\ z & 5 & 6 \end{vmatrix}$ **42.** $\begin{vmatrix} x & y & 1 \\ 3 & 4 & 1 \\ 2 & 5 & 1 \end{vmatrix}$

43. $\begin{vmatrix} a & 1 & 1 \\ 1 & b & 1 \\ 1 & 1 & c \end{vmatrix}$ **44.** $\begin{vmatrix} a+b & 2 & 1 \\ 2 & a-b & 1 \\ 3 & 3 & 1 \end{vmatrix}$

45. Consider $\begin{vmatrix} x & L & d \\ 0 & d & L \\ 1 & 2 & \frac{x}{d} \end{vmatrix} = 0$ and $|x - L| = d$, where $d > 0$

Absolute value bars

Is there any connection between these two expressions? Discuss their equivalency. Illustrate your conjecture with at least one numerical example.

7.6 Properties of Determinants

Because of the properties of determinants, a process of simplification can be developed to help evaluate determinants, especially those of large order.

Property I If two rows or two columns of a determinant are identical, the value of the determinant is zero.

Example 1 Expand about the first column.

$$\begin{vmatrix} 1 & 2 & 2 \\ 0 & 1 & 1 \\ -3 & 0 & 0 \end{vmatrix} = \begin{vmatrix} 1 & 1 \\ 0 & 0 \end{vmatrix} + (-3)\begin{vmatrix} 2 & 2 \\ 1 & 1 \end{vmatrix}$$

These columns are identical

$$= 0 + (-3)(0)$$
$$= \mathbf{0}$$

□

Property II If one row or column has all zero entries, the value of the determinant is zero.

Example 2 Expand about the second column.

$$\begin{vmatrix} 1 & 0 & -3 \\ 2 & 0 & 2 \\ 3 & 0 & 1 \end{vmatrix} = -0\begin{vmatrix} 2 & 2 \\ 3 & 1 \end{vmatrix} + 0\begin{vmatrix} 1 & -3 \\ 3 & 1 \end{vmatrix} - 0\begin{vmatrix} 1 & -3 \\ 2 & 2 \end{vmatrix}$$

$$= 0 + 0 - 0$$
$$= \mathbf{0}$$

□

Property III If each entry in one row or one column is multiplied by a constant, the value of the determinant is multiplied by that constant.

Example 3

$$\begin{vmatrix} 1 & 0 & 4 \\ -1 & 2 & 0 \\ 3 & 1 & 1 \end{vmatrix} = 1\begin{vmatrix} 2 & 0 \\ 1 & 1 \end{vmatrix} - 0 + 4\begin{vmatrix} -1 & 2 \\ 3 & 1 \end{vmatrix} \quad \text{Expand about the first row.}$$

$$= 1(2) - 0 + 4(-7)$$
$$= 2 - 0 - 28$$
$$= \mathbf{-26}$$

Multiply this column by 2

$$\begin{vmatrix} 2 & 0 & 4 \\ -2 & 2 & 0 \\ 6 & 1 & 1 \end{vmatrix} = 2\begin{vmatrix} 2 & 0 \\ 1 & 1 \end{vmatrix} - 0 + 4\begin{vmatrix} -2 & 2 \\ 6 & 1 \end{vmatrix} \quad \text{Expand about the first row.}$$

$$= 2(2) - 0 + 4(-14)$$
$$= 4 - 0 - 56$$
$$= \mathbf{-52}$$

Thus, multiplying the first column by 2 doubled the value of the determinant. □

But the key property is Property IV, which will have the greatest use.

Property IV If a multiple of the entries of one row (column) is added to corresponding entries of another row (column), respectively, the value of the determinant is unchanged.

Example 4

$$\begin{vmatrix} 1 & 0 & 4 \\ -1 & 2 & 0 \\ 3 & 1 & 1 \end{vmatrix} = \begin{vmatrix} 1 & 0 & 4+(-4) \\ -1 & 2 & 0+(4) \\ 3 & 1 & 1+(-12) \end{vmatrix} = \begin{vmatrix} 1 & 0 & 0 \\ -1 & 2 & 4 \\ 3 & 1 & -11 \end{vmatrix}$$

$$C_1 \quad C_2 \quad C_3 - 4C_1$$

$$= 1\begin{vmatrix} 2 & 4 \\ 1 & -11 \end{vmatrix} - 0 + 0$$
$$= -22 - 4$$
$$= \mathbf{-26} \quad \text{Compare with Example 3.}$$ □

Example 5

$$\begin{vmatrix} 1 & 0 & 4 \\ -1 & 2 & 0 \\ 3 & 1 & 1 \end{vmatrix} = \begin{vmatrix} 1-1 & 0+2 & 4+0 \\ -1 & 2 & 0 \\ 3-3 & 1+6 & 1+0 \end{vmatrix} \begin{matrix} R_1 + R_2 \\ R_2 \\ R_3 + 3R_2 \end{matrix}$$

$$= \begin{vmatrix} 0 & 2 & 4 \\ -1 & 2 & 0 \\ 0 & 7 & 1 \end{vmatrix}$$

$$= 0 - (-1)\begin{vmatrix} 2 & 4 \\ 7 & 1 \end{vmatrix} + 0$$
$$= 0 + (-26) + 0$$
$$= \mathbf{-26}$$ □

By the use of Property IV, you may introduce entries of 0 into the determinant to make expansion simpler. In Examples 4 and 5, you were able to create a row and a column, respectively, that contained all 0s except one. Expanding about that row or column meant evaluating only one 2×2 determinant to find the value of a 3×3 determinant. You may use this strategy to reduce any nth-order determinant to a single $(n - 1)$th-order determinant.

Example 6 Expand by first obtaining two zeros in one row or column.

Solution

$$\begin{vmatrix} 1 & 2 & 3 \\ 4 & -1 & -2 \\ 3 & 0 & 1 \end{vmatrix} = \begin{vmatrix} 1-3(3) & 2 & 3 \\ 4-3(-2) & -1 & -2 \\ 3-3(1) & 0 & 1 \end{vmatrix} = \begin{vmatrix} -8 & 2 & 3 \\ 10 & -1 & -2 \\ 0 & 0 & 1 \end{vmatrix}$$
$$\quad C_1 \quad C_2 \quad C_3 \qquad C_1 - 3C_3 \quad C_2 \quad C_3$$

$$= 0 - 0 + 1\begin{vmatrix} -8 & 2 \\ 10 & -1 \end{vmatrix}$$
$$= 1(8 - 20)$$
$$= \mathbf{-12}$$ □

Using the properties illustrated above, you can reduce a determinant of any order to a single 2×2 determinant (times some constant) as in the examples below, which illustrate the procedure.

Example 7

Pivot (circled -1 in row 1, column 3)

$$\begin{vmatrix} 3 & -2 & -1 & 2 \\ 4 & 1 & 2 & -3 \\ -9 & -5 & 7 & -8 \\ 1 & 5 & 3 & -2 \end{vmatrix} = \begin{vmatrix} 0 & 0 & -1 & 0 \\ 10 & -3 & 2 & 1 \\ 12 & -19 & 7 & 6 \\ 10 & -1 & 3 & 4 \end{vmatrix}$$
$$C_1 \quad C_2 \quad C_3 \quad C_4 \qquad C_1 + 3C_3 \quad C_2 - 2C_3 \quad C_3 \quad C_4 + 2C_3$$

New pivot for 3×3 (circled 1 in row 1, column 3)

$$= (-1)\begin{vmatrix} 10 & -3 & 1 \\ 12 & -19 & 6 \\ 10 & -1 & 4 \end{vmatrix}$$

$$= (-1)\begin{vmatrix} 10 & -3 & 1 \\ -48 & -1 & 0 \\ -30 & 11 & 0 \end{vmatrix} \begin{matrix} R_1 \\ R_2 - 6R_1 \\ R_3 - 4R_1 \end{matrix}$$

$$= (-1)(1)\begin{vmatrix} -48 & -1 \\ -30 & 11 \end{vmatrix}$$

$$= (-1)(1)(-528 - 30)$$

$$= \mathbf{558}$$

□

Example 8

$$\begin{vmatrix} 2 & -1 & 2 & -1 \\ 1 & 1 & -1 & 2 \\ 1 & 2 & 1 & -3 \\ 1 & 3 & 4 & 7 \end{vmatrix} = \begin{vmatrix} 2 & -3 & 4 & -5 \\ 1 & 0 & 0 & 0 \\ 1 & 1 & 2 & -5 \\ 1 & 2 & 5 & 5 \end{vmatrix}$$

$C_1 \quad C_2 \quad C_3 \quad C_4 \qquad C_1 \quad C_2 - C_1 \quad C_3 + C_1 \quad C_4 - 2C_1$

$$= -(1)\begin{vmatrix} -3 & 4 & -5 \\ 1 & 2 & -5 \\ 2 & 5 & 5 \end{vmatrix}$$

$$= -(1)(5)\begin{vmatrix} -3 & 4 & -1 \\ 1 & 2 & -1 \\ 2 & 5 & 1 \end{vmatrix}$$

$$= -5\begin{vmatrix} -3 & 4 & -1 \\ 1 & 2 & -1 \\ 2 & 5 & 1 \end{vmatrix}$$

$$= -5\begin{vmatrix} -1 & 9 & 0 \\ 3 & 7 & 0 \\ 2 & 5 & 1 \end{vmatrix} \begin{matrix} R_1 + R_3 \\ R_2 + R_3 \\ R_3 \end{matrix}$$

$$= -5(1)\begin{vmatrix} -1 & 9 \\ 3 & 7 \end{vmatrix}$$

$$= -5(-7 - 27)$$

$$= -5(-34)$$

$$= \mathbf{170}$$

□

Proceed slowly in your own work, because most errors occur in simple arithmetic. Use row or column labeling to ensure that you can recheck your work or search for an error.

If one or more of the entries of a determinant is a variable and the determinant is equal to another determinant or a constant, the equation can be solved for the value or values of the variable.

Example 9 Solve for x:

$$\begin{vmatrix} 1 & -3 & 1 \\ -1 & x & -2 \\ 2 & -1 & 3 \end{vmatrix} = 0 \qquad \text{Expand about the second row}$$

$$-(-1)\begin{vmatrix} -3 & 1 \\ -1 & 3 \end{vmatrix} + x\begin{vmatrix} 1 & 1 \\ 2 & 3 \end{vmatrix} - (-2)\begin{vmatrix} 1 & -3 \\ 2 & -1 \end{vmatrix} = 0$$

$$-(-1)(-8) + x(1) - (-2)(5) = 0$$

$$-8 + x + 10 = 0$$

$$\mathbf{x = -2}$$

□

Many useful equations and formulas can be stated in determinant form, as in the case of the equation of the line which is given next.

DETERMINANT EQUATION OF A LINE

The *equation of the line* through (x_1, y_1) and (x_2, y_2) is

$$\begin{vmatrix} x & y & 1 \\ x_1 & y_1 & 1 \\ x_2 & y_2 & 1 \end{vmatrix} = 0$$

Example 10 Find the equation of the line passing through $(2, 3)$ and $(-4, 1)$.

Solution

$$\begin{vmatrix} x & y & 1 \\ 2 & 3 & 1 \\ -4 & 1 & 1 \end{vmatrix} = 0$$

Expanding about the first row.

$$x\begin{vmatrix} 3 & 1 \\ 1 & 1 \end{vmatrix} - y\begin{vmatrix} 2 & 1 \\ -4 & 1 \end{vmatrix} + 1\begin{vmatrix} 2 & 3 \\ -4 & 1 \end{vmatrix} = 0$$

$$2x - 6y + 14 = 0$$

or

$$\mathbf{y = \tfrac{1}{3}x + \tfrac{7}{3}}$$

To check, first find the slope,

$$m = \frac{1-3}{-4-2} = \frac{-2}{-6} = \frac{1}{3}$$

and then substitute $(2, 3)$ into the point–slope formula:

$$y - 3 = \tfrac{1}{3}(x - 2)$$

$$= \tfrac{1}{3}x - \tfrac{2}{3}$$

$$\mathbf{y = \tfrac{1}{3}x + \tfrac{7}{3}}$$

□

Example 11 Find the equation of the line containing $\left(\frac{1}{2}, \frac{1}{4}\right)$ and $\left(1, -\frac{1}{2}\right)$.

Solution

$$\begin{vmatrix} x & y & 1 \\ \frac{1}{2} & \frac{1}{4} & 1 \\ 1 & -\frac{1}{2} & 1 \end{vmatrix} = 0$$

You can expand the determinant as it is, with the fractions, but instead we will use Property III to clear the fractions from the determinant:

$$\frac{1}{2}\begin{vmatrix} x & 2y & 1 \\ \frac{1}{2} & \frac{1}{2} & 1 \\ 1 & -1 & 1 \end{vmatrix} = \frac{1}{2} \cdot \frac{1}{2}\begin{vmatrix} x & 2y & 1 \\ 1 & 1 & 2 \\ 1 & -1 & 1 \end{vmatrix} = 0$$

Factor out $\frac{1}{2}$ from Row 2.

Factor out $\frac{1}{2}$ from Column 2.

Now, expand about the first row:

$$\frac{1}{4}\left(x\begin{vmatrix} 1 & 2 \\ -1 & 1 \end{vmatrix} - 2y\begin{vmatrix} 1 & 2 \\ 1 & 1 \end{vmatrix} + 1\begin{vmatrix} 1 & 1 \\ 1 & -1 \end{vmatrix}\right) = 0$$

Multiply both sides by 4, and expand the 2×2 determinants:

$$x(3) - 2y(-1) + 1(-2) = 0$$

$$3x + 2y - 2 = 0$$

or

$$\mathbf{y = -\tfrac{3}{2}x + 1}$$

□

Problem Set 7.6

A *Expand the determinants in Problems 1–20 and find the value of each.*

1. $\begin{vmatrix} 1 & 2 & 0 \\ 0 & -1 & 2 \\ 2 & 3 & 1 \end{vmatrix}$ **2.** $\begin{vmatrix} 2 & 0 & -1 \\ 1 & 2 & 0 \\ 0 & -1 & 2 \end{vmatrix}$ **3.** $\begin{vmatrix} 1 & -2 & 1 \\ 3 & 0 & 2 \\ -2 & 1 & 0 \end{vmatrix}$

4. $\begin{vmatrix} 2 & 4 & 1 \\ 3 & -1 & 2 \\ 5 & 2 & 1 \end{vmatrix}$ **5.** $\begin{vmatrix} 3 & -2 & -3 \\ -2 & 2 & 1 \\ -3 & 1 & 2 \end{vmatrix}$ **6.** $\begin{vmatrix} 2 & 3 & 7 \\ 3 & 4 & 10 \\ 5 & 6 & 16 \end{vmatrix}$

7. $\begin{vmatrix} -1 & 3 & 2 \\ 3 & 4 & -6 \\ -2 & 5 & 4 \end{vmatrix}$ **8.** $\begin{vmatrix} 2 & -3 & 1 \\ 3 & -2 & -3 \\ -3 & 1 & 2 \end{vmatrix}$ **9.** $\begin{vmatrix} 3 & 2 & 4 \\ 3 & 1 & 1 \\ -1 & -2 & 3 \end{vmatrix}$

10. $\begin{vmatrix} 1 & 2 & 4 \\ 1 & -1 & -3 \\ 1 & -2 & -3 \end{vmatrix}$ **11.** $\begin{vmatrix} 2 & -3 & 1 \\ -2 & 2 & 1 \\ -3 & 1 & 2 \end{vmatrix}$ **12.** $\begin{vmatrix} 2 & -1 & 3 \\ 1 & 2 & 1 \\ 3 & 1 & -3 \end{vmatrix}$

13. $\begin{vmatrix} 2 & -3 & 1 \\ 3 & -2 & -3 \\ -2 & 2 & 1 \end{vmatrix}$ **14.** $\begin{vmatrix} -2 & -2 & 3 \\ 3 & 3 & 2 \\ 2 & 2 & 3 \end{vmatrix}$ **15.** $\begin{vmatrix} 2 & 3 & 3 \\ 3 & 2 & 1 \\ -2 & 1 & 2 \end{vmatrix}$

16. $\begin{vmatrix} 0 & -1 & 2 \\ -1 & 2 & -1 \\ -11 & 9 & 6 \end{vmatrix}$ **17.** $\begin{vmatrix} 1 & 0 & -3 \\ 4 & 3 & 1 \\ 9 & -12 & 5 \end{vmatrix}$ **18.** $\begin{vmatrix} -1 & 2 & 0 \\ 3 & 1 & 4 \\ 7 & 4 & 13 \end{vmatrix}$

19. $\begin{vmatrix} 0 & a & b \\ -a & 0 & c \\ -b & -c & 0 \end{vmatrix}$ **20.** $\begin{vmatrix} 0 & b & b \\ a & 0 & c \\ a & c & 0 \end{vmatrix}$

In Problems 21–30 find the solution set of the determinant equations.

21. $\begin{vmatrix} 3 & 1 & 2 \\ 2 & 0 & 1 \\ 1 & x & -2 \end{vmatrix} = 5$ **22.** $\begin{vmatrix} 4 & x & 0 \\ -1 & -2 & 1 \\ 2 & 4 & 3 \end{vmatrix} = 0$

23. $\begin{vmatrix} 2 & 2 & 3 \\ 0 & 2 & 1 \\ 1 & x & 0 \end{vmatrix} = 12$ **24.** $\begin{vmatrix} 3 & 1 & -2 \\ 0 & x & 1 \\ 4 & -2 & 1 \end{vmatrix} = 21$

25. $\begin{vmatrix} 2 & 0 & -3 \\ -1 & x & 2 \\ 3 & 2 & -1 \end{vmatrix} = 5$ **26.** $\begin{vmatrix} -4 & 2 & x \\ 2 & 0 & 1 \\ 0 & 4 & 2 \end{vmatrix} = 32$

27. $\begin{vmatrix} x & 1 & -1 \\ 5 & 2 & 4 \\ 2x & 3 & 5 \end{vmatrix} = 0$ **28.** $\begin{vmatrix} 4 & -1 & 2 \\ 7 & 1 & 3 \\ x & 2 & 3x \end{vmatrix} = 60$

29. $\begin{vmatrix} 2x & x & 2 \\ 1 & 0 & 2 \\ -1 & 4 & 1 \end{vmatrix} = 65$ **30.** $\begin{vmatrix} 2 & 5 & -1 \\ 1 & 3x & 3 \\ 1 & x & 1 \end{vmatrix} = 0$

Find the equation of the line containing the points in Problems 31–36.

31. $(0, -2)$, $(-3, 0)$ **32.** $(0, 4)$, $(-1, 0)$ **33.** $(3, -1)$, $(-1, 2)$

34. $(-1, 2)$, $(2, 4)$ **35.** $\left(-\frac{1}{2}, 4\right)$, $(1, 5)$ **36.** $\left(\frac{1}{3}, -2\right)$, $(1, 1)$

B *In Problems 37–40 find the solution set of the determinant equations.*

37. $\begin{vmatrix} x & 2 & 3 \\ 1 & x & 3 \\ 1 & -1 & 1 \end{vmatrix} = 5$ **38.** $\begin{vmatrix} -1 & 4 & x \\ 2 & x & 1 \\ 3 & 5 & 1 \end{vmatrix} = -3$

39. $\begin{vmatrix} x & 1 & 1 \\ 3 & 2 & -2 \\ 1 & 3 & 2x \end{vmatrix} = 9$ **40.** $\begin{vmatrix} 1 & 2 & -1 \\ x & x & 1 \\ 2 & 3 & 0 \end{vmatrix} = x$

Evaluate the determinants in Problems 41–50.

41. $\begin{vmatrix} -2 & 3 & -1 & 1 \\ 4 & -2 & 1 & 2 \\ 0 & 0 & 0 & -3 \\ -6 & -1 & 5 & 4 \end{vmatrix}$ **42.** $\begin{vmatrix} 3 & -2 & 0 & -3 \\ -2 & 2 & 0 & 1 \\ -3 & 1 & 0 & 2 \\ 1 & -3 & -2 & 3 \end{vmatrix}$

43. $\begin{vmatrix} -2 & 2 & 3 & 3 \\ 0 & 0 & 1 & 0 \\ 2 & -2 & 3 & 3 \\ -3 & 3 & 2 & 2 \end{vmatrix}$

44. $\begin{vmatrix} 0 & -1 & 2 & -3 \\ 1 & 0 & -3 & -1 \\ -2 & 3 & 0 & 7 \\ 3 & 4 & -6 & 9 \end{vmatrix}$

45. $\begin{vmatrix} 1 & 2 & 0 & 3 \\ 0 & -1 & 2 & -1 \\ 2 & 3 & 1 & 5 \\ 1 & 0 & 0 & 1 \end{vmatrix}$

46. $\begin{vmatrix} 2 & 0 & -1 & 2 \\ 1 & 2 & 0 & 3 \\ 0 & -1 & 2 & -1 \\ 3 & 0 & 1 & 3 \end{vmatrix}$

47. $\begin{vmatrix} 4 & 5 & -2 & -3 \\ 0 & -3 & 1 & 0 \\ 3 & 2 & -1 & 1 \\ -1 & -4 & 3 & -2 \end{vmatrix}$

48. $\begin{vmatrix} 1 & 0 & -1 & 2 \\ 1 & -1 & 2 & -1 \\ 15 & -11 & 9 & 6 \\ 7 & 0 & 4 & 3 \end{vmatrix}$

49. $\begin{vmatrix} 2 & 0 & 2 & 0 \\ -1 & 1 & 0 & -3 \\ -7 & 4 & 3 & 1 \\ 3 & 9 & -12 & 5 \end{vmatrix}$

50. $\begin{vmatrix} -5 & 3 & -9 & 8 \\ 1 & -2 & 4 & -3 \\ 4 & -4 & 5 & -7 \\ -1 & 1 & -2 & 3 \end{vmatrix}$

Problems 51–60 will require one of the following formulas:

Area *The area of a triangle with vertices at (x_1, y_1), (x_2, y_2), and (x_3, y_3) is A:*

$$A = \left| \frac{1}{2} \begin{vmatrix} x_1 & y_1 & 1 \\ x_2 & y_2 & 1 \\ x_3 & y_3 & 1 \end{vmatrix} \right|$$

Absolute value bars

Collinearity *If the three points (x_1, y_1), (x_2, y_2), and (x_3, y_3) lie on the same line, then*

$$\begin{vmatrix} x_1 & y_1 & 1 \\ x_2 & y_2 & 1 \\ x_3 & y_3 & 1 \end{vmatrix} = 0$$

This works because the area of "a triangle" formed by collinear points would be 0.

Concurrence *The three lines with slopes m_1, m_2, m_3 and with y-intercepts b_1, b_2, b_3 intersect at a single point, if*

$$\begin{vmatrix} m_1 & b_1 & 1 \\ m_2 & b_2 & 1 \\ m_3 & b_3 & 1 \end{vmatrix} = 0$$

51. Find the area of $\triangle ACE$, where $A = (1, 1)$, $C = (-2, -3)$, and $E = (11, -3)$.

52. Find the area of $\triangle DAM$, where $D = (-3, 12)$, $A = (5, 6)$, and $M = (-3, -9)$.

53. Verify that $(1, 1)$, $(0, -2)$, and $(-1, -5)$ lie on the same line.

54. Determine whether $(-1, 7)$, $(5, -1)$, and $(2, 3)$ are collinear.

55. Verify that $y = 3x - 5$, $y = -2x + 5$, and $y = x - 1$ represent the same line.

56. Show that $\frac{1}{2}x + y = 4$, $2x + y + 2 = 0$, and $x - y + 10 = 0$ represent the same line.

57. Find the area of parallelogram *RECT*, where $R = (-8, 0)$, $E = (0, 15)$, $C = (12, 10)$, and $T = (4, -5)$.

58. What is the area of parallelogram *RHOM*, where $R = (-7, 7)$, $H = (-12, -5)$, $O = (0, 0)$, and $M = (5, 12)$?

59. Find the area of quadrilateral *QUAD*, where $Q = (0, 0)$, $U = (-1, 2)$, $A = (2, 5)$, and $D = (4, 1)$.

60. What is the area of the quadrilateral *FOUR*, where $F = (0, 4)$, $O = (3, 0)$, $U = (0, -2)$, and $R = (-4, 0)$?

C 61. Prove that

$$\begin{vmatrix} x & y & 1 \\ x_1 & y_1 & 1 \\ x_2 & y_2 & 1 \end{vmatrix} = 0$$

represents an equation of the line through the points (x_1, y_1) and (x_2, y_2).

62. Prove Property I for any determinant.
63. Prove Property II for any determinant.
64. Prove Property III for any determinant.

65. Show that: $\begin{vmatrix} a & b & b & b \\ a & b & a & a \\ b & b & a & b \\ a & a & a & b \end{vmatrix} = (a-b)^4$

66. Show that: $\begin{vmatrix} a & b & c & d \\ c & b & a & d \\ c & d & a & b \\ a & d & c & b \end{vmatrix} = 0$

7.7 Determinant Solution of Systems

Mathematics is so diverse and yet so compact and interrelated! Graphing; the addition, substitution, and Gauss–Jordan methods; matrix equations; and now a determinant method are many diverse approaches to the same problem: the solution of simultaneous linear equations.

Recall the system of equations represented by an augmented matrix:

$$\left[\begin{array}{cc|c} a_{11} & a_{12} & b_1 \\ a_{21} & a_{22} & b_2 \end{array}\right]$$

Solve for x by the addition method:

$$\begin{array}{r} a_{22} \\ -a_{12} \end{array}\begin{cases} a_{11}x + a_{12}y = b_1 \\ a_{21}x + a_{22}y = b_2 \end{cases}$$

$$+\begin{cases} a_{11}a_{22}x + a_{12}a_{22}y = a_{22}b_1 \\ -a_{12}a_{21}x - a_{12}a_{22}y = -a_{12}b_2 \end{cases}$$

$$a_{11}a_{22}x - a_{12}a_{21}x + 0 = a_{22}b_1 - a_{12}b_2$$

$$(a_{11}a_{22} - a_{12}a_{21})x = a_{22}b_1 - a_{12}b_2$$

$$x = \frac{a_{22}b_1 - a_{12}b_2}{a_{11}a_{22} - a_{12}a_{21}} \qquad a_{11}a_{22} - a_{12}a_{21} \neq 0$$

$$x = \frac{\begin{vmatrix} b_1 & a_{12} \\ b_2 & a_{22} \end{vmatrix}}{\begin{vmatrix} a_{11} & a_{12} \\ a_{21} & a_{22} \end{vmatrix}}$$

And similarly,

$$y = \frac{\begin{vmatrix} a_{11} & b_1 \\ a_{21} & b_2 \end{vmatrix}}{\begin{vmatrix} a_{11} & a_{12} \\ a_{21} & a_{22} \end{vmatrix}}$$

The denominator in both cases is the determinant formed by the coefficients of x and y, called the **determinant of the system,** D. The x column of D is replaced by the constants to obtain the numerator for the x solution. This numerator is denoted by D_x. The symbol D_y then denotes D with the y column replaced by the constants. If the system is now represented by a matrix of its coefficients, the solution may be written in the following condensed form. The solution of a system using determinants in this way is called **Cramer's Rule,** named after its Swiss inventor, Gabriel Cramer (1704–1752).

CRAMER'S RULE

If $\left[\begin{array}{cc|c} a_{11} & a_{12} & b_1 \\ a_{21} & a_{22} & b_2 \end{array}\right]$ is the augmented matrix of a linear system in x and y, then

$$x = \frac{D_x}{D} \quad \text{and} \quad y = \frac{D_y}{D} \qquad D \neq 0$$

Example 1 Solve the system for x and y: $\begin{cases} 4x - 8y = 17 \\ 12x + 16y = -9 \end{cases}$

Solution First, form the augmented matrix:

$$\left[\begin{array}{cc|c} 4 & -8 & 17 \\ 12 & 16 & -9 \end{array}\right]$$

$$x = \frac{D_x}{D} = \frac{\begin{vmatrix} 17 & -8 \\ -9 & 16 \end{vmatrix}}{\begin{vmatrix} 4 & -8 \\ 12 & 16 \end{vmatrix}} = \frac{200}{160} = \frac{5}{4}$$

$$y = \frac{D_y}{D} = \frac{\begin{vmatrix} 4 & 17 \\ 12 & -9 \end{vmatrix}}{160} = \frac{-240}{160} = -\frac{3}{2}$$

Thus, $(x, y) = \left(\frac{5}{4}, -\frac{3}{2}\right)$. □

Cramer's rule is general and can be applied to a linear system of any size, n equations in n variables. Stated for a system of three equations in three variables it would be as follows:

CRAMER'S RULE

If $\left[\begin{array}{ccc|c} a_{11} & a_{12} & a_{13} & b_1 \\ a_{21} & a_{22} & a_{23} & b_2 \\ a_{31} & a_{32} & a_{33} & b_3 \end{array}\right]$ is the augmented matrix of a linear system in x, y, and z.

$$x = \frac{D_x}{D} \qquad y = \frac{D_y}{D} \qquad z = \frac{D_z}{D} \qquad D \neq 0$$

In this statement, the matrix represents

$$\begin{cases} a_{11}x + a_{12}y + a_{13}z = b_1 \\ a_{21}x + a_{22}y + a_{23}z = b_2 \\ a_{31}x + a_{32}y + a_{33}z = b_3 \end{cases}$$

and $D = \begin{vmatrix} a_{11} & a_{12} & a_{13} \\ a_{21} & a_{22} & a_{23} \\ a_{31} & a_{32} & a_{33} \end{vmatrix}$

with the constants b_1, b_2, and b_3 replacing the coefficient columnns in D_x, D_y, and D_z. The rule would appear very similar written for any size system. The solution would, of course, involve larger-order determinants for larger systems. It should be sufficient to illustrate the rule for a system of three equations.

Example 2

$$\begin{cases} x - 2y - 3z = 3 \\ x + y - z = 2 \\ 2x - 3y - 5z = 5 \end{cases}$$

Solution

$$\left[\begin{array}{ccc|c} 1 & -2 & -3 & 3 \\ 1 & 1 & -1 & 2 \\ 2 & -3 & -5 & 5 \end{array}\right]$$

$$x = \frac{D_x}{D} = \frac{\begin{vmatrix} 3 & -2 & -3 \\ 2 & 1 & -1 \\ 5 & -3 & -5 \end{vmatrix}}{\begin{vmatrix} 1 & -2 & -3 \\ 1 & 1 & -1 \\ 2 & -3 & -5 \end{vmatrix}} = \frac{\begin{vmatrix} -3 & -5 & -3 \\ 0 & 0 & -1 \\ -5 & -8 & -5 \end{vmatrix}}{\begin{vmatrix} -2 & -5 & -3 \\ 0 & 0 & -1 \\ -3 & -8 & -5 \end{vmatrix}}$$

$$= \frac{-(-1)\begin{vmatrix} -3 & -5 \\ -5 & -8 \end{vmatrix}}{-(-1)\begin{vmatrix} -2 & -5 \\ -3 & -8 \end{vmatrix}} = \frac{24 - 25}{16 - 15} = \frac{-1}{1} = -1$$

$$y = \frac{D_y}{D} = \frac{\begin{vmatrix} 1 & 3 & -3 \\ 1 & 2 & -1 \\ 2 & 5 & -5 \end{vmatrix}}{1} = \frac{\begin{vmatrix} 1 & 3 & -3 \\ 0 & -1 & 2 \\ 0 & -1 & 1 \end{vmatrix}}{1}$$

$$= 1\begin{vmatrix} -1 & 2 \\ -1 & 1 \end{vmatrix} = -1 + 2 = 1$$

$$z = \frac{D_z}{D} = \frac{\begin{vmatrix} 1 & -2 & 3 \\ 1 & 1 & 2 \\ 2 & -3 & 5 \end{vmatrix}}{1} = \frac{\begin{vmatrix} 1 & -2 & 3 \\ 0 & 3 & -1 \\ 0 & 1 & -1 \end{vmatrix}}{1}$$

$$= \begin{vmatrix} 3 & -1 \\ 1 & -1 \end{vmatrix} = -2$$

Actually, once x and y are known, z may be found by substitution in one of the original equations.

Thus, $\boldsymbol{(x, y, z) = (-1, 1, -2)}$. □

The circle $x^2 + y^2 + Ax + By + C = 0$ is determined by three points. If you substitute each ordered pair into the equation, you obtain a system of three equations in A, B, and C, which can be solved for the coefficients of the desired equation.

Example 3 Find the equation of the circle determined by the points (0, 1), (1, 0), and (2, 1).

Solution Substitute values into $x^2 + y^2 + Ax + By + C = 0$:

$$\begin{aligned} (0,1)&: \quad 0^2 + 1^2 + A \cdot 0 + B \cdot 1 + C = 0 \\ (1,0)&: \quad 1^2 + 0^2 + A \cdot 1 + B \cdot 0 + C = 0 \\ (2,1)&: \quad 2^2 + 1^2 + A \cdot 2 + B \cdot 1 + C = 0 \end{aligned} \quad \text{or} \quad \begin{cases} \phantom{2A +{}} B + C = -1 \\ A \phantom{{}+ B} + C = -1 \\ 2A + B + C = -5 \end{cases}$$

Solve the system:

$$\left[\begin{array}{ccc|c} 0 & 1 & 1 & -1 \\ 1 & 0 & 1 & -1 \\ 2 & 1 & 1 & -5 \end{array}\right]$$

$$D = \begin{vmatrix} 0 & 1 & 1 \\ 1 & 0 & 1 \\ 2 & 1 & 1 \end{vmatrix} = \begin{vmatrix} 0 & 1 & 0 \\ 1 & 0 & 1 \\ 2 & 1 & 0 \end{vmatrix} = -\begin{vmatrix} 0 & 1 \\ 2 & 1 \end{vmatrix} = 2$$

$$D_A = \begin{vmatrix} -1 & 1 & 1 \\ -1 & 0 & 1 \\ -5 & 1 & 1 \end{vmatrix} = \begin{vmatrix} -1 & 1 & 0 \\ -1 & 0 & 1 \\ -5 & 1 & 0 \end{vmatrix} = -\begin{vmatrix} -1 & 1 \\ -5 & 1 \end{vmatrix} = -4$$

Similarly, $D_B = -4$ and $D_C = 2$. Therefore,

$$A = \frac{D_A}{D} = \frac{-4}{2} = -2 \qquad B = \frac{D_B}{D} = -2 \qquad C = \frac{D_C}{D} = 1$$

The equation of the circle is $x^2 + y^2 - 2x - 2y + 1 = 0$. □

Example 4 Solve for x and y: $\begin{cases} \frac{2}{x} - \frac{6}{y} = 1 \\ \frac{4}{x} - \frac{3}{y} = 1 \end{cases}$

Solution First, let $a = 1/x$ and $b = 1/y$. Then the equations become

$$\begin{cases} 2a - 6b = 1 \\ 4a - 3b = 1 \end{cases}$$

Now employ Cramer's Rule:

$$a = \frac{\begin{vmatrix} 1 & -6 \\ 1 & -3 \end{vmatrix}}{\begin{vmatrix} 2 & -6 \\ 4 & -3 \end{vmatrix}} \qquad b = \frac{\begin{vmatrix} 2 & 1 \\ 4 & 1 \end{vmatrix}}{\begin{vmatrix} 2 & -6 \\ 4 & -3 \end{vmatrix}}$$

$$a = \frac{3}{18} \qquad b = \frac{-2}{18}$$

$$a = \frac{1}{6} = \frac{1}{x} \qquad b = -\frac{1}{9} = \frac{1}{y}$$

Thus, the solution is $x = 6$ and $y = -9$, or simply $\{(6, -9)\}$ □

Problem Set 7.7

A *Use Cramer's Rule to solve the systems of equations in Problems 1–20.*

1. $\begin{cases} 3x - 2y = 0 \\ 2x + 3y = 13 \end{cases}$ **2.** $\begin{cases} 2x - 5y = 1 \\ 3x - 8y = 2 \end{cases}$ **3.** $\begin{cases} 2x - y = 1 \\ x + y = 1 \end{cases}$

4. $\begin{cases} 3x - y = 0 \\ x + 5y = 4 \end{cases}$ **5.** $\begin{cases} x - 3y = 3 \\ 3x + 4y = 4 \end{cases}$ **6.** $\begin{cases} 2x + y = 2 \\ 4x + 3y = 3 \end{cases}$

7. $\begin{cases} 3x + 2y = -4 \\ 4x + 3y = -1 \end{cases}$

8. $\begin{cases} 3x + 2y = 1 \\ 4x + 3y = 3 \end{cases}$

9. $\begin{cases} x + \frac{1}{2}y - \frac{1}{2} = 0 \\ \frac{1}{3}y - \frac{1}{2}x = \frac{3}{2} \end{cases}$

10. $\begin{cases} \frac{1}{4}x - \frac{1}{3}y - 1 = 0 \\ \frac{5}{6}x + \frac{1}{3}y + 1 = 0 \end{cases}$

11. $\begin{cases} x - y = 4 \\ y + z = -2 \\ x + z = 2 \end{cases}$

12. $\begin{cases} x - z = 3 \\ x + y = 5 \\ y + z = 2 \end{cases}$

13. $\begin{cases} 3x + y - z = 1 \\ 4y + 2z = -1 \\ x - 3z = 0 \end{cases}$

14. $\begin{cases} x - y + z = 2 \\ 2x + 2z = 3 \\ y + 3z = 2 \end{cases}$

15. $\begin{cases} x - 2z = 1 \\ 2x + 3y + 3\cdot = 0 \\ 4x - 3y - 4z = 3 \end{cases}$

16. $\begin{cases} x - 3y + 3z = 1 \\ 3x + 2y - 5z = 4 \\ 4x + y - 6z = 5 \end{cases}$

17. $\begin{cases} 2x - y - 2z = 5 \\ 4x + y + 3z = 1 \\ 8x - y + z = 5 \end{cases}$

18. $\begin{cases} 2x + 3y + z + 1 = 0 \\ x - y + 2z = 6 \\ 3x + 2y - z = 1 \end{cases}$

19. $\begin{cases} 2x + 3y + 3 = 0 \\ 3x + 2y + 2 = 5z \\ 4z - 3y = 8 \end{cases}$

20. $\begin{cases} 2x + y = z + 1 \\ 3x + 2 = y + z \\ z - 3 + 4x = 2y \end{cases}$

B *In Problems 21–26 find the equation of the circle determined by the given points.*

21. (0, 2), (2, 0), (4, 0)

22. (0, 2), (0, −4), (−1, −1)

23. (0, 1), (1, 2), (2, 5)

24. (−1, 1), (0, 2), (1, 1)

25. (−1, 2), (0, 1), (1, −2)

26. (0, −3), (1, 0), (2, 1)

Solve the systems in Problems 27–30 with Cramer's Rule. [Hint: Let $a = 1/x$ and $b = 1/y$.]

27. $\begin{cases} \dfrac{3}{x} - \dfrac{2}{y} = -1 \\ \dfrac{7}{x} - \dfrac{6}{y} = -2 \end{cases}$

28. $\begin{cases} \dfrac{2}{x} - \dfrac{3}{y} = 3 \\ \dfrac{1}{x} + \dfrac{2}{y} = 5 \end{cases}$

29. $\begin{cases} \dfrac{3}{x} + \dfrac{4}{y} = 6 \\ \dfrac{5}{x} + \dfrac{3}{y} = -1 \end{cases}$

30. $\begin{cases} \dfrac{5}{x} + \dfrac{1}{y} = 3 \\ \dfrac{1}{2x} - \dfrac{2}{y} = 1 \end{cases}$

C *Solve the systems in Problems 31–40 with Cramer's Rule.*

31. $\begin{cases} \dfrac{1}{x} - \dfrac{1}{y} = 1 \\ \dfrac{1}{x} + \dfrac{1}{z} = 1 \\ \dfrac{1}{y} - \dfrac{1}{z} = 1 \end{cases}$

32. $\begin{cases} \dfrac{1}{x} + \dfrac{1}{y} = 2 \\ \dfrac{1}{x} - \dfrac{1}{z} = 1 \\ \dfrac{1}{z} - \dfrac{1}{y} = 1 \end{cases}$

33. $\begin{cases} \dfrac{1}{x} + \dfrac{1}{y} + \dfrac{1}{z} = 0 \\ \dfrac{2}{x} - \dfrac{3}{y} - \dfrac{2}{z} = 5 \\ \dfrac{3}{x} - \dfrac{2}{y} + \dfrac{1}{z} = -5 \end{cases}$

34. $\begin{cases} \dfrac{3}{x} - \dfrac{4}{y} - \dfrac{2}{z} = -1 \\ \dfrac{1}{x} + \dfrac{2}{y} + \dfrac{1}{z} = -2 \\ \dfrac{3}{y} + \dfrac{1}{2z} - \dfrac{2}{x} = 1 \end{cases}$

35. $\begin{cases} \dfrac{1}{x} - \dfrac{3}{z} = 0 \\ \dfrac{4}{y} + \dfrac{2}{z} = -1 \\ \dfrac{3}{x} + \dfrac{1}{y} - \dfrac{1}{z} = 1 \end{cases}$

36. $\begin{cases} \dfrac{1}{y} + \dfrac{3}{z} = 2 \\ \dfrac{2}{x} + \dfrac{2}{z} = 3 \\ \dfrac{1}{x} - \dfrac{1}{y} + \dfrac{1}{z} = 2 \end{cases}$

37. $\begin{cases} w + x + y = 0 \\ 2x + y + z = 0 \\ w + 4y - z = 3 \\ w - y - z = 0 \end{cases}$

38. $\begin{cases} w + x + y = 0 \\ w + x + z = 6 \\ w + 2y - z = 1 \\ 3x + y + 2z = 0 \end{cases}$

39. $\begin{cases} 2x + 2y - z + w = -2 \\ 3x - y + 3z = 0 \\ 3y - 2z - w = 9 \\ x + z - 2w = 9 \end{cases}$

40. $\begin{cases} x + y + z + w = -3 \\ 2x + 3z - 4w = 6 \\ 4z - 3w - 2y = 0 \\ 5y - 2z + 4w = -1 \end{cases}$

41. If r_1, r_2, r_3, and r_4 are the four roots of the equation $x^4 = 1$, then show that

$$\begin{vmatrix} r_1 & r_2 & r_3 & r_4 \\ r_2 & r_3 & r_4 & r_1 \\ r_3 & r_4 & r_1 & r_2 \\ r_4 & r_1 & r_2 & r_3 \end{vmatrix} = 0$$

7.8 Summary and Review

Describing biological or management processes very often involves the relationship of large numbers of variables. In a natural ecosystem, for example, it may be necessary to record the interaction of more than a hundred plant and animal species. Matrices form a useful, simple mathematical framework for such work. Determinants are defined as functions of square matrices and simplify the solution of linear systems of any order.

CHAPTER OBJECTIVES

At the conclusion of your study of this chapter, you should be able to:

1. Solve linear and certain nonlinear systems of two equations in two unknowns and characterize the solution as consistent, inconsistent, or dependent
2. Perform matrix addition, subtraction, and multiplication, as well as scalar multiplication when the matrices are conformable for the operation
3. Represent a linear system of equations as an augmented matrix or as a matrix equation

4. Use elementary row operations to determine the inverse of a matrix if it exists
5. Use the Gauss–Jordan method, matrix equation with inverse, or Cramer's Rule to solve a linear system, when applicable
6. Use systems, matrices, and determinants to solve certain applied problems
7. Expand a determinant of order 2 or use minors and cofactors to expand a determinant of order 3 or larger

Problem Set 7.8

CONCEPT PROBLEMS *Supply the word or words that make the statements in Problems 1–30 complete and correct.*

[7.1] **1.** The solution to a system of equations is the ______ of the solutions of the individual equations.

[7.1] **2.** The addition method requires that you manipulate the equations so that the coefficients of one of the variables are ______ .

[7.1] **3.** The ______ method requires that one of the equations be solved for one of the variables.

[7.1] **4.** If the solution set of a system is empty, the equations are ______ .

[7.1] **5.** If the solution set of a system is infinite, the equations are ______ .

[7.2] **6.** A matrix of dimension $m \times n$ has ______ rows and ______ columns.

[7.2] **7.** A matrix with the same number of rows and columns is called a(n) ______ matrix.

[7.2] **8.** Two matrices are equal if and only if they have the same ______ and ______ entries are equal.

[7.2] **9.** Two matrices are ______ for addition if they have the same dimension.

[7.2] **10.** To multiply a matrix by a real number (scalar), each ______ is multiplied by that number.

[7.2] **11.** If $A_{r \times s}$ and $B_{p \times q}$ are ______ for multiplication, then $s =$ ______ .

[7.2] **12.** If A is a matrix and $AI = IA = A$, then I is a(n) ______ matrix.

[7.2] **13.** If $AB = BA = I$, then B is the ______ of A for multiplication and is written ______ .

[7.2] **14.** $A \sim B$ if A can be changed into B by a finite number of ______ operations.

[7.3] **15.** The ______ method reduces an augmented matrix to the row-reduced form.

[7.3] **16.** In the row-reduced form of the augmented matrix $[A|B]$, A is a(n) ______ .

[7.3] **17.** The row-reduced form of a matrix has ______ on the main diagonal and ______ directly above and below.

[7.4] **18.** A(n) ______ matrix has an inverse.

[7.4] **19.** To solve $AX = B$, multiply both sides by ______ on the left, if it exists.

[7.5] **20.** The domain of the determinant function is the set of ______ .

[7.5] **21.** The order of a determinant is the number of ______ and ______ .

[7.5] **22.** The ______ of an entry in a determinant is obtained by ruling out all entries in its row and column.

[7.5] **23.** The ______ of an entry is its ______ with the proper sign affixed.

[7.6] **24.** If two rows or two columns of a determinant are ______ , its value is zero.

[7.6] **25.** The value of a determinant is zero if all the entries of a row or column are ______ .

[7.6] **26.** If a multiple of the entries of one row of a determinant is added to ________ , the value of the determinant is ________ .

[7.7] **27.** Cramer's Rule is a scheme for solving ________ .

[7.7] **28.** Using Cramer's Rule, D is called the ________ .

[7.7] **29.** Using Cramer's Rule, D_x is obtained by ________ .

[7.7] **30.** A system has a solution only if the determinant of the system is ________ .

REVIEW PROBLEMS

[7.1] *Solve each system in Problems 31–36 by the method indicated.*

31. By graphing:
$$\begin{cases} 2x - y = 2 \\ 3x - 2y = 1 \end{cases}$$

32. By addition:
$$\begin{cases} x + 3y = 3 \\ 4x - 6y = -6 \end{cases}$$

33. By substitution:
$$\begin{cases} x + y = 1 \\ x^2 + y^2 = 13 \end{cases}$$

34. By any suitable method:
$$\begin{cases} y = 4x - 8 \\ y = (x-2)^2 \end{cases}$$

35. By any suitable method:
$$\begin{cases} x^2 + y^2 = 25 \\ 7x - y = 25 \end{cases}$$

36. By any suitable method:
$$\begin{cases} y = 2x^2 - 12x + 18 \\ y + 6x = x^2 + 10 \end{cases}$$

[7.2] *Perform the indicated matrix operations, if possible, in Problems 37–42, where*

$$A = \begin{bmatrix} 1 & 0 \\ 2 & -1 \end{bmatrix} \quad B = \begin{bmatrix} 2 & -1 & 0 \\ 1 & 0 & 1 \end{bmatrix} \quad C = \begin{bmatrix} 2 & 0 \\ 1 & 2 \\ -1 & 1 \end{bmatrix} \quad D = \begin{bmatrix} 0 & 1 & 0 \\ -1 & 0 & 0 \\ 0 & 1 & -1 \end{bmatrix}$$

37. $CB - 3D$ **38.** $2A + BC$ **39.** BDC

40. ABC **41.** BAD **42.** CAB

[7.3] *Represent each system in Problems 43–50 as an augmented matrix, and solve by the Gauss–Jordan method.*

43. $\begin{cases} 2x + y = 1 \\ 3x + 2y = 4 \end{cases}$ **44.** $\begin{cases} 2x + y = 1 \\ 5x + 2y = 1 \end{cases}$ **45.** $\begin{cases} 2x + 3y = 1 \\ 4x - 6y = -5 \end{cases}$

46. $\begin{cases} 2x + y = 1 \\ x + z = 1 \\ 2z - y = 1 \end{cases}$ **47.** $\begin{cases} 2x + y = 1 \\ x + y + z = 0 \\ y + 2z = 0 \end{cases}$ **48.** $\begin{cases} x + y + z = 2 \\ x + 2y - 2z = 1 \\ x + y + 3z = 4 \end{cases}$

49. $\begin{cases} 2x + y = 1 \\ w + y - 3z = 0 \\ x + y - z = 0 \\ w + x - z = -1 \end{cases}$ **50.** $\begin{cases} w + x + y + z = 1 \\ w - 3x - y - z = 1 \\ w + 2x + y - z = -1 \\ w - x - y - 3z = -1 \end{cases}$

[7.3] **51.** Find the equation of the parabola of the form $y = ax^2 + bx + c$ containing the points (1, 1), (2, 0), and (4, 4).

[7.4] *Find the inverse of each matrix in Problems 52–57.*

52. $\begin{bmatrix} 1 & 2 \\ 3 & 4 \end{bmatrix}$ **53.** $\begin{bmatrix} -3 & 2 \\ 2 & -1 \end{bmatrix}$ **54.** $\begin{bmatrix} \frac{2}{3} & \frac{1}{2} \\ -\frac{3}{2} & -\frac{1}{2} \end{bmatrix}$

55. $\begin{bmatrix} 1 & 1 & 0 \\ 0 & 1 & -1 \\ 0 & 1 & 1 \end{bmatrix}$ **56.** $\begin{bmatrix} 7 & -3 & -3 \\ -1 & 1 & 0 \\ -1 & 0 & 1 \end{bmatrix}$ **57.** $\begin{bmatrix} 1 & 2 & 3 \\ 2 & 3 & 4 \\ 1 & 2 & 1 \end{bmatrix}$

[7.4] *Represent each system in Problems 58–63 in a matrix equation, and solve using an inverse matrix. Note that the required inverses were found in Problems 52–57.*

58. $\begin{cases} x + 2y = 3 \\ 3x + 4y = 3 \end{cases}$ **59.** $\begin{cases} 2y - 3x = 1 \\ 2x - y = 4 \end{cases}$

60. $\begin{cases} 2x + y = 3 \\ -\frac{3}{2}x - \frac{1}{2}y = 1 \end{cases}$ **61.** $\begin{cases} x + y = -1 \\ y - z = 1 \\ y + z = -2 \end{cases}$

62. $\begin{cases} 7x - 3y - 3z = -4 \\ y - x = 1 \\ z - x = 4 \end{cases}$ **63.** $\begin{cases} x + 2y + 3z = 1 \\ 2x + 3y + 4z = 2 \\ x + 2y + z = 2 \end{cases}$

[7.4] **64.** *Chemistry* Two commercial preparations, I and II, contain two ingredients, A and B, in the following proportions:

	A	B
I	70%	30%
II	40%	60%

How many grams of each product must be mixed to obtain 60 g of a preparation that contains the ingredients in equal parts?

[7.4] **65.** *Business* A manufacturer of auto accessories uses three basic parts, A, B, and C, in its three products, I, II, and III, in the following proportions:

	A	B	C
I	2	1	1
II	2	2	1
III	3	2	2

The inventory shows 1250 of part A, 900 of part B, and 750 of part C on hand. How many of each product may be manufactured from the parts available?

[7.5] *Evaluate the determinants in Problems 66–79.*

66. $\begin{vmatrix} 7 & 5 \\ -3 & 4 \end{vmatrix}$ **67.** $\begin{vmatrix} -2 & 4 \\ 9 & 3 \end{vmatrix}$ **68.** $\begin{vmatrix} 6 & 5 \\ 9 & 8 \end{vmatrix}$

69. $\begin{vmatrix} .3 & .2 \\ .7 & .8 \end{vmatrix}$ **70.** $\begin{vmatrix} 12 & -11 \\ -13 & 14 \end{vmatrix}$ **71.** $\begin{vmatrix} \frac{1}{2} & \frac{1}{2} \\ \frac{1}{4} & \frac{3}{4} \end{vmatrix}$

72. $\begin{vmatrix} -1 & 1 & 0 \\ 1 & 0 & 2 \\ 3 & 1 & 0 \end{vmatrix}$ **73.** $\begin{vmatrix} 1 & 1 & -1 \\ 3 & -1 & 0 \\ 2 & -3 & 4 \end{vmatrix}$ **74.** $\begin{vmatrix} 1 & 1 & 1 \\ 2 & 1 & 1 \\ 1 & 1 & 2 \end{vmatrix}$

75. $\begin{vmatrix} -2 & 3 & 1 \\ 2 & -4 & 0 \\ 0 & 1 & 3 \end{vmatrix}$ **76.** $\begin{vmatrix} 3 & -2 & 0 \\ 5 & 4 & 1 \\ 3 & 0 & 1 \end{vmatrix}$ **77.** $\begin{vmatrix} 2 & 3 & -1 \\ 1 & 1 & 1 \\ 0 & 2 & -1 \end{vmatrix}$

78. $\begin{vmatrix} 0 & 1 & -1 & 1 \\ -1 & 3 & 0 & 1 \\ -1 & 0 & 2 & 1 \\ 2 & 1 & 1 & 0 \end{vmatrix}$

79. $\begin{vmatrix} 1 & 2 & -3 & -2 \\ 0 & 1 & 4 & -2 \\ 3 & -1 & 4 & 0 \\ 2 & 1 & 0 & 3 \end{vmatrix}$

[7.6] *Apply determinants to the solution of Problems 80–84.*

80. Find the area of $\triangle SIX$ if $S = (0, 0)$, $I = (3, 0)$, and $X = (3, -4)$.

81. Find the area of $\triangle NIL$ if $N = (1, -2)$, $I = (2, 1)$, and $L = (3, 4)$.

82. Show that the points $(1, -2)$, $(2, 1)$, and $(3, 4)$ are collinear.

83. Find the area of parallelogram $YAWN$, with vertices $Y = (-3, 3)$, $A = (-1, 6)$, $W = (5, -2)$, and $N = (3, -5)$.

84. Find the determinant equation of the line containing $(3, 1)$ and $\left(2, -\frac{1}{2}\right)$. Rewrite this equation in the slope–intercept form.

[7.7] *Solve each system in Problems 85–90 using Cramer's Rule.*

85. $\begin{cases} 7x + 5y = 1 \\ -3x + 4y = 18 \end{cases}$

86. $\begin{cases} 2x + 3y = 1 \\ 4x - 6y = -5 \end{cases}$

87. $\begin{cases} x + 2y = 3 \\ 3x + 4y = 3 \end{cases}$

88. $\begin{cases} 3x - 2y + 17 = 0 \\ 5x + 4y + z = 8 \\ 3x + z + 2 = 0 \end{cases}$

89. $\begin{cases} x + y + z = 2 \\ x + 2y - 2z = 1 \\ x + y + 3z = 4 \end{cases}$

90. $\begin{cases} x_1 + 2x_2 - x_4 = 0 \\ x_1 - 3x_3 - x_4 = 10 \\ x_2 - x_1 + x_3 = -2 \\ 2x_4 - x_1 = 3 \end{cases}$

CHAPTER 8

Sequences, Series, and the Binomial Theorem

Johann Bernoulli (1667–1748)

. . . but [Johann Bernoulli] was quite aware of the importance of his theorem, for he boasts that by means of it he calculated . . . the sum of the 10th powers of the first thousand integers, and found it [the number] to be 91,409,924,241,424,243,424,241,924,242,500

George Chrystal
Algebra, Part II,
Edinburgh, 1879

Historical Note

The remarkable Bernoulli family of Switzerland produced at least eight noted mathematicians over three generations. Two brothers, Jakob (1654–1705) and Johann, were bitter rivals. In spite of their disagreements, they maintained continuous communication with each other and with Leibniz. The brothers were extremely influential advocates of the newly born calculus. Jakob is credited with the development of polar coordinates and authored the first book devoted entirely to probability. Johann was the most prolific of the clan, and was responsible for the discovery of Bernoulli numbers, Bernoulli polynomials, the lemniscate of Bernoulli, the Bernoulli equation, the Bernoulli theorem, and the Bernoulli distribution. He was jealous and cantankerous; he tossed a son (Daniel) out of the house for winning an award he had expected to win himself. He was, however, the premier teacher of his time. Euler was his most famous pupil and was a close lifelong friend of his sons. The first calculus textbook was actually a collection of his lecture notes, published in 1696. Three of his sons, Nicholaus (1695–1736), Daniel (1700–1782), and Johann II (1710–1790), were noted mathematicians. Daniel was awarded the coveted prize of the French Academy no less than ten times, and his brother Johann three times. Daniel and Nicholaus served at the St. Petersburg Academy in Russia, and were responsible for Euler coming there in 1727. A nephew, Nicholaus II (1687–1759), was Chair of Mathematics at Padua, a position once held by Galileo. Johann II's sons Johann III (1744–1807), Daniel II (1751–1834), and Jakob II (1759–1789), Daniel II's son Christoph (1782–1863), and Christoph's son Johann Gustav (1811–1863) continued the dynasty through the 19th century.

Curiously, Jakob's father opposed his study of mathematics and science and tried to force him into theology. He chose the motto *Invito patre sidera verso,* which means "I study the stars against my father's will."

Chapter Overview

Sequences of letters, symbols, or numbers are the basis of pattern recognition. Sequences are defined mathematically in this chapter, and several special sequences are studied. The sum of the numbers of a sequence—a series—is similarly explored, and special notation is introduced to simplify the work. The limit of a series as the number of terms increases is of particular interest, and will be the subject of study in the calculus. Finally, the sum of the terms of a binomial raised to a power, $(a + b)^n$, is considered. The result, the Binomial Theorem, has wide application, particularly in probability and statistics.

8.1 Infinite Sequences

Proofs and patterns are both important in mathematics, and in this chapter they are blended together in the study of special kinds of functions called *sequences*, or *progressions*. Sequences of numbers are frequently encountered in mathematics. For instance, the numbers or *terms*

$$1, 3, 5, 7, 9, 11$$
$$1, 2, 3, \ldots, 99, 100$$

form sequences. These are *finite sequences* because there is a first and a last number. On the other hand, the sequences

$$0, 2, 4, 6, 8, \ldots$$
$$\ldots, -4, -3, -2, -1, 0$$

are *infinite sequences* because the three dots indicate that there is no last number. If the pattern of the sequence is not random, but specified so that successive terms are unique, then the rule for forming a sequence is a function.

FINITE SEQUENCE

A **finite sequence** is a function whose domain is the subset of counting numbers,

$$\{1, 2, 3, \ldots, n\}$$

INFINITE SEQUENCE

An **infinite sequence** is a function whose domain is the set of counting numbers,

$$N = \{1, 2, 3, \ldots, n, \ldots\}$$

Perhaps the simplest of the many special types of sequences is the sequence of counting numbers themselves:

$1, 2, 3, 4, 5, 6, \ldots$

This sequence is formed by adding 1 to each term to obtain the next term. Your assignments in this course may well have been a sequence of problems identified by their numbers. "Do the odd-numbered problems from 1 to 19, inclusive," that is, problems

$1, 3, 5, 7, 9, 11, 13, 15, 17, 19$

Like the counting numbers, this sequence is formed by adding the same number to each term—in this case 2. Sequences obtained in this way are called *arithmetic sequences*.

ARITHMETIC SEQUENCE

> An **arithmetic sequence** is a sequence whose consecutive terms differ by the same real number, called the **common difference.**

Example 1 Fill in the blank: $1, 4, 7, 10, 13, \_\_, \ldots$

Solution

$$4 - 1 = 3$$
$$7 - 4 = 3$$
$$10 - 7 = 3$$
$$13 - 10 = 3$$

First, verify that there is a common difference. If a number has been added in each case, then subtracting consecutive terms will identify that number.

The common difference is 3.

To find subsequent terms in an arithmetic sequence, simply add the common difference to the preceding term. The next term is

$$13 + 3 = \mathbf{16}$$

□

Example 2 Fill in the blank: $20, 14, 8, 2, -4, -10, \_\_, \ldots$

Solution The common difference is -6. Be careful: Do not just check the first one or two differences, they must *all* be -6 in order for the sequence of numbers to be an arithmetic sequence. The next term is found by adding the common difference:

$$-10 + (-6) = \mathbf{-16}$$

□

Example 3 Fill in the blank: $a_1, a_1 + d, a_1 + 2d, a_1 + 3d, a_1 + 4d, \_\_, \ldots$

Solution The common difference is d, and the next term is

$$(a_1 + 4d) + d = \boldsymbol{a_1 + 5d}$$

□

Another common type of sequence is the *geometric sequence*. If, instead of adding the same number, each term is *multiplied* by that number to obtain successive terms, a geometric sequence is formed.

GEOMETRIC SEQUENCE

> A **geometric sequence** is a sequence whose consecutive terms have the same quotient, called the **common ratio.**

If the sequence is geometric, the number obtained by dividing any term into the following term of that sequence will be the same nonzero number. No term of a geometric sequence may be zero.

Example 4 Fill in the blank: 2, 4, 8, 16, 32, __ , . . .

Solution

$$\frac{4}{2} = 2$$

First, verify that there is a common ratio. If a number has been multiplied in each case, then dividing consecutive terms will identify the number.

$$\frac{8}{4} = 2$$

$$\frac{16}{8} = 2$$

$$\frac{32}{16} = 2$$

The common ratio is 2.

To find subsequent terms in a geometric sequence, multiply the common ratio and the preceding term. The next term is

$$32(2) = \mathbf{64}$$ □

Example 5 Fill in the blank: $10, 5, \frac{5}{2}, \frac{5}{4}, \frac{5}{8}$, __ , . . .

Solution The common ratio is $\frac{1}{2}$ (be sure to check *each* ratio). The next term is

$$\frac{5}{8}\left(\frac{1}{2}\right) = \mathbf{\frac{5}{16}}$$ □

Example 6 Fill in the blank: $g_1, g_1r, g_1r^2, g_1r^3, g_1r^4$, __ , . . .

Solution The common ratio is r, and the next term is

$$(g_1r^4)r = \mathbf{g_1r^5}$$ □

Even though our attention will be focused on arithmetic and geometric sequences, it is important that you realize that there can be other types of sequences. The sequences in Examples 7–9 are neither arithmetic nor geometric, but each does have a pattern.

Example 7 Fill in the blanks: 1, 1, 2, 3, 5, 8, 13, __ , __ , __ , __ , . . .

Solution First, check to see if the sequence is arithmetic:

$$1 - 1 = 0$$
$$2 - 1 = 1$$

Not the same, so not arithmetic.

Next, check to see if the sequence is geometric:

$$\frac{1}{1} = 1$$
$$\frac{2}{1} = 2$$

Not the same, so not geometric.

Look for another pattern. Notice

$$1 + 1 = 2$$
$$1 + 2 = 3$$
$$2 + 3 = 5$$
$$3 + 5 = 8$$
$$5 + 8 = 13$$

This sequence is called the *Fibonacci sequence,* named after the 13th century mathematician Leonardo Fibonacci. The sequence has a surprising number of applications.

It appears that each term is obtained by adding the two preceding terms:

$$8 + 13 = 21$$
$$13 + 21 = \mathbf{34}$$
$$21 + \mathbf{34} = \mathbf{55}$$
$$\mathbf{34} + \mathbf{55} = 89$$

The next four terms are 21, 34, 55, and 89. □

Example 8 Fill in the blanks: 9, 6, 4, 3, __, __, __, __, . . .

Solution First, check to see if the sequence is arithmetic:

$$6 - 9 = -3$$
$$4 - 6 = -2$$

Not the same, so not arithmetic.

Next, check to see if it is geometric:

$$\frac{6}{9} = \frac{2}{3}$$
$$\frac{4}{6} = \frac{2}{3}$$
$$\frac{3}{4} = \frac{3}{4}$$

Not the same, so not geometric.

Look for another pattern. Notice the differences:

$$6 - 9 = -3$$
$$4 - 6 = -2$$
$$3 - 4 = -1$$

It looks like these differences are increasing by 1. Since this pattern fits all the given information, it is acceptable.

The pattern is:

$$9 + (-3) = 6$$
$$6 + (-2) = 4$$
$$4 + (-1) = 3$$
$$3 + (0) = 3$$
$$3 + (1) = 4$$
$$4 + (2) = 6$$
$$6 + (3) = 9$$

The next four terms are 3, 4, 6, and 9. □

Example 9 Fill in the blanks: 1, 2, 1, 1, 2, 1, 1, ___, ___, ___, ___, . . .

Solution This sequence is neither arithmetic nor geometric. As you study this pattern, you may notice that the pattern of the first few terms need not specify a unique sequence.

FIRST POSSIBLE PATTERN: one 1; 2; two 1s; 2; three 1s; 2; four 1s; . . .

In this case, the next four terms are 1, 2, 1, and 1.

SECOND POSSIBLE PATTERN: 1, 2, 1 repeated over and over.

In this case, the next four terms are 2, 1, 1, and 2. □

Example 9 shows that if a finite number of terms of a sequence are given, it cannot be assumed that the sequence is uniquely determined. In order to specify a sequence completely, a formula is usually given. This formula, called the **general term,** employs a new notation for functions which is used in connection with sequences. Remember that the domain is the set of counting numbers, so we might define a sequence by

$$s(n) = 3n - 2 \qquad \text{where } n \text{ is } 1, 2, 3, 4, \ldots$$

Thus,

$$s(1) = 3(1) - 2 = 1$$
$$s(2) = 3(2) - 2 = 4$$
$$s(3) = 3(3) - 2 = 7$$
$$\vdots$$

However, instead of writing $s(1)$ it is customary to write s_1; in place of $s(2)$ use s_2; and so on. In place of $s(n)$ write s_n, the *general term*. Thus, s_{15} means the 15th term of the sequence and you can find s_{15} in the same fashion as if you used the notation $s(15)$:

$$s_{15} = 3(15) - 2 = 43$$

Examples 10–14 show how to generate the first few terms of the sequence when you are given the general term.

Example 10 If $s_n = 26 - 6n$, find s_1, s_2, s_3, and s_4.

Solution

$$s_1 = 26 - 6(1) = 20$$
$$s_2 = 26 - 6(2) = 14$$
$$s_3 = 26 - 6(3) = 8$$
$$s_4 = 26 - 6(4) = 2$$

The sequence is: 20, 14, 8, 2, . . . □

Example 11 If $s_n = (-1)^n n^2$, find the first four terms.

Solution

$$s_1 = (-1)^1(1)^2 = -1$$
$$s_2 = (-1)^2(2)^2 = 4$$

$s_3 = (-1)^3(3)^2 = -9$

$s_4 = (-1)^4(4)^2 = 16$

The sequence is: $-1, 4, -9, 16, \ldots$ □

Example 12 If $s_n = s_{n-1} + s_{n-2}$, $n \geq 3$, where $s_1 = 1$ and $s_2 = 1$, find s_3 and s_4.

Solution

$s_1 = 1$ Given.

$s_2 = 1$ Given.

$s_3 = s_2 + s_1$

$= 1 + 1$ By substitution.

$= 2$

$s_4 = s_3 + s_2$

$= 2 + 1$

$= 3$

The sequence is: 1, 1, 2, 3, . . . (Compare with Example 7.) □

Example 13 If $s_n = 2n$, find the first four terms.

Solution $s_1 = 2, \quad s_2 = 4, \quad s_3 = 6, \quad s_4 = 8$

The sequence is: 2, 4, 6, 8, . . . □

Example 14 If $s_n = 2n + (n - 1)(n - 2)(n - 3)(n - 4)$, find the first four terms.

Solution

$s_1 = 2(1) + 0 = 2$

$s_2 = 2(2) + 0 = 4$

$s_3 = 2(3) + 0 = 6$

$s_4 = 2(4) + 0 = 8$

The sequence is: 2, 4, 6, 8, . . . □

Notice in Examples 13 and 14 that the first four terms of each sequence found from the general term are the same; however, the fifth terms are different. For Example 13, $s_5 = 10$, but for Example 14,

$$s_5 = 2(5) + (5 - 1)(5 - 2)(5 - 3)(5 - 4)$$
$$= 10 + (4)(3)(2)(1)$$
$$= 34$$

Problem Set 8.1

A ***a.*** *Classify the sequences in Problems 1–30 as arithmetic, geometric, both, or neither.*

b. *If arithmetic, give d; if geometric, give r; if neither, state a pattern using your own words.*

c. *Supply the next term.*

1. 2, 4, 6, ___ , . . .
2. 2, 4, 8, ___ , . . .
3. 5, 15, 25, ___ , . . .
4. 1, 5, 25, ___ , . . .
5. 9, 3, 1, ___ , . . .
6. 1, 3, 9, ___ , . . .
7. 3, 4, 2, 5, 1, ___ , . . .
8. 3, 4, 7, 11, 18, ___ , . . .
9. 21, 20, 18, 15, 11, ___ , . . .
10. 8, 6, 7, 5, 6, 4, ___ , . . .
11. 2, 5, 8, 11, 14, ___ , . . .
12. 3, 6, 12, 24, 48, ___ , . . .
13. 5, −15, 45, −135, 405, ___ , . . .
14. 10, 10, 10, ___ , . . .
15. −8, −8, −8, ___ , . . .
16. 5, −5, −15, −25, −35, ___ , . . .
17. 2, 4, 6, 10, ___ , . . .
18. 3, 6, 9, 15, ___ , . . .
19. 97, 86, 75, 64, ___ , . . .
20. 100, 99, 97, 94, 90, ___ , . . .
21. 1, 8, 27, 64, 125, ___ , . . .
22. 8, 12, 18, 27, ___ , . . .
23. $3^2, 3^5, 3^8, 3^{11}$, ___ , . . .
24. $4^5, 4^4, 4^3, 4^2$, ___ , . . .
25. 3, 7, 3, 7, 7, 3, 7, 7, 7, ___ , . . .
26. 5, 2, 5, 5, 2, 2, 5, 5, ___ , . . .
27. $\frac{1}{2}, \frac{1}{3}, \frac{2}{3}, \frac{1}{4}, \frac{3}{4}, \frac{1}{5}, \frac{2}{5}, \frac{3}{5}, \frac{4}{5}, \frac{1}{6}$, ___ , . . .
28. $\frac{1}{10}, \frac{1}{5}, \frac{3}{10}, \frac{2}{5}, \frac{1}{2}$, ___ , . . .
29. $\frac{4}{3}, 2, 3, 4\frac{1}{2}$, ___ , . . .
30. $\frac{7}{12}, \frac{2}{3}, \frac{3}{4}, \frac{5}{6}$, ___ , . . .

B *a.* *Find the first three terms of the sequences whose nth terms are given in Problems 31–50.*

b. *Classify the sequence as arithmetic (give d), geometric (give r), both, or neither.*

31. $s_n = 4n - 3$
32. $s_n = -3 + 3n$
33. $s_n = 10n$
34. $s_n = 2 - n$
35. $s_n = 7 - 3n$
36. $s_n = 10 - 10n$
37. $s_n = \dfrac{2}{n}$
38. $s_n = 1 + \dfrac{1}{n}$
39. $s_n = \dfrac{n-1}{n+1}$
40. $s_n = \frac{1}{2}n(n+1)$
41. $s_n = \frac{1}{6}n(n+1)(2n+1)$
42. $s_n = \frac{1}{4}n^2(n+1)^2$
43. $s_n = 6(2)^{n-1}$
44. $s_n = \dfrac{10}{2^{n-1}}$
45. $s_n = 3(2)^{1-n}$
46. $s_n = (-1)^n$
47. $s_n = (-1)^n(n+1)$
48. $s_n = (-1)^{n-1}n$
49. $s_n = -5$
50. $s_n = \frac{2}{3}$

Find the term in Problems 51–60.

51. Find the 15th term of the sequence $s_n = 4n - 3$.
52. Find the 12th term of the sequence $s_n = 10^n - 1$.
53. Find the 102nd term of the sequence $s_n = -3 + 3n$.
54. Find the 69th term of the sequence $s_n = 7 - 3n$.
55. Find the 8th term of the sequence $s_n = 3(2)^{1-n}$.

56. Find the 10th term of the sequence $s_n = \dfrac{10}{2^{n-1}}$.

57. Find the 49th term of the sequence $s_n = 1 + \dfrac{(-1)^n}{n}$.

58. Find the 20th term of the sequence $s_n = (-1)^n(n + 1)$.

59. Find the 3rd term of the sequence $s_n = (-1)^{n+1}5^{n+1}$.

60. Find the 2nd term of the sequence $s_n = (-1)^{n-1}7^{n-1}$.

C **61.** Find the first five terms of the sequence where $s_1 = 2$ and $s_n = 3s_{n-1}$, $n \geq 2$.

62. Find the first five terms of the sequence where $s_1 = 3$ and $s_n = \frac{1}{3}s_{n-1}$, $n \geq 2$.

63. Find the first five terms of the sequence where $s_1 = 1$, $s_2 = 1$, and $s_n = s_{n-1} + s_{n-2}$, $n \geq 3$.

64. Find the first five terms of the sequence where $s_1 = 1$, $s_2 = 2$, and $s_n = s_{n-1} + s_{n-2}$, $n \geq 3$.

65. Find an 8th term of the sequence: 1, 3, 4, 7, 11, 18, 29, . . .

66. Find a 5th term of the sequence: 225, 625, 1225, 2025, . . .

In Problems 67–69 fill in the blanks so that

— , 8, — , — , 27, — , . . .

is each of the following:

67. An arithmetic sequence.

68. A geometric sequence.

69. A sequence that is neither arithmetic nor geometric, for which you are able to write a general term.

8.2 Arithmetic Sequences and Series

Let us now concentrate on *arithmetic* sequences and, for convenience, denote the terms of an arithmetic sequence by $a_1, a_2, a_3, a_4, \ldots, a_n, \ldots$. The a will remind us that the sequence is *a*rithmetic. Recall that an arithmetic sequence has a common difference between successive terms, denoted by d.

$$a_n - a_{n-1} = d \qquad \text{for every } n > 1.$$

If this is written as

$$a_n = a_{n-1} + d$$

and $n = 2$, then

$$a_2 = a_1 + d \qquad \text{One difference } d \text{ is added for the second term.}$$

If $n = 3, 4, \ldots$, then

$$a_3 = a_2 + d$$
$$= (a_1 + d) + d$$

$= a_1 + 2d$ $\quad$ $2d$ for the 3rd term

$a_4 = a_3 + d$

$= (a_1 + 2d) + d$

$= a_1 + 3d$ $\quad$ $3d$ for the 4th term

$\vdots$

$\mathbf{a_n = a_1 + (n - 1)d}$ $\quad$ In general, $(n - 1)d$ for the nth term.

This formula gives us the nth term when the first term and common difference are known.

GENERAL TERM OF AN ARITHMETIC SEQUENCE

> For an arithmetic sequence $a_1, a_2, a_3, a_4, \ldots, a_n, \ldots$ with common difference d:
>
> $\mathbf{a_n = a_1 + (n - 1)d}$ $\quad$ for every $n \geqslant 1$

Example 1 Find an expression for the general term of the arithmetic sequence: 18, 14, 10, 6, . . .

Solution Here, $a_1 = 18$. The common difference is $d = 14 - 18 = -4$. Next, use the formula:

$$a_n = 18 + (n - 1)(-4)$$
$$= 18 - 4n + 4$$
$$\mathbf{a_n = 22 - 4n}$$

□

Example 2 $a_1 = 5$, $d = 3$, $a_{10} = ?$

Solution Find a_n, where $n = 10$.

$$a_n = a_1 + (n - 1)d$$
$$a_{10} = 5 + (10 - 1)3$$
$$a_{10} = 5 + 27$$
$$\mathbf{a_{10} = 32}$$

□

Example 3 $a_1 = -4$, $a_{12} = 51$, $d = ?$

Solution Find the common difference.

$$a_n = a_1 + (n - 1)d$$
$$a_{12} = a_1 + (12 - 1)d$$
$$51 = -4 + 11d$$
$$55 = 11d$$
$$\mathbf{5 = d}$$

□

Example 4 $a_1 = 43, a_{10} = 7, a_{25} = ?$

Solution First, find the common difference.

$$a_{10} = a_1 + (10 - 1)d$$
$$7 = 43 + 9d$$
$$-36 = 9d$$
$$-4 = d$$

Now, find the 25th term.

$$a_{25} = 43 + 24(-4)$$
$$a_{25} = 43 - 96$$
$$\mathbf{a_{25} = -53}$$

□

If the terms of a sequence are added, the expression is called a **series.** Since an infinite sequence contains infinitely many terms, a finite sequence is considered first, and the indicated sum of its terms is called a **finite series.**

DEFINITION OF A FINITE SERIES

The indicated sum of the terms of a finite sequence $s_1, s_2, s_3, \ldots, s_n$ is called **finite series** and is denoted by

$$s_1 + s_2 + s_3 + \cdots + s_n$$

In Examples 5–7 consider the sequence 1, 4, 7, 10, ... Since this sequence is arithmetic, the sum is denoted by A_n. That is,

$$A_n = a_1 + a_2 + a_3 + \cdots + a_n$$

Example 5 $A_1 = \mathbf{1}$ □

Example 6 $A_4 = 1 + 4 + 7 + 10$

$= \mathbf{22}$ □

Example 7 $A_{10} = 1 + 4 + 7 + 10 + 13 + 16 + 19 + 22 + 25 + 28$

$= \mathbf{145}$ Brute force addition □

There is something bothersome about this given solution to Example 7. Suppose you needed to find A_{50} or A_{100}? Consider the following alternate solution to Example 7:

$$A_{10} = 1 + 4 + 7 + \cdots + 22 + 25 + 28$$

Also,

$$A_{10} = 28 + 25 + 22 + \cdots + 7 + 4 + 1$$

Add these two equations term-by-term:

$$A_{10} + A_{10} = (1 + 28) + (4 + 25) + (7 + 22) + \cdots + (22 + 7) + (25 + 4) + (28 + 1)$$
$$2A_{10} = 29 + 29 + 29 + \cdots + 29 + 29 + 29$$

Notice that the sums are *all* equal to the sum of the first and last terms. The only additional information necessary to find the sum *easily* is the *number of terms*. Since this is A_{10}, there are 10 terms. Thus,

$$\begin{aligned} 2A_{10} &= 10(29) \\ A_{10} &= \frac{10(29)}{2} \\ &= 145 \end{aligned}$$

Example 8 Find A_{100} for the series considered in Example 7 using the procedure outlined above.

Solution The first term is 1; the 100th term is found by using the formula

$$\begin{aligned} a_n &= a_1 + (n - 1)d \\ a_{100} &= 1 + 99(3) && \text{Since } n = 100, a_1 = 1; \text{and } d = 3 \\ &= 298 \\ A_{100} &= \frac{100(1 + 298)}{2} \\ &= \mathbf{14{,}950} \end{aligned}$$

□

In general, consider the **sum** A_n of the first n terms of any arithmetic sequence:

$$\begin{array}{lllllll} A_n = a_1 & + a_2 & + a_3 & + \cdots + a_{n-2} & + a_{n-1} & + a_n \\ \updownarrow & \updownarrow & \updownarrow & \updownarrow & \updownarrow & \updownarrow \\ A_n = a_n & + a_{n-1} & + a_{n-2} & + \cdots + a_3 & + a_2 & + a_1 \end{array}$$

Simply reverse the order of the terms.

If these equations are added term-by-term as shown, you obtain

$$2A_n = (a_1 + a_n) + (a_2 + a_{n-1}) + (a_3 + a_{n-2}) + \cdots + (a_{n-2} + a_3) + (a_{n-1} + a_2) + (a_n + a_1)$$

Now, you can rewrite each of the quantities within parentheses as follows:

$$\begin{aligned} a_1 + a_n &= a_1 + [a_1 + (n - 1)d] \\ &= 2a_1 + (n - 1)d \\ a_2 + a_{n-1} &= (a_1 + d) + [a_1 + (n - 2)d] \\ &= 2a_1 + (n - 1)d \longleftarrow && \text{Notice that this is the same result as for } a_1 + a_n. \\ a_3 + a_{n-2} &= (a_1 + 2d) + [a_1 + (n - 3)d] \\ &= 2a_1 + (n - 1)d \longleftarrow && \text{We got the same result again!} \end{aligned}$$

There is a total of n such sums, so

$$2A_n = n[2a_1 + (n-1)d]$$

$$A_n = \frac{n}{2}[2a_1 + (n-1)d]$$

Or, since $2a_1 + (n-1)d = a_1 + [a_1 + (n-1)d] = a_1 + a_n$, you obtain

$$A_n = \frac{n}{2}(a_1 + a_n)$$

which is equivalent to n times the average of the first and last terms:

$$A_n = n\underbrace{\left(\frac{\overset{\text{First term}}{a_1} + \overset{\text{Last term}}{a_n}}{2}\right)}_{\text{Average of first and last terms}}$$

ARITHMETIC SERIES

> For an arithmetic sequence $a_1, a_2, a_3, a_4, \ldots, a_n$ with common difference d, the sum of the arithmetic series $a_1 + a_2 + a_3 + \cdots + a_n$ is
>
> $$\boldsymbol{A_n = \frac{n}{2}(a_1 + a_n)} \qquad \text{or, equivalently,} \qquad \boldsymbol{A_n = \frac{n}{2}[2a_1 + (n-1)d]}$$

Example 9 Find the sum of the first 100 even integers.

Solution The finite sequence is 2, 4, 6, . . . , 200; $a_1 = 2$, $a_{100} = 200$, and

$$A_{100} = \frac{100}{2}(2 + 200) \qquad \text{Use } A_n = \frac{n}{2}(a_1 + a_n) \text{ since the first and last terms are known.}$$

$$= 100(101)$$

$$= \mathbf{10{,}100}$$

□

Alternate Solution The last term need not be used, since it is known that $n = 100$, $a_1 = 2$, and $d = 2$. You can use the alternate formula

$$A_n = \frac{n}{2}[2a_1 + (n-1)d] \qquad \text{Since the first term and common difference are known.}$$

$$A_{100} = \frac{100}{2}[2(2) + (99)(2)]$$

$$= 50(4 + 198)$$

$$= \mathbf{10{,}100}$$

□

Example 10 Find the sum of the first 50 terms of the arithmetic sequence whose first term is -10 and whose common difference is 4.

Solution $a_1 = -10$, $n = 50$, and $d = 4$

$$A_n = \frac{n}{2}[2a + (n - 1)d]$$

$$A_{50} = \frac{50}{2}[2(-10) + 49(4)]$$

$$= 25(-20 + 196)$$

$$= \mathbf{4400}$$

□

Example 11 Find A_{20} when $a_1 = -10$ and $a_9 = -66$.

Solution First, find d and then a_{20}.

$$a_9 = a_1 + (9 - 1)d$$

$$-66 = -10 + (9 - 1)d$$

$$-56 = 8d$$

$$-7 = d$$

$$a_{20} = a_1 + (20 - 1)d$$

$$a_{20} = -10 + (19)(-7)$$

$$a_{20} = -143$$

Now, find the sum.

$$A_{20} = \frac{20}{2}(a_1 + a_{20})$$

$$A_{20} = \frac{20}{2}(-10 - 143)$$

$$A_{20} = 10(-153)$$

$$A_{20} = \mathbf{-1530}$$

□

Problem Set 8.2

A *In each of Problems 1–10 write out the first four terms of the arithmetic sequence whose first element is the given a_1 and whose common difference is the given d. Also write an expression for the general term.*

1. $a_1 = 5, \quad d = 4$

2. $a_1 = 85, \quad d = 3$

3. $a_1 = -15, \quad d = 8$

4. $a_1 = -22, \quad d = 7$

5. $a_1 = 100, \quad d = -5$

6. $a_1 = 20, \quad d = -4$

7. $a_1 = \frac{1}{2}, \quad d = \frac{3}{2}$

8. $a_1 = \frac{5}{3}, \quad d = -\frac{2}{3}$

9. $a_1 = 5, \quad d = x$

10. $a_1 = x, \quad d = y$

Find a_1 and d for each of the arithmetic sequences in Problems 11–20.

11. 3, 7, 11, 15, . . .

12. 2, 7, 12, 17, . . .

13. 6, 11, 16, . . .

14. 35, 46, 57, . . .

15. $-8, -1, 6, \ldots$

16. $-1, 1, 3, \ldots$

17. 151, 142, 133, . . .

18. 76, 89, 102, . . .

19. $x, 2x, 3x, \ldots$

20. $x - 5b, x - 3b, x - b$

B *Use the formula $a_n = a_1 + (n - 1)d$ to find an expression for the general term of each of the arithmetic sequences in Problems 21–30. Notice that you found a_1 and d for each of these sequences in Problems 11–20.*

21. 3, 7, 11, 15, . . .

22. 2, 7, 12, 17, . . .

23. 6, 11, 16, . . .

24. 35, 46, 57, . . .

25. $-8, -1, 6, \ldots$

26. $-1, 1, 3, \ldots$

27. 151, 142, 133, . . .

28. 76, 89, 102, . . .

29. $x, 2x, 3x, \ldots$

30. $x - 5b, x - 3b, x - b$

Find the missing quantities in Problems 31–50 for the given arithmetic sequences.

31. $a_1 = 6, \quad d = 5; \quad a_{20} = ?$

32. $a_1 = 35, \quad d = 11; \quad a_{10} = ?$

33. $a_1 = -30, \quad d = 4; \quad a_{12} = ?$

34. $a_1 = 19, \quad d = -8; \quad a_{10} = ?$

35. $a_1 = -5, \quad a_{30} = -63; \quad d = ?$

36. $a_1 = 4, \quad a_6 = 24; \quad d = ?$

37. $a_1 = \frac{1}{2}, \quad a_7 = \frac{13}{2}; \quad d = ?$

38. $a_1 = -\frac{7}{3}, \quad a_{12} = \frac{4}{3}; \quad d = ?$

39. $a_1 = 35, \quad d = 11; \quad A_{10} = ?$

40. $a_1 = -7, \quad d = -2; \quad A_{100} = ?$

41. $a_1 = 17, \quad a_{10} = -55; \quad A_{10} = ?$

42. $a_1 = -29, \quad a_8 = 20; \quad A_8 = ?$

43. $a_1 = 4, \quad a_6 = 24; \quad A_{15} = ?$

44. $a_1 = -5, \quad a_{30} = -63; \quad A_{10} = ?$

45. $a_1 = -8, \quad a_5 = 12; \quad A_{10} = ?$

46. $a_1 = 12, \quad a_5 = -8; \quad A_{10} = ?$

47. $a_1 = 100, \quad a_{11} = 30; \quad A_{25} = ?$

48. $a_1 = 125, \quad a_6 = 90; \quad A_{19} = ?$

49. $a_1 = -1, \quad a_7 = 2; \quad A_{100} = ?$

50. $a_1 = 2, \quad a_7 = -1; \quad A_{100} = ?$

51. Find the sum of the first 20 terms of the arithmetic sequence whose first term is 100 and whose common difference is 50.

52. Find the sum of the first 50 terms of the arithmetic sequence whose first term is -15 and whose common difference is 5.

53. Find the sum of the even integers between 41 and 99.

54. Find the sum of the odd integers between 100 and 80.

55. Find the sum of the odd integers between 48 and 136.

56. Find the sum of the first n odd integers.

57. Find the sum of the first n even integers.

58. How many blocks would be needed to build a stack like the one shown at the right if the bottom row had 28 blocks?

59. Repeat Problem 58 if the bottom row has 87 blocks.

60. Repeat Problem 58 if the bottom row has 100 blocks.

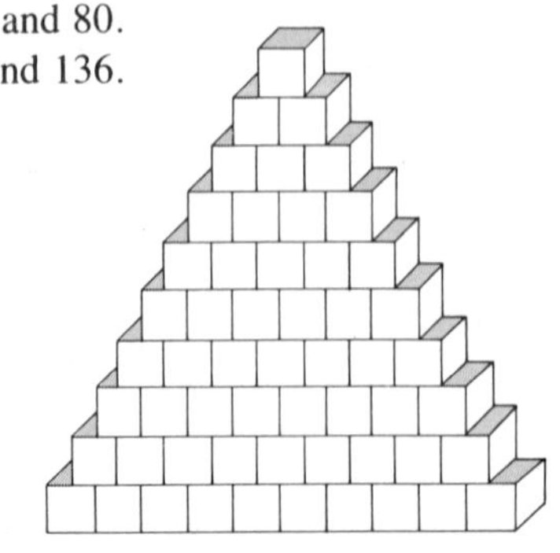

The following is needed in Problems 61–70.

Consider the arithmetic sequence s_1, x, s_2. The number x can be found by considering the system

$$\begin{cases} x = s_1 + d \\ x = s_2 - d \end{cases}$$

$$2x = s_1 + s_2 \qquad \text{By adding}$$

$$x = \frac{s_1 + s_2}{2}$$

Here, x is called the **arithmetic mean** between s_1 and s_2. Find the arithmetic mean between each of the given pairs of numbers.

61. 1, 8 **62.** 1, 7 **63.** $-5, 3$ **64.** 80, 88 **65.** 40, 56

66. 4, 20 **67.** 4, 15 **68.** $\frac{1}{2}, \frac{1}{3}$ **69.** $-10, -2$ **70.** $-\frac{2}{3}, \frac{4}{5}$

A sequence $s_1, s_2, \ldots, s_n$ is a ***harmonic sequence*** *if and only if its reciprocals form an arithmetic sequence. In Problems 71–75 which sequences are harmonic?*

71. $1, \frac{1}{2}, \frac{1}{3}, \frac{1}{4}, \frac{1}{5}, \ldots$ **72.** $\frac{1}{2}, \frac{1}{5}, \frac{1}{8}, \frac{1}{11}, \frac{1}{14}, \ldots$

73. $2, \frac{2}{3}, \frac{2}{5}, \frac{2}{7}, \ldots$ **74.** $\frac{1}{5}, -\frac{1}{5}, \frac{1}{15}, \frac{1}{25}, \ldots$

75. $\frac{3}{4}, \frac{1}{2}, \frac{1}{3}, \frac{2}{9}, \ldots$

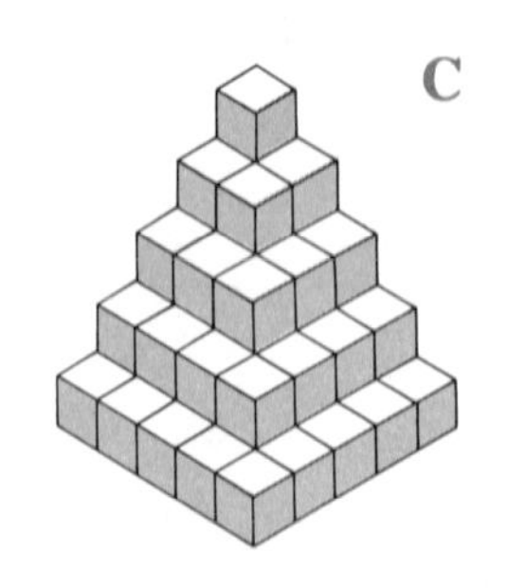

C **76.** **a.** How many blocks are there in the solid figure shown in the margin?
b. How many blocks are there in a similar figure with 50 layers?

77. *Business* Suppose you were hired for a job paying \$21,000 per year and were given the following options:
OPTION *A*: Annual salary increase of \$1440
OPTION *B*: Semiannual salary increase of \$360
Which is the better option?

78. Repeat Problem 77 for the following options:
OPTION *C*: Quarterly salary increase of \$90
OPTION *D*: Monthly salary increase of \$10

79. What are the differences in the amounts earned in the first year from options *A–D* in Problems 77 and 78?

80. Repeat Problem 79 for the first 2 years.

81. Write the arithmetic series for the total amount of money earned in 10 years under the options described in Problem 77:
a. OPTION *A* **b.** OPTION *B*

8.3 Geometric Sequences and Series

Let us now turn our attention to *geometric* sequences. For convenience, we denote the terms of a geometric sequence by $g_1, g_2, g_3, g_4, \ldots, g_n$. The g will remind us that the sequence is geometric. Recall that a geometric sequence has a common ratio, denoted by r:

$$\frac{g_n}{g_{n-1}} = r \qquad \text{or} \qquad g_n = g_{n-1}r, \qquad \text{for} \quad n > 1$$

We use a procedure that is similar to the development of the general term of an arithmetic sequence, and write

$$g_2 = g_1 r$$
$$g_3 = g_2 r = (g_1 r)r \;= g_1 r^2$$
$$g_4 = g_3 r = (g_1 r^2)r = g_1 r^3$$
$$\vdots$$
$$\mathbf{g_n = g_1 r^{n-1}}$$

Notice that the power of the ratio is just 1 less than the number of the term.

In general, the nth term contains r^{n-1}.

This formula gives the nth term when the first term and the common ratio are known.

GENERAL TERM OF GEOMETRIC SEQUENCE

For a geometric sequence $g_1, g_2, g_3, g_4, \ldots, g_n, \ldots$ with common ratio r,

$\mathbf{g_n = g_1 r^{n-1}}$ for every $n > 1$

Example 1 Write the general term for the geometric sequence: 200, 400, 800, . . .

Solution $g_1 = 200, r = \dfrac{400}{200} = 2$

$$g_n = g_1 r^{n-1}$$
$$= 200(2)^{n-1}$$
$$= 2 \cdot 10^2 \cdot 2^{n-1}$$
$$\mathbf{g_n = 2^n \cdot 10^2}$$

□

Example 2 $g_1 = 3, r = 2, g_{10} = ?$

Solution Find g_n, where $n = 10$.

$$g_n = g_1 r^{n-1}$$
$$g_{10} = 3 \cdot 2^{10-1}$$
$$= 3 \cdot 2^9$$
$$= 3 \cdot 512$$
$$= \mathbf{1536}$$

□

Example 3 $g_1 = 486, g_6 = 2, r = ?$

Solution Find the common ratio.

$$g_6 = g_1 r^5$$
$$2 = 486 \cdot r^5$$

$$\frac{1}{243} = r^5$$

$$\left(\frac{1}{3}\right)^5 = r^5$$

$$\mathbf{\frac{1}{3} = r}$$

□

Example 4 $g_1 = \frac{16}{27}, g_6 = -\frac{9}{2}, g_9 = ?$

Solution First, find the common ratio:

$$g_6 = g_1 r^5$$

$$-\frac{9}{2} = \frac{16}{27} r^5$$

$$-\frac{243}{32} = r^5$$

$$\left(-\frac{3}{2}\right)^5 = r^5$$

$$-\frac{3}{2} = r$$

Now, use $r = -\frac{3}{2}$ to find the 9th term:

$$g_9 = g_1 r^8$$

$$= \frac{16}{27}\left(-\frac{3}{2}\right)^8$$

$$= \frac{2^4}{3^3} \cdot \frac{3^8}{2^8}$$

$$= \frac{3^5}{2^4}$$

$$\mathbf{g_9 = \frac{243}{16}}$$

□

Now consider the sum G_n of the first n terms:

$$G_n = g_1 + g_1 r + g_1 r^2 + g_1 r^3 + \cdots + g_1 r^{n-1}$$

Multiply both sides by r:

$$rG_n = g_1 r + g_1 r^2 + g_1 r^3 + g_1 r^4 + \cdots + g_1 r^n$$

Notice that, except for the first and last terms, all the terms in the expressions for G_n and rG_n are the same, so that

$$G_n - rG_n = g_1 - g_1 r^n$$

Now solve for G_n:

$$(1 - r)G_n = g_1(1 - r^n)$$

$$G_n = \frac{g_1(1 - r^n)}{1 - r} \qquad r \neq 1$$

GEOMETRIC SERIES

For a geometric series $g_1 + g_2 + g_3 + g_4 + \cdots + g_n$ with common ratio $r \neq 1$,

$$\mathbf{G_n = \frac{g_1(1 - r^n)}{1 - r}}$$

Example 5 Find the sum of the first six terms of the geometric sequence with $g_1 = -3$ and $r = 2$.

Solution

$$G_n = \frac{g_1(1 - r^n)}{1 - r}$$

$$G_6 = \frac{(-3)(1 - 2^6)}{1 - 2}$$

$$= \frac{(-3)(-63)}{-1}$$

$$\mathbf{G_6 = -189}$$

□

Example 6 Find the sum of the first ten terms of the geometric sequence with $g_1 = \frac{1}{2}$ and $r = \frac{1}{2}$.

Solution

$$G_n = \frac{g_1(1 - r^n)}{1 - r}$$

$$G_{10} = \frac{\frac{1}{2}\left[1 - \left(\frac{1}{2}\right)^{10}\right]}{1 - \frac{1}{2}}$$

$$= 1 - \left(\frac{1}{2}\right)^{10}$$

$$= 1 - \frac{1}{1024}$$

$$\mathbf{G_{10} = \frac{1023}{1024}}$$

□

Example 7 $g_1 = \frac{9}{4}, g_4 = -\frac{1}{12}, G_7 = ?$

Solution First, find r:

$$g_4 = g_1 r^3$$

$$-\frac{1}{12} = \frac{9}{4} r^3$$

$$-\frac{1}{27} = r^3$$

$$-\frac{1}{3} = r$$

Now use $r = -\frac{1}{3}$ to find G_n for $n = 7$:

$$G_7 = \frac{g_1(1 - r^7)}{1 - r}$$

$$= \frac{\frac{9}{4}\left[1 - \left(-\frac{1}{3}\right)^7\right]}{1 - \left(-\frac{1}{3}\right)}$$

$$= \frac{\frac{9}{4}\left[1 + \frac{1}{2187}\right]}{\frac{4}{3}}$$

$$= \frac{27}{16}\left[\frac{2188}{2187}\right]$$

$$\mathbf{G_7 = \frac{547}{324}}$$

□

Example 8 $128 - 192 + 288 - \cdots + 1458 = ?$

Solution First, find the general term:

$$g_n = g_1 r^{n-1}$$

We need g_1 and r; $g_1 = 128$ by inspection, and

$$r = \frac{-192}{128} = -\frac{3}{2}.$$

Thus,

$$g_n = 128\left(-\frac{3}{2}\right)^{n-1}$$

Then determine the number of terms:

$$128\left(-\frac{3}{2}\right)^{n-1} = 1458$$

$$\left(-\frac{3}{2}\right)^{n-1} = \frac{1458}{128} = \frac{729}{64}$$

$$= \frac{3^6}{2^6}$$

$$\left(-\frac{3}{2}\right)^{n-1} = \left(-\frac{3}{2}\right)^6$$

$$n - 1 = 6$$

$$n = 7$$

Finally, compute the sum of the series:

$$G_n = \frac{g_1(1 - r^n)}{1 - r}$$

$$G_7 = \frac{128\left[1 - \left(-\frac{3}{2}\right)^7\right]}{1 - \left(-\frac{3}{2}\right)}$$

$$= \frac{2^7\left[1 + \dfrac{3^7}{2^7}\right]}{\frac{5}{2}}$$

$$= 2^7\left[\frac{2^7 + 3^7}{2^7}\right] \cdot \frac{2}{5}$$

$$= [2^7 + 3^7] \cdot \frac{2}{5}$$

$$= [2315] \cdot \frac{2}{5}$$

$$\mathbf{G_7 = 926}$$

□

Problem Set 8.3

A *In each of Problems 1–10 write out the first four terms of the geometric sequence whose first element is g_1 and whose common ratio is r. Also, write an expression for the general term.*

1. $g_1 = 5, \quad r = 3$

2. $g_1 = -12, \quad r = 3$

3. $g_1 = 1, \quad r = -2$

4. $g_1 = 1, \quad r = 2$

5. $g_1 = -15, \quad r = \frac{1}{5}$

6. $g_1 = 625, \quad r = -\frac{1}{5}$

7. $g_1 = 54, \quad r = -\frac{2}{3}$

8. $g_1 = 128, \quad r = \frac{3}{2}$

9. $g_1 = 8, \quad r = x$

10. $g_1 = x, \quad r = y$

Find g_1 and r for each of the geometric sequences in Problems 11–20.

11. 3, 6, 12, . . .

12. 7, 14, 28, . . .

13. $1, \frac{1}{2}, \frac{1}{4}, \ldots$

14. 100, 50, 25, . . .

15. 3125, −625, 125, . . .

16. $3, -1, \frac{1}{3}, -\frac{1}{9}, \ldots$

17. 100, 150, 225, . . .

18. 625, 250, 100, . . .

19. $x, x^2, x^3, \ldots$

20. $xyz, xy, \ldots$

B *Use the formula $g_n = g_1r^{n-1}$ to find an expression for the general term of each of the geometric sequences in Problems 21–30. Notice that you found g_1 and r for each of these sequences in Problems 11–20.*

21. 3, 6, 12, . . .

22. 7, 14, 28, . . .

23. $1, \frac{1}{2}, \frac{1}{4}, \ldots$

24. 100, 50, 25, . . .

25. 3125, −625, 125, . . .

26. $3, -1, \frac{1}{3}, -\frac{1}{9}, \ldots$

27. 100, 150, 225, . . .

28. 625, 250, 100, . . .

29. $x, x^2, x^3, \ldots$

30. $xyz, xy, \ldots$

Find the missing quantities in Problems 31–50 for the given geometric sequences.

31. $g_1 = 6, \quad r = 3; \quad g_5 = ?$

32. $g_1 = 5, \quad r = 2; \quad g_7 = ?$

33. $g_1 = 1024, \quad r = \frac{1}{2}; \quad g_9 = ?$

34. $g_1 = 100, \quad r = \frac{1}{10}; \quad g_{10} = ?$

35. $g_1 = 27, \quad g_4 = -1; \quad g_7 = ?$

36. $g_1 = 72, \quad g_4 = -\frac{1}{6}; \quad g_6 = ?$

37. $g_1 = 8, \quad g_3 = \frac{3}{2}; \quad g_5 = ?$

38. $g_1 = 9, \quad g_4 = \frac{8}{3}; \quad g_7 = ?$

39. $g_1 = 6, \quad r = 3; \quad G_5 = ?$

40. $g_1 = 5, \quad r = 2; \quad G_7 = ?$

41. $g_1 = 1024, \quad r = \frac{1}{2}; \quad G_9 = ?$

42. $g_1 = 100, \quad r = \frac{1}{10}; \quad G_{10} = ?$

43. $g_1 = 512, \quad r = \frac{3}{2}; \quad G_8 = ?$

44. $g_1 = 729, \quad r = -\frac{4}{3}; \quad G_7 = ?$

45. $g_1 = \frac{1}{3}, \quad r = -\frac{1}{3}; \quad G_5 = ?$

46. $g_1 = \frac{1}{2}, \quad r = -\frac{1}{2}; \quad G_6 = ?$

47. $g_1 = \frac{1}{8}, \quad g_8 = -16; \quad G_6 = ?$

48. $g_1 = 81, \quad g_5 = 16; \quad G_7 = ?$

49. $g_1 = 12, \quad g_4 = -\frac{4}{9}; \quad G_8 = ?$

50. $g_1 = 9, \quad g_4 = \frac{8}{3}; \quad G_9 = ?$

Find the sum of each series in Problems 51–60.

51. $5 + 15 + 45 + \cdots + 3645$

52. $-12 - 36 - 108 - \cdots - 2916$

53. $6561 + 4374 + 2916 + \cdots + 256$

54. $6561 - 4374 + 2916 - \cdots + 256$

55. $3125 + 2500 + 2000 + \cdots + 1024$

56. $3125 - 2500 + 2000 - \cdots - 1024$

57. $4096 - 3072 + 2304 - \cdots + 729$

58. $128 + 192 + 288 + \cdots + 2187$

59. $8 + 8x + 8x^2 + \cdots + 8x^{10}$

60. $x + xy + xy^2 + \cdots + xy^{20}$

61. A chain letter requires each person receiving the letter to send copies to six friends. If each person does as directed by the chain letter, what is the total mailed with five mailings? [*Hint:* The first mailing consists of six letters.]

62. Answer Problem 61 for 15 mailings.

63. According to the 1980 census, the U.S. population is 226,504,825. If everyone follows the directions in the chain letter (Problem 61), how many mailings would be necessary to include the *entire* U.S. population? [*Hint:* Assume that nobody receives a letter more than once.]

64. *Social science* According to the 1980 census, the population of Hawaii is about 965,000. If the population increases 12% every 5 years, what will the population be in the year 2000? (A 12% increase every 5 years means that the population each 5 years is 112% of the previous total.)

65. *Biology* A culture of bacteria increases by 100% every 24 hours. If the original culture contains one million bacteria, find the number of bacteria present after 10 days.

66. Use Problem 65 to find a formula for the number of bacteria present after d days.

C *The following is needed for Problems 67–71:*

Consider the geometric sequence g_1, x, g_2. The number x can be found by considering the equations

$$\frac{x}{g_1} = r \quad \text{and} \quad \frac{g_2}{x} = r$$

So,

$$\frac{x}{g_1} = \frac{g_2}{x}$$

$$x^2 = g_1 g_2$$

This equation has two solutions:

$$x = \sqrt{g_1 g_2} \quad \text{and} \quad x = -\sqrt{g_1 g_2}$$

If g_1 and g_2 are both positive, then $\sqrt{g_1 g_2}$ is called the ***geometric mean*** *of g_1 and g_2.*

If g_1 and g_2 are both negative, then $-\sqrt{g_1 g_2}$ is the geometric mean. Find the geometric mean of each of the given pairs of numbers.

67. 1, 8 **68.** 2, 8 **69.** $-5, -3$ **70.** $-10, -2$ **71.** 4, 20

72. Find three distinct numbers whose sum is 9 so that these numbers form an arithmetic sequence and their squares form a geometric sequence.

8.4 Infinite Series

The sum of the first n terms of a geometric series has been investigated. Sometimes it is also possible to determine the sum of an entire infinite geometric series. Suppose an infinite series

$$s_1 + s_2 + s_3 + s_4 + \cdots$$

is denoted by S. The **partial sums** for S are defined by

$$\begin{aligned} S_1 &= s_1 \\ S_2 &= s_1 + s_2 \\ S_3 &= s_1 + s_2 + s_3 \\ &\vdots \\ S_{10} &= s_1 + s_2 + s_3 + s_4 + s_5 + s_6 + s_7 + s_8 + s_9 + s_{10} \\ &\vdots \end{aligned}$$

Consider the infinite geometric series with $g_1 = \frac{1}{2}$ and $r = \frac{1}{2}$. The first few partial sums can be found as follows:

$$\begin{aligned} G_1 &= \frac{1}{2} \\ G_2 &= \frac{1}{2} + \frac{1}{4} = \frac{3}{4} \\ G_3 &= \frac{1}{2} + \frac{1}{4} + \frac{1}{8} = \frac{7}{8} \\ &\vdots \\ G_{10} &= \frac{1}{2} + \frac{1}{4} + \frac{1}{8} + \cdots + \frac{1}{1024} = \frac{1023}{1024} \end{aligned}$$

Does this series have a sum if you add *all* its terms? It does seem that as you take more terms of the series, the sum is closer and closer to 1. Graphically, you can see that if the terms are laid end-to-end as lengths on a number line, as shown in Figure 8.1, each term is half the remaining distance to 1.

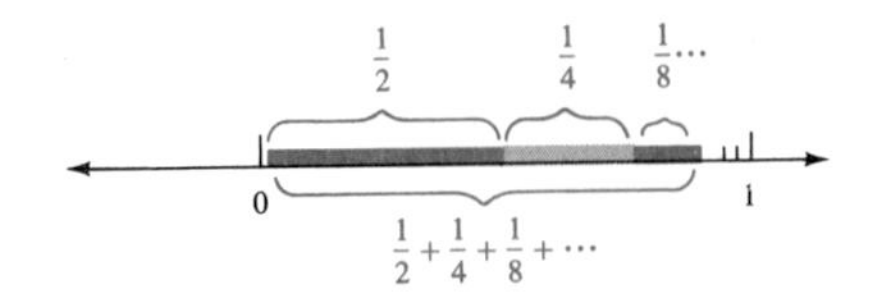

Figure 8.1

That is, $\frac{1}{2} + \frac{1}{4} = \frac{3}{4}$, and the next term is $\frac{1}{8}$, just half of the remaining $\frac{1}{4}$ distance to 1. However, the remaining length is shrinking quickly $\left(\frac{1}{2}, \frac{1}{4}, \frac{1}{8}, \frac{1}{16}, \frac{1}{32}, \ldots\right)$ and as n becomes large, $\left(\frac{1}{2}\right)^n$ will become and remain as close to zero as you please. We say that zero is the **limit** of $\left(\frac{1}{2}\right)^n$ as n increases without bound. Symbolically,

$$\lim_{n \to \infty} \left(\frac{1}{2}\right)^n = 0$$

You can reinforce this concept if you have a calculator. To calculate $\left(\frac{1}{2}\right)^n = .5^n$ for larger and larger values of n, see the steps shown in Table 8.1. It appears that the value of f is getting closer and closer to 0.

Table 8.1

n	$f(n) = .5^n$ (approximate value)
1	.5
10	9.77×10^{-4}
100	7.89×10^{-31}
1000	0*

*Perhaps your calculator shows ERROR for this step, meaning (in this case) that the number is too small (too close to 0) to calculate.

For each step, press

.5 y^x n =

Particular value for n. For example, for $n = 10$ the display is

.00097656
(or 9.7656 -04)

In general, it may be said that

$$\lim_{n \to \infty} r^n = 0 \qquad \text{if } |r| < 1$$

That is, any number r between -1 and 1 raised to large powers will approach 0 as a limit.

Now return to the original question of the sum of *all* the terms of a geometric series. Does G_n have a limit as n becomes large without bound?

$$G_n = \frac{g_1(1 - r^n)}{1 - r}$$

$$= \frac{g_1}{1 - r}(1 - r^n)$$

Notice that as n becomes large, r^n decreases and approaches 0 if $|r| < 1$. That is,

$$\lim_{n \to \infty} G_n = \lim_{n \to \infty} \left[\frac{g_1}{1 - r}(1 - r^n)\right]$$

$$= \frac{g_1}{1 - r}(1 - 0)$$

$$= \frac{g_1}{1 - r}$$

Thus, in this example, with $g_1 = \frac{1}{2}$ and $r = \frac{1}{2}$,

$$\lim_{n \to \infty} G_n = \frac{\frac{1}{2}}{1 - \frac{1}{2}}$$

$$= \frac{\frac{1}{2}}{\frac{1}{2}}$$

$$= 1$$

In general, we denote the limit of the sum of an infinite geometric series by $\lim_{n \to \infty} G_n$ and call this limit G.

INFINITE GEOMETRIC SERIES

If $G_n = g_1 + g_1 r + g_1 r^2 + g_1 r^3 + \cdots$, then

$$G = \lim_{n \to \infty} G_n = \frac{g_1}{1 - r} \qquad \text{for } |r| < 1$$

If $|r| \geq 1$, the infinite geometric series has no sum.

Example 1 $\frac{1}{3} + \frac{1}{9} + \frac{1}{27} + \frac{1}{81} + \cdots = ?$

Solution

$$G = \frac{g_1}{1 - r}$$

$$G = \frac{\frac{1}{3}}{1 - \frac{1}{3}} \qquad g_1 = \tfrac{1}{3} \text{ and } r = \tfrac{1}{3} \text{ by inspection.}$$

$$= \frac{\frac{1}{3}}{\frac{2}{3}}$$

$$= \mathbf{\frac{1}{2}}$$

□

Example 2 $2 + 4 + 6 + 8 + \cdots = ?$

Solution G does not exist; that is, this infinite geometric series has no sum because $r = 2$, which is greater than 1. □

Example 3 $100 + 50 + 25 + \cdots = ?$

Solution

$$G = \frac{g_1}{1 - r}$$

$$G = \frac{100}{1 - \frac{1}{2}} \qquad g_1 = 100 \text{ and } r = \tfrac{1}{2} \text{ by inspection.}$$

$$= \mathbf{200}$$

□

Example 4 $1 - \frac{2}{3} + \frac{4}{9} - \frac{8}{27} + - + \cdots = ?$

Solution

$$G = \frac{g_1}{1 - r}$$

$$G = \frac{1}{1 - \left(-\frac{2}{3}\right)} \qquad g_1 = 1 \text{ and } r = -\tfrac{2}{3}.$$

$$= \frac{1}{\frac{5}{3}}$$

$$= \mathbf{\frac{3}{5}}$$

□

Example 5 $-5 + 10 - 20 + \cdots = ?$

Solution Here, $g_1 = -5$ and $r = -2$. Since $|r| \geq 1$, this infinite series does not have a sum.

□

Infinite geometric series can be used to represent repeating decimals as quotients of two integers. For example, the repeating decimal .4444. . . is often denoted by $.\overline{4}$ and is a rational number. As such, it can be written as the quotient of two integers. In order to write it in this form, consider it written as an infinite series:

$$\begin{aligned} .4444\ldots &= .4 + .04 + .004 + .0004 + \cdots \\ &= .4 + .4(.1) + .4(.01) + .4(.001) + \cdots \\ &= .4 + .4(.1) + .4(.1)^2 + .4(.1)^3 + \cdots \end{aligned}$$

Notice that $g_1 = .4$ with $r = .1$, so

$$G = \frac{g_1}{1 - r}$$

$$\lim_{n \to \infty} G_n = \frac{.4}{1 - .1}$$

$$= \frac{.4}{.9}$$

$$= \frac{4}{9}$$

Thus, **.4444. . .** $= \mathbf{\frac{4}{9}}$.

Calculator check: 4  9 = DISPLAY: .444444444444

Example 6 Express .727272. . . as the quotient of two integers.

Solution

$$\begin{aligned} .727272\ldots &= .72 + .0072 + .000072 + \cdots \\ &= .72 + .72(.01) + .72(.0001) + \cdots \\ &= .72 + .72(.01) + .72(.01)^2 + \cdots \end{aligned}$$

Thus, $g_1 = .72$, $r = .01$, and

$$\begin{aligned} \lim_{n \to \infty} G_n &= \frac{.72}{1 - .01} \\ &= \frac{.72}{.99} \\ &= \frac{72}{99} \\ &= \frac{8}{11} \end{aligned}$$

Remember, you are trying to write .727272. . . as a fraction, not a decimal. If you divide to put the answer in decimal form, you will simply return to the form you started with.

Thus, **.727272. . .** $= \mathbf{\frac{8}{11}}$. □

Example 7 Express 1.2345345345. . . as the quotient of two integers.

Solution

$$\begin{aligned} 1.2345345345\ldots &= 1.2 + .0345 + .0000345 + .0000000345 + \cdots \\ 1.2\overline{345} &= 1.2 + .0345 + .0345(.001) + .0345(.001)^2 + \cdots \end{aligned}$$

Here, we have 1.2 added to an infinite geometric series with $g_1 = .0345$ and $r = .001$. First, find the series.

$$\begin{aligned} .0\overline{345} &= .0345 + .0345(.001) + .0345(.001)^2 + \cdots \\ &= \frac{.0345}{1 - .001} \\ &= \frac{.0345}{.999} \\ &= \frac{345}{9990} \\ .0\overline{345} &= \frac{23}{666} \end{aligned}$$

Now, adding 1.2 and the series $.0\overline{345}$, we obtain

$$\begin{aligned} 1.2\overline{345} &= 1.2 + \frac{23}{666} \\ &= \frac{12}{10} + \frac{23}{666} \\ &= \frac{6}{5} + \frac{23}{666} \end{aligned}$$

$$= \frac{3996}{3330} + \frac{115}{3330}$$

$$= \mathbf{\frac{4111}{3330}}$$

□

Example 8 The path of each swing of a pendulum bob is .85 as long as the path of the previous swing (after the first). If the path of the first swing is 30 cm long, how far does the bob travel before eventually coming to rest?

Solution

$$\begin{pmatrix}\text{TOTAL}\\ \text{DISTANCE}\end{pmatrix} = 30 + 30(.85) + 30(.85)^2 + \cdots$$

$$= \frac{30}{1 - .85}$$

$$= \frac{30}{.15}$$

On a calculator, press: 30 ÷ (1 − .85) =

$$= 200$$

The pendulum bob travels 200 cm.

□

Problem Set 8.4

A *If possible, find the sum of the infinite geometric series in Problems 1–20.*

1. $1 + \frac{1}{2} + \frac{1}{4} + \cdots$

2. $1 + \frac{1}{3} + \frac{1}{9} + \cdots$

3. $1 + \frac{3}{4} + \frac{9}{16} + \cdots$

4. $1 + \frac{3}{2} + \frac{9}{4} + \cdots$

5. $1000 + 500 + 250 + \cdots$

6. $100 + 50 + 25 + \cdots$

7. $-20 + 10 - 5 + \cdots$

8. $-45 - 15 - 5 - \cdots$

9. $\frac{1}{3} + \frac{1}{9} + \frac{1}{27} - \cdots$

10. $\frac{1}{4} + \frac{1}{16} + \frac{1}{64} + \cdots$

11. $\frac{1}{3} - \frac{1}{9} + \frac{1}{27} - \cdots$

12. $\frac{1}{4} - \frac{1}{16} + \frac{1}{64} - \cdots$

13. $128 + 64 + 32 + \cdots$

14. $243 + 81 + 27 + \cdots$

15. $128 + 96 + 72 + \cdots$

16. $243 + 162 + 108 + \cdots$

17. $\frac{3}{4} + \frac{1}{2} + \frac{1}{3} + \cdots$

18. $\frac{4}{9} + \frac{1}{3} + \frac{1}{4} + \cdots$

19. $\frac{3}{4} - \frac{1}{2} + \frac{1}{3} - \cdots$

20. $\frac{4}{9} - \frac{1}{3} + \frac{1}{4} - \cdots$

Represent the repeating decimals in Problems 21–40 as quotients of two integers by considering an infinite geometric series.

21. $.\overline{3}$

22. $.\overline{5}$

23. $.\overline{6}$

24. $.\overline{27}$

25. $.\overline{18}$

26. $.\overline{54}$

27. $.\overline{923}$

28. $.\overline{418}$

29. $1.\overline{34}$

30. $2.\overline{45}$

31. $.1\overline{6}$

32. $.08\overline{3}$

33. $.\overline{225}$

34. $.0\overline{225}$

35. $2.\overline{252}$

36. $22.\overline{522}$

37. $.\overline{132}$

38. $.00\overline{132}$

39. $1.\overline{321}$

40. $13.\overline{213}$

B *If possible, find the sum of the infinite geometric series in Problems 41–60.*

41. $\frac{16}{27} + \frac{8}{9} + \frac{4}{3} + \cdots$

42. $\frac{625}{256} + \frac{125}{64} + \frac{25}{16} + \cdots$

43. $\frac{27}{32} - \frac{9}{16} + \frac{3}{8} - \cdots$

44. $\frac{27}{32} + \frac{9}{8} + \frac{3}{2} + \cdots$

45. $.3 + .03 + .003 + \cdots$

46. $.9 - .09 + .009 - \cdots$

47. $1 + (1.08)^{-1} + (1.08)^{-2} + \cdots$

48. $1 + (1.10)^{-1} + (1.10)^{-2} + \cdots$

49. $5.03\overline{1}$

50. $2.2\overline{534}$

51. $.\overline{571428}$

52. $.\overline{714285}$

53. $2 + \sqrt{2} + 1 + \cdots$

54. $3 + \sqrt{3} + 1 + \cdots$

55. $9\sqrt{2} - 6 + 2\sqrt{2} - \cdots$

56. $4\sqrt{3} - 6 + 3\sqrt{3} - \cdots$

57. $(1 + \sqrt{2}) + 1 + (-1 + \sqrt{2}) + \cdots$

58. $(\sqrt{2} - 1) + 1 + (\sqrt{2} + 1) + \cdots$

59. $1 + (1 - \sqrt{2}) + (3 - 2\sqrt{2}) + \cdots$

60. $1 + (1 + \sqrt{2}) + (3 + 2\sqrt{2}) + \cdots$

61. *Physics* A pendulum is swung 20 cm and allowed to swing free until it eventually comes to rest. Each subsequent swing of the bob of the pendulum is 90% as far as the preceding swing. How far will the bob travel before coming to rest?

62. *Physics* The initial swing of the bob of a pendulum is 25 cm. If each swing of the bob is 75% of the preceding swing, how far does the bob travel before eventually coming to rest?

63. *Physics* A flywheel is brought to a speed of 375 revolutions per minute (rpm) and allowed to slow and eventually come to rest. If, in slowing, it rotates three-fourths as fast each subsequent minute, how many revolutions will the wheel make before returning to rest?

64. *Physics* A rotating flywheel is allowed to slow to a stop from a speed of 500 rpm. While slowing, each minute it rotates two-thirds as many times as in the preceding minute. How many revolutions will the wheel make before coming to rest?

65. *Physics* A new type of superball advertises that it will rebound $\frac{9}{10}$ of its original height. If it is dropped from a height of 10 ft, how far will the ball travel before coming to rest?

66. *Physics* A tennis ball is dropped from a height of 10 ft. If the ball rebounds $\frac{2}{3}$ of its height on each bounce, how far will the ball travel before coming to rest?

67. *Business* Suppose that a piece of machinery costing \$10,000 depreciates 20% of its present value each year. That is, during the first year \$10,000(.20) = \$2000 is depreciated. The second year's depreciation is \$8000(.2) = \$1600, since the value the second year is \$10,000 − \$2000 = \$8000. The third year's depreciation is \$6400(.2) = \$1280. If the depreciation is calculated this way indefinitely, what is the total depreciation?

C **68.** The sum of an infinite geometric series is 30. Each term is exactly 4 times the sum of the remaining terms. Determine the series with these properties.

69. Each term is 5 times the sum of all the terms that follow it. Determine the infinite geometric series if the sum of the series is 20.

70. A unit square *ABCD* is cut out of construction paper. Another square *EFGH* of side $\frac{1}{2}$ is cut out of paper and placed on top of *ABCD*, as shown in the figure in

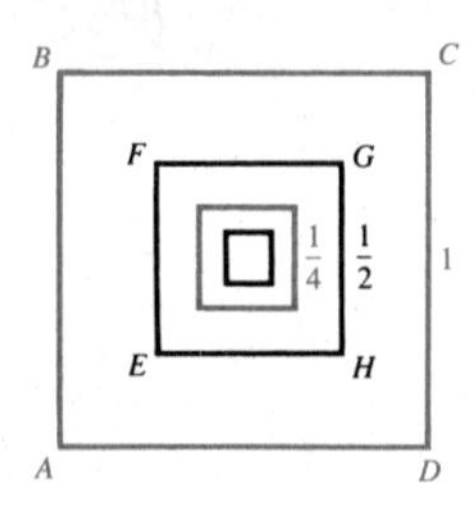

the margin. Next, squares of side $\frac{1}{4}, \frac{1}{8}, \frac{1}{16}, \ldots$ are cut out. What is the total area of all the squares?

71. Repeat Problem 70 for square $ABCD$ of side length a, $EFGH$ of side length $a/2$, and so on.

72. Consider a sequence of equilateral triangles as shown in the figure where the sides of $\triangle ABC$ are 1, the sides of $\triangle ADE$ and $\triangle EFB$ are $\frac{1}{2}$, and so on. If this process is repeated forever, what is the total perimeter of the triangles and what is the total area of the triangles?

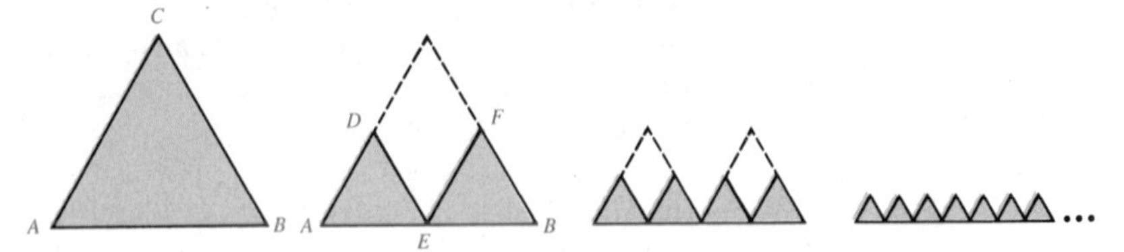

Recall that the height of an equilateral triangle is $\frac{1}{2}a\sqrt{3}$, where a is the length of the side.

73. An equilateral triangle of side a is cut out of paper, as shown in figure a. Next, three equilateral triangles, each of side $a/3$, are cut out and placed in the middle of each side of the first triangle, as shown in figure b. Then 12 equilateral triangles, each of side $a/9$, are placed halfway along each of the sides of this figure, as shown in figure c. Figure d shows the result of adding 48 equilateral triangles, each of side $a/27$, to the previous figure. The curve produced by repeating this procedure indefinitely is called the *snowflake curve*. Find:

a. Perimeter of the snowflake curve

b. Area of the snowflake curve

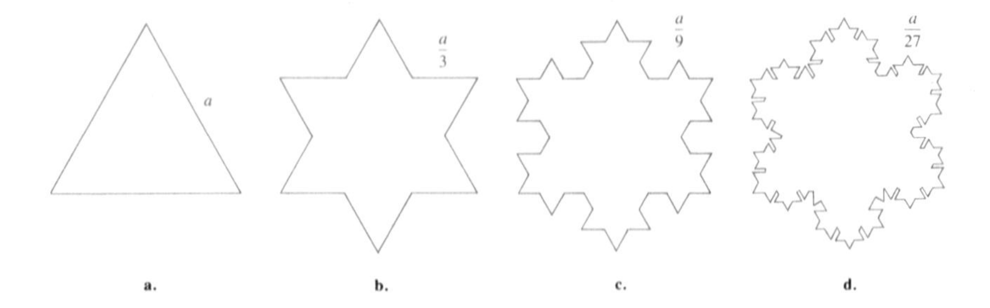

The following is needed for Problems 74–79:

Business *If \$1 is invested for n years at r percent interest, the amount present, A, is given by*

$$A = (1 + r)^n$$

If the interest rate is 8.06% and the period is 20 years, the amount is

$$A = (1.0806)^{20}$$
$$\approx 4.71$$

If the interest is compounded semiannually, then the amount is given by

$$A = \left(1 + \frac{r}{2}\right)^{2n}$$

For the above example,

$$A = \left(1 + \frac{.0806}{2}\right)^{40}$$
$$\approx 4.86$$

74. Write a formula for the amount present if it is compounded s times a year.
75. Calculate the amount present if it is compounded quarterly.
76. Calculate the amount present if it is compounded monthly.
77. Calculate the amount present if it is compounded daily. (1 year = 365 days.)
78. Calculate the amount present if it is compounded every minute. (The number of minutes in a year is $60 \cdot 24 \cdot 365$.)
79. Does there appear to be a limit to the amount present as s becomes large without limit?

8.5 Summation, Factorial, and Pascal

Recall that a finite sequence is a function whose domain is the set of counting numbers less than some given number. Consider the function $s(k) = 2k$ and $N = \{1, 2, 3, 4\}$, for which the following table may be compiled:

k	$s(k) = 2k$
1	2
2	4
3	6
4	8

The sum of this finite sequence is

$$2 + 4 + 6 + 8$$

This sum may be denoted by the Greek letter Σ, called *sigma:*

$$\sum_{k=1}^{4} 2k = 2 + 4 + 6 + 8$$

This is the last natural number in the domain. (the upper index 4)

This is the function that is evaluated. ($2k$)

This is the first natural number in the domain. ($k = 1$)

The Greek letter sigma means you evaluate the function for each number in the domain and *add* the results.

Thus,

$$\sum_{k=1}^{4} 2k = 20$$

Example 1 Let $s(k) = 2k + 1$ and $N = \{3, 4, 5, 6\}$.

k	$s(k) = 2k + 1$
3	7
4	9
5	11
6	13

First natural number in domain: $k = 3$ → 3

Last natural number in domain: $k = 6$ → 6

The sigma means you add these values.

$$\sum_{k=3}^{6} (2k + 1) = 7 + 9 + 11 + 13$$

(9: $k = 4$; 11: $k = 5$; 13: $k = 6$)

This is obtained by letting $k = 3$ and evaluating $2k + 1$.

The sum of this series is 40. □

Example 2

$$\sum_{k=1}^{5} (3k - 2) = 1 + 4 + 7 + 10 + 13$$

This is an arithmetic series; use the formula

$$A_n = \frac{n}{2}(a_1 + a_n)$$

$$A_5 = \frac{5}{2}(1 + 13) \qquad a_1 = 1, a_5 = 13, n = 5$$

$$= 35$$

Thus,

$$\sum_{k=1}^{5} (3k - 2) = 35$$ □

Example 3

$$\sum_{k=1}^{20} (15 - 2k) = 13 + 11 + 9 + \cdots + (-23) + (-25)$$

$$= \frac{20}{2}(13 - 25) \qquad a_1 = 13, a_{20} = -25, n = 20$$

$$= 10(-12)$$

$$= \mathbf{-120}$$ □

Example 4

$$A_n = a_1 + a_2 + \cdots + a_n \qquad \text{Arithmetic series}$$

$$= \sum_{k=1}^{n} a_k$$

$$= \frac{n}{2}(a_1 + a_n) \qquad \text{Formula from page 334}$$ □

Example 5

$$\sum_{k=1}^{5} 3^{1-k} = 3^0 + 3^{-1} + 3^{-2} + 3^{-3} + 3^{-4}$$

$$= 1 + \frac{1}{3} + \frac{1}{9} + \frac{1}{27} + \frac{1}{81}$$

This is a geometric series; use the formula

$$G_n = \frac{g_1(1 - r^n)}{1 - r}, \qquad r \neq 1$$

$$G_5 = \frac{1\left(1 - \frac{1}{243}\right)}{1 - \frac{1}{3}} \qquad r = \tfrac{1}{3}, n = 5$$

$$= \frac{3}{2}\left(\frac{242}{243}\right)$$

$$= \frac{121}{81}$$

Thus,

$$\boldsymbol{\sum_{k=1}^{5} 3^{1-k} = \frac{121}{81}}$$

□

Example 6

$$G_n = g_1 + g_1 r + g_1 r^2 + \cdots + g_1 r^{n-1} \qquad \text{Geometric series}$$

$$= \sum_{k=1}^{n} g_1 r^{k-1}$$

$$= \boldsymbol{\frac{g_1(1 - r^n)}{1 - r}, \qquad r \neq 1} \qquad \text{Formula from page 339}$$

□

The sigma notation is used to indicate certain types of sums. Similarly, there is a special notation for certain types of products. To indicate the *product* of the first n natural numbers, the symbol $\boldsymbol{n!}$ is used; this is called ***n* factorial.**

DEFINITION OF FACTORIAL

> $n!$ is called ***n* factorial** and is defined by $\boldsymbol{n! = 1 \cdot 2 \cdot 3 \cdot \cdots \cdot (n - 1) \cdot n}$ for n a natural number. Also, $0! = 1$.

Example 7 Find the factorial of each whole number up to 10.

$0! = 1$

$1! = 1$

$2! = 1 \cdot 2 = 2$

$3! = 1 \cdot 2 \cdot 3 = 6$

$$\begin{aligned}
4! &= 1\cdot 2\cdot 3\cdot 4 = 24\\
5! &= 1\cdot 2\cdot 3\cdot 4\cdot 5 = 120\\
6! &= 1\cdot 2\cdot 3\cdot 4\cdot 5\cdot 6 = 720\\
7! &= 1\cdot 2\cdot 3\cdot 4\cdot 5\cdot 6\cdot 7 = 5040\\
8! &= 1\cdot 2\cdot 3\cdot 4\cdot 5\cdot 6\cdot 7\cdot 8 = 40{,}320\\
9! &= 1\cdot 2\cdot 3\cdot\cdots\cdot 8\cdot 9 = 362{,}880\\
10! &= 1\cdot 2\cdot 3\cdot\cdots\cdot 9\cdot 10 = 3{,}628{,}800
\end{aligned}$$

□

Example 8

$$\begin{aligned}
5! - 4! &= 120 - 24\\
&= 96
\end{aligned}$$

□

Example 9

$$\begin{aligned}
(5 - 4)! &= 1!\\
&= 1
\end{aligned}$$

Note: $5! - 4! \neq (5 - 4)!$

□

Example 10

$$\begin{aligned}
\frac{8!}{4!} &= \frac{8\cdot 7\cdot 6\cdot 5\cdot \not{4}\cdot \not{3}\cdot \not{2}\cdot \not{1}}{\not{4}\cdot \not{3}\cdot \not{2}\cdot \not{1}}\\
&= 8\cdot 7\cdot 6\cdot 5\\
&= 1680
\end{aligned}$$

Note: $\frac{8!}{4!} \neq 2!$

□

Example 11

$$\begin{aligned}
(2\cdot 3)! &= 6!\\
&= 720
\end{aligned}$$

□

Example 12

$$\begin{aligned}
2!3! &= (2\cdot 1)\cdot(3\cdot 2\cdot 1)\\
&= 12
\end{aligned}$$

Note: $(2\cdot 3)! \neq 2!3!$

□

Example 13

$$\begin{aligned}
\frac{10!}{8!} &= \frac{10\cdot 9\cdot \not{8}!}{\not{8}!}\\
&= 90
\end{aligned}$$

Note:

$$\begin{aligned}
10 &= 10\cdot 9!\\
&= 10\cdot 9\cdot 8!\\
&= 10\cdot 9\cdot 8\cdot 7!\\
&\ \vdots
\end{aligned}$$

□

Example 14

$$\begin{aligned}
\frac{8!}{3!(8-3)!} &= \frac{8!}{3!5!}\\
&= \frac{8\cdot 7\cdot \not{6}\cdot \not{5}!}{(\not{3}\cdot \not{2}\cdot 1)\cdot \not{5}!}\\
&= 56
\end{aligned}$$

□

Example 14 illustrates a formula using factorials that will be needed in the next section. It is called the *binomial coefficient 8, 3* and is denoted by $\binom{8}{3}$.

BINOMIAL COEFFICIENT n, r

The symbol $\binom{n}{r}$ is defined for $0 \leq r \leq n$:

$$\binom{n}{r} = \frac{n!}{r!(n-r)!}$$

$\binom{n}{r}$ is read **"the binomial coefficient n, r."**

Example 15

$$\binom{52}{2} = \frac{n!}{r!(n-r)!}, \quad \text{where } n = 52 \text{ and } r = 2$$

$$= \frac{52!}{2!(52-2)!}$$

$$= \frac{52!}{2!50!}$$

$$= \frac{52 \cdot 51 \cdot \cancel{50!}}{2 \cdot 1 \cdot \cancel{50!}}$$

$$= 1326$$

□

Example 16

$$\binom{n}{n} = \frac{n!}{n!(n-n)!}$$

$$= \frac{n!}{n!0!}$$

$$= 1$$

□

Example 17

$$\binom{n}{n-1} = \frac{n!}{(n-1)![n-(n-1)]!}$$

$$= \frac{n!}{(n-1)!1!}$$

$$= \frac{n \cdot \cancel{(n-1)!}}{\cancel{(n-1)!}}$$

$$= n$$

□

For small values of n and r, there is an easier way to find $\binom{n}{r}$ than using the formula. Consider

$$\binom{0}{0} = 1$$

$$\binom{1}{0} = 1 \quad \binom{1}{1} = 1$$

$$\binom{2}{0} = 1 \quad \binom{2}{1} = 2 \quad \binom{2}{2} = 1$$

$$\binom{3}{0} = 1 \quad \binom{3}{1} = 3 \quad \binom{3}{2} = 3 \quad \binom{3}{3} = 1$$

$$\vdots$$

This pattern is called **Pascal's Triangle;** it is illustrated further in Figure 8.2.

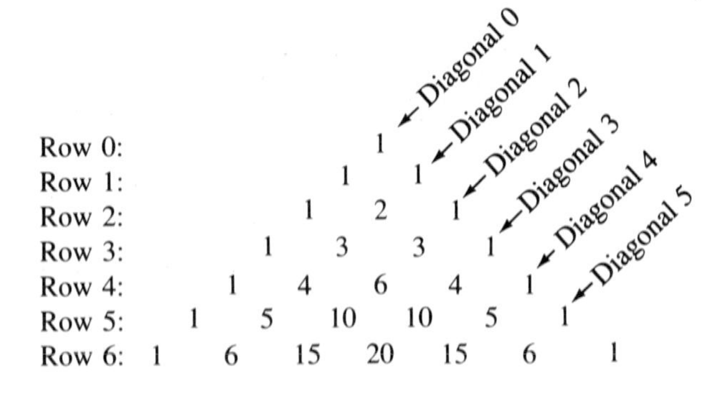

Figure 8.2
Pascal's Triangle

There are many interesting relationships associated with this pattern, but we are concerned with an expression representing the entries of this pattern. Do you see how to generate additional rows of the triangle?

1. Each row begins and ends with a 1.
2. Notice that we have begun counting the rows with Row 0. This is because after Row 0, the second entry in the row is the same as the row number. Thus, Row 7 would begin 1, 7, ...
3. The triangle is symmetric about the middle. This means that the entries of each row are the same at the beginning and the end. Thus, Row 7 ends with . . . , 7, 1.
4. To find new entries we can simply add the two entries just above in the preceding row. Thus, Row 7 is found by looking at Row 6:

Row 6: 1 6 15 20 15 6 1
Row 7: 1 7 21 35 35 21 7 1

This property, stated in symbols, is

$$\binom{n-1}{r} + \binom{n-1}{r-1} = \binom{n}{r}$$

You are asked to prove this in Problem 70 of Problem Set 8.5. Notice that $\binom{n}{r}$ is found in the nth row and rth diagonal. When counting rows and diagonals in Pascal's Triangle, remember to begin counting with 0 rather than 1.

Example 18 Find $\binom{4}{3}$ from Pascal's Triangle.

Solution Look in the fourth row, third diagonal; the entry is 4, so

$$\binom{4}{3} = 4$$

□

Example 19 Find $\binom{7}{5}$ from Pascal's Triangle.

Solution Look in the seventh row; count across (beginning with 0) to the fifth entry; it is 21, so

$$\binom{7}{5} = 21$$

□

Problem Set 8.5

A *Evaluate the expressions in Problems 1–40.*

1. $\sum_{k=1}^{5} (3k - 1)$ **2.** $\sum_{k=1}^{4} (4k - 1)$ **3.** $\sum_{n=0}^{6} (2n + 1)$ **4.** $\sum_{k=1}^{5} (3k + 2)$

5. $\sum_{k=2}^{6} k$ **6.** $\sum_{k=4}^{10} k$ **7.** $\sum_{m=1}^{4} m^2$ **8.** $\sum_{k=3}^{8} k^2$

9. $\sum_{k=1}^{6} 2(3)^k$ **10.** $\sum_{k=1}^{4} 10\left(\frac{1}{4}\right)^k$ **11.** $\sum_{k=1}^{5} 5\left(\frac{1}{2}\right)^k$ **12.** $\sum_{k=1}^{5} 3(2)^{k-1}$

13. $9!$ **14.** $10!$ **15.** $11!$ **16.** $12!$

17. $6! - 4!$ **18.** $7! - 3!$ **19.** $(6 - 4)!$ **20.** $(7 - 3)!$

21. $\frac{9!}{7!}$ **22.** $\frac{10!}{6!}$ **23.** $\frac{11!}{7!}$ **24.** $\frac{12!}{10!}$

25. $\frac{9!}{2!7!}$ **26.** $\frac{10!}{4!6!}$ **27.** $\frac{12!}{3!(12 - 3)!}$ **28.** $\frac{52!}{3!(52 - 3)!}$

29. $\binom{8}{0}$ **30.** $\binom{8}{1}$ **31.** $\binom{5}{4}$ **32.** $\binom{8}{2}$

33. $\binom{52}{48}$ **34.** $\binom{52}{3}$ **35.** $\binom{8}{3}$ **36.** $\binom{6}{3}$

37. $\binom{8}{4}$ **38.** $\binom{5}{5}$ **39.** $\binom{1000}{0}$ **40.** $\binom{1000}{1}$

B *Write the expressions in Problems 41–56 using summation notation.*

41. $\frac{1}{2} + \frac{1}{4} + \frac{1}{8} + \cdots + \frac{1}{128}$

42. $\frac{1}{3} + \frac{1}{12} + \frac{1}{48} + \cdots + \frac{1}{12{,}288}$

43. $2 + 4 + 6 + \cdots + 100$

44. $5 + 15 + 45 + 135 + 405$

45. $1 + 6 + 36 + 216 + 1296$

46. $2 + 6 + 18 + \cdots + 1458$

47. $1 + 11 + 21 + \cdots + 101$

48. $1 - 10 + 100 - \cdots + 1{,}000{,}000$

49. $\frac{1}{2} + \frac{5}{6} + \frac{7}{6} + \cdots + \frac{5}{2}$

50. $\frac{7}{3} + \frac{13}{6} + 2 + \cdots + \frac{1}{3}$

51. $\frac{1}{3} + \frac{1}{9} + \frac{1}{27} + \frac{1}{81} + \cdots + \frac{1}{2187}$

52. $1 + \frac{3}{4} + \frac{9}{16} + \cdots + \frac{243}{1024}$

53. $1 - \frac{2}{3} + \frac{4}{9} - \frac{8}{27} + \cdots + \frac{64}{729}$

54. $1 + \frac{3}{2} + \frac{9}{4} + \cdots + \frac{729}{64}$

55. $8 - 4\sqrt{2} + 4 - \cdots + \frac{1}{2}$

56. $2\sqrt{3} + 3 + \frac{3\sqrt{3}}{2} + \cdots + \frac{27}{16}$

Write out the expressions in Problems 57–60 without summation notation.

57. $\sum_{j=1}^{r} j$ **58.** $\sum_{k=1}^{n} k$ **59.** $\sum_{j=1}^{r} a_j b_j$ **60.** $\sum_{k=0}^{n} ka^k$

C **61.** If $\sum_{k=1}^{4} xk = 700$, find x.

62. If $\sum_{j=0}^{2} x^j = 3$, find x.

63. If $\sum_{k=3}^{5} (kx + y) = -3$ and $\sum_{k=2}^{4} (kx + y) = -6$, find x and y.

64. If $\sum_{j=1}^{r} a_j b_j$ and you let $b_j = k$, show that $\sum_{j=1}^{r} ka_j = k \sum_{j=1}^{r} a_j$.

65. If $\sum_{j=1}^{r} a_j b_j$ and you let $a_j = 1$ and $b_j = k$, show that $\sum_{j=1}^{r} k = kr$.

66. Show that $\sum_{k=1}^{n} (a_k + b_k) = \sum_{k=1}^{n} a_k + \sum_{k=1}^{n} b_k$.

67. Examine Pascal's Triangle carefully, look for patterns, and answer the following questions:

a. What is the second number in the 100th row?

b. What is the next-to-last number in the 200th row?

68. Which rows of Pascal's Triangle contain only odd numbers?

69. Consider $11^0, 11^1, 11^2, 11^3, \ldots$, and explain how the powers of 11 are related to Pascal's Triangle.

70. Show that $\binom{n-1}{r-1} + \binom{n-1}{r} = \binom{n}{r}$.

71. Show that $\binom{n}{r} = \binom{n}{n-r}$. This shows the symmetry that you may have noticed when looking at Pascal's Triangle.

72. Show that $\binom{13}{3}\binom{13}{2} = \binom{13}{6}\binom{13}{1}$.

73. Is it true in general that $\binom{n}{a}\binom{n}{b} = \binom{n}{ab}\binom{n}{1}$? (See Problem 72.)

8.6 Binomial Theorem

In mathematics it is frequently necessary to find $(a + b)^n$. If n is very large, direct calculation is tedious, so an easy pattern is sought that not only will help find $(a + b)^n$ but also will allow you to find any given term in that expansion.

Consider the first few powers of $(a + b)$, which are found by direct multiplication:

$$\begin{aligned}
(a + b)^0 &= 1 \\
(a + b)^1 &= 1 \cdot a + 1 \cdot b \\
(a + b)^2 &= 1 \cdot a^2 + 2 \cdot ab + 1 \cdot b^2 \\
(a + b)^3 &= 1 \cdot a^3 + 3 \cdot a^2b + 3 \cdot ab^2 + 1 \cdot b^3 \\
(a + b)^4 &= 1 \cdot a^4 + 4 \cdot a^3b + 6 \cdot a^2b^2 + 4 \cdot ab^3 + 1 \cdot b^4 \\
(a + b)^5 &= 1 \cdot a^5 + 5 \cdot a^4b + 10 \cdot a^3b^2 + 10 \cdot a^2b^3 + 5 \cdot ab^4 + 1 \cdot b^5
\end{aligned}$$

If the coefficients are ignored and attention is focused on the variables, a pattern can be seen:

$$\begin{aligned}
&(a + b)^1: \quad a \quad b \\
&(a + b)^2: \quad a^2 \quad ab \quad b^2 \\
&(a + b)^3: \quad a^3 \quad a^2b \quad ab^2 \quad b^3 \\
&(a + b)^4: \quad a^4 \quad a^3b \quad a^2b \quad ab^3 \quad b^4 \\
&(a + b)^5: \quad a^5 \quad a^4b \quad a^3b \quad a^2b \quad ab \quad b^5
\end{aligned}$$

Do you see the pattern? From left to right, the powers of a decrease and the powers of b increase. Notice that the sum of the exponents for each term is the same as the original power. The general pattern is

$$(a + b)^n: \quad a^n \quad a^{n-1}b \quad a^{n-2}b^2 \ldots a^{n-r}b^r \ldots a^2b^{n-2} \quad ab^{n-1} \quad b^n$$

Next, consider the coefficients:

$$\begin{array}{lccccccccccc}
(a + b)^0: & & & & & & 1 & & & & & \\
(a + b)^1: & & & & & 1 & & 1 & & & & \\
(a + b)^2: & & & & 1 & & 2 & & 1 & & & \\
(a + b)^3: & & & 1 & & 3 & & 3 & & 1 & & \\
(a + b)^4: & & 1 & & 4 & & 6 & & 4 & & 1 & \\
(a + b)^5: & 1 & & 5 & & 10 & & 10 & & 5 & & 1 \\
 & & & & & & \vdots & & & & &
\end{array}$$

Do you see the pattern? The coefficients are the numbers in Pascal's Triangle. Using the notation for Pascal's Triangle introduced in the last section,

$$(a + b)^n = \binom{n}{0}a^n + \binom{n}{1}a^{n-1}b + \binom{n}{2}a^{n-2}b^2 + \cdots + \binom{n}{r}a^{n-r}b^r + \cdots + \binom{n}{n-2}a^2b^{n-2} + \binom{n}{n-1}ab^{n-1} + \binom{n}{n}b^n$$

A more compact form, called the **Binomial Theorem,** is obtained using summation notation.

BINOMIAL THEOREM

For any positive integer n,

$$(a + b)^n = \sum_{k=0}^{n} \binom{n}{k} a^{n-k}b^k \qquad \text{where} \qquad \binom{n}{k} = \frac{n!}{k!(n-k)!}$$

The proof of the Binomial Theorem is by mathematical induction, which will be discussed in Chapter 10.

Example 1 Find $(x + y)^8$.

Solution For smaller powers (say 8 or less), use Pascal's Triangle to obtain the coefficients in the expansion.

$$(x + y)^8 = x^8 + 8x^7y + 28x^6y^2 + 56x^5y^3 + 70x^4y^4 + 56x^3y^5 + 28x^2y^6 + 8x^1y^7 + y^8 \quad \square$$

Example 2 Find $(x - 2y)^4$.

Solution In this example $a = x$ and $b = -2y$, and the coefficients are found in Pascal's Triangle.

$$\begin{aligned}(x - 2y)^4 &= 1\cdot x^4 + 4\cdot x^3(-2y) + 6\cdot x^2(-2y)^2 + 4\cdot x(-2y)^3 + 1\cdot(-2y)^4 \\ &= x^4 - 8x^3y + 24x^2y^2 - 32xy^3 + 16y^4\end{aligned} \quad \square$$

Example 3 Find $(3x + 5)^5$.

Solution Here, $a = 3x$ and $b = 5$.

$$\begin{aligned}(3x + 5)^5 &= 1(3x)^5 + 5(3x)^4(5) + 10(3x)^3(5^2) + 10(3x)^2(5^3) + 5(3x)(5^4) + 1(5^5) \\ &= 1(3^5x^5) + 5(3^4x^4)(5) + 10(3^3x^3)(5^2) + 10(3^2x^2)5^3 + 5(3x)(5^4) + 1(5^5) \\ &= 243x^5 + 2025x^4 + 6750x^3 + 11{,}250x^2 + 9375x + 3125\end{aligned}$$

$\square$

Example 4 Find $(a + b)^{15}$.

Solution The power is rather large, so use the Binomial Theorem.

$$\begin{aligned}(a + b)^{15} &= \binom{15}{0}a^{15} + \binom{15}{1}a^{14}b + \binom{15}{2}a^{13}b^2 + \cdots + \binom{15}{14}ab^{14} + \binom{15}{15}b^{15} \\ &= \frac{15!}{0!15!}a^{15} + \frac{15!}{1!14!}a^{14}b + \frac{15!}{2!13!}a^{13}b^2 + \cdots + \frac{15!}{14!1!}ab^{14} + \frac{15!}{15!0!}b^{15} \\ &= a^{15} + 15a^{14}b + 105a^{13}b^2 + \cdots + 15ab^{14} + b^{15}\end{aligned} \quad \square$$

Example 5 Find the coefficient of the term x^2y^{10} in the expansion of $(x + 2y)^{12}$.

Solution $n = 12$, $k = 10$, $a = x$, and $b = 2y$; thus, the term is

$$\binom{12}{10}x^2(2y)^{10} = \frac{12!}{10!2!}(2^{10})x^2y^{10}$$
$$= 66(1024)x^2y^{10}$$

The coefficient is $66(1024) =$ **67,584.** □

Example 6 Find the coefficient of x^7 in the expansion of $(2x - 3)^{10}$.

Solution $n = 10$, $k = 3$, $a = 2x$, and $b = -3$; thus,

$$\binom{10}{3}(2x)^7(-3)^3 = (120)(2^7x^7)(-3)^3$$
$$= -414{,}720x^7$$

Notice that even for relatively small powers, the coefficient can be very large indeed.

The coefficient is **−414,720.** □

Problem Set 8.6

A *In Problems 1–10 expand using the Binomial Theorem.*

1. $(x + 1)^4$ **2.** $(x + 1)^8$ **3.** $(x - 1)^5$ **4.** $(x - 1)^9$
5. $(x - y)^6$ **6.** $(x + y)^5$ **7.** $(x + 2)^5$ **8.** $(x - 2)^6$
9. $(x - 3)^5$ **10.** $(x + 4)^4$

Find the coefficient of the given term in the expansion of the given binomial in Problems 11–20.

11. a^5b^6 in $(a - b)^{11}$ **12.** a^4b^7 in $(a + b)^{11}$
13. $x^{10}y^4$ in $(x + y)^{14}$ **14.** $x^{10}y^5$ in $(x - y)^{15}$
15. x^{12} in $(x - 1)^{16}$ **16.** y^8 in $(y + 1)^{12}$
17. r^5 in $(r + 2)^9$ **18.** s^5 in $(s - 2)^{10}$
19. a^7b in $(a - 2b)^8$ **20.** a^4b^4 in $(a + 2b)^8$

B *Find the first four terms in the expansion of the given binomial in Problems 21–30.*

21. $(x - y)^{15}$ **22.** $(x + 2y)^{16}$ **23.** $(x + \sqrt{2})^8$ **24.** $(x - 2y)^{12}$
25. $(x - 3y)^{10}$ **26.** $(x + \sqrt{3})^9$ **27.** $(ab - 2b)^{15}$ **28.** $(rs - 3t)^{13}$
29. $(z^2 + 5k)^{11}$ **30.** $(z^3 - k^2)^7$

C *Find the last three terms in the expansion of the given binomial in Problems 31–40.*

31. $(x^{2/3} + y^{1/3})^{15}$ **32.** $(2x^{1/5} + 3y^{3/5})^{10}$
33. $(z^{3/4} - 2x^{1/4})^{20}$ **34.** $(2z^2 + \sqrt{y})^7$
35. $(\sqrt{p} - 2\sqrt{q})^8$ **36.** $(x\sqrt{x} - y\sqrt{y})^5$
37. $(2m^{-1} + \frac{1}{3}m^{-2})^9$ **38.** $(3r - 2r^{-1})^{12}$

39. $(q^{-2} - \sqrt{2}r^{-1})^6$

40. $(xy^{-1} - 2y^{-2})^{11}$

41. Show that $\binom{n}{0} + \binom{n}{1} + \binom{n}{2} + \cdots + \binom{n}{n-1} + \binom{n}{n} = 2^n$. This says that the sum of the entries of the nth row of Pascal's Triangle is 2^n.

42. Show that $\sum_{j=0}^{n} (-1)^j \binom{n}{j} = 0$ for every integer n.

43. Show that $\sum_{j=0}^{n} 2^j \binom{n}{j} = 3^n$ for every positive integer n.

8.7 Summary and Review

One of the most important skills to acquire in your study of mathematics is the ability to recognize and use patterns to generalize from simple cases to more complicated cases.

Two patterns examined in this chapter—arithmetic and geometric progressions—occur frequently and form the cornerstone for the study of additional sequences in more advanced mathematics.

The Binomial Theorem, along with the associated notation, is a powerful result that lends itself to a wide variety of applications.

Indeed, the ideas of this chapter provide a transition from elementary mathematics to the more advanced mathematics in calculus. The ideas of sequences, series, limits, and the Binomial Theorem all play an important role in the study of calculus.

CHAPTER OBJECTIVES

After studying this chapter, you should be able to:

1. Classify a given sequence as arithmetic, geometric, or neither
2. Given the general term, find specific terms
3. Find the sum of the first n terms of an arithmetic or geometric sequence
4. Find the sum of an infinite geometric series if $|r| < 1$
5. Solve word problems involving arithmetic and geometric sequences and series
6. Use summation and factorial notation in representing sequences and series
7. Evaluate expressions involving factorials, summation notation, and Pascal's Triangle
8. Use the Binomial Theorem to expand binomials
9. Find a given term in a binomial expansion

Problem Set 8.7

CONCEPT PROBLEMS

Fill in the word or words to make the statements in Problems 1–20 complete and correct.

[8.1] **1.** A(n) ________ is a function whose domain is the subset of the set of counting numbers, $\{1, 2, 3, \ldots, n\}$.

[8.1] **2.** A(n) ________ is a function whose domain is the set of counting numbers.

[8.1] **3.** An arithmetic sequence is a sequence that has a common ______ between successive terms.

[8.1] **4.** A formula that completely specifies a sequence is called a(n) ______ .

[8.2] **5.** The indicated sum of the terms of a finite sequence is called a(n) ______ .

[8.2] **6.** For an arithmetic sequence with first and last terms a_1 and a_n, respectively, the sum is $A_n =$ ______ .

[8.3] **7.** For a geometric sequence with first term g_1 and common ratio r, the general term $g_n =$ ______ .

[8.3] **8.** For a geometric sequence with general term g_n and common ratio r, the sum is $G_n =$ ______ .

[8.3] **9.** Each successive term in a geometric sequence is obtained by ______ the preceding term by a nonzero constant.

[8.4] **10.** A geometric series has a(n) ______ as the number of terms increases without bound if $|r| < 1$.

[8.4] **11.** An infinite geometric series has no sum if ______ .

[8.4] **12.** If G is the sum of an infinite geometric series with first term g_1 and ratio r, then $G =$ ______ , where $|r|$ ______ .

[8.5] **13.** The notation $\sum_{k=1}^{n} s_k$ means the ______ of the terms s_k from $k =$ ______ to $k =$ ______ .

[8.5] **14.** The product of the first n natural numbers is called n ______ .

[8.5] **15.** $\binom{n}{r}$ indicates the number in the ______ row and ______ diagonal of Pascal's Triangle.

[8.5] **16.** $n! =$ ______

[8.5] **17.** $\binom{n}{r} =$ ______

18. The symbol $\binom{n}{r}$ is read "the ______ of n, r."

[8.6] **19.** Using sigma notation, for any positive integer n, $(a + b)^n =$ ______ .

[8.6] **20.** The numerical coefficients in the expansion of $(a + b)^n$ are the numbers in ______ .

REVIEW PROBLEMS [8.1, 8.2, 8.3]

Classify each sequence in Problems 21–30 as arithmetic, geometric, or neither. Find an expression for the general term if it is an arithmetic or geometric sequence. If it is neither, give the next two terms.

21. 1, 11, 121, 1331, . . .

22. 1, 11, 111, 1111, . . .

23. 7, 10, 13, 16, . . .

24. 54, 18, 6, 2, . . .

25. 1, 4, 9, 16, . . .

26. 1, 11, 21, 31, . . .

27. $-\frac{1}{2}, -\frac{1}{4}, -\frac{1}{8}, \ldots$

28. $\frac{1}{1}, \frac{3}{4}, \frac{5}{9}, \frac{7}{16}, \ldots$

29. $\frac{1}{2}, \frac{2}{3}, \frac{3}{4}, \ldots$

30. $\frac{1}{2}, \frac{3}{4}, 1, \frac{5}{4}, \frac{3}{2}, \ldots$

[8.2] *Find the missing quantities in Problems 31–40.*

31. $a_1 = 50, \quad d = -5; \quad a_{10} = ?$

32. $a_1 = 4, \quad d = 3, \quad a_{12} = ?$

33. $a_1 = 50, \quad d = -5; \quad A_5 = ?$

34. $a_1 = 3, \quad a_7 = 33; \quad d = ?$

35. $a_1 = 2, \quad a_{10} = 20; \quad a_7 = ?$

36. $a_1 = 3, \quad a_7 = 33; \quad a_5 = ?$

37. $a_1 = 2, \quad a_{10} = 20; \quad d = ?$

38. $a_1 = .1, \quad a_9 = .9; \quad d = ?$

39. $a_1 = 2, \quad a_{10} = 20; \quad A_{10} = ?$

40. $a_1 = .1, \quad a_9 = .9; \quad A_9 = ?$

[8.3] *Find the missing quantities in Problems 41–50.*

41. $g_1 = 5, \quad r = 2; \quad g_{10} = ?$

42. $g_1 = 5, \quad r = \frac{1}{2}; \quad g_{10} = ?$

43. $g_1 = 5, \quad r = 2; \quad G_5 = ?$

44. $g_1 = 512, \quad g_6 = -\frac{1}{2}; \quad G_5 = ?$

45. $g_1 = 50{,}000, \quad g_6 = 16; \quad g_7 = ?$

46. $g_1 = 64{,}000, \quad g_{10} = -125; \quad g_6 = ?$

47. $g_1 = 512, \quad g_{10} = 1; \quad r = ?$

48. $g_1 = 384, \quad g_4 = 162; \quad r = ?$

49. $g_1 = 512, \quad g_{10} = 1; \quad G_{10} = ?$

50. $g_1 = 384, \quad g_4 = 162; \quad G_4 = ?$

[8.1] *Find the first four terms of the given sequences in Problems 51–60.*

51. $a_n = 2 + 3n$

52. $a_n = 5n - 2$

53. $g_n = \dfrac{2^{n+1}}{40}$

54. $g_n = \dfrac{10^5}{2^{n-1}}$

55. $s_n = \dfrac{(-1)^n}{n}$

56. $s_n = (-1)^{n-1}n$

57. $s_n = \dfrac{n-1}{n+1}$

58. $s_n = \dfrac{(-1)^n}{n!}$

59. $s_1 = 1, \quad s_2 = 2, \quad s_n = s_{n-1} + s_{n-2}, \quad n \geqslant 3$

60. $s_1 = 1, \quad s_2 = -5, \quad s_n = s_{n-1} + s_{n-2}, \quad n \geqslant 3$

[8.2, 8.3, 8.5] *In Problems 61–70 find the sum of the series, if possible.*

61. $2 + 4 + 6 + \cdots + 512$

62. $2 + 4 + 8 + \cdots + 512$

63. $486 + 324 + 216 + \cdots + 64$

64. $3125 + 1250 + 500 + \cdots + 32$

65. $\frac{243}{256} + \frac{81}{128} + \frac{27}{64} + \cdots + \frac{1}{27}$

66. $\frac{2}{3} - \frac{4}{9} + \frac{8}{27} - \cdots$

67. $100 + 50 + 25 + \cdots$

68. $100 + 20 + 4 + \cdots$

69. $128 + 192 + 288 + \cdots$

70. $144 + 96 + 64 + \cdots$

[8.5] *Evaluate each expression in Problems 71–80.*

71. $8! - 4!$

72. $(8 - 4)!$

73. $\dfrac{8!}{4!}$

74. $\dbinom{8}{4}!$

75. $\dbinom{8}{4}$

76. $\dbinom{7}{3}$

77. $\dbinom{15}{0}$

78. $\dbinom{15}{14}$

79. $\dbinom{p}{q}$

80. $\dbinom{r+1}{s-1}$

[8.2, 8.3, 8.5] *In Problems 81–90 write out the given expressions without using summation notation.*

81. $\sum_{k=1}^{10} k$ **82.** $\sum_{k=1}^{6} (-k)$ **83.** $\sum_{k=1}^{8} (2k - 3)$

84. $\sum_{k=1}^{9} (9 - 2k)$ **85.** $\sum_{k=1}^{5} 2(3)^{k-1}$ **86.** $\sum_{k=1}^{5} 3(2)^{k-1}$

87. $\sum_{k=1}^{5} n(n + 1)$ **88.** $\sum_{k=1}^{4} \frac{n}{n + 1}$ **89.** $\sum_{k=0}^{n} \binom{n}{k}$

90. $\sum_{k=0}^{n} a^{n-k}b^k$

[8.5] *Write the expressions in Problems 91–95 using summation notation.*

91. $2 + 4 + 6 + 8 + 10$ **92.** $256 + 128 + 64 + \cdots + 2 + 1$

93. $a_1 + a_2 + a_3 + \cdots + a_{n-2} + a_{n-1} + a_n$

94. $g_1r + g_1r^2 + g_1r^3 + g_1r^4 + \cdots + g_1r^n$

95. $(x + y)^n$

[8.4] **96.** Find $2.\overline{18}$ as the quotient of two integers by considering an infinite geometric series.

[8.6] **97.** Find:

a. $(x - y)^5$ **b.** $(2x + y)^5$

c. The coefficient of x^8y^4 in the expansion of $(x + 2y)^{12}$

[8.2, 8.3] **98.** *Sociology* Suppose someone tells you she has traced her family tree back 10 generations. What is the minimum number of people on her family tree if there were no intermarriages?

[8.3] **99.** *Biology* A certain bacterium divides into two bacteria every 20 minutes. If there are 1024 bacteria in the culture now, how many will there be in 24 hours, assuming that no bacteria die? Leave your answer in exponential form.

[8.4] **100.** *Physics* A pendulum is swung 125 cm and allowed to swing free until it eventually comes to rest. Each subsequent swing of the bob of the pendulum is 80% as far as the preceding swing. How far will the bob travel before coming to rest?

CHAPTER 9

Combinatorics and Probability

Blaise Pascal
(1623–1662)

Pierre de Fermat
(1601–1665)

The mathematical theory of probability is a science which aims at reducing to calculation, where possible, the amount of credence due to propositions or statements, or to the occurrence of events, future or past, more especially as contingent or dependent upon other propositions or events the probability of which is known.

M. W. Crofton,
"Probability," Encyclopedia Britannica, 9th ed., Cambridge, England, at the University Press, New York, 1875–1889

Historical Note

The mathematical theory of probability arose in France in the 17th century when a gambler, Chevalier de Méré, became interested in adjusting the stakes so that he could be certain of winning if he played long enough. In 1654 he sent some problems to Blaise Pascal who in turn sent them to Pierre de Fermat; together they developed the first theory of probability. From these beginnings, the theory of probability developed into one of the most important branches of mathematics. Today it has applications in business, economics, genetics, insurance, medicine, physics, psychology, and the social sciences.

Chapter Overview

This chapter applies some of the algebraic concepts considered earlier. Probability is a method of mathematically considering events that are not certain. It is one of the best examples of mathematical modeling that can be considered at this level. If you continue your study in mathematics, you will work with mathematical modeling a great deal.

9.1 Permutations

Consider a set

$$A = \{a, b, c, d, e\}$$

Remember, when set symbols are used, the order in which the elements are listed is not important. (See Appendix A for a review of sets.) Suppose now that you wish to select elements from A by taking them in a certain order. The selected elements are enclosed in parentheses to signify order and are called an **arrangement** of elements of A. For example, if the elements a and b are selected from A, then there are two different ordered pairs, or arrangements.

$$(a, b) \qquad \text{and} \qquad (b, a)$$

Remember that when parentheses are used, the order in which the elements are listed *is* important. This example shows ordered pairs, but you could also select an **ordered triple** such as (d, c, a), from A. These arrangements are said to be selected *without repetitions,* since a symbol cannot be used twice in the same arrangement.

Example 1 List the arrangements of the elements a, c, and d selected from A.

Solution

$$(a, c, d), \quad (a, d, c), \quad (c, a, d), \quad (c, d, a), \quad (d, a, c), \quad (d, c, a)$$

□

If an arbitrary finite set S has n elements and r elements are selected from S (where $r \leq n$), then an arrangement without repetitions of the r selected elements is called a *permutation*.

PERMUTATION

A **permutation** of r elements of a set S with n elements is an ordered arrangement of those r elements selected without repetitions.

Example 2 How many permutations of 2 elements can be selected from a set of 6 elements?

Solution Let $B = \{a, b, c, d, e, f\}$ and select 2 elements.

$$\left.\begin{matrix} (a,b), & (a,c), & (a,d), & (a,e), & (a,f) \\ (b,a), & (b,c), & (b,d), & (b,e), & (b,f) \\ (c,a), & (c,b), & (c,d), & (c,e), & (c,f) \\ (d,a), & (d,b), & (d,c), & (d,e), & (d,f) \\ (e,a), & (e,b), & (e,c), & (e,d), & (e,f) \\ (f,a), & (f,b), & (f,c), & (f,d), & (f,e) \end{matrix}\right\}$$

There are 30 permutations of 2 elements selected from a set of 6 elements

□

Example 2 brings up two difficulties. The first is the lack of notation for the phrase,

"*the number of permutations of 2 elements selected from a set of 6 elements,*"

and the second is the inadequacy of relying on direct counting, especially if the sets are very large.

NOTATION FOR PERMUTATIONS

$_nP_r$ is a symbol used to denote the **number of permutations** of r elements selected from a set of n elements.

Example 2 can now be shortened by writing

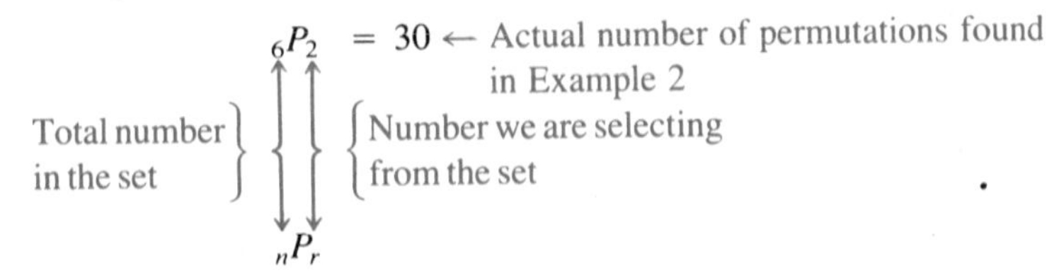

Next, to find a formula for $_nP_r$, we turn to a general result called the **Fundamental Counting Principle.**

FUNDAMENTAL COUNTING PRINCIPLE

If task A can be performed in m ways, and, after task A is performed, a second task B can be performed in n ways, then task A followed by task B can be performed in $m \cdot n$ ways.

Example 3 Select 2 elements from a set of 6 elements, and compute $_6P_2$ using the Fundamental Counting Principle rather than direct listing (as was done in Example 2).

Solution

Number of choices for first component	Number of choices for second component
6	5

Thus, from the Fundamental Counting Principle,

$$_6P_2 = 6 \cdot 5$$
$$= \mathbf{30}$$ □

Example 4 $_7P_3 = \underbrace{7 \cdot 6 \cdot 5}_{\text{Three factors}} = \mathbf{210}$ □

Example 5 $_{10}P_4 = \underbrace{10 \cdot 9 \cdot 8 \cdot 7}_{\text{Four factors}} = \mathbf{5040}$ □

In general, the number of permutations of n objects taken r at a time is given by the following formula:

$$\mathbf{{}_nP_r = \underbrace{n \cdot (n-1) \cdot (n-2) \cdot \cdots \cdot (n-r+1)}_{r \text{ factors}}}$$

Using factorials, this formula for permutations can be written more simply. For example, ${}_6P_6 = 6!$ and, in general,

$${}_nP_n = n!$$

In order to write ${}_nP_r$ using factorial notation, notice

$$\begin{aligned}
{}_nP_r &= n(n-1)(n-2) \cdot \cdots \cdot (n-r+1) \qquad \textit{Note: } n-(r-1) = n-r+1 \\
&= n(n-1)(n-2) \cdot \cdots \cdot (n-r+1) \cdot \frac{(n-r)!}{(n-r)!} \\
&= \frac{n(n-1)(n-2) \cdot \cdots \cdot (n-r+1)(n-r)(n-r-1) \cdot \cdots \cdot 3 \cdot 2 \cdot 1}{(n-r)!} \\
&= \frac{n!}{(n-r)!}
\end{aligned}$$

This is the general formula for ${}_nP_r$.

PERMUTATION FORMULA

$$\mathbf{{}_nP_r = \frac{n!}{(n-r)!}}$$

Example 6

$$\begin{aligned}
{}_{10}P_2 &= \frac{10!}{(10-2)!} \\
&= \frac{10 \cdot 9 \cdot \cancel{8!}}{\cancel{8!}} \\
&= \mathbf{90}
\end{aligned}$$

□

Example 7

$$\begin{aligned}
{}_nP_0 &= \frac{n!}{(n-0)!} \\
&= \mathbf{1}
\end{aligned}$$

□

Example 8 Find the number of license plates possible in a state using only 3 letters, if none of the letters can be repeated. This is a permutation of 26 objects taken 3 at a time. Thus, the solution is given by

$${}_{26}P_3 = \underbrace{26 \cdot 25 \cdot 24}_{\text{Three factors}} = \mathbf{15{,}600}$$

□

Example 9 Repeat Example 8 if repetitions are allowed. Remember, permutations do not allow repetitions, so this is *not* a permutation. To answer the question, use the Fundamental Counting Principle:

$$26 \cdot 26 \cdot 26 = \mathbf{17{,}576}$$

□

Example 10 Find the number of arrangements of letters in the word MATH. This is a permutation of 4 objects taken 4 at a time.

$${}_4P_4 = 4 \cdot 3 \cdot 2 \cdot 1 = \mathbf{24}$$

□

Example 11 How many permutations are there of the letters in the word HATH? If you try to solve this problem as you did Example 10, the result is

$${}_4P_4 = 4! = 24$$

However, if you list the possibilities, you will find only *12 different permutations*. The difficulty here is that two of the letters in the word HATH are *indistinguishable*. If you label them as H_1ATH_2, you can find additional possibilities, such as

H_1ATH_2
H_2ATH_1
H_1AH_2T

If you complete this list, you will find $4! = 24$ possibilities. This means that, since there are *two* indistinguishable letters, you divide the total, $4!$, by 2 to find the result:

$$\frac{4!}{2} = \frac{24}{2} = \mathbf{12}$$

□

Example 12 How many permutations are there of the letters in the word ASSIST? There are 6 letters, and if you consider the letters as *distinguishable*, as in

$AS_1S_2IS_3T$

there are ${}_6P_6 = 6! = 720$ possibilities. However,

$AS_1S_2IS_3T$
$AS_1S_3IS_2T$
$AS_2S_1IS_3T$
$AS_2S_3IS_1T$
$AS_3S_1IS_2T$
$AS_3S_2IS_1T$

are all indistinguishable, so you must divide the total by $3! = 6$:

$$\frac{6!}{3!} = \frac{6 \cdot 5 \cdot 4 \cdot \cancel{3!}}{\cancel{3!}} = 120$$

There are **120 permutations of the letters in the word ASSIST.** □

Examples 11 and 12 suggest a general result.

NUMBER OF DISTINGUISHABLE PERMUTATIONS

> The number of **distinguishable permutations** of n objects, of which r are alike and the remaining objects are different from each other, is
>
> $$\frac{n!}{r!}$$

This result generalizes to include several subcategories.

Example 13 The number of permutations of the letters in the word

ATTRACT

is 7! (since there are 7 letters) divided by factorials of the number of subcategories of repeated letters:

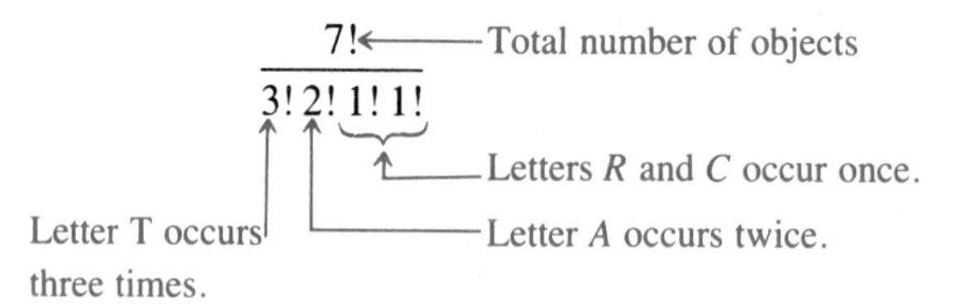

This number can now be simplified:

$$\frac{7 \cdot 6 \cdot 5 \cdot \overset{2}{\cancel{4}} \cdot \cancel{3!}}{\cancel{3!} \cdot \cancel{2}} = \mathbf{420}$$

GENERAL FORMULA FOR THE NUMBER OF DISTINGUISHABLE PERMUTATIONS

> The number of **distinguishable permutations** of n objects in which n_1 are of one kind, n_2 are of another kind, . . . , and n_k are of a further kind so that
>
> $$n = n_1 + n_2 + \cdots + n_k$$
>
> is given by the formula
>
> $$\frac{n!}{n_1!n_2! \cdot \cdots \cdot n_k!}$$

Example 14 What is the number of distinguishable permutations of the letters in the words COLLEGE ALGEBRA?

Solution

$$\frac{14!}{3!\ 3!\ 2!\ 2!\ 1!\ 1!\ 1!\ 1!} = \frac{\overset{7}{\cancel{14}} \cdot 13 \cdot \cancel{12} \cdot 11 \cdot \overset{5}{\cancel{10}} \cdot \overset{3}{\cancel{9}} \cdot 8!}{\cancel{3} \cdot \cancel{2} \cdot \cancel{3} \cdot \cancel{2} \cdot \cancel{2} \cdot \cancel{2}}$$ Cancel where possible.

$= 7 \cdot 13 \cdot 11 \cdot 5 \cdot 3 \cdot 8!$ Factored from answer.

$= \mathbf{605{,}404{,}800}$ Use a calculator if you do not want to leave your answer in factored form.

If you also consider the space between the words and its proper location (which would be necessary if you were programming this on a computer), then the number of possibilities is found by

$$\frac{15!}{3!\ 3!\ 2!\ 2!\ 1!\ 1!\ 1!\ 1!\ 1!} = 9{,}081{,}072{,}000$$

□

Problem Set 9.1

Evaluate each expression in Problems 1–30.

A

1. ${}_9P_1$	**2.** ${}_9P_2$	**3.** ${}_9P_3$	**4.** ${}_9P_4$	**5.** ${}_9P_0$
6. ${}_5P_4$	**7.** ${}_{52}P_3$	**8.** ${}_7P_2$	**9.** ${}_4P_4$	**10.** ${}_{100}P_1$
11. ${}_{12}P_5$	**12.** ${}_5P_3$	**13.** ${}_8P_4$	**14.** ${}_8P_0$	**15.** ${}_gP_h$
16. ${}_{92}P_0$	**17.** ${}_{52}P_1$	**18.** ${}_7P_5$	**19.** ${}_{16}P_3$	**20.** ${}_nP_4$
21. ${}_7P_3$	**22.** ${}_5P_5$	**23.** ${}_{50}P_{48}$	**24.** ${}_{25}P_1$	**25.** ${}_mP_3$
26. ${}_8P_3$	**27.** ${}_{12}P_0$	**28.** ${}_{10}P_2$	**29.** ${}_{11}P_4$	**30.** ${}_nP_5$

How many permutations are there in the words given in Problems 31–39?

31. HOLIDAY	**32.** ANNEX	**33.** ESCHEW
34. OBFUSCATION	**35.** MISSISSIPPI	**36.** CONCENTRATION
37. BOOKKEEPING	**38.** GRAMMATICAL	**39.** APOSIOPESIS

40. State the Fundamental Counting Principle.

B

41. In how many ways can a group of 5 people elect a president, vice-president, secretary, and treasurer?

42. In how many ways can a group of 15 people elect a president and a vice-president?

43. In how many ways can a group of 10 people elect a president, a vice-president, and a secretary?

44. How many outfits consisting of a skirt and a blouse can a woman select if she has three skirts and five blouses?

45. How many outfits consisting of a suit and a tie can a man select if he has two suits and eight ties?

46. In how many different ways can eight books be arranged on a shelf?

47. In how many ways can you select and read three books from a shelf of eight books?

48. In how many ways can a row of three contestants for a TV game show be selected from an audience of 362 people?

49. How many 7 digit telephone numbers are possible if the first 2 digits cannot be ones or zeros?

50. Foley's Village Inn offers the following menu in its restaurant:

Main Course	Dessert	Beverage
Prime rib	Ice cream	Coffee
Steak	Sherbet	Tea
Chicken	Cheesecake	Milk
Ham		Sanka
Shrimp		

In how many different ways can someone order a meal consisting of one choice from each category?

51. A typical Social Security number is 576-38-4459; the first digit cannot be zero. How many Social Security numbers are possible?

52. California license plates consist of 1 digit, followed by 3 letters, followed by 3 digits. How many such license plates are possible?

53. If a state issues license plates that consist of 1 letter followed by 5 digits, how many different plates are possible?

54. Repeat Problem 53 if the first letter cannot be O, Q, or I.

55. Repeat Problem 53 where repetition of digits is not allowed.

56. New York license plates consist of 3 letters followed by 3 digits. It is also known that 245 specific arrangements of 3 letters are not allowed because they are considered obscene. How many license plates are possible?

57. A certain lock has five tumblers, and each tumbler can assume six positions. How many different positions are there?

58. Tarot cards are used for telling fortunes, and in a reading, the arrangement of the cards is as important as the cards themselves. How many different readings are possible if 3 cards are selected from a set of 7 Tarot cards?

59. Texas license plates have 3 letters followed by 3 digits. If we assume that all such plates are possible, how many Texas license plates have a repeating letter?

60. Suppose you flip a coin and keep a record of the results. In how many ways could you obtain at least one head if you flip the coin six times?

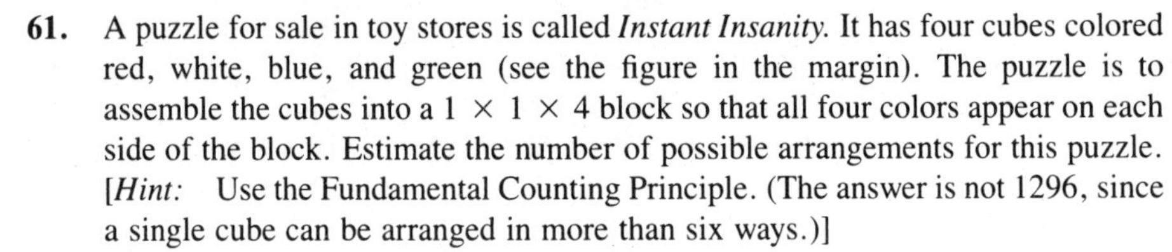

C **61.** A puzzle for sale in toy stores is called *Instant Insanity.* It has four cubes colored red, white, blue, and green (see the figure in the margin). The puzzle is to assemble the cubes into a $1 \times 1 \times 4$ block so that all four colors appear on each side of the block. Estimate the number of possible arrangements for this puzzle. [*Hint:* Use the Fundamental Counting Principle. (The answer is not 1296, since a single cube can be arranged in more than six ways.)]

62. How many triangles are there in each figure?

a. 

b. 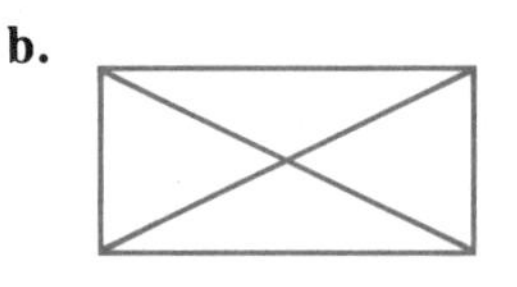

c. 

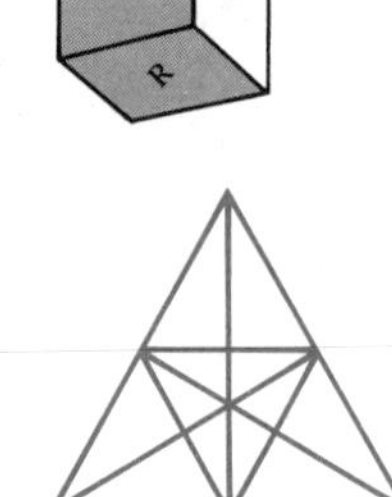

63. How many triangles are there in the figure shown in the margin?

9.2 Combinations

Again, consider the set

$$A = \{a, b, c, d, e\}$$

If two elements are selected from A in a certain order, they are represented by an *ordered pair* and the ordered pair is called a *permutation*. On the other hand, if two elements are selected from A *without regard to the order in which they are selected,* they are represented as a subset of A.

Example 1 Select two elements from A:

Permutations—Order Important

(a, b), (a, c), (a, d), (a, e)
(b, a), (b, c), (b, d), (b, e)
(c, a), (c, b), (c, d), (c, e)
(d, a), (d, b), (d, c), (d, e)
(e, a), (e, b), (e, c), (e, d)

There are 20 permutations—order is important.

Notation: ${}_5P_2 = 20$

Subsets—Order Not Important

$\{a, b\}$, $\{a, c\}$, $\{a, d\}$, $\{a, e\}$
$\{b, c\}$, $\{b, d\}$, $\{b, e\}$
$\{c, d\}$, $\{c, e\}$
$\{d, e\}$

Do not list $\{b, a\}$ since $\{b, a\} = \{a, b\}$.

There are 10 subsets; notice that set notation is used. □

A name and some notation are needed for listing different subsets of a given set.

DEFINITION OF COMBINATION

A **combination** of r elements of a finite set S is a subset of S that contains r distinct elements.

Remember, when listing the elements of a subset the order in which those elements are listed is not important. A notation similar to that used for permutations is used to denote the number of combinations.

NOTATION FOR COMBINATIONS

${}_nC_r$ is a symbol used to denote the **number of combinations** of r elements selected from a set of n elements ($r \leq n$).

The formula for the number of permutations leads directly to a formula for the number of combinations. Since each subset of r elements has $r!$ permutations of its members,

$$ {}_nP_r = {}_nC_r \cdot r! \qquad \text{so} \qquad {}_nC_r = \frac{{}_nP_r}{r!} = \frac{n!}{r!(n-r)!} $$

COMBINATION FORMULA

$$ {}_nC_r = \frac{n!}{r!(n-r)!} $$

To find ${}_nC_r$ you may either use this formula or, for small numbers, proceed as follows:

1. Write ${}_nP_r$ in factored form.
2. Divide by $r!$.

3. Simplify the result.

Example 2

$$_{10}C_3 = \frac{_{10}P_3}{3!} = \frac{10 \cdot 9 \cdot 8}{3 \cdot 2 \cdot 1} = \mathbf{120}$$ □

Example 3

$$_nC_0 = \frac{n!}{0!(n-0)!} = \mathbf{1}$$ □

Example 4

$$_{m-1}C_2 = \frac{(m-1)!}{2!(m-1-2)!} = \frac{(m-1)(m-2)(m-3)!}{2 \cdot 1 \cdot (m-3)!} = \frac{\mathbf{(m-1)(m-2)}}{\mathbf{2}}$$ □

Example 5 In how many ways can a club of 5 members select a 3 person committee?

Solution

$$_5C_3 = \frac{5 \cdot 4 \cdot 3}{3!} = \mathbf{10}$$ The order is not important. □

Example 6 Find the number of 5 card hands that can be drawn from an ordinary deck of cards.

Solution

$$_{52}C_5 = \frac{_{52}P_5}{5!} = \frac{52 \cdot 51 \cdot 50 \cdot 49 \cdot 48}{5 \cdot 4 \cdot 3 \cdot 2 \cdot 1} = \mathbf{2{,}598{,}960}$$ □

Example 7 In how many ways can a diamond flush be drawn in poker? (A diamond flush is a hand of 5 diamonds.) This is a combination of 13 objects (diamonds) taken 5 at a time. Thus, the solution is given by

Solution

$$_{13}C_5 = \frac{_{13}P_5}{5!} = \frac{13 \cdot 12 \cdot 11 \cdot 10 \cdot 9}{5!} = \mathbf{1287}$$ □

Example 8 In how many ways can a flush be drawn in poker?

Solution Begin with the Fundamental Counting Principle:

$$\underbrace{4}_{\text{Number of suits}} \cdot \underbrace{1287}_{\text{Number of ways of drawing a flush in a particular suit (from Example 7)}} = \mathbf{5148}$$ □

Compare the formulas for ${}_nC_r$ and $\binom{n}{r}$, and you will see that combinations are the same as the binomial coefficients. The relationship is used in the following example.

Example 9 If a set has 10 elements, find the number of distinct subsets of all sizes.

Solution Number of subsets of

Size 0: ${}_{10}C_0$
Size 1: ${}_{10}C_1$
. .
. .
. .
Size 9: ${}_{10}C_9$
Size 10: ${}_{10}C_{10}$

$$\text{TOTAL: } {}_{10}C_0 + {}_{10}C_1 + {}_{10}C_2 + \cdots + {}_{10}C_9 + {}_{10}C_{10}$$

$$= \binom{10}{0} + \binom{10}{1} + \binom{10}{2} + \cdots + \binom{10}{9} + \binom{10}{10}$$

Now, consider $(1 + 1)^{10}$ using the Binomial Theorem:

$$(1 + 1)^{10} = \binom{10}{0}1^{10} \cdot 1^0 + \binom{10}{1}1^9 \cdot 1^1 + \binom{10}{2}1^8 \cdot 1^2 + \cdots$$

$$+ \binom{10}{9}1^1 \cdot 1^9 + \binom{10}{10}1^0 \cdot 1^{10}$$

$$= \binom{10}{0} + \binom{10}{1} + \binom{10}{2} + \cdots + \binom{10}{9} + \binom{10}{10}$$

Thus, **the total number of subsets is $(1 + 1)^{10} = 2^{10}$.** □

In practice, you are usually required to decide whether a given counting problem is a permutation or a combination (or neither) before you find the solution. Remember, with a permutation the *order is important;* with a combination the *order is not important*.

PERMUTATION AND COMBINATION

> A **permutation** of a set of objects is an arrangement of certain of these objects in a *definite order and without repetitions.*
>
> A **combination** of a set of objects is a collection of certain of these objects *without regard to their order.*

Determine whether you would use a permutation, a combination, or neither to find the given number in Examples 10–13.

Example 10 What is the number of license plates possible in Florida if each license plate consists of 3 letters followed by 3 digits, and we add the condition that repetition of letters or digits is not permitted?

Solution This is a permutation problem since the *order* in which the elements are arranged is important. That is, CWB072 and BCW072 are different plates. The number is found by using permutations along with the Fundamental Counting Principle:

$$\begin{aligned} {}_{26}P_3 \cdot {}_{10}P_3 &= 26 \cdot 25 \cdot 24 \cdot 10 \cdot 9 \cdot 8 \leftarrow \text{This answer is acceptable if you do not have a calculator.} \\ &= \mathbf{11{,}232{,}000} \end{aligned}$$

□

Example 11 Find the number of 3 letter "words" that can be formed using 3 letters from among $\{m, a, t, h\}$.

Solution Finding these words is also a permutation problem since *mat* is different from *tam*.

$$\begin{aligned} {}_4P_3 &= 4 \cdot 3 \cdot 2 \\ &= \mathbf{24} \end{aligned}$$

□

Example 12 Find the number of bridge hands (13 cards) consisting of 6 hearts, 4 spades, and 3 diamonds.

Solution The order in which the cards were received is unimportant, so finding the number of bridge hands is a combination problem. Also use the Fundamental Counting Principle:

$$\underbrace{{}_{13}C_6}_{\text{Number of ways of obtaining 6 hearts}} \cdot \underbrace{{}_{13}C_4}_{\text{Number of ways of obtaining 4 spades}} \cdot \underbrace{{}_{13}C_3}_{\text{Number of ways of obtaining 3 diamonds}}$$

$$= \frac{13!}{6!(13-6)!} \cdot \frac{13!}{4!(13-4)!} \cdot \frac{13!}{3!(13-3)!}$$

$$= \frac{13!13!13!}{6!7!4!9!3!10!} \leftarrow \text{This answer is acceptable if you do not have a calculator.}$$

$$= \mathbf{350{,}904{,}840}$$

□

Example 13 A club with 42 members wants to elect a president, a vice-president, and a treasurer. From the other members, an advisory committee of 5 people is to be selected. In how many ways can this be done?

Solution This is both a permutation and a combination problem, with the final result calculated by using the Fundamental Counting Principle:

$$\underbrace{{}_{42}P_3}_{\text{Number of ways of selecting officers}} \cdot \underbrace{{}_{39}C_5}_{\text{Number of ways of selecting committee}}$$

$$= 42 \cdot 41 \cdot 40 \cdot \frac{39!}{5!(39-5)!}$$

$= \mathbf{39{,}658{,}142{,}160}$ Found with a calculator □

Example 14 How many subsets are there of the members of a club of 42 people?

Solution This is the sum of a large number of combinations, since ${}_{42}C_1$ is the number of subsets of one person; ${}_{42}C_2$ is the number of subsets of two persons, and so on. The number you are looking for is

$${}_{42}C_0 + {}_{42}C_1 + {}_{42}C_2 + \cdots + {}_{42}C_{41} + {}_{42}C_{42}$$

From Pascal's Triangle (Problem 41 of Problem Set 8.6), this is $\mathbf{2^{42}}$. This answer is acceptable. On a calculator, it is given approximately: $\mathbf{4.398046 \times 10^{12}}$. □

Problem Set 9.2

A *Evaluate each expression in Problems 1–30.*

1. ${}_9C_1$	**2.** ${}_9C_2$	**3.** ${}_9C_3$	**4.** ${}_9C_4$	**5.** ${}_9C_0$
6. ${}_5C_4$	**7.** ${}_{52}C_3$	**8.** ${}_7C_2$	**9.** ${}_4C_4$	**10.** ${}_{100}C_1$
11. ${}_7C_3$	**12.** ${}_5C_5$	**13.** ${}_{50}C_{48}$	**14.** ${}_{25}C_1$	**15.** ${}_gC_h$
16. ${}_7C_5$	**17.** ${}_8C_0$	**18.** ${}_{10}C_2$	**19.** ${}_{12}C_5$	**20.** ${}_nC_4$
21. ${}_{92}C_0$	**22.** ${}_{52}C_1$	**23.** ${}_4C_3$	**24.** ${}_{16}C_3$	**25.** ${}_nC_5$
26. ${}_{28}C_1$	**27.** ${}_7C_5$	**28.** ${}_7C_7$	**29.** ${}_{12}C_2$	**30.** ${}_mC_n$

Use Pascal's Triangle to evaluate the expressions in Problems 31–35.

31. ${}_6C_5$	**32.** ${}_7C_3$	**33.** ${}_8C_6$	**34.** ${}_7C_5$	**35.** ${}_8C_2$

36. A bag contains 12 pieces of candy. In how many ways can 5 pieces be selected?

37. If the Senate is to form a new committee of 5 members, in how many different ways can the committee be chosen if all 100 senators are available to serve on this committee?

38. In how many ways can 3 aces be drawn from a deck of cards?

39. In how many ways can 2 kings be drawn from a deck of cards?

40. In how many ways can 4 aces be drawn from a deck of cards?

41. In how many ways can a heart flush be obtained? (A heart flush is a hand of 5 hearts.)

42. In how many ways can a spade flush be obtained? (A spade flush is a hand of 5 spades.)

43. In how many ways can a full house of 3 aces and 2 kings be obtained?

44. In how many ways can a full house of 3 jacks and a pair of 2s be obtained?

B **45.** Explain the difference between a permutation and a combination and illustrate with everyday examples.

In Problems 46–67 decide whether you would use a permutation, a combination, or neither. Next, write the solution using permutation or combination notation, if possible, and finally, answer the question asked.

46. How many different arrangements are there of the letters in the word CORRECT?

47. At Mr. Furry's Dance Studio, every man must dance the last dance. If there are five men and eight women, in how many ways can dance couples be formed for the last dance?

48. Martin's Ice Cream Store sells sundaes with chocolate, strawberry, butterscotch, or marshmallow toppings, nuts, and whipped cream. If you can choose exactly three of these extras, how many possible sundaes are there?

49. Five people are to dine together at a rectangular table, but the host cannot decide on a seating arrangement. In how many ways can the guests be seated, assuming fixed chair positions?

50. In how many ways can 3 hearts be drawn from a deck of cards?

51. A shipment of 100 TV sets is received. Six sets are to be chosen at random and tested for defects. In how many ways can the six sets be chosen?

52. A night watchman visits 15 offices every night. To prevent others from knowing when he will be at a particular office, he varies the order of his visits. In how many ways can this be done?

53. A certain manufacturing process calls for six chemicals to be mixed. One liquid is to be poured into the vat, and then the others are to be added in turn. All possible combinations must be tested to see which gives the best results. How many tests must be performed?

54. There are three boys and three girls at a party. In how many ways can they be seated in a row if they can sit down four at a time?

55. There are three boys and three girls at a party. In how many ways can they be seated in a row if they want to sit alternating boy, girl, boy, girl?

56. If there are ten people in a club, in how many ways can they choose a dishwasher and a bouncer?

57. How many subsets can be formed from a set of 100 elements?

58. In how many ways can you be dealt 2 cards from an ordinary deck of 52 cards?

59. In how many ways can five taxi drivers be assigned to six cars?

60. How many arrangements are there of the letters in the word GAMBLE?

61. A student is asked to answer 10 out of 12 questions on an exam. In how many ways can the questions to be answered be selected?

62. In how many ways can a group of seven choose a committee of four?

63. In how many ways can seven books be arranged on a bookshelf?

64. In how many ways can you choose two books to read from a bookshelf containing seven books?

65. A certain mathematics test consists of ten questions. In how many ways can the test be answered if the possible answers are *true* and *false*?

66. Answer the question in Problem 65 if the possible answers are *true*, *false*, and *maybe*.

67. Answer the question in Problem 65 if the possible answers are (a), (b), (c), (d), and (e); that is, the test is multiple-choice.

C **68.** A club consists of 17 men and 19 women. In how many ways can they choose a president, vice-president, treasurer, and secretary, along with an advisory committee of 6 people?

69. In Problem 68, how many ways can the selection be made if two of the officers must be women?

70. In a bridge tournament, eight couples played bridge seven times. The people changed groupings every time they played, and, as it turned out, no man and woman were ever partners or opponents more than once. How were the couples distributed if no husband and wife were ever partners?

71. Prove: ${}_nC_r = {}_nC_{n-r}$

9.3 Definition of Probability

A **probability function** is a function that assigns to an event a number representing the likelihood that the event will occur. However, before a precise definition can be given, some preliminary terminology is necessary. The **sample space** of an experiment is the set of *all* possible outcomes, while an **event** is simply a subset of the sample space. For example, if you simultaneously toss a coin and roll a die, the sample space is

$$\{1H, 1T, 2H, 2T, 3H, 3T, 4H, 4T, 5H, 5T, 6H, 6T\}$$

where H represents heads and T represents tails. An event might be *obtaining a 5*, which is the subset

$$\{5H, 5T\}$$

If the sample space can be divided into **mutually exclusive** (*if one event occurs, the others in the sample space S cannot occur*) and **equally likely** outcomes, then the probability of an event can be defined. Consider an experiment in which a coin is tossed. A suitable sample space is

$$S = \{\text{Heads, Tails}\}$$

Suppose you wish to consider the event of *obtaining heads;* call it event A:

$$A = \{\text{Heads}\}$$

In this book the probability function is denoted by P, so the probability of A is denoted by $P(A)$. Notice that the outcomes in this sample space are *mutually exclusive*. If each outcome in the sample space is *equally likely*, the probability of A is found by

$$P(A) = \frac{\text{Number of successful results}}{\text{Number of possible results}}$$

A *successful* result is a result that corresponds to the probability you are seeking—in this case, {Heads}. Since you can obtain a head in only one way (success), and the total number of possible outcomes is 2 (number of elements in the sample space), then the probability of heads as given by this definition is

$$P(\text{Heads}) = P(A) = \tfrac{1}{2}$$

This, of course, corresponds to your experience of flipping coins. We now give a definition of probability.

PROBABILITY OF AN EVENT THAT CAN OCCUR IN ANY ONE OF n MUTUALLY EXCLUSIVE AND EQUALLY LIKELY WAYS

> If an experiment can occur in any of n, $n \geq 1$, mutually exclusive and equally likely ways and if s of these ways are considered favorable, then the probability of the event E, denoted by $P(E)$, is
>
> $$P(E) = \frac{s}{n} = \frac{\textbf{Number of outcomes favorable to } E}{\textbf{Number of all possible outcomes}}$$

This probability function P has the following properties, which follow directly from the definition:

1. $P(E) \geq 0$, since both $s \geq 0$ and $n \geq 1$.
2. $P(E) \leq 1$, since $s \leq n$. (Remember, n represents the total of *all* possible outcomes.)
3. If $s = 0$, then $P(E) = 0$, which means that a success *cannot* occur.
4. If $s = n$, then $P(E) = 1$, which means that a success *must* occur.
5. In summary,

$$0 \leq P(E) \leq 1$$

The closer $P(E)$ is to 0, the less likely is the event E; the closer $P(E)$ is to 1, the more likely is the event E. If the event E *must* occur (it is certain), then $P(E) = 1$; and if the event E cannot occur, then $P(E) = 0$.

Example 1 Suppose a single card is selected from an ordinary deck of 52 cards. Find the requested probabilities.

a. $P(\text{Heart})$. There are 52 elements in the sample space and 13 of these are hearts, or successes. Therefore,

$$P(\text{Heart}) = \frac{13}{52} = \frac{1}{4}$$

Probability fractions should be reduced or be presented in decimal form.

b. $P(\text{Ace}) = \frac{4}{52} = \frac{1}{13}$

c. $P(\text{Heart or ace}) = \frac{16}{52} = \frac{4}{13}$

13 hearts + 3 *additional* aces (Be careful not to count the ace of hearts twice.)

d. $P(\text{Heart and ace}) = \frac{1}{52}$

The ace of hearts is the only such card.

e. $P(\text{Ace or } 2) = \frac{8}{52}$

$= \mathbf{\frac{2}{13}}$

f. $P(\text{Ace and } 2) = \frac{0}{52}$ There is no way of drawing a single card and obtaining an ace and a 2!

$= \mathbf{0}$ □

FINDING THE PROBABILITY OF AN EVENT

In summary, to find the probability of some event:

1. Describe and identify the sample space, and then count the number of elements (these should be equally likely). Call this number n.
2. Count the number of occurrences that interest you; call this the *number of successes* and denote it by s.
3. Compute the probability of the event: $P(E) = \frac{s}{n}$.

Remember that this scheme works only if the event occurs in n mutually exclusive and equally likely ways.

In carrying out Step 1 of the above procedure, it is important that you make sure the elements in the sample space are equally likely. Consider the following three examples.

Example 2 Roll a single die. The sample space is $\{1, 2, 3, 4, 5, 6\}$. Find $P(2)$.

Solution $P(2) = \mathbf{\frac{1}{6}}$ ←Number of elements in the sample space □

Example 3 Roll a pair of dice. The sample space is $\{2, 3, 4, 5, 6, 7, 8, 9, 10, 11, 12\}$. Find $P(2)$, where $P(2)$ means that the dice add up to 2.

Solution $P(2) \neq \frac{1}{11}$ ←Number of elements in the sample space

If you can find the probability in Example 2 by counting the elements in the sample space, why not do the same for this example? You cannot, because the outcomes in the sample space for this example are *not equally likely*. Study Figure 9.1.

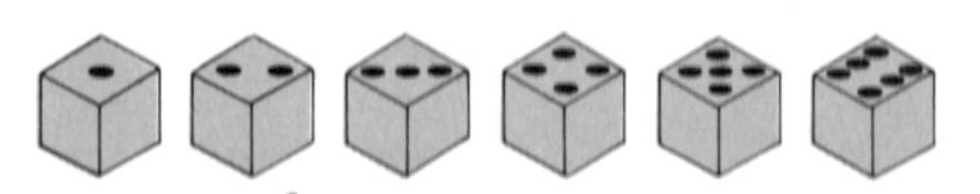

6 equally likely possibilities

Figure 9.1a
Sample Space for One Die

Sample space for tossing a pair of dice:

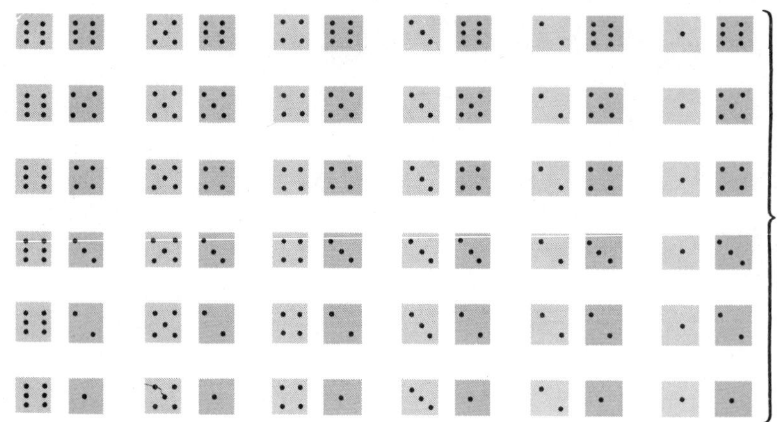

Figure 9.1b
Sample Space Two Dice

There are 36 equally likely events. Thus,

$$P(2) = \frac{1}{36} \quad \begin{array}{l}\leftarrow \text{Number of outcomes in Figure 9.1 totaling 2} \\ \leftarrow \text{Total number of outcomes in Figure 9.1}\end{array}$$

□

Example 4 Roll a pair of dice. What is the probability of *craps* (rolling a sum of 2, 3, or 12)?

Solution

$$P(\text{Craps}) = \frac{4}{36} \quad \begin{array}{l}\leftarrow \text{4 outcomes (see Figure 9.1)} \\ \leftarrow \text{36 possible outcomes}\end{array}$$

$$= \frac{1}{9}$$

□

Examples 2–4 illustrate the importance of setting up a sample space of equally likely possibilities. If you do, then the probability of an event can be found by counting successes and dividing by the number of equally likely possibilities, as outlined in the procedure given above.

Example 5 Assume that a jar contains 3 red marbles, 2 black marbles, and 5 green marbles. The experiment consists of drawing a single marble. Find the requested probabilities.

a. $P(\text{Red})$. A possible sample space is {Red, Green, Black}, but these possibilities are not equally likely. (Why?) Consider the following sample space:

$\{R_1, R_2, R_3, B_1, B_2, G_1, G_2, G_3, G_4, G_5\}$

where R_1, R_2, R_3 represent red marbles; B_1, B_2 represent black marbles; and G_1, . . . , G_5 represent green marbles. Then the sample space, as now described, consists of equally likely outcomes, so this is an appropriate model. Thus,

$$P(\text{Red}) = \frac{\text{Number of favorable outcomes}}{\text{Number of all possible outcomes}} = \frac{3}{10}$$

b. $P(\text{Black or green}) = \frac{7}{10}$

c. $P(\text{Black and green}) = \frac{0}{10} = \mathbf{0}$

The number of favorable outcomes is 0—you cannot draw out a marble that is both black and green. □

If the fractions necessary to represent a probability are very complicated, it is customary to represent them in decimal form (rounded to some degree of accuracy). *For the remainder of this section, it is assumed that you have a calculator.*

Example 6 What is the probability of being dealt a flush in poker?

Solution A flush is 5 cards from a deck of 52 cards in which all of them are the same suit (hearts, diamonds, clubs, or spades).

$$P(\text{Flush}) = \frac{\text{Number of ways of obtaining a flush}}{\text{Number of possible poker hands}}$$

$$= \frac{{}_4C_1 \cdot {}_{13}C_5}{{}_{52}C_5}$$

$_4C_1$: Number of kinds of flushes (there are four)

$_{13}C_5$: Number of flushes of a particular kind; this is 1287 from Example 7 on page 376

$_{52}C_5$: Number of ways of drawing 5 cards from 52 cards; this is 2,598,960 from Example 6 on page 376

$$= \frac{4 \cdot 1287}{2{,}598{,}960}$$

$$\approx \mathbf{.00198}$$

This is about two chances in a thousand. □

Example 7 Suppose that in an assortment of 20 electronic calculators, there are 5 with defective switches. Write the requested probabilities as decimals to the nearest hundredth.

a. If a machine is selected at random, what is the probability that it has a defective switch? The solution is given by

$$\frac{\text{Number of ways of selecting 1 defective from the 5}}{\text{Number of ways of selecting 1 machine from the 20}} = \frac{{}_5C_1}{{}_{20}C_1} = \frac{5}{20} = \mathbf{.25}$$

b. If 2 machines are selected at random, what is the probability that they both have defective switches? The solution is given by

$$\frac{\text{Number of ways of selecting 2 defectives from the 5}}{\text{Number of ways of selecting 2 machines from the 20}} = \frac{{}_5C_2}{{}_{20}C_2} = \frac{\frac{5 \cdot 4}{2}}{\frac{20 \cdot 19}{2}} = \frac{5 \cdot 4}{20 \cdot 19} \approx \mathbf{.05}$$

c. If 3 machines are selected at random, what is the probability that exactly 1 has a defective switch? In this problem use the Fundamental Counting Principle to determine the number of successes. Picking exactly 1 machine with a defective switch:

$$\frac{\begin{pmatrix}\text{Number of ways of picking} \\ \text{1 defective from the 5}\end{pmatrix} \cdot \begin{pmatrix}\text{Number of ways of picking the} \\ \text{other 2 from the remaining 15}\end{pmatrix}}{(\text{Number of ways of picking 3 from 20})} = \frac{{}_5C_1 \cdot {}_{15}C_2}{{}_{20}C_3}$$

$$= \frac{\dfrac{5 \cdot 15 \cdot 14}{1 \cdot 2 \cdot 1}}{\dfrac{20 \cdot 19 \cdot 18}{3 \cdot 2 \cdot 1}}$$

$$= \frac{5 \cdot 15 \cdot 7}{20 \cdot 19 \cdot 3}$$

$$= \frac{35}{76}$$

$$\approx .46$$ □

Problem Set 9.3

A *For Problems 1–12 suppose a single card is drawn from an ordinary deck of 52 cards. Find the probabilities and write your answers as reduced fractions.*

1. $P(2)$
2. $P(\text{Spade})$
3. $P(2 \text{ or spade})$
4. $P(2 \text{ and spade})$
5. $P(\text{King})$
6. $P(\text{Queen})$
7. $P(\text{King or queen})$
8. $P(\text{King and queen})$
9. $P(\text{King and spade})$
10. $P(\text{Heart or club})$
11. $P(\text{Not a face card})$
12. $P(\text{King or spade})$

For Problems 13–22 suppose a single die is rolled. Find the probabilities. Write your answers as reduced fractions.

13. $P(4)$
14. $P(4 \text{ or } 5)$
15. $P(\text{Even})$
16. $P(\text{Less than } 5)$
17. $P(4 \text{ and } 5)$
18. $P(5 \text{ or } 6)$
19. $P(1, 2, 3, \text{ or } 4)$
20. $P(\text{Odd or even})$
21. $P(\text{Not a } 5)$
22. $P(\text{Not a } 7)$

For Problems 23–28 suppose an urn contains 6 red balls, 4 white balls, and 3 black balls. One ball is drawn at random. Find the probabilities. Write your answers as reduced fractions.

23. $P(\text{Red})$
24. $P(\text{White})$
25. $P(\text{Black})$
26. $P(\text{Black or white})$
27. $P(\text{Red or white})$
28. $P(\text{Black and white})$

Randomly choose a number between 1 and 100, inclusive. Find the probabilities in Problems 29–33. Write your answers in decimal form.

29. $P(5)$
30. $P(\text{Less than } 5)$
31. $P(\text{Greater than } 5)$
32. $P(\text{Less than 5 or greater than } 5)$
33. $P(\text{Multiple of } 5)$

B *For Problems 34–40 suppose a pair of dice are rolled. (See Figure 9.1.) Find the probabilities. Note: P(4) means that the sum of the dice is 4. Write your answers as reduced fractions.*

34. $P(4)$
35. $P(6)$
36. $P(\text{Even})$
37. $P(\text{Greater than } 5)$
38. $P(\text{Greater than } 7)$
39. P(4 or 5 on *one* of the dice but not the other)
40. P(4 on the first die and 5 on the second die)
41. The game of craps is popular in gambling casinos. Two dice are tossed, and various amounts are paid according to the outcome. If a 7 or 11 occurs on the first roll, the player wins. What is the probability of winning on the first roll?
42. In the game of craps, a pair of 1s is called *snake eyes*. What is the probability of losing a dice game by rolling snake eyes on the first roll?

It is known that a company has 10 pieces of machinery with no defects, 4 with minor defects, and 2 with major defects. Thus, each of the 16 machines falls into one of these three categories. The inspector is about to come, and it is her policy to choose one machine and check it. Find the probabilities in Problems 43–45.

43. P(No defects)
44. P(No major defects)
45. P(No minor defects)

C *The remaining problems in this problem set require a calculator. Find the probabilities as decimals rounded to the nearest hundredth.*

For Problems 46–48 assume that the inspector described in Problems 43–45 selects two machines at random.

46. P(Both defective)
47. P(One with no defects and one with major defect)
48. P(Both nondefective)

For Problems 49–52 assume that the inspector described in Problems 43–45 selects three machines at random. Find the probabilities.

49. P(All have major defects)
50. P(All have minor defects)
51. P(All are nondefective)
52. P(One of each type)

For Problems 53–58 suppose 2 balls are drawn at random from the urn described for Problems 23–28. Find the probabilities.

53. P(Both red)
54. P(Both white)
55. P(Both black)
56. P(1 red and 1 white)
57. P(1 black and 1 white)
58. P(1 red and 1 black)

For Problems 59–65 suppose 3 balls are drawn at random from the urn described for Problems 23–28. Find the probabilities.

59. P(3 red)
60. P(3 white)
61. P(3 black)
62. P(2 red and 1 white)
63. P(2 white and 1 black)
64. P(2 black and 1 red)
65. P(1 of each color)

66. Find the probability of obtaining a royal flush (ace, king, queen, jack, and 10 of one suit), correct to eight decimal places.

67. Find the probability of obtaining a full house of 3 aces and a pair of 2s, correct to eight decimal places.

68. The historical note at the beginning of this chapter tells of Chevalier de Méré's sending some problems to Blaise Pascal. Chevalier de Méré used to bet that he could get at least one 6 in four rolls of a die. He also bet that, in 24 tosses of a pair of dice, he would get at least one 12. He found that he won more often than he lost with the first bet, but not with the second. He did not know why, so he wrote to Pascal seeking the probabilities of these events. What are the probabilities for winning in these two games?

9.4 Calculated Probabilities

There are many probabilities that can be calculated from the probabilities of known events. These probabilities are often denoted by using set notation. You can review sets and set operations in Appendix A, but here is a quick review:

$\varnothing$	Empty set	$A \cup B$	Union of sets A and B
U	Universal set	$A \cap B$	Intersection of sets A and B
		$\overline{A}$	Complement of a set A

Probabilistic Statement	**Set Notation**
Events A and B	A, B
A and B are mutually exclusive	$A \cap B = \varnothing$
A and B occur	$A \cap B$
A or B occurs	$A \cup B$
A does not occur	$\overline{A}$
Neither A nor B occurs	$\overline{A \cup B}$, or (equivalently) $\overline{A} \cap \overline{B}$
A and B are equally likely	$P(A) = P(B)$
A is more likely than B	$P(A) > P(B)$
A is less likely than B	$P(A) < P(B)$

An easily calculated probability often involves **complementary events.** Events are complementary if they are *mutually exclusive*, and together make up the entire sample space. The complement of any event E is denoted by $\overline{E}$, and

$$P(E) + P(\overline{E}) = 1$$

or, in a more useful form,

$$P(E) = 1 - P(\overline{E})$$

which is used when it is easier to find $P(\overline{E})$ than it is to find $P(E)$.

Example 1 Find the probability (to the nearest hundredth) that a poker hand has at least 1 ace.

Solution Direct calculation involves finding the probabilities of having 1 ace, 2 aces, 3 aces, or 4 aces in a hand. But,

$$\text{if} \quad E = \{\text{At least 1 ace}\}, \quad \text{then} \quad \overline{E} = \{\text{Not at least 1 ace}\}$$
$$= \{\text{No aces}\}$$

It is easier to find $P(\overline{E})$ than it is to find $P(E)$.

$$P(\overline{E}) = P(\text{No aces}) = \frac{{}_{48}C_5}{{}_{52}C_5} \quad \begin{matrix} \leftarrow \text{5 out of 48 cards} \\ \leftarrow \text{5 out of 52 cards} \end{matrix}$$

$$\approx .6588 \qquad \text{Using a calculator}$$

Thus,

$$\begin{aligned} P(E) &= 1 - P(\overline{E}) \\ &\approx 1 - .6588 \\ &\approx \mathbf{.34} \end{aligned}$$

□

Frequently when you compute the probability of an event, you have additional information that may alter the sample space. Denote the probability of an event E *given* that an event F has occurred by $P(E|F)$. This notation means that rather than simply computing $P(E)$, you reevaluate $P(E)$ in light of the information that F has occurred. That is, you consider the *altered sample space*. Compare Examples 2 and 3.

Example 2 What is the probability of a family with two children having two boys?

Solution Let E = {Two boys}.

Sample Space

BB ← Success
BG
GB
GG

} Four possibilities

$$P(E) = \frac{1}{4}$$

□

Example 3 Suppose you already know that the family described in Example 2 has at least one boy. Now, answer the question in Example 2.

Solution Let F = {At least one is a boy}.

Sample Space

BB ← Success
BG
GB
~~GG~~ ← This is crossed out because we have additional information that the family has at least one boy. This is called *altering the sample space*.

$$P(E|F) = \frac{1}{3}$$

□

It is not always practical to list the sample space and reduced sample space, as was done in Examples 2 and 3, so the following definition is often used:

CONDITIONAL PROBABILITY

The **conditional probability** that an event E has occurred given that the event F has occurred is denoted by $P(E|F)$ and defined by

$$\mathbf{P(E|F) = \frac{P(E \cap F)}{P(F)}} \qquad \text{provided} \quad P(F) \neq 0$$

Example 4 A tire manufacturer found that 10% of the tires produced had cosmetic defects and 2% had both cosmetic and structural defects. What is the probability that one tire selected at random is structurally defective if it is known that it has a cosmetic defect?

Solution Let C = {Cosmetic defect on tire} and let S = {Structural defect on tire}. Find $P(S|C)$.

$$P(S|C) = \frac{P(S \cap C)}{P(C)} = \frac{\frac{2}{100}}{\frac{10}{100}} = \mathbf{\frac{1}{5}}$$

□

Sometimes, more complicated models can be built by breaking the events being considered into simpler ones using the words *and* (*intersection*) or *or* (*union*).

INTERSECTION: $P(E \cap F) = P(E) \cdot P(F)$ provided E and F are independent
UNION: $P(E \cup F) = P(E) + P(F) - P(E \cap F)$

Events E and F are **independent** if the occurrence of one of these events in no way affects the occurrence of the other. Also, notice that if E and F are mutually exclusive, then $E \cap F = \varnothing$ and $P(E \cap F) = 0$, so $P(E \cup F) = P(E) + P(F)$.

Example 5 Suppose a die is rolled twice. Let

A = {First toss is a prime} B = {First toss is a 3}
C = {Second toss is a 2} D = {Second toss is a 3}

A sample space for one toss is S = {1, 2, 3, 4, 5, 6}, so the sample space for this experiment has 36 possible outcomes, as shown in Figure 9.1, page 384. By considering this sample space, we can count the successes directly, using Figure 9.1:

$$P(A) = \frac{18}{36} = \frac{1}{2}$$

$$P(B) = \frac{6}{36} = \frac{1}{6} \qquad P(A \cap B) = \frac{6}{36} = \frac{1}{6}$$

$$P(C) = \frac{6}{36} = \frac{1}{6} \qquad P(A \cap C) = \frac{3}{36} = \frac{1}{12} \qquad P(B \cap C) = \frac{1}{36}$$

$$P(D) = \frac{6}{36} = \frac{1}{6} \qquad P(A \cap D) = \frac{3}{36} = \frac{1}{12} \qquad P(B \cap D) = \frac{1}{36} \qquad P(C \cap D) = 0$$

Now, use the formula for intersection where possible:

$P(A \cap B) \neq P(A) \cdot P(B)$ — Because A and B are not independent.

$$\begin{aligned} P(A \cap C) &= P(A) \cdot P(C) \\ &= \frac{1}{2} \cdot \frac{1}{6} \\ &= \frac{1}{12} \end{aligned}$$

Because A and C are independent.

Similarly, $P(A \cap D) = P(A) \cdot P(D)$, $P(B \cap C) = P(B) \cdot P(C)$, and $P(B \cap D) = P(B) \cdot P(D)$, but $P(C \cap D) \neq P(C) \cdot P(D)$ because C and D are not independent (obtaining a 2 on the second toss means that a 3 *cannot* be obtained on the same toss). □

Example 6 Find the indicated unions of the events defined in Example 5.

$$\begin{aligned} P(A \cup B) &= P(A) + P(B) - P(A \cap B) \\ &= \frac{1}{2} + \frac{1}{6} - \frac{1}{6} \\ &= \mathbf{\frac{1}{2}} \end{aligned}$$

This is the probability that a prime or a 3 is obtained on the first toss.

$$\begin{aligned} P(A \cup C) &= P(A) + P(C) - P(A \cap C) \\ &= \frac{1}{2} + \frac{1}{6} - \frac{1}{12} \\ &= \mathbf{\frac{7}{12}} \end{aligned}$$

This is the probability that the first toss is a prime or the second toss is a 2.

$$\begin{aligned} P(B \cup D) &= P(B) + P(D) - P(B \cap D) \\ &= \frac{1}{6} + \frac{1}{6} - \frac{1}{36} \\ &= \mathbf{\frac{11}{36}} \end{aligned}$$

This is the probability that the first toss is a 3 or the second toss is a 3.

$$\begin{aligned} P(C \cup D) &= P(C) + P(D) - P(C \cap D) \\ &= \frac{1}{6} + \frac{1}{6} - 0 \\ &= \mathbf{\frac{1}{3}} \end{aligned}$$

This is the probability that the second toss is a 2 or a 3. Notice that C and D are mutually exclusive. □

Example 7 Find the indicated conditional probabilities for the events defined in Example 5.

$$\begin{aligned} P(A|C) &= \frac{P(A \cap C)}{P(C)} \\ &= \frac{\frac{1}{12}}{\frac{1}{6}} \end{aligned}$$

This is the probability of a prime on the first toss, given that the second toss is a 2. Notice that the knowledge that event C occurred had no effect on the occurrence of event A.

$$= \mathbf{\frac{1}{2}}$$

$$P(A|B) = \frac{P(A \cap B)}{P(B)} = \frac{\frac{1}{6}}{\frac{1}{6}} = \mathbf{1}$$

This is the probability of a prime on the first toss given that the first toss is a 3. Notice that the probability is 1—a certainty—since 3 *is* a prime.

$$P(B|A) = \frac{P(B \cap A)}{P(A)} = \frac{\frac{1}{6}}{\frac{1}{2}} = \mathbf{\frac{1}{3}}$$

This is the probability of a 3 on the first toss given that the first toss is a prime. Notice that $P(A|B)$ is not the same as $P(B|A)$.

□

SUMMARY OF PROBABILITY FORMULAS

1. $P(E) = \frac{s}{n}$, where event E can occur in n mutually exclusive and equally likely ways, and s of them are considered favorable
2. $P(S) = 1$, where S is the sample space
3. $P(\emptyset) = 0$
4. $P(E) = 1 - P(\overline{E})$
5. $P(E|F) = \frac{P(E \cap F)}{P(F)}$, provided $P(F) \neq 0$
6. $P(E \text{ and } F) = P(E \cap F) = P(E) \cdot P(F)$, provided E and F are independent
7. $P(E \text{ or } F) = P(E \cup F) = P(E) + P(F) - P(E \cap F)$
8. $P(E \text{ or } F) = P(E \cup F) = P(E) + P(F)$, provided E and F are mutually exclusive

Example 8 Suppose a coin is tossed and a die is simultaneously rolled. Let $T = \{\text{Tail is tossed}\}$, and $F = \{4 \text{ is rolled}\}$. Find the indicated probabilities.

a. $P(T \cap F) = P(T) \cdot P(F) = \frac{1}{2} \cdot \frac{1}{6} = \mathbf{\frac{1}{12}}$

This is the probability of a tail and a 4. Notice that T and F are independent.

b. $P(T \cup F) = P(T) + P(F) - P(T \cap F) = \frac{1}{2} + \frac{1}{6} - \frac{1}{12} = \mathbf{\frac{7}{12}}$

This is the probability of a tail or a 4.

□

Example 9 Suppose 2 cards are drawn from a deck of 52 cards. Let

$$S_1 = \{\text{Draw a spade on the first draw}\}$$
$$H_1 = \{\text{Draw a heart on the first draw}\}$$
$$H_2 = \{\text{Draw a heart on the second draw}\}$$

To draw **with replacement** means that the first card is drawn, the result noted, and then it is replaced before the second card is drawn. To draw **without replacement** means that the first card is drawn, the result noted, and then a second card is drawn without replacing the first card. Find the indicated probabilities.

a. $P(S_1 \cap H_2)$, *with replacement.* The events S_1 and H_2 are independent since the cards are drawn with replacement.

$$\begin{aligned} P(S_1 \cap H_2) &= P(S_1) \cdot P(H_2) \\ &= \frac{1}{4} \cdot \frac{1}{4} \\ &= \mathbf{\frac{1}{16}} \end{aligned}$$

b. $P(S_1 \cap H_2)$, *without replacement.* In this experiment, the events are not independent since the probability of drawing the second card depends on what was drawn on the first card. This problem is equivalent to drawing 2 cards from a deck of cards. The order in which the cards are drawn is important since the question specifies that the *first* card is a spade and the *second* card is a heart; thus, this is a permutation problem.

$$\begin{aligned} P(S_1 \cap H_2) &= \frac{{}_{13}P_1 \cdot {}_{13}P_1}{{}_{52}P_2} \\ &= \frac{13 \cdot 13}{52 \cdot 51} \\ &= \mathbf{\frac{13}{204}} \end{aligned}$$

c. What is the probability of drawing 2 hearts *with replacement?* The events H_1 and H_2 are independent.

$$\begin{aligned} P(H_1 \cap H_2) &= P(H_1) \cdot P(H_2) \\ &= \frac{1}{4} \cdot \frac{1}{4} \\ &= \mathbf{\frac{1}{16}} \end{aligned}$$

d. What is the probability of drawing 2 hearts *without replacement?* The events H_1 and H_2 are not independent, but the order in which the cards are drawn is not important since the question simply asks for 2 hearts; thus, this is a combination problem.

$$P(2\text{ hearts}) = \frac{_{13}C_2}{_{52}C_2} = \frac{\dfrac{13 \cdot 12}{2}}{\dfrac{52 \cdot 51}{2}} = \mathbf{\frac{1}{17}}$$

□

Problem Set 9.4

A *Suppose that events A, B, and C are all independent such that*

$$P(A) = \frac{1}{2} \qquad P(B) = \frac{1}{3} \qquad P(C) = \frac{1}{6}$$

Find the probabilities in Problems 1–15.

1. $P(\overline{A})$	**2.** $P(\overline{B})$	**3.** $P(\overline{C})$
4. $P(A \cap B)$	**5.** $P(A \cap C)$	**6.** $P(B \cap C)$
7. $P(A \cup B)$	**8.** $P(A \cup C)$	**9.** $P(B \cup C)$
10. $P(\overline{A \cap B})$	**11.** $P(\overline{A \cap C})$	**12.** $P(\overline{B \cap C})$
13. $P(\overline{A \cup B})$	**14.** $P(\overline{A \cup C})$	**15.** $P(\overline{B \cup C})$

16. What is the probability that a family with three children has exactly two boys?
17. What is the probability in Problem 16 if it is known that at least one of the children is a boy?
18. What is the probability that a family with three children has exactly one girl?
19. What is the probability in Problem 18 if it is known that at least one of them is a girl?
20. What is the probability that a family with three children has exactly two girls?
21. What is the probability in Problem 20 if it is known that at least one of them is a girl?
22. What is the probability that a family with three children has at least one girl?
23. What is the probability in Problem 22 if it is known that all three are girls?
24. What is the probability of flipping a coin four times and obtaining exactly 3 heads?
25. What is the probability in Problem 24 if it is known that there are at least 2 heads?
26. What is the probability of flipping a coin four times and obtaining exactly 2 heads?
27. What is the probability in Problem 26 if it is known that at least 1 head has turned up?
28. What is the probability of flipping a coin four times and obtaining exactly 3 tails?
29. What is the probability in Problem 28 if it is known that at least 1 tail has turned up?

In Problems 30–33 suppose a coin is tossed twice. Find the requested probabilities.

30. $P(2 \text{ heads})$ **31.** $P(2 \text{ tails})$
32. $P(1 \text{ head and } 1 \text{ tail})$ **33.** $P(1 \text{ head or } 1 \text{ tail})$

Suppose A, B, and C are independent events such that $P(A) = \frac{1}{2}$, $P(B) = \frac{1}{3}$, *and* $P(C) = \frac{5}{6}$. *Find the probabilities in Problems 34–45.*

34. $P(\overline{A})$ **35.** $P(\overline{B})$ **36.** $P(\overline{C})$ **37.** $P(A \cap B)$
38. $P(A \cap C)$ **39.** $P(B \cap C)$ **40.** $P(A \cup B)$ **41.** $P(A \cup C)$
42. $P(B \cup C)$ **43.** $P(\overline{A \cap C})$ **44.** $P(\overline{A \cup B})$ **45.** $P(\overline{B \cup C})$

B *In Problems 46–60 suppose a coin is tossed and simultaneously a die is rolled. Let*

$$H = \{\text{A head is tossed}\} \quad S = \{\text{A 6 is rolled}\} \quad E = \{\text{An even number is rolled}\}$$

Find the requested probabilities.

46. $P(H)$ **47.** $P(S)$ **48.** $P(E)$ **49.** $P(H \cap S)$
50. $P(H \cap E)$ **51.** $P(S \cap E)$ **52.** $P(S \cup E)$ **53.** $P(H \cup E)$
54. $P(H \cup S)$ **55.** $P(H|S)$ **56.** $P(S|H)$ **57.** $P(S|E)$
58. $P(E|S)$ **59.** $P(H|E)$ **60.** $P(E|H)$

61. The probability of tossing a coin four times and obtaining 4 heads in a row is $P(4H) = \frac{1}{2} \cdot \frac{1}{2} \cdot \frac{1}{2} \cdot \frac{1}{2} = \frac{1}{16}$. What is the probability of tossing a coin and obtaining a head if we know that heads have occurred on the previous four flips of the coin?

62. One "system" used by roulette players is to watch a game of roulette until a large number of reds occur in a row (say, 10). After 10 successive reds, they reason that black is "due" to occur and begin to bet large sums on black. Where is the fallacy in the reasoning of this "system betting"?

63. The Eureka Light Bulb Company has found that 5% of the bulbs it manufactures have defective filaments and 3% have both defective filaments and defective workmanship. What is the probability of defective workmanship in a particular bulb if you know it has a defective filament?

64. At Learnalot University a survey of students taking both College Algebra and Statistics found that 25% dropped College Algebra, 30% dropped Statistics, and 10% dropped both courses. If a person dropped College Algebra, what is the probability that the person also dropped Statistics?

C *In Problems 65–70 assume a box has 5 red cards and 3 black cards and 2 cards are drawn with replacement. Find the requested probabilities.*

65. $P(2 \text{ red cards})$ **66.** $P(2 \text{ black cards})$
67. $P(1 \text{ red and } 1 \text{ black card})$
68. $P(\text{Red on first draw and black on second draw})$
69. $P(1 \text{ red or } 1 \text{ black card})$
70. $P(\text{Red on first draw or black on second draw})$

Problems 71–76 repeat the experiment of Problems 65–70, except that cards are drawn without replacement.

71. P(2 red cards)

72. P(2 black cards)

73. P(1 red and 1 black card)

74. P(Red on first draw and black on second draw)

75. P(1 red or 1 black card)

76. P(Red on first draw or black on second draw)

77. A game of solitaire called *Tournament* involves two ordinary decks of cards shuffled together. Eight cards are dealt face up. In the game, it is desirable to have an ace or a king turned up. What is the probability that you will turn up at least 1 ace or 1 king in the first 8 cards?

78. Show that if A and B are independent and have no common elements, then at least one of them is impossible.

79. Show that if A and B are independent events, then so are $\overline{A}$ and $\overline{B}$.

9.5 Summary and Review

In this chapter we have introduced you to some of the notions of counting and probability.

CHAPTER OBJECTIVES

After studying this chapter, you should be able to:

1. State and apply the Fundamental Counting Principle
2. Distinguish between a permutation and a combination
3. Use and understand the notation for permutations and combinations
4. Use the formulas $_nP_r = \dfrac{n!}{(n-r)!}$ and $_nC_r = \dfrac{_nP_r}{r!}$
5. Define probability
6. Calculate simple probabilities using the definition
7. Calculate conditional probabilities
8. Calculate probabilities of combined events

Problem Set 9.5

CONCEPT PROBLEMS

Fill in the word or words necessary to make the statements in Problems 1–20 complete and correct.

[9.1] **1.** A permutation of q elements of a set S with m elements is ____________________.

[9.1] **2.** The notation $_qP_m$ means ____________________.

[9.1] **3.** $_qP_m =$ ________

[9.1] **4.** The Fundamental Counting Principle states: ____________________.

[9.1] **5.** The number of distinguishable permutations of m objects, of which q_1 are one kind, q_2 another kind, . . . , and q_k a further kind so that $q_1 + q_2 + \cdots + q_k = m$, is given by the formula ____________________.

[9.2] **6.** A combination of q elements of a set S with m elements is ____ .

[9.2] **7.** The notation ________ means q elements selected from a set of m elements without regard to the order selected.

[9.2] **8.** ${}_qC_r$ = ________

[9.2] **9.** If certain elements are selected from a set in a definite order, it is called a(n) ________ ; and if they are selected without order, it is called a(n) ________ .

[9.3] **10.** The definition of probability is: ____ .

[9.3] **11.** $P(S)$ = ________ , where S is the sample space.

[9.3] **12.** $P(\varnothing)$ = ________

[9.4] **13.** If E and F are complementary events, then $P(E)$ = ________________ .

[9.4] **14.** $P(E \mid F)$ = ________

[9.4] **15.** $P(E \cap F)$ = ________ , provided ________ .

[9.4] **16.** $P(E \cup F)$ = ________

[9.4] **17.** $P(E \cap F)$ = ________ if E and F are independent.

[9.4] **18.** $P(E \cup F)$ = ________ if E and F are independent.

[9.4] **19.** To draw a card with replacement means ________ .

[9.4] **20.** To draw a card without replacement means ________ .

REVIEW PROBLEMS

[9.1] *How many permutations are there in the words given in Problems 21–26?*

21. APPLE **22.** COMPUTER **23.** MATHEMATICS
24. REPEAT **25.** CURRICULUM **26.** TENNESSEE

[9.1, 9.2] *Compute the numbers requested in Problems 27–41.*

27. ${}_5P_3$ **28.** ${}_7P_5$ **29.** ${}_{12}P_8$ **30.** ${}_nP_3$ **31.** ${}_jP_h$
32. ${}_5C_3$ **33.** ${}_7C_5$ **34.** ${}_{12}C_8$ **35.** ${}_nC_3$ **36.** ${}_jC_h$
37. ${}_4P_2$ **38.** ${}_2P_0$ **39.** ${}_7C_6$ **40.** ${}_8P_8$ **41.** ${}_6C_0$

[9.1] **42.** The Sticky Widget Company has 12 directors, and at the stockholders' meeting a chairperson, vice-chairperson, secretary, and treasurer are to be elected from the board. In how many different ways can this be done?

[9.2] **43.** A committee of 4 is to be selected from a board of 12 directors. In how many ways can this be done?

[9.2] *The executive board of the Sticky Widget Company needs to pick a committee to study labor relations. Three members of the board have excellent qualifications, seven members could be rated as mediocre, and the remaining two are not qualified. In how many ways can the company select a committee of three as specified in Problems 44–48?*

44. They all have excellent qualifications.
45. One has excellent qualifications, and two are mediocre.
46. One has excellent qualifications, one is mediocre, and one is not qualified.
47. All three are mediocre.
48. All three are not qualified.

[9.3] *Suppose a card is drawn from a deck of 52 cards. Find the probabilities in Problems 49–54.*

49. $P(7)$ **50.** $P(\text{Club})$ **51.** $P(\text{Face card})$
52. $P(7 \text{ or club})$ **53.** $P(7 \text{ and club})$ **54.** $P(7 \text{ and face card})$

[9.3] *Suppose 2 cards are drawn from a deck of 52 cards at the same time. Find the probabilities in Problems 55–59 as decimals to the nearest thousandth.*

55. $P(\text{Both 7s})$ **56.** $P(\text{Both clubs})$ **57.** $P(7 \text{ and club})$
58. $P(7 \text{ of clubs and 6 of hearts})$ **59.** $P(7 \text{ of clubs and face card})$

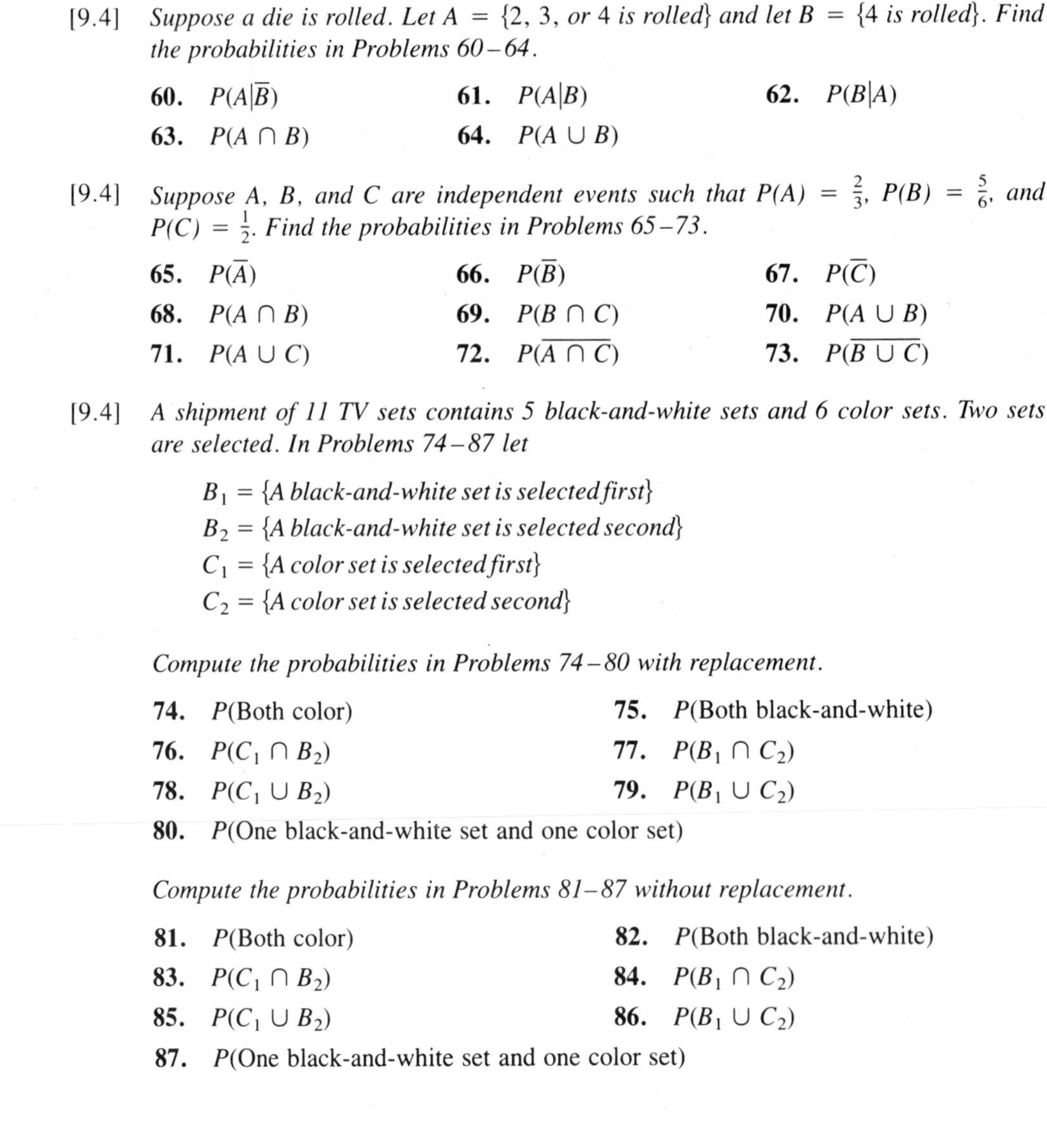

[9.4] *Suppose a die is rolled. Let* $A = \{2, 3, \text{ or } 4 \text{ is rolled}\}$ *and let* $B = \{4 \text{ is rolled}\}$. *Find the probabilities in Problems 60–64.*

60. $P(A|\overline{B})$ **61.** $P(A|B)$ **62.** $P(B|A)$
63. $P(A \cap B)$ **64.** $P(A \cup B)$

[9.4] *Suppose A, B, and C are independent events such that* $P(A) = \frac{2}{3}$, $P(B) = \frac{5}{6}$, *and* $P(C) = \frac{1}{2}$. *Find the probabilities in Problems 65–73.*

65. $P(\overline{A})$ **66.** $P(\overline{B})$ **67.** $P(\overline{C})$
68. $P(A \cap B)$ **69.** $P(B \cap C)$ **70.** $P(A \cup B)$
71. $P(A \cup C)$ **72.** $P(\overline{A \cap C})$ **73.** $P(\overline{B \cup C})$

[9.4] *A shipment of 11 TV sets contains 5 black-and-white sets and 6 color sets. Two sets are selected. In Problems 74–87 let*

$$B_1 = \{\text{A black-and-white set is selected first}\}$$
$$B_2 = \{\text{A black-and-white set is selected second}\}$$
$$C_1 = \{\text{A color set is selected first}\}$$
$$C_2 = \{\text{A color set is selected second}\}$$

Compute the probabilities in Problems 74–80 with replacement.

74. $P(\text{Both color})$ **75.** $P(\text{Both black-and-white})$
76. $P(C_1 \cap B_2)$ **77.** $P(B_1 \cap C_2)$
78. $P(C_1 \cup B_2)$ **79.** $P(B_1 \cup C_2)$
80. $P(\text{One black-and-white set and one color set})$

Compute the probabilities in Problems 81–87 without replacement.

81. $P(\text{Both color})$ **82.** $P(\text{Both black-and-white})$
83. $P(C_1 \cap B_2)$ **84.** $P(B_1 \cap C_2)$
85. $P(C_1 \cup B_2)$ **86.** $P(B_1 \cup C_2)$
87. $P(\text{One black-and-white set and one color set})$

CHAPTER 10

Mathematical Proof

George Boole
(1815–1864)

In mathematics we see the conscious logical activity of our mind in its purest and most perfect form; here is made manifest to us all the labor and the great care with which it progresses, the precision which is necessary to determine exactly the source of the established general theorems, and the difficulty with which we form and comprehend abstract conceptions; but we also learn to have confidence in the certainty, breadth, and fruitfulness of such intellectual labor.

Benjamin Pierce,
American Journal of Mathematics,
Vol. 4, 1881

Historical Note

Logic began to flourish during the classical Greek period. Aristotle (384–322 B.C.) was the first person to study the subject systematically, and he and many other Greeks searched for universal truths that were irrefutable. The logic of this period, referred to as *Aristotelian logic,* is still used today and is based on the *syllogism* (discussed in this chapter).

The second great period for logic came with the use of symbols to simplify complicated logical arguments. This was first done when the great mathematician Gottfried Leibniz, at the age of 14, attempted to reform Aristotelian logic. However, the world took little notice of Leibniz' logic, and it was not until George Boole (1815–1864) completed his book, *An Investigation of the Laws of Thought,* that logic entered its third and most important period. Boole considered various mathematical operations by separating them from the other commonly used symbols. This idea was popularized by Bertrand Russell (1872–1970) and Alfred North Whitehead (1861–1947) in their monumental *Principia Mathematica*. In this work, they began with a few assumptions and three undefined terms and built a system of symbolic logic. From this, they then formally developed the theorems of arithmetic and mathematics.

*Section 10.5 does not require Sections 10.1–10.4.

Chapter Overview
This chapter forms the bridge between elementary and advanced mathematics. Basic algebra courses (including this course) are concerned mainly with the mechanics of manipulation, solving equations, and graphing certain types of functions and curves. More advanced algebra courses (usually called *abstract algebra*) are concerned with structure and with proof. This chapter introduces you to some of these more advanced concepts.

10.1 Axioms of the Real Numbers

A **mathematical system** is a set with at least one defined operation and some developed properties. In building any mathematical system, certain fundamental terms must necessarily remain undefined. For example, words to describe sorting objects into similar groupings are an essential part of every language: a *herd* of cattle, a *flock* of birds, a track *team,* a stamp *collection,* or a *set* of dishes. All these grouping words serve the same purpose; and in mathematics, the word *set* is used to refer to any collection of objects. However, any attempt to define **set** would be circular or would require the acceptance of other undefined terms, so the concept of *set* is thus accepted as an undefined term. For this text we also assume the existence of a set of numbers called the set of **real numbers, R.**

To understand the notions of a mathematical system and the nature of proof, it is first necessary to understand some fundamental ideas of logic. When the term *logic* is used in this book it refers to **deductive logic.** This is a type of reasoning that accepts no conclusions except those that are inescapable. For example, the simplest type of deductive structure is the **syllogism,** which consists of three statements: the first two are called **premises** (or **hypotheses**) and the last is called the **conclusion:**

1. All who read the *Times* are well informed. Premise
2. Mr. Smith reads the *Times*. Premise
3. Therefore, Mr. Smith is well informed. Conclusion

If you accept the premises, then you *must* accept the conclusion. If the premises *imply* the conclusion, then the argument is **valid.** You must remember that we are not interested in the truth of the premises or of the conclusion, but only whether the conclusion *follows from* the premises. For this example, it means that we are not concerned about whether Mr. Smith really reads the *Times;* we do not care whether

Times readers are well or ill informed. We care only that the conclusion follows *if* the premises are *assumed*.

It is often easier to show that a given argument is not valid than to show that it is valid. An argument may be shown to be valid by proving it true for *all* possibilities. An argument may be shown to be false by finding just *one* example for which it is false. An example that disproves a statement is called a **counterexample.**

Find a counterexample to disprove the statements in Examples 1 and 2.

Example 1 All primes are odd.

Counterexample 2 is prime and it is not odd; therefore, not all primes are odd. □

Example 2 For all values of a and b, $\sqrt{a^2 + b^2} = a + b$.

Counterexample Let $a = 3$ and $b = 4$; then $\sqrt{a^2 + b^2} = \sqrt{9 + 16} = 5$, and $5 \neq 3 + 4$. Therefore, $\sqrt{a^2 + b^2} \neq a + b$. □

The mathematical structure that will be built up in this chapter is concerned with the set of real numbers and the proofs will be more complex than simple syllogisms, but both syllogisms and proofs have two important parts:

1. Premises (or agreements) that are made at the beginning and serve as a point of departure.
2. Conclusions drawn from these premises.

Thus, in building a mathematical structure, you must accept some *undefined terms* (set, set of real numbers), some noncontradictory axioms or *premises*, and a logical system of reasoning, which leads to *conclusions* drawn from the premises.

In Chapter 1, we assumed certain properties of equality, which we must now state more formally for completeness.

AXIOMS OF EQUALITY

For $a, b, c \in \mathbf{R}$,

AXIOM E1	REFLEXIVE:	$a = a$
AXIOM E2	SYMMETRIC:	If $a = b$, then $b = a$.
AXIOM E3	TRANSITIVE:	If $a = b$ and $b = c$, then $a = c$.
AXIOM E4	SUBSTITUTION:	If $a = b$, then a may be replaced throughout by b (or b by a) in any statement without changing the truth or falsity of the statement.

We also repeat the assumed axioms of the real numbers for clarity and reference. Axioms F1–F11 are also called the **field axioms.**

FIELD AXIOMS

AXIOMS F1–F11

CLOSURE:	For any $a, b \in \mathbf{R}$, F1: $(a + b) \in \mathbf{R}$	F2: $ab \in \mathbf{R}$
COMMUTATIVE:	For any $a, b \in \mathbf{R}$, F3: $a + b = b + a$	F4: $ab = ba$
ASSOCIATIVE:	For any $a, b, c \in \mathbf{R}$, F5: $(a + b) + c = a + (b + c)$	F6: $(ab)c = a(bc)$
IDENTITY:	There exist two (unequal) numbers 0 and 1 such that for every $a \in \mathbf{R}$, F7: $a + 0 = a$	F8: $a \cdot 1 = a$
INVERSE:	F9: For each $a \in \mathbf{R}$, there exists exactly one element $(-a) \in \mathbf{R}$ such that $a + (-a) = 0$. F10: For each $a \in \mathbf{R}$, $a \neq 0$, there exists exactly one element $(1/a) \in \mathbf{R}$ such that $a(1/a) = 1$.	
DISTRIBUTIVE:	For any $a, b, c \in \mathbf{R}$, F11: $a(b + c) = ab + ac$	

Now that our mathematical system has some undefined terms along with some premises (or axioms), the next step is to illustrate a mathematical proof of a new statement (or conclusion) that is inescapable.

Example 3 Axiom F7 says $a + 0 = a$ and not $0 + a = a$. Prove: $0 + a = a$

For clarity, the premises and conclusions will be carefully stated in a two-column form.

Proof GIVEN states the premises, and TO PROVE states the intended conclusion.

GIVEN: $a \in \mathbf{R}$ TO PROVE: $0 + a = a$

Statements	Reasons
1. $a + 0 = a$	1. Identity (Axiom F7)
2. $0 + a = a + 0$	2. Commutative (Axiom F3)
$\therefore$3. $0 + a = a$	3. Transitive (Axiom E3), or substitution of Step 2 into Step 1 (Axiom E4)

↑
$\therefore$ is a symbol meaning *therefore*. It usually denotes the conclusion of a proof.

□

An analysis of the proof in Example 3 shows that it consists of a chain of syllogisms. Each step of the proof is a syllogism in which the conclusion is listed in the column headed Statements; one premise is expressed as the reason, and the other premise is implied.

Example 4 Write the proof in Example 3 in syllogism form.

Proof GIVEN: $a \in \mathbf{R}$ TO PROVE: $0 + a = a$

	Premise 1 (Stated)	Premise 2 (Implied)	Conclusion
1.	Axiom F7: There exists a number 0 such that for every $a \in \mathbf{R}$, $a + 0 = a$.	$a \in \mathbf{R}$	$a + 0 = a$
2.	Axiom F3: For any $a, b \in \mathbf{R}$, $a + b = b + a$.	$a, 0 \in \mathbf{R}$ $b = 0$	$a + 0 = 0 + a$
∴3.	Axiom E4: If $a = b$, then a may be replaced by b in any statement without changing its truth or falsity.	If $a + 0 = a$ (Step 1) and $a + 0 = 0 + a$ (Step 2), then substitute $0 + a$ for $a + 0$ to obtain $0 + a = a$.	$0 + a = a$

□

The assumptions and derived theorems form a logical structure. This algebra course was one such logical structure, but there are other possible algebras that begin with different initial assumptions. Although it is not our intent to give a rigorous treatment of the algebra in this book, we will develop some of the more important theorems of algebra in the next three sections, but before we do, consider an important mathematical connective called **implication.**

Most mathematical theorems are stated in what is called an *if–then statement* since it is very convenient to reason from this form. For example, the statement *if* $a = b$, *then* $a + c = b + c$ is an if–then statement. The *if* part (GIVEN) is called the *hypothesis* (the given part), and the *then* part (TO PROVE) is called the *conclusion* (the part to be proven). The symbol $p \rightarrow q$ is used to denote a proposition expressed in the form if p, then q.

The *if* part of an implication need not be stated first. All the following statements have the same meaning:

IMPLICATION

Implication	Example
If p, then q	If you are 18, then you can vote.
$p \rightarrow q$	Age 18 implies that you can vote.
q, if p	You can vote if you are 18.
p, only if q	You are 18 only if you can vote.
All p are q	All 18-year-olds can vote.

One of the most often used principles of logic is called **direct reasoning,** which is also sometimes called a *syllogism, direct proof,* or *modus ponens.*

DIRECT REASONING

Direct Reasoning

MAJOR PREMISE	$p \rightarrow q$
MINOR PREMISE	p
CONCLUSION	$\therefore q$

Example 5 If you read the *Times,* then you are well informed.
You read the *Times*.

Therefore, you are well informed. □

Example 6 If you play chess, then you are intelligent.
You play chess.

Therefore, you are intelligent. □

Example 7 If you are a logical person, then you will understand this example.
You are a logical person.

Therefore, you understand this example. □

Problem Set 10.1

A

1. In what ways does a mathematical definition differ from a dictionary definition?
2. Why must some mathematical terms remain undefined?
3. Distinguish among an undefined term, an axiom, and a theorem.
4. What is the difference between truth and validity?

Write each of the statements in Problems 5–10 in the if–then format.

5. "Everything happens to everybody sooner or later if there is time enough" (G. B. Shaw).
6. "We are not weak if we make a proper use of those means which the God of Nature has placed in our power" (Patrick Henry).
7. "A useless life is an early death" (Goethe).
8. "All work is noble" (Thomas Carlyle).
9. "Everything's got a moral if only you can find it" (Lewis Carroll).
10. "You can go to your friend's house only if you have finished your homework" (Karl Smith).
11. Write the following statement in several equivalent ways: If you like mathematics, then you are intelligent.
12. Write the following statement in several equivalent ways: All elephants are big.

B *Identify the axiom illustrated by each statement in Problems 13–22.*

13. $-a + [-(-a)] + a = [-(-a)] + (-a) + a$
14. $[-(-a)] + (-a) + a = [-(-a)] + [(-a) + a]$
15. $a[b + (-b)] = a \cdot 0$
16. $ac \cdot \frac{1}{bd} \cdot 1 \cdot 1 = ac \cdot \frac{1}{bd} \cdot \left(b \cdot \frac{1}{b}\right) \cdot \left(d \cdot \frac{1}{d}\right); \quad b, d \neq 0$
17. If $a \in \mathbf{R}$, then $aa \in \mathbf{R}$.
18. If $a = 0$, then $ab = 0 \cdot b$.
19. If $a + (-b) = c$ and $a + (-b) = a - b$, then $a - b = c$.

20. $\left(c \cdot \frac{b}{d}\right) \cdot d = c \cdot \left(\frac{b}{d} \cdot d\right); \quad d \neq 0$

21. $a \cdot 1 \cdot \frac{d}{bc} = a \cdot \frac{d}{bc}$

22. $[-(-a)] + 0 = -(-a)$

Disprove the statements in Problems 23–31 by finding a counterexample.

23. A number is prime if it is divisible by itself and 1.

24. For all values of a and b, $\sqrt{a^2 - b^2} = a - b$.

25. A number is prime if its only divisors are itself and 1.

26. For all values of x and y, $(x^{1/2} + y^{1/2})^2 = x + y$.

27. If $ax^2 + bx + c = 0$, then $x = \frac{-b \pm \sqrt{b^2 - 4ac}}{2a}$.

28. For all values of x, $\sqrt{x^2} = x$.

29. For all values of x and y, $\frac{x^3 - y^3}{x - y} = x^2 - y^2$.

30. For all values of x and y, $2^x \cdot 3^y = 6^{x+y}$.

31. For all values of x and y, $\frac{6^x}{2^y} = 3^{x-y}$.

In Problems 32–37 form a valid conclusion using direct reasoning.

32. If you can learn mathematics, then you are intelligent.
You can learn mathematics.

33. If I am idle, then I become lazy.
I am idle.

34. If a^2 is even, then a is even.
a^2 is even.

35. All squares have equal diagonals.
$ABCD$ is a square.

36. If you go to college, then you will get a good job.
You go to college.

37. If I can earn enough money this summer, I will attend college in the fall.
I can earn enough money this summer.

C 38. Consider the following proof that $1 \cdot a = a$:

GIVEN: $a \in \mathbf{R}$ TO PROVE: $1 \cdot a = a$

Statements	Reasons
1. $a \cdot 1 = a$	1. Identity (Axiom F8)
2. $1 \cdot a = a \cdot 1$	2. Commutative (Axiom F4)
∴3. $1 \cdot a = a$	3. Substitution of Step 2 into Step 1 (Axiom E4)

Rewrite this proof in syllogism form.

10.2 Mathematical Proof

In this section some of the theorems used in this book are proved. In Chapter 2 the solution of linear equations was accomplished by performing operations on both sides of an equation. These operations are justified by the following theorems.

Theorem 1 **Addition Law of Equality.** If $a = b$, then $a + c = b + c$.

Proof GIVEN: $a, b, c \in \mathbf{R}$, $a = b$ TO PROVE: $a + c = b + c$

Statements	Reasons
1. $a, b, c \in \mathbf{R}$	1. Given
2. $(a + c) \in \mathbf{R}$	2. Closure (Axiom F1)
3. $a + c = a + c$	3. Reflexive (Axiom E1)
4. $a = b$	4. Given
∴5. $a + c = b + c$	5. Substitution of Step 4 into Step 3 (Axiom E4)

A necessary assumption for all the proofs in this chapter is that the variables represent real numbers. Thus, while this fact is explicitly stated in the GIVEN, it is usually omitted from the statement of the theorems. Notice that Steps 1 and 2 are part of the implied premise in Step 3, which says that

$$a + c = a + c$$

Compare this with Axiom E1, which says that if $a \in \mathbf{R}$, then

$$a = a$$

Thus, before Axiom E1 can be applied to $a + c = a + c$, it is necessary to determine that $a + c$ *is* a real number.

The order in which the steps of a proof are presented can also be varied somewhat. For Theorem 1, it is possible to include Step 4 as part of Step 1.

Theorem 2 **Multiplication Law of Equality.** If $a = b$, then $ac = bc$.

Example 1 Fill in the blanks in the proof of Theorem 2.

Proof GIVEN: **a.** ________ **b.** ________ TO PROVE: $ac = bc$

Statements	Reasons
1. $a, b, c \in \mathbf{R}$	1. **c.** ________
2. $ac \in \mathbf{R}$	2. **d.** ________
3. $ac = ac$	3. **e.** ________
4. **f.** ________	4. Given
∴5. $ac = bc$	5. Substitution of Step 4 into Step 3 (Axiom E4)

Answers **a.** $a, b, c \in \mathbf{R}$ **b.** $a = b$ **c.** Given **d.** Closure (Axiom F2) **e.** Reflexive (Axiom E1) **f.** $a = b$ □

DEFINITION OF SUBTRACTION AND DIVISION

If $a, b \in \mathbf{R}$,

$a - b$ means $a + (-b)$

If $a, b \in \mathbf{R}$, $b \neq 0$,

$\frac{a}{b}$ means $a \div b = a \cdot \frac{1}{b}$

Example 2 Prove: If $a = b$, then $a/c = b/c$, $c \neq 0$.

Proof GIVEN: $a, b, c \in \mathbf{R}$ TO PROVE: $\frac{a}{c} = \frac{b}{c}$

$c \neq 0$

$a = b$

Statements	Reasons
1. $a, b, c \in \mathbf{R}$, $c \neq 0$	1. Given
2. $(1/c) \in \mathbf{R}$	2. Inverse, $c \neq 0$ (Axiom F10)
3. $a \cdot (1/c) = b \cdot (1/c)$	3. Theorem 2
∴4. $a/c = b/c$	4. Definition of division □

Direct proof, which was introduced in the last section, is suitable only for rather short proofs. In order to tie together longer chains of reasoning a second principle of proof is needed, called **logical transitivity.**

LOGICAL TRANSITIVITY

PREMISE $p \to q$
PREMISE $q \to r$
CONCLUSION $\therefore p \to r$

Example 3 If you attend class, then you will pass the course.
If you pass the course, then you will graduate.

Therefore, if you attend class, then you will graduate. □

Transitivity can be extended so that a chain of several if–then sentences is connected together.

Example 4 If you graduate, then you will get a good job.
If you get a good job, then you will meet the right people.
If you meet the right people, then you will become well known.

Therefore, if you graduate, then you will become well known. □

Transitivity is needed for the proof of the following theorem in order to tie together the premise with the conclusion:

Theorem 3 **Zero Multiplication Law.** $a \cdot 0 = 0$

Proof GIVEN: $a \in \mathbf{R}$ TO PROVE: $a \cdot 0 = 0$

Statements	Reasons
1. $a \in \mathbf{R}$	1. Given
2. $a + 0 = a$	2. Identity (F7)
3. $a(a + 0) = aa$	3. Theorem 2
4. $aa + a \cdot 0 = aa$	4. Distributive (F11)
5. $aa \in \mathbf{R}$	5. Closure (F2)
6. $(-aa) \in \mathbf{R}$	6. Inverse (F9)
7. $(aa + a \cdot 0) + (-aa) = aa + (-aa)$	7. Step 4, Theorem 1
8. $(a \cdot 0 + aa) + (-aa) = aa + (-aa)$	8. Commutative (F3)
9. $a \cdot 0 + [(aa) + (-aa)] = aa + (-aa)$	9. Associative (F5)
10. $a \cdot 0 + 0 = 0$	10. Inverse (F9)
$\therefore$11. $a \cdot 0 = 0$	11. Identity (F7)

The proof of Theorem 3 can be divided into separate statements:

If $a \in \mathbf{R}$, then $a + 0 = a$.
If $a + 0 = a$, then $a(a + 0) = aa$.
If $a(a + 0) = aa$, then $aa + a \cdot 0 = aa$.

.
.
.

If $a \cdot 0 + [(aa + (-aa)] = aa + (-aa)$, then $a \cdot 0 + 0 = 0$.
If $a \cdot 0 + 0 = 0$, then $a \cdot 0 = 0$.

Thus, the premise (first step), $a + 0 = a$, implies the conclusion (last step), $a \cdot 0 = 0$.

The next two theorems, which are needed in order to develop the algebraic structure for this book, will provide you with some additional examples of the proper format for proofs.

Theorem 4 **Cancellation Law for Addition.** If $a + c = b + c$, then $a = b$.

Proof GIVEN: $a, b, c \in \mathbf{R}$ TO PROVE: $a = b$
$a + c = b + c$

Statements	Reasons
1. $a, b, c \in \mathbf{R}$	1. Given
2. $a + c = b + c$	2. Given

Statements	Reasons
3. $(-c) \in \mathbf{R}$	3. Inverse (F9)
4. $[a + c] + (-c) = [b + c] + (-c)$	4. Step 2, Theorem 1
5. $a + [c + (-c)] = b + [c + (-c)]$	5. Associative (F5)
6. $a + 0 = b + 0$	6. Inverse (F9)
$\therefore$7. $a = b$	7. Identity (F7)

Theorem 5 **Cancellation Law for Multiplication.** If $ac = bc$, $c \neq 0$, then $a = b$.

The proof is similar to that of Theorem 4 and is left for the reader.

Notice that Theorems 1 and 4 seem very similar. This is because Theorem 4 is the *converse* of Theorem 1.

CONVERSE

The **converse** of a statement

if p, then q is the statement if q, then p

A given statement may be true and its converse false. For example, it is true that

if *ABCD* is a square, then *ABCD* is a rectangle

but the statement

if *EFGH* is a rectangle, then *EFGH* is a square

is false. In the case of Theorem 1, the converse is also true and was proved as a separate theorem. In logic and mathematics the two ideas are often combined into one statement by using the words *if and only if*.

IF AND ONLY IF

The statement **p if and only if q**, abbreviated **p iff q**, means that both the following statements are true:

if p, then q and if q, then p

Other translations for the words *if and only if* are also sometimes used:

1. p if and only if q.
2. q if and only if p.
3. If p then q, and conversely.
4. If q then p, and conversely.
5. p is necessary and sufficient for q.

The proof of a theorem using any of these phrases requires two parts: first, prove one statement and then prove its converse. Theorem 6 illustrates this type of proof. Theorem 6 is called the *Factoring Theorem;* this theorem makes it possible to solve polynomial equations by factoring. For example, $(x - 3)(2x + 5) = 0$ is solved by setting $x - 3 = 0$ or $2x + 5 = 0$ to obtain $x = 3$ or $x = -\frac{5}{2}$.

Theorem 6 **Factoring Theorem.** $ab = 0$ if and only if $a = 0$ or $b = 0$.

Part 1 If $ab = 0$, then $a = 0$ or $b = 0$.

Proof

GIVEN: $a, b \in \mathbf{R}$, $ab = 0$ TO PROVE: $a = 0$ or $b = 0$

Statements	Reasons
1. $a, b \in \mathbf{R}$, $ab = 0$	1. Given
2. Let $a \neq 0$.	2. If $a = 0$, the theorem is true. Thus, this proof considers the case where $a \neq 0$.
3. $a \cdot 0 = 0$	3. Theorem 3
4. $ab = a \cdot 0$	4. Substitution of Step 3 into Step 1 (E4)
5. $ba = 0 \cdot a$	5. Commutative (F4)
$\therefore$6. $b = 0$	6. Theorem 5, $a \neq 0$

Part 2 If $a = 0$ or $b = 0$, then $ab = 0$.

GIVEN: $a, b \in \mathbf{R}$, $a = 0$ or $b = 0$ TO PROVE: $ab = 0$

Statements	Reasons
1. $a, b \in \mathbf{R}$	1. Given
2. If $a = 0$, then $ab = 0 \cdot b$.	2. Substitution
3. If $ab = 0 \cdot b$, then $ab = b \cdot 0$.	3. Commutative (F4)
4. If $b = 0$, then $ab = a \cdot 0$.	4. Substitution
5. If both $a = b = 0$, then $ab = 0 \cdot 0$.	5. Substitution
6. $a = 0$, $b = 0$, or both	6. Given
$\therefore$7. $ab = 0$	7. Steps 3, 4, 5, and Theorem 3

Problem Set 10.2

A *State the converse of the propositions in Problems 1–10. If the proposition is not in if–then form, restate it before writing the converse.*

1. If you break the law, then you will go to jail.
2. I will go with you if your car is air-conditioned.
3. You are happy if the sun shines.
4. I will go on Saturday if I get paid.
5. If your car is not air-conditioned, you will not have any friends.
6. If you brush your teeth with Smiles toothpaste, then you will have fewer cavities.
7. All cockroaches are ugly.
8. All politicians are liars.

9. Your grandchildren will not survive if we do not stop the pollution of our environment.

10. If $a = b$ and $c = d$, then $a + c = b + d$.

Use the principle of logical transitivity to form a valid conclusion for Problems 11–16 using all the statements for each argument.

11. If I eat that piece of pie, I will get fat.
If I get fat, I will not be able to jog a mile.
If I am not able to jog a mile, then I will be unhealthy.

12. If a nail is lost, then a shoe is lost.
If a shoe is lost, then a horse is lost.
If a horse is lost, then a rider is lost.
If a rider is lost, then a battle is lost.
If a battle is lost, then a kingdom is lost.

13. All snarks are fribbles.
All fribbles are ugly.

14. All trebbles are frebbles.
All frebbles are expensive.

15. If the World Health Organization sends pesticide to Borneo, then it will be used to kill mosquitoes.
If pesticide is used to kill mosquitoes, then bodies of the mosquitoes will become contaminated.
If the mosquitoes' bodies become contaminated, then the roaches that eat them will have an accumulation of pesticide in their bodies.
If the roaches have an accumulation of pesticide in their bodies, then the lizards that eat the roaches will also become contaminated.
If the lizards become contaminated, then the cats that eat the lizards will die from a buildup of the pesticide in their bodies.
If the cats die, then the rats will go unchecked in Borneo.
If the rats go unchecked in Borneo, then there is a real danger of plague.

16. If the World Health Organization sends pesticide to Borneo, then it will kill a type of parasite that feeds on caterpillars.
If the parasite that feeds on caterpillars is killed, then the caterpillars will multiply in the huts where they live.
If the caterpillars multiply, then they will eat away too much of the roof thatching.
If the caterpillars eat away too much of the roof thatching, then the roofs will cave in.

For the propositions given in Problems 17–21 identify p and q and write the proposition in the form if p then q, if q then p, or p if and only if q.

17. The judge will sentence you if and only if the prosecution can prove your guilt.

18. $x \cdot y = 0$ if and only if $x = 0$ or $y = 0$.

19. $x^2 = 16$ if $x = 4$.

20. A polygon is a square only if it is a rectangle.

21. If $\triangle ABC$ is a right triangle, then $a^2 + b^2 = c^2$ and conversely.

B *In Problems 22–78 fill in the blanks to complete the proofs. Theorems marked by an asterisk are used in subsequent theorems.*

***Theorem 10.2.1** Axiom F11 is a left-distributive law. Prove the right-distributive law:

$$(a + b)c = ac + bc.$$

Proof GIVEN: $a, b \in \mathbf{R}$ TO PROVE: $(a + b)c = ac + bc$

Statements	Reasons
1. $a, b \in \mathbf{R}$	1. Given
2. $(a + b)c = c(a + b)$	2. ______ **22.**
3. $c(a + b) = ca + cb$	3. ______ **23.**
4. $(a + b)c = ca + cb$	4. ______ **24.**
5. $ca + cb = ac + bc$	5. ______ **25.**
∴6. ______ **26.**	6. ______ **27.**

***Theorem 10.2.2** The opposite of a real number is unique: If $a + b = 0$, then $a = -b$.

Proof GIVEN: $a, b \in \mathbf{R}$, $a + b = 0$ TO PROVE: $a = -b$

Statements	Reasons
1. $b \in \mathbf{R}$	1. Given
2. ______ **28.**	2. Given
3. $-b \in \mathbf{R}$	3. ______ **29.**
4. $a + b + (-b) = 0 + (-b)$	4. Addition Law of Equality (Theorem 1)
5. $a + [b + (-b)] = 0 + (-b)$	5. ______ **30.**
6. ______ **31.**	6. Inverse (F9)
∴7. ______ **32.**	7. Identity (F7)

***Theorem 10.2.3** The reciprocal of a real number is unique: If $ab = 1$, then $b = 1/a$ $(a, b \neq 0)$.

Proof GIVEN: ______ **33.** ______ **34.** TO PROVE: ______ **35.**

Statements	Reasons
1. $a \in \mathbf{R}, a \neq 0$	1. Given
2. ______ **36.**	2. Given
3. $1/a \in \mathbf{R}$	3. ______ **37.**
4. $ab(1/a) = 1(1/a)$	4. ______ **38.**
5. $ba(1/a) = 1(1/a)$	5. ______ **39.**

Statements	Reasons
6. ____40.____	6. Associative (F6)
7. ____41.____	7. Inverse (F10)
∴8. ____42.____	8. Identity (F8)

***Theorem 10.2.4** Prove: $(a - b) + b = a$

Proof GIVEN: $a, b \in \mathbf{R}$ TO PROVE: $(a - b) + b = a$

Statements	Reasons
1. $a, b \in \mathbf{R}$	1. Given
2. $(a - b) + b = (a - b) + b$	2. ____43.____
3. $= [a + (-b)] + b$	3. ____44.____
4. ____45.____	4. Associative (F5)
5. ____46.____	5. Inverse (F9)
6. ____47.____	6. Identity (F7)
∴7. ____48.____	7. Transitive (E3)

Theorem 10.2.5 Prove: $(3a + b) + (2a - b) = 5a$

Proof GIVEN: ____49.____ TO PROVE: ____50.____

Statements	Reasons
1. $a, b \in \mathbf{R}$	1. Given
2. $(3a + b) + (2a - b)$ $= (3a + b) + (2a - b)$	2. ____51.____
3. $= (3a + b) + [2a + (-b)]$	3. ____52.____
4. $= 3a + 2a + b + (-b)$	4. ____53.____
5. $= (3a + 2a) + [b + (-b)]$	5. ____54.____
6. ____55.____	6. Inverse (F9)
7. ____56.____	7. Identity (F7)
8. $= (3 + 2)a$	8. ____57.____
9. $= 5a$	9. Closure (F1)
∴10. $(3a + b) + (2a - b) = 5a$	10. Transitive (E3)

***Theorem 10.2.6** Prove: $(ab)\frac{1}{a} \cdot \frac{1}{b} = 1$

Proof GIVEN: ____58.____ TO PROVE: ____59.____

Statements	Reasons
1. $a, b \in \mathbf{R}, a \neq 0, b \neq 0$	1. Given
2. **60.** ______	2. Reflexive (E1)
3. $= a \cdot \frac{1}{a} \cdot b \cdot \frac{1}{b}$	3. **61.** ______
4. $= \left(a \cdot \frac{1}{a}\right) \cdot \left(b \cdot \frac{1}{b}\right)$	4. **62.** ______
5. **63.** ______	5. **64.** ______
6. **65.** ______	6. **66.** ______
∴7. $ab\left(\frac{1}{a} \cdot \frac{1}{b}\right) = 1$	7. **67.** ______

Theorem 10.2.7 If $b + 2a = 0$, then $a = -b/2$.

Proof GIVEN: **68.** ______ **69.** ______ TO PROVE: **70.** ______

Statements	Reasons
1. $a, b \in \mathbf{R}$	1. Given
2. $b + 2a = 0$	2. **71.** ______
3. $2a = -b$	3. **72.** ______
4. **73.** ______	4. Multiplication Law of Equality (Theorem 2)
5. $\frac{1}{2}(2a) = \frac{1}{2}(-b)$	5. Commutative (F4)
6. **74.** ______	6. Definition of division
7. $\left(\frac{1}{2} \cdot 2\right)a = \frac{-b}{2}$	7. **75.** ______
8. **76.** ______	8. Inverse F(10)
∴9. **77.** ______	9. **78.** ______

C **79.** If $a = 1/b$, prove: $b = 1/a \quad (a, b \neq 0)$

80. Prove: $(3a - b) + (4a + b) = 7a$

81. Prove: $(2b - 3a) + (3a - 2b) = 0$

82. Prove Theorem 3 by assuming Theorem 4 has been previously proved.

83. Prove: $\frac{1}{b}(ba) = a \quad (b \neq 0)$

***84.** Prove: $x = \frac{a}{b}$ if and only if $a = bx$

85. Prove: $a - 2b = 0$ if and only if $b = a/2$

86. **a.** Which of the field axioms F1–F11 are satisfied by the set $\{0, 1\}$?
b. Let $1 + 1 = 0$ and answer the question of part a.

87. Prove: $-0 = 0$

10.3 Real Number Theorems

To build your ability to construct proofs, it is necessary for you to see many examples of proofs, as well as practice the construction of proofs on your own. This section will give you the opportunity for this practice as we change our emphasis from the nature of proof to the development of several real number theorems needed to lay the necessary foundation for the results used throughout this text.

The first theorems of this section are concerned with the properties of opposites.

Theorem 7 **Double Negative Theorem.** $-(-a) = a$ for any real number a.

Proof GIVEN: $a \in \mathbf{R}$ TO PROVE: $-(-a) = a$

Statements	Reasons
1. $a \in \mathbf{R}$*	1. Given*
2. $-a \in \mathbf{R}$	2. Inverse (F9)
3. There exists $-(-a) \in \mathbf{R}$ such that $-a + [-(-a)] = 0$.	3. Inverse (F9)
4. $-a + [-(-a)] + a = 0 + a$	4. Theorem 1
5. $[-(-a)] + (-a) + a = a + 0$	5. Commutative (F3)
6. $[-(-a)] + [(-a) + a] = a + 0$	6. Associative (F5)
7. $[-(-a)] + 0 = a + 0$	7. Inverse (F9)
∴8. $-(-a) = a$	8. Identity (F7)

Theorem 8 **Properties of Opposites.** If $a, b \in \mathbf{R}$, then:

1. $a(-b) = -(ab)$
2. $(-a)b = -(ab)$
3. $(-a)(-b) = ab$
4. $(-1)a = -a$

Proof *Part 1* GIVEN: $a, b \in \mathbf{R}$ TO PROVE: $a(-b) = -(ab)$

Statements	Reasons
1. $a[b + (-b)] = ab + a(-b)$	1. Distributive (F11)
2. $a \cdot 0 = ab + a(-b)$	2. Inverse (F9)
3. $0 = ab + a(-b)$	3. Zero Multiplication Law (Theorem 3)
4. $ab + a(-b) = 0$	4. Symmetric (E2)
5. $a(-b) + ab = 0$	5. Commutative (E4)
∴6. $a(-b) = -(ab)$	6. Opposite of a real number is unique (Theorem 10.2.2)

*For the remainder of the theorems in this book, the step in the proofs asserting that the variables are real numbers will be assumed.

Part 2 Interchange a and b in part 1 and apply the commutative property. The proof is left as a problem.

Part 3 Replace a by $-a$ in part 1; the result follows. The proof is left as a problem.

Part 4 Let $b = 1$ in part 1 to obtain

$$a(-1) = -(a \cdot 1)$$
$$(-1)a = -(a \cdot 1)$$
$$(-1)a = -a$$

The proof is left as a problem.

When reading Theorem 8 it is necessary to be careful to read $-a$ as "the opposite of a." We can, however, easily relate Theorem 8 to positive and negative numbers. In Chapter 1 we assumed that you were familiar with the notions of positive and negative numbers, but if you review these concepts from an advanced standpoint you need a precise definition.

POSITIVE NUMBERS

Order Axiom There exists a subset of **R**, called the **positive numbers,** with the properties:

O1. If $a \in \mathbf{R}$, then exactly one of the following is true:
i. $a = 0$ ii. a is positive iii. $-a$ is positive

O2. The sum of two positive numbers is positive.

O3. The product of two positive numbers is positive.

Property O1 is used so frequently it is given a name, **trichotomy.** That is, when the trichotomy property is referred to, it is the property we have numbered O1.

The Order Axiom defines a subset of the reals called the *positives*. The following definition gives a notation of an element of this subset and defines the negative numbers:

NEGATIVE NUMBERS

For a real number a:

If a is positive, then we write $a > 0$.
If $-a$ is positive, then a is **negative,** and we write $a < 0$.

From Theorem 8 and the Order Axiom, the rules for multiplication of integers can be derived. For example, we will show that the product of a positive and a negative is negative.

Let p and q be positive numbers; then from Theorem 8, part 1,

$$\underset{\substack{\uparrow \\ \text{Negative}}}{(-p)} \cdot \underset{\substack{\uparrow \\ \text{Positive}}}{q} = \underset{\substack{\uparrow \\ \text{Negative}}}{\underbrace{-(pq)}}$$

INEQUALITY SYMBOLS

a is **greater than** b, written $a > b$, if and only if $a - b$ is positive.
a is **less than** b, written $a < b$, if and only if $b > a$.
a is **greater than or equal to** b, written $a \geqslant b$, if and only if $a > b$ or $a = b$.
a is **less than or equal to** b, written $a \leqslant b$, if and only if $a < b$ or $a = b$.

The first theorem dealing with inequalities proves the converse of the statement in the definition of negative numbers.

Theorem 9 If $a < 0$, then $-a$ is positive.

Proof GIVEN: $a \in \mathbf{R}$, $a < 0$ TO PROVE: $-a$ is positive

Statements	Reasons
1. $a < 0$	1. Given
2. $0 > a$	2. Definition of less than
3. $0 - a$ is positive.	3. Definition of greater than
4. $0 - a = 0 + (-a)$	4. Definition of subtraction
5. $0 + (-a) = -a$	5. Identity (F7)
6. $0 - a = -a$	6. Transitive (E3); Steps 4 and 5
∴7. $-a$ is positive.	7. Substitution (E4); Step 6 into Step 3

Using the definition of the inequality symbols, you can now prove some properties of inequality that are related to the theorems of equality of the previous section.

Theorem 10 **Addition Law of Inequality.** If $a > b$, then

$$a + c > b + c \quad \text{for all real numbers } a, b, \text{ and } c$$

Proof GIVEN: $a, b, c \in \mathbf{R}$, $a > b$ TO PROVE: $a + c > b + c$

Statements	Reasons
1. $a > b$	1. Given
2. $a - b > 0$	2. Definition of greater than
3. $(a + 0) - b > 0$	3. Identity (F7)
4. $c + (-c) = 0$	4. Inverse (F9)
5. $a + [c + (-c)] - b > 0$	5. Substitute Step 4 into Step 3 (E4)
6. $(a + c) + [(-c) - b] > 0$	6. Associative (F5)
7. $(a + c) + [-c + (-b)] > 0$	7. Definition of subtraction
8. $(a + c) + [(-1)c + (-1)b] > 0$	8. Property of Opposites (Theorem 8)
9. $(a + c) + (-1)(c + b) > 0$	9. Distributive (F11)

Statements	Reasons
10. $(a + c) - (c + b) > 0$	10. Definition of subtraction
11. $a + c > c + b$	11. Definition of greater than
∴12. $a + c > b + c$	12. Commutative (F3)

Theorem 11 **Multiplication Law of Inequality.** For all real numbers a, b, and c:

1. If $a > b$ and $c > 0$, then $ac > bc$.
2. If $a > b$ and $c < 0$, then $ac < bc$.

Proof *Part 1* GIVEN: $a, b, c \in \mathbf{R}$, $a > b, c > 0$ TO PROVE: $ac > bc$

Statements	Reasons
1. $a > b, c > 0$	1. Given
2. $a - b > 0$	2. Definition of greater than
3. $(a - b)c > 0$	3. The product of positives is positive (Order Axiom).
4. $ac - bc > 0$	4. Distributive (Theorem 10.2.1)
∴5. $ac > bc$	5. Definition of greater than

Part 2 The proof is similar to part 1 and is left as a problem.

Theorems 10 and 11 also hold for $<$, $\leq$, and $\geq$. These symbols will be assumed to be included when referring to Theorems 10 and 11.

The rest of the theorems of this section are concerned with the properties of absolute value, which are summarized in Table 10.1. The first six properties follow directly from the definition.

ABSOLUTE VALUE

The **absolute value** of a, written $|a|$, is defined by

$$|a| = \begin{cases} a & \text{if } a > 0 \\ 0 & \text{if } a = 0 \\ -a & \text{if } a < 0 \end{cases}$$

Theorem 12 $|a| \geq 0$

Proof GIVEN: $a \in \mathbf{R}$ TO PROVE: $|a| \geq 0$

Statements	Reasons
1. $a > 0$, $a = 0$, or $a < 0$	1. Trichotomy (O1)
2. If $a > 0$, then $\|a\| = a$.	2. Definition of absolute value
3. $\|a\| > 0$	3. Substitution, Step 2
4. If $a < 0$, then $\|a\| = -a$.	4. Definition of absolute value
5. $-a$ is positive.	5. Theorem 9
6. $\|a\|$ is positive.	6. Substitution (Steps 4 and 5)
7. $\|a\| > 0$	7. Definition of negative numbers
8. If $a = 0$, then $\|a\| = 0$.	8. Definition of absolute value
9. $a \geqslant 0$	9. Definition of greater than or equal to and Steps 3 and 8
∴10. $\|a\| \geqslant 0$ (all cases)	10. Steps 3, 7, and 9

The proofs of Properties 2–6 listed in Table 10.1 are similar to the proof of Theorem 12, and all rely on the definition of absolute value. You are asked to prove these in Problem Set 10.3. Theorem 13 shows a proof for Property 7. Remember, $a = \pm b$ means $a = b$ or $a = -b$.

Table 10.1 Properties of Absolute Value

Let a be a real number. **Property**	**Comment**
1. $\|a\| \geqslant 0$	1. Absolute value is nonnegative.
2. $\|-a\| = \|a\|$	2. The absolute value of a number and the absolute value of its opposite are equal.
3. $\|a\|^2 = a^2$	3. If an absolute value is squared, the absolute value sign can be dropped.
4. $\|ab\| = \|a\|\,\|b\|$	4. This property tells how to multiply absolute values. The absolute value of a product is the product of the absolute values.
5. $\left\|\frac{a}{b}\right\| = \frac{\|a\|}{\|b\|}$ $\quad b \neq 0$	5. This property tells how to divide absolute values. The absolute value of a quotient is the quotient of the absolute values.
6. $-\|a\| \leqslant a \leqslant \|a\|$	6. Any number a is between + and − the absolute value of that number, inclusive.
7. $\|a\| = \|b\|$ if and only if $a = \pm b$	7. This property tells how to solve absolute value equations.
8. $\|a\| < b$ if and only if $-b < a < b$ for $b > 0$	8. These properties tell how to solve
9. $\|a\| > b$ if and only if either $a > b$ or $a < -b$ for $b > 0$.	9. absolute value inequalities. Property 8 also holds for $\leqslant$ and Property 9 holds for $\geqslant$. The condition $b > 0$ is necessary because of Property 1.
10. $\|a + b\| \leqslant \|a\| + \|b\|$	10. This property is called the **triangle inequality** and is an important property of absolute value.
11. $\|a - b\| \geqslant \|a\| - \|b\|$	11. This is a property of absolute value of a difference. Also, $\|a - b\| \leqslant \|a\| + \|b\|$.

Theorem 13 **Absolute Value Equations.** For all real numbers a and b,

$$|a| = |b| \quad \text{if and only if} \quad a = \pm b$$

Proof *Part 1* If $|a| = |b|$, then $a = \pm b$.

GIVEN: $a, b \in \mathbf{R}$ TO PROVE: $a = b$ or $a = -b$
$|a| = |b|$

Statements	Reasons
1. $\lvert a\rvert = \lvert b\rvert$	1. Given
2. $\lvert a\rvert \lvert a\rvert = \lvert b\rvert \lvert a\rvert$	2. Multiplication Law of Equality (Theorem 2)
3. $\lvert a\rvert \lvert a\rvert = \lvert b\rvert \lvert b\rvert$	3. Substitution of Step 1 into Step 2 (E4)
4. $\lvert a\rvert^2 = \lvert b\rvert^2$	4. Definition of exponent
5. $a^2 = b^2$	5. Property 3 of absolute value (Table 10.1)
6. $a^2 + (-b^2) = b^2 + (-b^2)$	6. Addition Law of Equality (Theorem 1)
7. $a^2 + (-b^2) = 0$	7. Inverse (F9)
8. $a^2 - b^2 = 0$	8. Definition of subtraction
9. $(a - b)(a + b) = 0$	9. Factoring (see discussion following proof)
10. $a - b = 0$ or $a + b = 0$	10. Factoring Theorem (Theorem 6)
11. $a + (-b) = 0$ or $a + b = 0$	11. Definition of subtraction
12. $a = -(-b)$ or $a = -b$	12. Opposite of a real number (Theorem 10.2.2)
∴13. $a = b$ or $a = -b$	13. Double Negative (Theorem 7)

Part 2 If $a = \pm b$, then $|a| = |b|$.

GIVEN: $a, b \in \mathbf{R}$ TO PROVE: $|a| = |b|$
$a = b$ or $a = -b$

The proof of this follows directly from the definition of absolute value and is left as a problem.

In the proof of Theorem 13 we used *factoring* as a reason. This, of course, is not proper since factoring is not a proved result. But it might be instructive to show how this factoring step follows from the properties that have been developed.

$$\begin{aligned}
(a - b)(a + b) &= (a - b)a + (a - b)b && \text{Distributive (F11)}\\
&= a^2 - ba + ab - b^2 && \text{Distributive (F11)}\\
&= a^2 + (-ba) + (ab) - b^2 && \text{Definition of subtraction}\\
&= a^2 + [(-ba) + (ab)] - b^2 && \text{Associative (F5)}\\
&= a^2 + [(-ab) + (ab)] - b^2 && \text{Commutative (F3)}\\
&= a^2 + 0 - b^2 && \text{Inverse (F9)}\\
&= a^2 - b^2 && \text{Identity (F7)}
\end{aligned}$$

Theorem 14 **Absolute Value Inequalities.** For real numbers a and b where $b > 0$,

$$|a| < b \quad \text{if and only if} \quad -b < a < b$$

Proof *Part 1* If $|a| < b,\ b > 0$, then $-b < a < b$.

GIVEN: $a, b \in \mathbf{R}$ TO PROVE: $-b < a$ and $a < b$
$b > 0$
$|a| < b$

Statements	Reasons
1. $\|a\| < b$	1. Given
2. $\|a\| = \pm a$	2. Definition of absolute value
Case i: $\|a\| = a$	
∴3. $a < b$	3. Substitution into Step 1
Case ii: $\|a\| = -a$	
4. $-a < b$	4. Substitution into Step 1 (E4)
5. $-a + (a - b) < b + (a - b)$	5. Addition Law of Inequality (Theorem 10)
6. $-a + [a + (-b)] < b + [a + (-b)]$	6. Definition of subtraction
7. $[-a + a] + (-b) < a + [b + (-b)]$	7. Commutative (F3) and Associative (F5)
8. $0 + (-b) < a + 0$	8. Inverse (F9)
∴9. $-b < a$	9. Identity (F7)

Part 2 If $-b < a < b,\ b > 0$, then $|a| < b$.

GIVEN: $a, b \in \mathbf{R}$ TO PROVE: $|a| < b$
$b > 0$
$-b < a$ and $a < b$

The steps of part 1 can be reversed to arrive at the desired result.

Theorems 13 and 14 can serve as examples for other absolute value proofs you may encounter.

Problem Set 10.3

A *In Problems 1–73 fill in the blanks to complete the proofs. Theorems marked by an asterisk are used in subsequent theorems.*

Theorem 10.3.1 Prove (part 2 of Theorem 8): $(-a)b = -(ab)$

Proof GIVEN: $a, b \in \mathbf{R}$ TO PROVE: $(-a)b = -(ab)$

Statements	Reasons
1. $[a + (-a)]b = ab + (-a)b$	1. **1.** ____________
2. **2.** ____________	2. Inverse (F9)
3. **3.** ____________	3. Zero Multiplication Law (Theorem 3)
4. $ab + (-a)b = 0$	4. **4.** ____________
5. $(-a)b + ab = 0$	5. **5.** ____________
∴6. **6.** ____________	6. **7.** ____________

***Theorem 10.3.2** Prove (part 4 of Theorem 8): $(-1)a = -a$

Proof GIVEN: $a, b \in \mathbf{R}$ TO PROVE: $(-1)a = -a$

Statements	Reasons
1. $a(-b) = -(ab)$	1. **8.**
2. If $b = 1$, then $a(-1) = -(a \cdot 1)$.	2. **9.**
3. **10.**	3. Commutative (F4)
∴4. **11.**	4. **12.**

***Theorem 10.3.3** Prove: If $a < 0$, then $-a > 0$.

Proof GIVEN: $a \in \mathbf{R}$, $a < 0$ TO PROVE: $-a > 0$

Statements	Reasons
1. $a < 0$	1. **13.**
2. $(-1) < 0$	2. Definition of a negative
3. $a \cdot (-1) > 0 \cdot (-1)$	3. **14.**
4. $(-1) \cdot a > (-1) \cdot 0$	4. **15.**
5. $(-1) \cdot a > 0$	5. **16.**
∴6. $-a > 0$	6. **17.**

Theorem 10.3.4 Prove (part 2 of Theorem 11): If $a > b$ and $c < 0$, then $ac < bc$.

Proof GIVEN: **18.** **19.** TO PROVE: **20.**

Statements	Reasons
1. $a > b, c < 0$	1. **21.**
2. $a - b > 0$	2. **22.**
3. $(-c) > 0$	3. **23.**
4. $(-c)(a - b) > 0$	4. **24.**
5. $-(-c)(a - b) < 0$	5. **25.**
6. $c(a - b) < 0$	6. **26.**
7. **27.**	7. Distributive (F11)
8. **28.**	8. Commutative (F3)
∴9. **29.**	9. Definition of less than

***Theorem 10.3.5** Prove the transitive property of order: If $a > b$ and $b > c$, then $a > c$.

Proof GIVEN: **30.** **31.** TO PROVE: **32.**

Statements	Reasons
1. $a > b,\ b > c$	1. **33.** ______
2. $a - b > 0,\ b - c > 0$	2. **34.** ______
3. $(a - b) + (b - c) > 0$	3. **35.** ______
4. $[a + (-b)] + [b + (-c)] > 0$	4. **36.** ______
5. $[a + (-c)] + [(-b) + b] > 0$	5. Repeated use of Associative (F5) and Commutative (F3)
6. $[a + (-c)] + 0 > 0$	6. **37.** ______
7. $a + (-c) > 0$	7. **38.** ______
8. $a - c > 0$	8. **39.** ______
$\therefore$9. $a > c$	9. **40.** ______

Theorem 10.3.6 Prove: If $a > 1$, then $a^2 > 1$.

Proof GIVEN: **41.** ______ **42.** ______ TO PROVE: **43.** ______

Statements	Reasons
1. $a > 1$	1. **44.** ______
2. $a - 1 > 0$	2. **45.** ______
3. $a(a - 1) > 0$	3. **46.** ______
4. $a^2 - a > 0$	4. Distributive (F11) and definition of exponent
5. $a^2 > a$	5. **47.** ______
$\therefore$6. $a^2 > 1$	6. **48.** ______

Theorem 10.3.7 Prove (Property 2 of absolute value): $|-a| = |a|$

Proof GIVEN: **49.** ______ TO PROVE: **50.** ______

Statements	Reasons
Case i: $a > 0$	
1. $-a < 0$	1. **51.** ______
2. $\lvert a\rvert = a,\ \lvert -a\rvert = -(-a)$	2. **52.** ______
3. $\lvert a\rvert = a,\ \lvert -a\rvert = a$	3. **53.** ______
$\therefore$4. $\lvert -a\rvert = \lvert a\rvert$	4. **54.** ______
Case ii: $a = 0$	
5. $\lvert a\rvert = 0,\ \lvert -a\rvert = 0$	5. **55.** ______
$\therefore$6. $\lvert -a\rvert = \lvert a\rvert$	6. **56.** ______
Case iii: $a < 0$	
7. $-a > 0$	7. **57.** ______
8. $\lvert a\rvert = -a,\ \lvert -a\rvert = -a$	8. **58.** ______
$\therefore$9. $\lvert -a\rvert = \lvert a\rvert$	9. **59.** ______

Theorem 10.3.8 Prove (Property 3 of absolute value): $|a|^2 = a^2$

Proof GIVEN: **60.** ________ TO PROVE: **61.** ________

Statements	Reasons
1. $\|a\|\,\|a\| = \|a\|\,\|a\|$	1. **62.** ________
Case i: $a > 0$	
2. $\|a\| = a$	2. **63.** ________
3. $\|a\|\,\|a\| = aa$	3. **64.** ________
∴4. $\|a\|^2 = a^2$	4. Definition of exponent
Case ii: $a = 0$	
5. $\|a\| = 0$	5. **65.** ________
6. $\|a\|\,\|a\| = 0 \cdot 0$	6. **66.** ________
7. $\|a\|\,\|a\| = aa$	7. **67.** ________
∴8. **68.** ________	8. Definition of exponent
Case iii: $a < 0$	
9. $\|a\| = -a$	9. **69.** ________
10. **70.** ________	10. Substitution of Step 9 into Step 1
11. **71.** ________	11. Property of Opposites (Theorem 8)
∴12. **72.** ________	12. **73.** ________

B **74.** Prove part 3 of Theorem 8: $(-a)(-b) = ab$

***75.** Prove: $a(b - c) = ab - ac$
(This is the left-distributive law for multiplication over subtraction.)

76. Prove: $(a - b)c = ac - bc$
(This is the right-distributive law for multiplication over subtraction.)

77. Prove: $(-a) + (-b) = -(a + b)$

78. If $a = b$, prove: $-a = -b$

***79.** Prove: $-(a - b) = b - a$

80. Prove: $x - a(-b) = x + ab$

81. Prove part 2 of Theorem 13: If $a = \pm b$, then $|a| = |b|$.

***82.** If $a - b > 0$, prove: $|a - b| = |b - a|$

***83.** If $a - b < 0$, prove: $|a - b| = |b - a|$

***84.** If $a - b = 0$, prove: $|a - b| = |b - a|$

85. Prove: $|a - b| = |b - a|$

86. Prove the factoring property of a perfect square:

$$a^2 + 2ab + b^2 = (a + b)^2$$

87. Prove the factoring property of a perfect square:

$$a^2 - 2ab + b^2 = (a - b)^2$$

C **88.** What is wrong with the following "proof" that $0 = 1$?

Let $x = 0$. Then $x(x-1) = 0(x-1)$.	Multiplication Law of Equality (Theorem 2)
$x(x-1) = 0$	Zero Multiplication Law (Theorem 3)
$x = 0, \quad x - 1 = 0$	Factoring Theorem (Theorem 6)
$x = 0, \quad x = 1$	Opposite of a real number (Theorem 10.2.2)
$\therefore 0 = 1$	Transitive (E3)

89. Prove Property 6 of absolute value: $-|a| \leq a \leq |a|$

***90.** Prove Property 4 of absolute value: $|ab| = |a|\,|b|$

91. Prove Property 5 of absolute value: $\left|\frac{a}{b}\right| = \frac{|a|}{|b|}$

10.4 Methods of Proof

The previous examples of proof have used one of two important proof forms:

DIRECT PROOF

$$\begin{array}{l} p \rightarrow q \\ p \\ \hline \therefore q \end{array}$$

TRANSITIVE PRINCIPLE

$$\begin{array}{l} p \rightarrow q \\ q \rightarrow r \\ \hline \therefore p \rightarrow r \end{array}$$

Sometimes a third method of proof, called **indirect proof,** is useful. Other names given to indirect proof are *proof by contradiction* and *modus tollens*. Let the symbol $\sim p$ represent the **negation of** p and $\sim q$ represent the **negation of** q. Then:

INDIRECT PROOF FORM

PREMISE:	$p \rightarrow q$	If you know that p implies q and that q is false, then you can conclude that p is also false.
PREMISE:	$\sim q$	
CONCLUSION:	$\therefore \sim p$	

Example 1 If you receive an A on the final, then you will pass the course.
You did not pass the course.

Therefore, you did not receive an A on the final. □

Example 2 If the cat takes the rat, then the rat will take the cheese.
The rat does not take the cheese.

Therefore, the cat does not take the rat. □

Example 3 If you received an A on the test, then I am Napoleon.
I am not Napoleon.

Therefore, you did not receive an A on the test. □

Example 4 If $\sqrt{2}$ is rational, then $\sqrt{2}$ can be written as p/q where $p, q \in \mathbf{Z}$.
$\sqrt{2}$ cannot be written as p/q with $p, q \in \mathbf{Z}$.

Therefore, $\sqrt{2}$ is not rational. □

Example 4 leads to an important theorem in algebra proving the existence of an irrational number. This proof is included as an example of an indirect proof.

Theorem 15 $\sqrt{2}$ is not rational.

Indirect Proof Assume $\sqrt{2}$ is rational. Then $\sqrt{2} = p/q$ where p and q are integers, $q \neq 0$, and let p/q be reduced. (If it is not reduced, write it in reduced form.)

Statements	Reasons
1. $\sqrt{2} = p/q$, where p/q is reduced.	1. Given
2. $2 = p^2/q^2$	2. Square both sides.
3. $2q^2 = p^2$	3. Multiply both sides by q^2.
4. p^2 is even.	4. Definition of an even number
5. p is even.	5. See Theorem 10.4.2, Problem Set 10.4.
6. Let $p = 2k$, k an integer.	6. Definition of an even number
7. $2q^2 = (2k)^2$	7. Substitution of Step 6 into Step 3.
8. $2q^2 = 4k^2$	8. Squaring
9. $q^2 = 2k^2$	9. Multiply both sides by $\frac{1}{2}$.
10. q^2 is even.	10. Definition of an even number
11. q is even.	11. See Theorem 10.4.2, Problem Set 10.4.
12. p/q is not reduced.	12. Step 5 and Step 11; p and q each have a factor of 2.

But Step 12 contradicts Step 1. Therefore, the assumption that $\sqrt{2}$ is rational is incorrect, and $\sqrt{2}$ is not rational.

Problem Set 10.4

A *Name the type of reasoning illustrated by the arguments in Problems 1–10.*

1. If I inherit \$1000, I will buy you a cookie.
I inherit \$1000.
Therefore, I will buy you a cookie.

2. If you climb the highest mountain, you will feel great.
If you feel great, then you are happy.
Therefore, if you climb the highest mountain, then you are happy.

3. If $b = 0$, then $a = 0$.
$a \neq 0$.
Therefore, $b \neq 0$.

4. If $a \cdot b = 0$, then $a = 0$ or $b = 0$.
$a \cdot b = 0$.
Therefore, $a = 0$ or $b = 0$.

5. If a^2 is even, then a is even.
a is odd. (Assume that a is a counting number.)
Therefore, a^2 is odd.

6. If a^2 is even, then a is even.
a^2 is even.
Therefore, a is even.

7. All prunes are organic.
All organic foods are healthy.
Therefore, all prunes are healthy.

8. If I am lazy, then I am not rewarded.
If I am not rewarded, then I do not attend to my duties.
Therefore, if I am lazy, then I do not attend to my duties.

9. If I am lazy, then I do not attend to my duties.
I am lazy.
Therefore, I do not attend to my duties.

10. If you go to college, then you will get a good job.
If you get a good job, then you will make a lot of money.
You go to college.
Therefore, you will make a lot of money.

11. Give an example of direct reasoning.

12. Give an example of indirect reasoning.

In Problems 13–22 some of the arguments illustrate a valid form of reasoning, but others do not. If the problem illustrates a valid argument, name the type of reasoning used.

13. If you understand logic, then you will enjoy this sort of problem.
You do not understand logic.
Therefore, you will not enjoy this sort of problem.

14. If you understand a problem, it is easy.
The problem is not easy.
Therefore, you do not understand the problem.

15. If I do not get a raise in pay, I will quit.
I do not get a raise.
Therefore, I quit.

16. If I do not get a raise in pay, I will quit.
I quit.
Therefore, I did not get a raise in pay.

17. If Fermat's Last Theorem is ever proved, then my life is complete.
My life is not complete.
Therefore, Fermat's Last Theorem is not proved.

18. If a triangle is a right triangle, then the square of the hypotenuse is equal to the sum of the squares of the other two sides.
In $\triangle ABC$ the square of the hypotenuse is equal to the sum of the squares of the other two sides.
Therefore, $\triangle ABC$ is a right triangle.

19. If a triangle is a right triangle, then it is not equiangular.
$\triangle ABC$ is an equiangular triangle.
Therefore, $\triangle ABC$ is not a right triangle.

20. If a triangle is isosceles, then it has two sides of equal length.
$\triangle ABC$ has two sides of equal length.
Therefore, $\triangle ABC$ is isosceles.

21. Only if a whole number is divisible by 3 is it divisible by 6.
36 is divisible by 3.
Therefore, 36 is divisible by 6.

22. $x = 2$ only if $x^2 = 4$.
$x^2 = 4$
Therefore, $x = 2$.

B *In Problems 23–33 fill in the blanks to complete the proof.*

Theorem 10.4.1 The square of an odd integer is an odd integer. [*Hint:* An odd number is a number that can be written as $2k + 1$, where k is some integer.]

Proof GIVEN: a is an odd integer. TO PROVE: a^2 is an odd integer.

Statements	Reasons
1. Let $a = 2k + 1$ (k any integer).	1. Definition of an odd integer
2. $a^2 = a \cdot a$	2. Definition of squares
3. $a^2 = (2k + 1)(2k + 1)$	3. ____ **23.** ____
4. $= (2k + 1) \cdot 2k + (2k + 1) \cdot 1$	4. ____ **24.** ____
5. $= 2k \cdot 2k + 1 \cdot 2k + (2k + 1) \cdot 1$	5. ____ **25.** ____
6. $= 2k \cdot 2k + 2k + 2k + 1$	6. ____ **26.** ____
7. $= 2(k \cdot 2k + k + k) + 1$	7. ____ **27.** ____
8. $k \cdot 2k + k + k$ is an integer.	8. ____ **28.** ____
∴9. a^2 is odd.	9. ____ **29.** ____

Theorem 10.4.2 If a is an integer and a^2 is even, then a is even.

Proof by Contradiction Assume a is an odd integer.

Statements	Reasons
1. a is odd.	1. ____ **30.** ____
2. ____ **31.** ____	2. Theorem 10.4.1
3. a^2 is even.	3. ____ **32.** ____
∴4. ____ **33.** ____	4. Contradiction in Steps 2 and 3; therefore, assumption is false.

34. Prove that if an integer is even, then its square is even.

35. Prove that the set of odd integers is closed for multiplication.

36. Prove that the set of even integers is closed for multiplication.

37. If p^2 is divisible by 3, prove that p is divisible by 3.

38. Prove that $\sqrt{3}$ is irrational.

39. Theorem 15 proved that $\sqrt{2}$ is irrational. In a similar manner show that $\frac{1}{2}$ is not an integer.

C **40.** A subset S of the real numbers is said to be *well-ordered* if each nonempty subset contains a smallest member. An axiom of the positive integers is called the *well-ordering principle:* the positive integers are well-ordered. Prove by contradiction that there is no integer between 0 and 1.

41. Give an example of a subset of the real numbers that is not well-ordered (see Problem 40).

42. Prove that a set S of positive integers that includes 1 and includes $n + 1$ whenever it includes n, also includes every positive integer (see Problem 40). [*Hint:* Let T be the nonempty set of all positive integers not in S; apply the well-ordering principle to T, assuming $T \neq 0$, and see what happens.

*10.5 Mathematical Induction

Mathematical induction is another important method of proof in mathematics that allows us to prove results involving the set of positive integers. Do not confuse mathematical induction with the scientific method or the inductive logic used in the experimental sciences. Mathematical induction is a form of deductive logic in which conclusions are inescapable.

The first step in establishing a result by mathematical induction is often the observation of a pattern. Let us begin with a simple example. Suppose you wish to know the sum of the first n odd integers. You could begin by looking for a pattern:

$$\begin{aligned} 1 &= 1 \\ 1 + 3 &= 4 \\ 1 + 3 + 5 &= 9 \\ 1 + 3 + 5 + 7 &= 16 \\ 1 + 3 + 5 + 7 + 9 &= 25 \end{aligned}$$

Do you see a pattern? It appears that the sum of the first n odd numbers is n^2, since we see that the sum of the first 3 odd numbers is 3^2, of the first 4 odd numbers is 4^2, and so on. You now wish to *prove* deductively that

$$1 + 3 + 5 + \cdots + \underbrace{(2n - 1)}_{\substack{\uparrow \\ n\text{th odd number}}} = n^2$$

is true for all positive integers n. How can you proceed? You can use a method called *mathematical induction*, which is used to prove certain propositions about the positive integers. The proposition is denoted by $P(n)$. For example, in the above problem, let

$$P(n) = 1 + 3 + 5 + \cdots + (2n - 1) = n^2$$

*This section does not require Sections 10.1–10.4.

This means

$$
\begin{array}{ll}
P(1): & 1 = 1^2 \\
P(2): & 1 + 3 = 2^2 \\
P(3): & 1 + 3 + 5 = 3^2 \\
P(4): & 1 + 3 + 5 + 7 = 4^2 \\
 & \vdots \\
P(100): & 1 + 3 + 5 + \cdots + 199 = 100^2 \\
 & \vdots \\
P(x-1): & 1 + 3 + 5 + \cdots + (2x-3) = (x-1)^2 \\
P(x): & 1 + 3 + 5 + \cdots + (2x-1) = x^2 \\
P(x+1): & 1 + 3 + 5 + \cdots + (2x+1) = (x+1)^2
\end{array}
$$

Now, you want to show that $P(n)$ is true for all n (n a positive integer).

MATHEMATICAL INDUCTION

Principle of Mathematical Induction (PMI) If a given proposition $P(n)$ is true for $P(1)$ and if the truth of $P(k)$ implies the truth of the proposition for $P(k + 1)$, then $P(n)$ is true for all positive integers.

This gives us the following procedure for proof by mathematical induction:

1. Prove $P(1)$ is true.
2. Assume $P(k)$ is true.
3. Prove $P(k + 1)$ is true.
4. Conclude $P(n)$ true for all positive integers n.

Students often have a certain uneasiness when they first use the Principle of Mathematical Induction as a method of proof. Suppose we use this principle with a stack of dominoes, as shown in the cartoon.

How can the man in the cartoon be certain of knocking over all the dominoes?

He would have to be able to knock over the first one. He would have to have the dominoes arranged so that *if* the kth domino falls, then the next one, the $(k + 1)$st, will also fall. That is, each domino is set up so that if it falls, it causes the next one to fall. We have set up a kind of "chain reaction" here. The first domino falls; this knocks over the next one (the second domino); the second one knocks over the next one (the third domino); the third one knocks over the next one; this continues until all the dominoes are knocked over.

Returning to the example to prove (for all positive integers n)

$$1 + 3 + 5 + \cdots + (2n - 1) = n^2$$

Step 1 Prove $P(1)$ true: $1 = 1^2$ is true.

Step 2 Assume $P(k)$ true: $1 + 3 + 5 + \cdots + (2k - 1) = k^2$

Step 3 Prove $P(k + 1)$ true.

TO PROVE: $1 + 3 + 5 + \cdots + [2(k + 1) - 1] = (k + 1)^2$
This is found by substituting $(k + 1)$ for n in the original statement you want to prove. Next, simplify the TO PROVE statement. This is so you will know when you are finished with Step 3 of the proof.

$$1 + 3 + 5 + \cdots + [2(k + 1) - 1] = (k + 1)^2$$
$$1 + 3 + 5 + \cdots + [2k + 2 - 1] = (k + 1)^2$$
$$1 + 3 + 5 + \cdots + (2k + 1) = (k + 1)^2$$

The procedure for Step 3 is to begin with the hypotheses (from Step 2) and *prove*
$1 + 3 + 5 + \cdots + (2k + 1) = (k + 1)^2$.

Statements	Reasons
1. $1 + 3 + 5 + \cdots + (2k - 1) = k^2$	1. By hypothesis (Step 2)
2. $1 + 3 + 5 + \cdots + (2k - 1) + (2k + 1) = k^2 + (2k + 1)$	2. Add $(2k + 1)$ to both sides.
3. $1 + 3 + 5 + \cdots + (2k - 1) + (2k + 1) = k^2 + 2k + 1$	3. Associative
4. $1 + 3 + 5 + \cdots + (2k - 1) + (2k + 1) = (k + 1)^2$	4. Factoring (distributive)

Step 4 The proposition is true for all positive integers by PMI.

Example 1 Prove or disprove: $2 + 4 + 6 + \cdots + 2n = n(n + 1)$

Proof *Step 1* Prove $P(1)$ true:
$2 \stackrel{?}{=} 1(1 + 1)$
$2 = 2$
It is true.

Step 2 Assume $P(k)$. HYPOTHESIS: $2 + 4 + 6 + \cdots + 2k = k(k + 1)$

Step 3 Prove $P(k + 1)$. TO PROVE: $2 + 4 + 6 + \cdots + 2(k + 1) = (k + 1)(k + 2)$

Statements	Reasons
1. $2 + 4 + 6 + \cdots + 2k = k(k + 1)$	1. Hypothesis (Step 2)
2. $2 + 4 + 6 + \cdots + 2k + 2(k + 1) = k(k + 1) + 2(k + 1)$	2. Add $2(k + 1)$ to both sides.
3. $= (k + 1)(k + 2)$	3. Factoring

Step 4 The proposition is true for all positive integers by PMI. □

Example 2 Prove or disprove: $n^3 + 2n$ is divisible by 3

Proof *Step 1* Prove $P(1)$: $1^3 + 2 \cdot 1 = 3$, which is divisible by 3.

Step 2 Assume $P(k)$. HYPOTHESIS: $k^3 + 2k$ is divisible by 3.

Step 3 Prove $P(k + 1)$. TO PROVE: $(k + 1)^3 + 2(k + 1)$ is divisible by 3.

Statements	Reasons
1. $(k + 1)^3 + 2(k + 1) = k^3 + 3k^2 + 3k + 1 + 2k + 2$	1. Distributive, associative, and commutative axioms
2. $= (3k^2 + 3k + 3) + (k^3 + 2k)$	2. Commutative and associative axioms
3. $= 3(k^2 + k + 1) + (k^3 + 2k)$	3. Distributive axiom
4. $3(k^2 + k + 1)$ is divisible by 3.	4. Definition of divisibility by 3
5. $k^3 + 2k$ is divisible by 3.	5. Hypothesis
6. $(k + 1)^3 + 2(k + 1)$ is divisible by 3.	6. Both terms are divisible by 3 and therefore the sum is divisible by 3.

Step 4 The proposition is true for all positive integers n by PMI. □

Example 3 shows that *even though* you make an assumption in Step 2, it is not going to help if the proposition is not true.

Example 3 Prove or disprove: $n + 1$ is prime

Proof *Step 1* Prove $P(1)$: $1 + 1 = 2$ is a prime.

Step 2 Assume $P(k)$. HYPOTHESIS: $k + 1$ is a prime.

Step 3 Prove $P(k + 1)$. TO PROVE: $(k + 1) + 1$ is a prime. This is not possible since $(k + 1) + 1 = k + 2$, which is not prime whenever k is an even positive integer.

Step 4 Any conclusions? You cannot conclude that the statement is false, only that induction does not work. But this statement is in fact false, and a counterexample is $n = 3$, since $n + 1 = 4$ is not prime. □

Example 4 Prove or disprove: $1 \cdot 2 \cdot 3 \cdot 4 \cdot \cdots \cdot n < 0$

Proof Students often slip into the habit of skipping either the first or second step in a proof by mathematical induction. This is dangerous, and it is important to check every step. Suppose a careless person did not verify the first step.

Step 2 Assume $P(k)$. HYPOTHESIS: $1 \cdot 2 \cdot 3 \cdot \cdots \cdot k < 0$

Step 3 Prove $P(k + 1)$. TO PROVE: $1 \cdot 2 \cdot 3 \cdot \cdots \cdot k \cdot (k + 1) < 0$

$1 \cdot 2 \cdot 3 \cdot \cdots \cdot k < 0$ by hypothesis.
$k + 1$ is positive since k is a positive integer. Then, since we know that the product of a negative and a positive is negative,*

$$\underbrace{1 \cdot 2 \cdot 3 \cdot \cdots \cdot k}_{\text{Negative}} \cdot \underbrace{(k + 1)}_{\text{Positive}} < 0$$

Step 3 is proved.

Step 4 The proposition is not true for all positive integers since the first step, $1 < 0$, does not hold. □

Problem Set 10.5

A

1. Prove: $1 + 2 + 3 + \cdots + n = \dfrac{n(n + 1)}{2}$ for all positive integers n

2. Prove: $1^2 + 2^2 + 3^2 + \cdots + n^2 = \dfrac{n(n + 1)(2n + 1)}{6}$ for all positive integers n

3. Prove: $\displaystyle\sum_{j=1}^{n} (2j - 1)^2 = \frac{n(2n - 1)(2n + 1)}{3}$ for all positive integers n

4. Prove: $\displaystyle\sum_{j=1}^{n} j^3 = \frac{n^2(n + 1)^2}{4}$ for all positive integers n

5. Prove: $2^2 + 4^2 + 6^2 + \cdots + (2n)^2 = \dfrac{2n(n + 1)(2n + 1)}{3}$ for all positive integers n

6. Prove: $1 \cdot 2 + 2 \cdot 3 + 3 \cdot 4 + \cdots + n(n + 1) = \dfrac{n(n + 1)(n + 2)}{3}$ for all positive integers n

7. Prove: $1 \cdot 3 + 2 \cdot 4 + 3 \cdot 5 + \cdots + n(n + 2) = \dfrac{n(n + 1)(2n + 7)}{6}$ for all positive integers n

8. Prove: $1 + r + r^2 + \cdots + r^n = \dfrac{r^{n+1} - 1}{r - 1}$ for all positive integers n

B

9. Prove: $n^5 - n$ is divisible by 5 for all positive integers n.

10. Prove: $n(n + 1)(n + 2)$ is divisible by 6 for all positive integers n.

---

*If you have studied Section 10.3, you should recognize this as part 2 of Theorem 8.

11. Prove: $(1 + n)^2 \geq 1 + n^2$ for all positive integers n.

12. Notice

$$\begin{aligned} 1^3 &= 1^2 \\ 1^3 + 2^3 &= 3^2 \\ 1^3 + 2^3 + 3^3 &= 6^2 \\ 1^3 + 2^3 + 3^3 + 4^3 &= 10^2 \end{aligned}$$

Make a conjecture based on the above pattern and then prove or disprove your conjecture.

13. Notice

$$\begin{aligned} 1 &= 1 \\ 1 + 4 &= 5 \\ 1 + 4 + 7 &= 12 \\ 1 + 4 + 7 + 10 &= 22 \end{aligned}$$

Make a conjecture based on the above pattern and then prove or disprove your conjecture.

Define $b^{n+1} = b^n \cdot b$ *and* $b^0 = 1$. *Use this definition to prove the properties of exponents in each of Problems 14–17 for all positive integers* n.

14. $b^m \cdot b^n = b^{m+n}$

15. $(b^m)^n = b^{mn}$

16. $(ab)^n = a^n b^n$

17. $\left(\dfrac{a}{b}\right)^n = \dfrac{a^n}{b^n}$

C

18. Prove: $\dfrac{1}{2} + \dfrac{1}{3} + \dfrac{1}{4} + \dfrac{1}{5} + \cdots + \dfrac{1}{2^n} < n$ for all positive integers

19. Prove: $2^n > n$ for all positive integers.

20. Prove the generalized distributive property:

$$a(b_1 + b_2 + b_3 + \cdots + b_n) = ab_1 + ab_2 + ab_3 + \cdots + ab_n$$

21. Prove the generalized triangle inequality:

$$|a_1 + a_2 + \cdots + a_n| \leq |a_1| + |a_2| + \cdots + |a_n|$$

22. Prove: $\dbinom{k}{r} + \dbinom{k}{r-1} = \dbinom{k+1}{r}$

Use the formula $\dbinom{n}{m} = \dfrac{n!}{m!(n-m)!}$ and not induction. You will need this result in Problem 23.

23. The Binomial Theorem can be proved for any positive integer n by using mathematical induction. Fill in the missing steps and reasons.

TO PROVE:

$$(a+b)^n = \sum_{j=0}^{n} \binom{n}{j} a^{n-j}b^j$$

$$= \binom{n}{0} a^n + \binom{n}{1} a^{n-1}b + \binom{n}{2} a^{n-2}b^2 + \cdots + \binom{n}{r} a^{n-r}b^r + \cdots + \binom{n}{n-1} ab^{n-1} + \binom{n}{n} b^n.$$

Step 1 Prove it true for $n = 1$.

a. Fill in these details.

Step 2 Assume it true for $n = k$.

b. Fill in the statement of the hypothesis.

Step 3 Prove it true for $n = k + 1$.

TO PROVE: $(a+b)^{k+1} = \sum_{j=0}^{k+1} \binom{k+1}{j} a^{k+1-j}b^j$

BY HYPOTHESIS:

c. Fill in the statement of the hypothesis.

d. Fill in these details; the final simplified form is

$$(a+b)^{k+1} = a^{k+1} + \cdots + \left[\binom{k}{r} + \binom{k}{r-1}\right] a^{k-r+1}b^r + \cdots + b^{k+1}$$

e. Use Problem 22 to complete the proof.

24. What is wrong with the following "proof" that $1 = 2$?

$$(a+b)^n = a^n + na^{n-1}b + \frac{n(n-1)}{2!}a^{n-2}b^2 + \cdots + nab^{n-1} + b^n$$

Let $n = 0$. Then	Binomial Theorem
$(a+b)^0 = a^0 + 0 + 0 + \cdots + 0 + b^0$	Substitution*
$1 = 1 + 0 + 0 + \cdots + 0 + 1$	Definition of zero exponent
$1 = 2$	

10.6 Summary and Review

In this chapter we have introduced you to the idea of formal mathematical proof. In one sense, it is a review of many parts of the course; in another sense, it bridges the gap between elementary algebra and abstract algebra. If you continue your mathematical studies through calculus, you will probably take a course in abstract algebra, which will allow you to reinterpret the results of this course in a much more general way.

*If you have covered Section 10.1, you should recognize this as Axiom E4.

CHAPTER OBJECTIVES

After studying this chapter, you should be able to:

1. Describe the differences among an axiom, definition, and theorem
2. Distinguish between validity and truth
3. Disprove theorems or statements by means of counterexamples
4. Restate propositions in if–then form
5. Distinguish among direct reasoning, indirect reasoning, and transitivity
6. Rewrite a statement using the words *if and only if* as a direct statement and its converse
7. Use direct proof to prove some elementary theorems
8. Use indirect proof to prove certain theorems
9. Use the equality axioms, field axioms, and rules of logic to prove certain real number theorems
10. State the Principle of Mathematical Induction, and identify propositions to which PMI applies
11. Prove certain theorems using the Principle of Mathematical Induction

Problem Set 10.6

CONCEPT PROBLEMS

[10.1] *In Problems 1–15 state the property named.*

1. Reflexive property for equality
2. Symmetric property for equality
3. Transitive property for equality
4. Substitution property for equality
5. Closure property for addition
6. Closure property for multiplication
7. Associative property for multiplication
8. Associative property for addition
9. Commutative property for addition
10. Commutative property for multiplication
11. Identity property for addition
12. Identity property for multiplication
13. Inverse property for multiplication
14. Inverse property for addition
15. Distributive property

Fill in the word or words necessary to make the statements in Problems 16–26 complete and correct.

[10.1] **16.** Direct reasoning states:
Major premise: ________
Minor premise: ________
Conclusion: ___________

[10.2] **17.** Transitive reasoning states:
Major premise: ________
Minor premise: ________
Conclusion: ___________

[10.2] **18.** The statement p if and only if q means ________ and ________ .

[10.2] **19.** $a - b$ means ________ .

[10.2] **20.** $\frac{a}{b}$ means ________ .

[10.3] **21.** The Order Axiom defines a set called the ________ with the property that if a is any real number, then exactly one of the following is true:
i. ________ ii. ________ iii. ________

[10.3] **22.** $a > b$ means ________ .

[10.3] **23.** $a \leqslant b$ means ________ .

[10.3] **24.** The triangle inequality is ________ .

[10.4] **25.** Indirect reasoning states:
Major premise: ________
Minor premise: ________
Conclusion: ________

[10.5] **26.** The Principle of Mathematical Induction states:
____

REVIEW PROBLEMS [10.1, 10.2] *Write each statement in Problems 27–32 in if–then format and then write its converse.*

27. $b^2 = 81$ if $b = 9$.

28. $b > 0$ only if $b^{1/2} = \sqrt{b}$.

29. $P^2 = Q$ implies $|P| = \sqrt{Q}$.

30. $b^h < b^x < b^k$ if $h < x < k$.

31. All mathematicians are strange.

32. $x^2 = 9$ if and only if $x = 3$ or $x = -3$.

[10,1, 10,2, 10.4] *Some of the arguments in Problems 33–43 illustrate a valid form of reasoning, but others do not. If the problem illustrates a valid argument, name the type of reasoning used.*

33.
$$\begin{array}{l} p \rightarrow q \\ \sim p \\ \hline \therefore \sim q \end{array}$$

34.
$$\begin{array}{l} p \rightarrow q \\ q \rightarrow r \\ \hline \therefore p \rightarrow r \end{array}$$

35.
$$\begin{array}{l} p \rightarrow q \\ \sim q \\ \hline \therefore \sim p \end{array}$$

36.
$$\begin{array}{l} p \rightarrow q \\ q \\ \hline \therefore p \end{array}$$

37.
$$\begin{array}{l} p \rightarrow q \\ p \\ \hline \therefore q \end{array}$$

38.
$$\begin{array}{l} p \rightarrow q \\ r \rightarrow q \\ \hline \therefore p \rightarrow r \end{array}$$

39. If wages go up, then prices will also go up.
Prices do not go up.
Therefore, wages did not go up.

40. If you go to college, then you will obtain a good job.
You do not go to college.
Therefore, you will not obtain a good job.

41. If wages go up, then retail prices will go up.
If inflation continues, then wages will go up.
If retail prices go up, then the economy will suffer.
Therefore, if inflation continues, then the economy will suffer.

42. If Albert recommends Eats Restaurant, then you can be sure that they serve good food.
Albert recommends Eats Restaurant.
Therefore, Eats Restaurant serves good food.

43. All trade union executives are labor leaders.
Mr. C. D. Money is a labor leader.
Therefore, Mr. Money is a union executive.

[10.2] *Fill in the blanks in the following proof for Problems 44–54.*

GIVEN: $a, b \in \mathbf{R}$ TO PROVE: $5a + 2b - (3a + b) = 2a + b$

	Statements		Reasons
1.	$5a + 2b - (3a + b) = 5a + 2b - (3a + b)$	1.	**44.**
2.	$= 5a + 2b + [-(3a + b)]$	2.	**45.**
3.	$= 5a + 2b + [(-1)(3a + b)]$	3.	**46.**
4.	$= 5a + 2b + (-1)(3a) + (-1)b$	4.	**47.**
5.	$= 5a + 2b + (-3a) + (-1)b$	5.	**48.**
6.	$= 5a + (-3a) + 2b + (-1)b$	6.	**49.**
7.	$= [5a + (-3a)] + [2b + (-1)b]$	7.	**50.**
8.	$= [5 + (-3)]a + [2 + (-1)]b$	8.	**51.**
9.	$= 2a + 1 \cdot b$	9.	**52.**
10.	$= 2a + b$	10.	**53.**
∴11.	$5a + 2b - (3a + b) = 2a + b$	11.	**54.**

[10.2] **55.** Prove: $ax + by + c = 0$ $(b \neq 0)$ only if $y = -\frac{a}{b}x - \frac{c}{b}$

[10.3] **56.** Prove: If $a < b$, then $a + c < b + c$.

[10.3] **57.** Prove or disprove: $|x^2| = x^2$

[10.3] **58.** Prove or disprove: $|x^3| = x^3$

[10.5] **59.** Prove: $1 + 2 + 3 + \cdots + n = \frac{1}{2}n(n + 1)$

[10.1] **60.** Which of the field axioms are satisfied by the set of numbers of the form $a + b\sqrt{2}$, where a and b are integers?

APPENDIX A Sets

One of the fundamental ideas of mathematics is that of *sets*. You are probably familiar with many sets. This appendix is designed to refresh your memory concerning some of the notation and ideas related to sets.

ELEMENTS OF A SET

A **set** is a collection of well-defined objects called **elements,** or **members,** of the set. A set is generally designated by a capital letter, and the elements are said to *belong* to a set. A set is *well defined* if there is an established criterion to determine whether a particular element belongs to a given set. Braces are used to contain elements of a set.

Sets may be defined by **roster,** which means the elements are listed; for example,

$A = \{5, 7, 9, 16\}$

$B = \{3, 6, 9, 12, \ldots\}$

Another method of defining sets is by **description;** for example,

$C = \{\text{Counting numbers less than } 10\}$

$D = \{\text{Real numbers between 5 and 7}\}$

A third way of representing sets is to use **set-builder notation,** as follows:

This is normally written simply as

$$S = \{x | 5 < x < 7\}$$

If S is a set and a is an element of S, write $a \in S$. If b is not an element of S, write $b \notin S$. The number of elements in a set is called the **cardinality** of a set. If the cardinality of a set is 0, then the set is empty and is denoted by $\{\ \}$, or $\varnothing$.

Relationships among Sets

SUBSET

> A set A is a **subset** of a set B, denoted by $A \subseteq B$, if every element of A is an element of B.

A **universal set,** U, containing all the elements under consideration in a given discussion should either be specified or implied.

Consider the following sets:

$U = \{1, 2, 3, 4, 5, 6, 7, 8, 9\}$

$A = \{2, 4, 6, 8\}$ $\qquad B = \{1, 3, 5, 7, 9\}$ $\qquad C = \{5, 7\}$

Here, $A \subseteq U$, $B \subseteq U$, $C \subseteq U$, and $C \subseteq B$, but C is not a subset of A since $5 \in C$ and $5 \notin A$.

A useful way to depict relationships among sets is to let the universal set be represented by a rectangle, with the sets in the universe represented by circular or oval-shaped regions, which are called **Venn diagrams** (after John Venn, 1834–1923). Venn diagrams for subset and equal relationships are shown in Figure A.1.

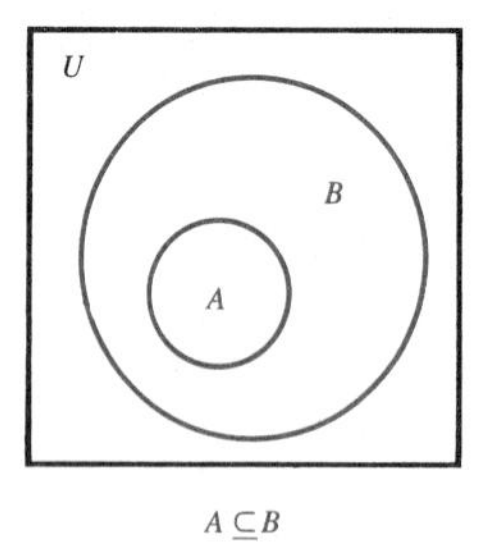

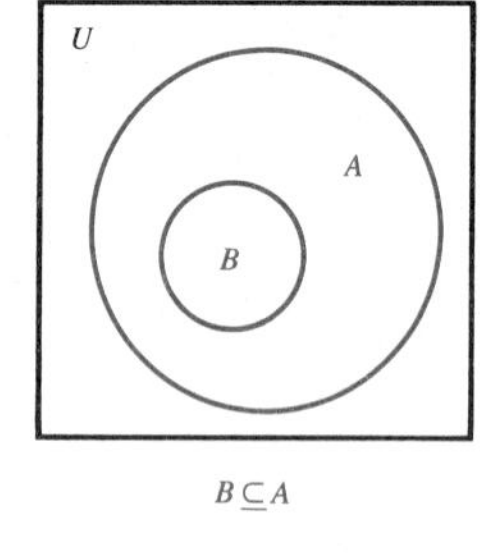

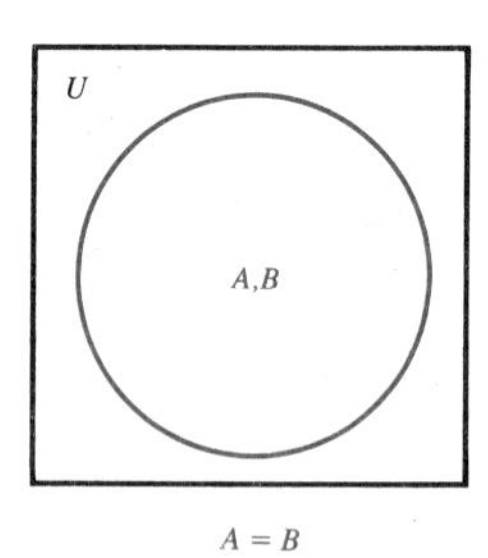

Figure A.1

We can also illustrate other relationships between two sets: A and B may have no elements in common, in which case they are **disjoint**, or they may be overlapping sets having some elements in common, as shown in Figure A.2.

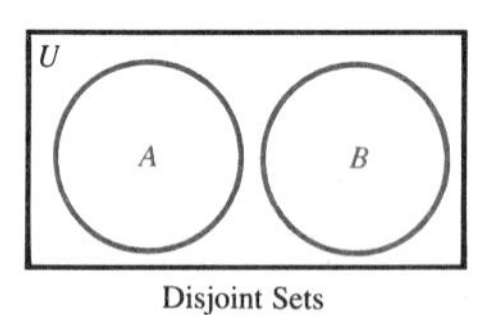

Disjoint Sets

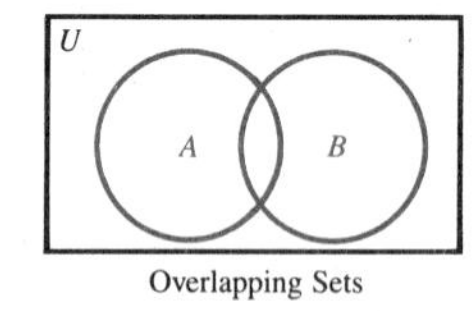

Overlapping Sets

Figure A.2
Venn Diagrams for Disjoint and Overlapping Sets

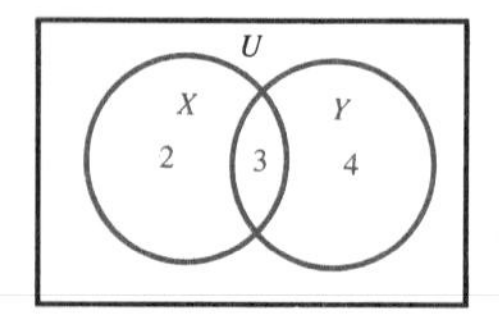

Figure A.3
Venn Diagram Showing Two General Sets

Sometimes you are given two sets X and Y, and you know nothing about the way they are related. In this situation draw a general Venn diagram, such as the one shown in Figure A.3. Notice that these circles divide the universe into four disjoint regions. When we draw the sets in this manner, we do not mean to imply that the only possibility is overlapping sets.

If $X \subseteq Y$, then Region 2 is empty.

If $Y \subseteq X$, then Region 4 is empty.

If $X = Y$, then Regions 2 and 4 are both empty.

If X and Y are disjoint, then Region 3 is empty.

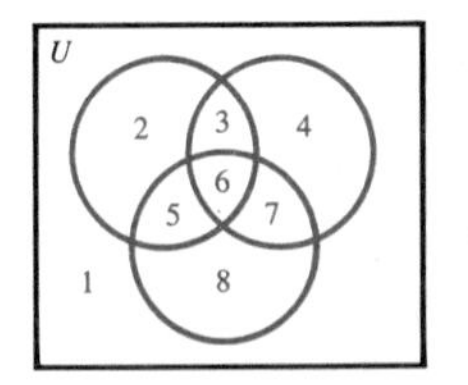

Figure A.4
Venn Diagram Showing Three General Sets

If three general sets are involved, the universe is partitioned into eight regions, as shown in Figure A.4.

Operations on Sets

There are three common operations that are performed on sets: *union, intersection,* and *taking the complement.*

UNION, INTERSECTION, AND COMPLEMENT

> Given sets A and B contained in a universe U:
>
> 1. The **union**, denoted by $A \cup B$, is the set consisting of all elements of A or B or both.
> 2. The **intersection**, denoted by $A \cap B$, is the set consisting of all elements common to A and B.
> 3. The **complement** of A, denoted by $\overline{A}$, is the set consisting of all elements in U that are not in the set A.

The Venn diagrams for these operations are shown in Figure A.5.

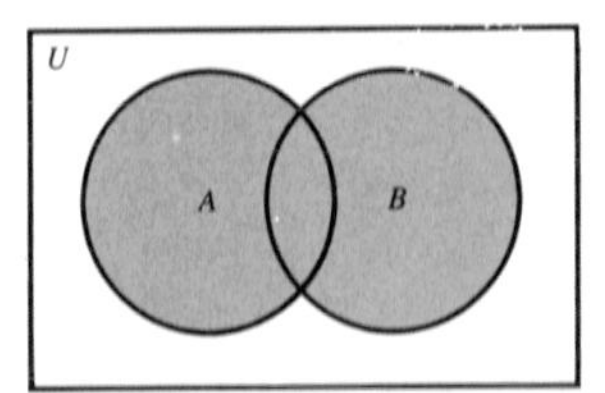

Venn diagram showing union as the shaded portion

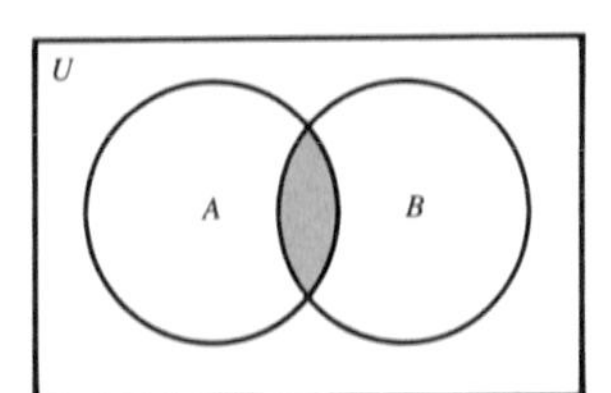

Venn diagram showing intersection as the shaded portion

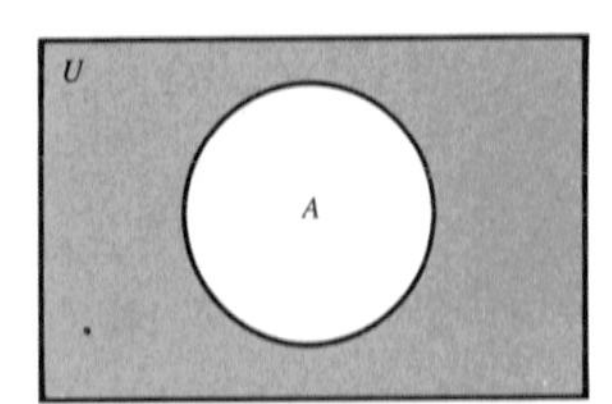

Venn diagram showing complement as the shaded portion

Figure A.5

For Examples 1–6, let

$U = \{m, a, t, h, i, s, f, u, n\}$

$A = \{m, a\}$ $\qquad B = \{t, h, i, s\}$ $\qquad C = \{i, s\}$ $\qquad D = \{n, u, t, s\}$

Example 1 $\overline{A \cup B}$. This is a combination of the operations of union and taking the complement. It is asking for the complement of A union B. First, find

$$A \cup B = \{m, a, t, h, i, s\}$$

Now,

$$\overline{A \cup B} = \overline{\{m, a, t, h, i, s\}} = \{\boldsymbol{f}, \boldsymbol{u}, \boldsymbol{n}\}$$

□

Example 2 $\overline{A} \cup \overline{B}$. This is asking for the union of the complements. First, find the complements:

$$\overline{A} = \{t, h, i, s, f, u, n\} \qquad \overline{B} = \{m, a, f, u, n\}$$

Then take their union:

$$\begin{aligned} \overline{A} \cup \overline{B} &= \{t, h, i, s, f, u, n\} \cup \{m, a, f, u, n\} \\ &= \{m, a, t, h, i, s, f, u, n\} \\ &= U \end{aligned}$$

Therefore, $\boldsymbol{\overline{A} \cup \overline{B} = U}$. □

Example 3 $\overline{A} \cap \overline{B}$. Find the complements (see Example 2) and then find the intersection:

$$\overline{A} \cap \overline{B} = \{t, h, i, s, f, u, n\} \cap \{m, a, f, u, n\} = \boldsymbol{\{f, u, n\}}$$ □

Compare Examples 1 and 2, and notice that $\overline{A \cup B} \neq \overline{A} \cup \overline{B}$. On the other hand, Examples 1 and 3 suggest a result that is true for any sets X and Y:

DE MORGAN'S LAWS

$$\overline{X \cup Y} = \overline{X} \cap \overline{Y} \quad \text{and} \quad \overline{X \cap Y} = \overline{X} \cup \overline{Y}$$

Example 4 $\overline{D} \cap B$. Find the intersection of the complement of D and B. First, find

$$\overline{D} = \{m, a, h, i, f\}$$

Then,

$$\overline{D} \cap B = \{m, a, h, i, f\} \cap \{t, h, i, s\} = \boldsymbol{\{h, i\}}$$ □

Example 5 Find $(A \cup C) \cap \overline{C}$, and draw the corresponding Venn diagram.

Solution Parentheses first: $A \cup C = \{m, a, i, s\}$
Then find $\overline{C}$: $\overline{C} = \{m, a, t, h, f, u, n\}$
Finally, consider their intersection:

$$\begin{aligned} (A \cup C) \cap \overline{C} &= \{m, a, i, s\} \cap \{m, a, t, h, f, u, n\} \\ &= \{m, a\} \\ &= \boldsymbol{A} \end{aligned}$$

The Venn diagram is shown in the margin. □

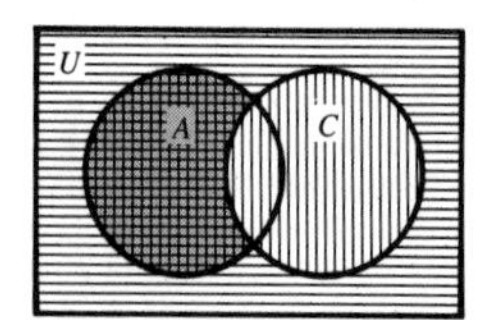

First, draw $A \cup C$ (vertical lines); then draw $\overline{C}$ (horizontal lines). The result is the part shown in color.

Example 6 Find the solution set and draw the Venn diagram for $\overline{A \cup B} \cap D$. This problem involves three sets, A, B, and D.

Solution

$$
\begin{aligned}
A \cup B &= \{m, a, t, h, i, s\} \\
\overline{A \cup B} &= \{f, u, n\} \\
\overline{A \cup B} \cap D &= \{f, u, n\} \cap \{n, u, t, s\} \\
&= \{\mathbf{n}, \mathbf{u}\}
\end{aligned}
$$

The Venn diagram for $\overline{A \cup B} \cap D$ is shown in the margin. □

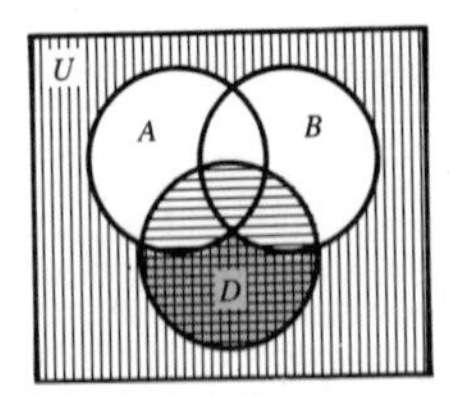

First, draw $\overline{A \cup B}$ (vertical lines); then draw D (horizontal lines). The result is the part shown in color.

Counting the Number of Elements in Sets

Suppose

$$U = \{1, 2, 3, 4, 5, 6, 7, 8, 9\}$$

$$A = \{2, 4, 6, 8\} \qquad B = \{1, 3, 5, 7, 9\} \qquad C = \{2, 3, 5, 7\}$$

If $n(X)$ represents the cardinality of the set X, then

$$n(U) = 9, \qquad n(A) = 4, \qquad n(B) = 5, \qquad n(C) = 4$$

Furthermore, by direct counting:

$$
\begin{aligned}
A \cap B &= \{\ \} && \text{so} \quad n(A \cap B) = 0 \\
B \cap C &= \{3, 5, 7\} && \text{so} \quad n(B \cap C) = 3 \\
A \cup B &= \{1, 2, 3, 4, 5, 6, 7, 8, 9\} && \text{so} \quad n(A \cup B) = 9 \\
B \cup C &= \{1, 2, 3, 5, 7, 9\} && \text{so} \quad n(B \cup C) = 6
\end{aligned}
$$

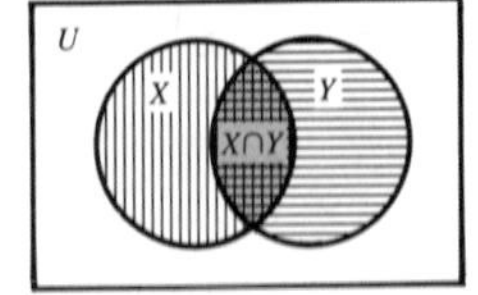

Figure A.6
Venn Diagram for the Number of Elements in the Union of Two Sets

However, if these sets were very large, the direct counting method would not be satisfactory. Consider the general Venn diagram shown in Figure A.6. Notice that if you count X and then count Y, the intersection (the part in color) has been counted twice.

NUMBER OF ELEMENTS IN THE UNION OF TWO SETS

$$n(X \cup Y) = \underbrace{n(X) + n(Y)}_{\text{The elements in the intersection are counted twice here}} - \underbrace{n(X \cap Y)}_{\text{This corrects for the "error" introduced by counting the elements in the intersection twice}}$$

Use the formula to find the number of elements in each union of the sets A, B, C (listed above) indicated in Examples 7 and 8.

Example 7

$$
\begin{aligned}
n(A \cup B) &= n(A) + n(B) - n(A \cap B) \\
&= 4 + 5 - 0 \\
&= \mathbf{9}
\end{aligned}
$$

□

Example 8

$$\begin{aligned} n(B \cup C) &= n(B) + n(C) - n(B \cap C) \\ &= 5 + 4 - 3 \\ &= \mathbf{6} \end{aligned}$$

□

Example 9 What is the millionth positive integer that is not the square or cube of an integer?

Solution Let

$$\begin{aligned} S &= \{\text{Perfect squares}\} \\ &= \{1, 4, 9, 16, 25, \ldots, 1{,}000{,}000\} \\ &= \{1^2, 2^2, 3^2, 4^2, 5^2, \ldots, 1000^2\} \\ n(S) &= 1000 \end{aligned}$$

Since the last element is 1000^2

Let

$$\begin{aligned} C &= \{\text{Perfect cubes}\} \\ &= \{1^3, 2^3, 3^3, 4^3, 5^3, \ldots, 100^3\} \\ n(C) &= 100 \end{aligned}$$

Then

$$\begin{aligned} S \cap C &= \{\text{Perfect sixth powers}\} \\ &= \{1^6, 2^6, 3^6, \ldots, 10^6\} \\ n(S \cap C) &= 10 \end{aligned}$$

Thus,

$$\begin{aligned} n(S \cup C) &= n(S) + n(C) - n(S \cap C) \\ &= 1000 + 100 - 10 \\ &= 1090 \end{aligned}$$

Since the next square and cube ($1001^2 = 1{,}002{,}001$ and $101^3 = 1{,}030{,}301$, respectively) are both greater than 1,001,090, you see that there are no additional numbers to be excluded between 1,000,000 and 1,001,090. Thus, the millionth number that is not a square or a cube is **1,001,090.**

□

APPENDIX **B**

Tables

Table I
Squares and Square Roots

n	n^2	$\sqrt{n}$	$\sqrt{10n}$	n	n^2	$\sqrt{n}$	$\sqrt{10n}$
1	1	1.000000	3.162278	51	2601	7.141428	22.583180
2	4	1.414214	4.472136	52	2704	7.211103	22.803509
3	9	1.732051	5.477226	53	2809	7.280110	23.021729
4	16	2.000000	6.324555	54	2916	7.348469	23.237900
5	25	2.236068	7.071068	55	3025	7.416198	23.452079
6	36	2.449490	7.745967	56	3136	7.483315	23.664319
7	49	2.645751	8.366600	57	3249	7.549834	23.874673
8	64	2.828427	8.944272	58	3364	7.615773	24.083189
9	81	3.000000	9.486833	59	3481	7.681146	24.289916
10	100	3.162278	10.000000	**60**	3600	7.745967	24.494897
11	121	3.316625	10.488088	61	3721	7.810250	24.698178
12	144	3.464102	10.954451	62	3844	7.874008	24.899799
13	169	3.605551	11.401754	63	3969	7.937254	25.099801
14	196	3.741657	11.832160	64	4096	8.000000	25.298221
15	225	3.872983	12.247449	65	4225	8.062258	25.495098
16	256	4.000000	12.649111	66	4356	8.124038	25.690465
17	289	4.123106	13.038405	67	4489	8.185353	25.884358
18	324	4.242641	13.416408	68	4624	8.246211	26.076810
19	361	4.358899	13.784049	69	4761	8.306624	26.267851
20	400	4.472136	14.142136	**70**	4900	8.366600	26.457513
21	441	4.582576	14.491377	71	5041	8.426150	26.645825
22	484	4.690416	14.832397	72	5184	8.485281	26.832816
23	529	4.795832	15.165751	73	5329	8.544004	27.018512
24	576	4.898979	15.491933	74	5476	8.602325	27.202941
25	625	5.000000	15.811388	75	5625	8.660254	27.386128
26	676	5.099020	16.124515	76	5776	8.717798	27.568098
27	729	5.196152	16.431677	77	5929	8.774964	27.748874
28	784	5.291503	16.733201	78	6084	8.831761	27.928480
29	841	5.385165	17.029386	79	6241	8.888194	28.106939
30	900	5.477226	17.320508	**80**	6400	8.944272	28.284271
31	961	5.567764	17.606817	81	6561	9.000000	28.460499
32	1024	5.656854	17.888544	82	6724	9.055385	28.635642
33	1089	5.744563	18.165902	83	6889	9.110434	28.809721
34	1156	5.830952	18.439089	84	7056	9.165151	28.982753
35	1225	5.916080	18.708287	85	7225	9.219544	29.154759
36	1296	6.000000	18.973666	86	7396	9.273618	29.325757
37	1369	6.082763	19.235384	87	7569	9.327379	29.495762
38	1444	6.164414	19.493589	88	7744	9.380832	29.664794
39	1521	6.244998	19.748418	89	7921	9.433981	29.832868
40	1600	6.324555	20.000000	**90**	8100	9.486833	30.000000
41	1681	6.403124	20.248457	91	8281	9.539392	30.166206
42	1764	6.480741	20.493902	92	8464	9.591663	30.331502
43	1849	6.557439	20.736441	93	8649	9.643651	30.495901
44	1936	6.633250	20.976177	94	8836	9.695360	30.659419
45	2025	6.708204	21.213203	95	9025	9.746794	30.822070
46	2116	6.782330	21.447611	96	9216	9.797959	30.983867
47	2209	6.855655	21.679483	97	9409	9.848858	31.144823
48	2304	6.928203	21.908902	98	9604	9.899495	31.304952
49	2401	7.000000	22.135944	99	9801	9.949874	31.464265
50	2500	7.071068	22.360680	**100**	10000	10.000000	31.622777
n	n^2	$\sqrt{n}$	$\sqrt{10n}$	n	n^2	$\sqrt{n}$	$\sqrt{10n}$

Table II
Powers of e

x	e^x	e^{-x}	x	e^x	e^{-x}	x	e^x	e^{-x}
0.00	1.000	1.000	0.50	1.649	0.607	1.00	2.718	0.368
0.01	1.010	0.990	0.51	1.665	0.600	1.01	2.746	0.364
0.02	1.020	0.980	0.52	1.682	0.595	1.02	2.773	0.361
0.03	1.031	0.970	0.53	1.699	0.589	1.03	2.801	0.357
0.04	1.041	0.961	0.54	1.716	0.583	1.04	2.829	0.353
0.05	1.051	0.951	0.55	1.733	0.577	1.05	2.858	0.350
0.06	1.062	0.942	0.56	1.751	0.571	1.06	2.886	0.346
0.07	1.073	0.932	0.57	1.768	0.566	1.07	2.915	0.343
0.08	1.083	0.923	0.58	1.786	0.560	1.08	2.945	0.340
0.09	1.094	0.914	0.59	1.804	0.554	1.09	2.974	0.336
0.10	1.105	0.905	0.60	1.822	0.549	1.10	3.004	0.333
0.11	1.116	0.896	0.61	1.840	0.543	1.11	3.034	0.330
0.12	1.127	0.887	0.62	1.859	0.538	1.12	3.065	0.326
0.13	1.139	0.878	0.63	1.878	0.533	1.13	3.096	0.323
0.14	1.150	0.869	0.64	1.896	0.527	1.14	3.127	0.320
0.15	1.162	0.861	0.65	1.916	0.522	1.15	3.158	0.317
0.16	1.174	0.852	0.66	1.935	0.517	1.16	3.190	0.313
0.17	1.185	0.844	0.67	1.954	0.512	1.17	3.222	0.310
0.18	1.197	0.835	0.68	1.974	0.507	1.18	3.254	0.307
0.19	1.209	0.827	0.69	1.994	0.502	1.19	3.287	0.304
0.20	1.221	0.819	0.70	2.014	0.497	1.20	3.320	0.301
0.21	1.234	0.811	0.71	2.034	0.492	1.21	3.353	0.298
0.22	1.246	0.803	0.72	2.054	0.487	1.22	3.387	2.295
0.23	1.259	0.795	0.73	2.075	0.482	1.23	3.421	0.292
0.24	1.271	0.787	0.74	2.096	0.477	1.24	3.456	0.289
0.25	1.284	0.779	0.75	2.117	0.472	1.25	3.490	0.287
0.26	1.297	0.771	0.76	2.138	0.468	1.26	3.525	0.284
0.27	1.310	0.763	0.77	2.160	0.463	1.27	3.561	0.281
0.28	1.323	0.756	0.78	2.182	0.458	1.28	3.597	0.278
0.29	1.336	0.748	0.79	2.203	0.454	1.29	3.633	0.275
0.30	1.350	0.741	0.80	2.226	0.449	1.30	3.669	0.273
0.31	1.363	0.733	0.81	2.248	0.445	1.31	3.706	0.270
0.32	1.377	0.726	0.82	2.270	0.440	1.32	3.743	0.267
0.33	1.391	0.719	0.83	2.293	0.436	1.33	3.781	0.264
0.34	1.405	0.712	0.84	2.316	0.432	1.34	3.819	0.262
0.35	1.419	0.705	0.85	2.340	0.427	1.35	3.857	0.259
0.36	1.433	0.698	0.86	2.363	0.423	1.36	3.896	0.257
0.37	1.448	0.691	0.87	2.387	0.419	1.37	3.935	0.254
0.38	1.462	0.684	0.88	2.411	0.415	1.38	3.975	0.252
0.39	1.477	0.677	0.89	2.435	0.411	1.39	4.015	0.249
0.40	1.492	0.670	0.90	2.460	0.407	1.40	4.055	0.247
0.41	1.507	0.664	0.91	2.484	0.403	1.41	4.096	0.244
0.42	1.522	0.657	0.92	2.509	0.399	1.42	4.137	0.242
0.43	1.537	0.651	0.93	2.535	0.395	1.43	4.179	0.239
0.44	1.553	0.644	0.94	2.560	0.391	1.44	4.221	0.237
0.45	1.568	0.638	0.95	2.586	0.387	1.45	4.263	0.235
0.46	1.584	0.631	0.96	2.612	0.383	1.46	4.306	0.232
0.47	1.600	0.625	0.97	2.638	0.379	1.47	4.349	0.230
0.48	1.616	0.619	0.98	2.664	0.375	1.48	4.393	0.228
0.49	1.632	0.613	0.99	2.691	0.372	1.49	4.437	0.225

x	e^x	e^{-x}	x	e^x	e^{-x}	x	e^x	e^{-x}
1.50	4.482	0.223	2.00	7.389	0.135	2.50	12.182	0.082
1.51	4.527	0.221	2.01	7.463	0.134	2.51	12.305	0.081
1.52	4.572	0.219	2.02	7.538	0.133	2.52	12.429	0.080
1.53	4.618	0.217	2.03	7.614	0.131	2.53	12.554	0.080
1.54	4.665	0.214	2.04	7.691	0.130	2.54	12.680	0.079
1.55	4.712	0.212	2.05	7.768	0.129	2.55	12.807	0.078
1.56	4.759	0.210	2.06	7.846	0.127	2.56	12.936	0.077
1.57	4.807	0.208	2.07	7.925	0.126	2.57	13.066	0.077
1.58	4.855	0.206	2.08	8.004	0.125	2.58	13.197	0.076
1.59	4.904	0.204	2.09	8.085	0.124	2.59	13.330	0.075
1.60	4.953	0.202	2.10	8.166	0.122	2.60	13.464	0.074
1.61	5.003	0.200	2.11	8.248	0.121	2.61	13.599	0.074
1.62	5.053	0.198	2.12	8.331	0.120	2.62	13.736	0.073
1.63	5.104	0.196	2.13	8.415	0.119	2.63	13.874	0.072
1.64	5.155	0.194	2.14	8.499	0.118	2.64	14.013	0.071
1.65	5.207	0.192	2.15	8.585	0.116	2.65	14.154	0.071
1.66	5.259	0.190	2.16	8.671	0.115	2.66	14.296	0.070
1.67	5.312	0.188	2.17	8.758	0.114	2.67	14.440	0.069
1.68	5.366	0.186	2.18	8.846	0.113	2.68	14.585	0.069
1.69	5.420	0.185	2.19	8.935	0.112	2.69	14.732	0.068
1.70	5.474	0.183	2.20	9.025	0.111	2.70	14.880	0.067
1.71	5.529	0.181	2.21	9.116	0.110	2.71	15.029	0.067
1.72	5.585	0.179	2.22	9.207	0.109	2.72	15.180	0.066
1.73	5.641	0.177	2.23	9.300	0.108	2.73	15.333	0.065
1.74	5.697	0.176	2.24	9.393	0.106	2.74	15.487	0.065
1.75	5.755	0.174	2.25	9.488	0.105	2.75	15.643	0.064
1.76	5.812	0.172	2.26	9.583	0.104	2.76	15.800	0.063
1.77	5.871	0.170	2.27	9.679	0.103	2.77	15.959	0.063
1.78	5.930	0.169	2.28	9.777	0.102	2.78	16.119	0.062
1.79	5.989	0.167	2.29	9.875	0.101	2.79	16.281	0.061
1.80	6.050	0.165	2.30	9.974	0.100	2.80	16.445	0.061
1.81	6.110	0.164	2.31	10.074	0.099	2.81	16.610	0.060
1.82	6.172	0.162	2.32	10.176	0.098	2.82	16.777	0.060
1.83	8.234	0.160	2.33	10.278	0.097	2.83	16.945	0.059
1.84	6.297	0.159	2.34	10.381	0.096	2.84	17.116	0.058
1.85	6.360	0.157	2.35	10.486	0.095	2.85	17.288	0.058
1.86	6.424	0.156	2.36	10.591	0.094	2.86	17.462	0.057
1.87	6.488	0.154	2.37	10.697	0.093	2.87	17.637	0.057
1.88	6.553	0.153	2.38	10.805	0.093	2.88	17.814	0.056
1.89	6.619	0.151	2.39	10.913	0.092	2.89	17.993	0.056
1.90	6.686	0.150	2.40	11.023	0.091	2.90	18.174	0.055
1.91	6.753	0.148	2.41	11.134	0.090	2.91	18.357	0.054
1.92	6.821	0.147	2.42	11.246	0.089	2.92	18.541	0.054
1.93	6.890	0.145	2.43	11.359	0.088	2.93	18.728	0.053
1.94	6.959	0.144	2.44	11.473	0.087	2.94	18.916	0.053
1.95	7.029	0.142	2.45	11.588	0.086	2.95	19.106	0.052
1.96	7.099	0.141	2.46	11.705	0.085	2.96	19.298	0.052
1.97	7.171	0.139	2.47	11.822	0.085	2.97	19.492	0.051
1.98	7.243	0.138	2.48	11.941	0.084	2.98	19.688	0.051
1.99	7.316	0.137	2.49	12.061	0.083	2.99	19.886	0.050
						3.00	20.086	0.050

Table III
Common Logarithms

N	0	1	2	3	4	5	6	7	8	9
1.0	.0000	.0043	.0086	.0128	.0170	.0212	.0253	.0294	.0334	.0374
1.1	.0414	.0453	.0492	.0531	.0569	.0607	.0645	.0682	.0719	.0755
1.2	.0792	.0828	.0864	.0899	.0934	.0969	.1004	.1038	.1072	.1106
1.3	.1139	.1173	.1206	.1239	.1271	.1303	.1335	.1367	.1399	.1430
1.4	.1461	.1492	.1523	.1553	.1584	.1614	.1644	.1673	.1703	.1732
1.5	.1761	.1790	.1818	.1847	.1875	.1903	.1931	.1959	.1987	.2014
1.6	.2041	.2068	.2095	.2122	.2148	.2175	.2201	.2227	.2253	.2279
1.7	.2304	.2330	.2355	.2380	.2405	.2430	.2455	.2480	.2504	.2529
1.8	.2553	.2577	.2601	.2625	.2648	.2672	.2695	.2718	.2742	.2765
1.9	.2788	.2810	.2833	.2856	.2878	.2900	.2923	.2945	.2967	.2989
2.0	.3010	.3032	.3054	.3075	.3096	.3118	.3139	.3160	.3181	.3201
2.1	.3222	.3243	.3263	.3284	.3304	.3324	.3345	.3365	.3385	.3404
2.2	.3424	.3444	.3464	.3483	.3502	.3522	.3541	.3560	.3579	.3598
2.3	.3617	.3636	.3655	.3674	.3692	.3711	.3729	.3747	.3766	.3784
2.4	.3802	.3820	.3838	.3856	.3874	.3892	.3909	.3927	.3945	.3962
2.5	.3979	.3997	.4014	.4031	.4048	.4065	.4082	.4099	.4116	.4133
2.6	.4150	.4166	.4183	.4200	.4216	.4232	.4249	.4265	.4281	.4298
2.7	.4314	.4330	.4346	.4362	.4378	.4393	.4409	.4425	.4440	.4456
2.8	.4472	.4487	.4502	.4518	.4533	.4548	.4564	.4579	.4594	.4609
2.9	.4624	.4639	.4654	.4669	.4683	.4698	.4713	.4728	.4742	.4757
3.0	.4771	.4786	.4800	.4814	.4829	.4843	.4857	.4871	.4886	.4900
3.1	.4914	.4928	.4942	.4955	.4969	.4983	.4997	.5011	.5024	.5038
3.2	.5051	.5065	.5079	.5092	.5105	.5119	.5132	.5145	.5159	.5172
3.3	.5185	.5198	.5211	.5224	.5237	.5250	.5263	.5276	.5289	.5302
3.4	.5315	.5328	.5340	.5353	.5366	.5378	.5391	.5403	.5416	.5428
3.5	.5441	.5453	.5465	.5478	.5490	.5502	.5514	.5527	.5539	.5551
3.6	.5563	.5575	.5587	.5599	.5611	.5623	.5635	.5647	.5658	.5670
3.7	.5682	.5694	.5705	.5717	.5729	.5740	.5752	.5763	.5775	.5786
3.8	.5798	.5809	.5821	.5832	.5843	.5855	.5866	.5877	.5888	.5899
3.9	.5911	.5922	.5933	.5944	.5955	.5966	.5977	.5988	.5999	.6010
4.0	.6021	.6031	.6042	.6053	.6064	.6075	.6085	.6096	.6107	.6117
4.1	.6128	.6138	.6149	.6160	.6170	.6180	.6191	.6201	.6212	.6222
4.2	.6232	.6243	.6253	.6263	.6274	.6284	.6294	.6304	.6314	.6325
4.3	.6335	.6345	.6355	.6365	.6375	.6385	.6395	.6405	.6415	.6425
4.4	.6435	.6444	.6454	.6464	.6474	.6484	.6493	.6503	.6513	.6522
4.5	.6532	.6542	.6551	.6561	.6571	.6580	.6590	.6599	.6609	.6618
4.6	.6628	.6637	.6646	.6656	.6665	.6675	.6684	.6693	.6702	.6712
4.7	.6721	.6730	.6739	.6749	.6758	.6767	.6776	.6785	.6794	.6803
4.8	.6812	.6821	.6830	.6839	.6848	.6857	.6866	.6875	.6884	.6893
4.9	.6902	.6911	.6920	.6928	.6937	.6946	.6955	.6964	.6972	.6981
5.0	.6990	.6998	.7007	.7016	.7024	.7033	.7042	.7050	.7059	.7067
5.1	.7076	.7084	.7093	.7101	.7110	.7118	.7126	.7135	.7143	.7152
5.2	.7160	.7168	.7177	.7185	.7193	.7202	.7210	.7218	.7226	.7235
5.3	.7243	.7251	.7259	.7267	.7275	.7284	.7292	.7300	.7308	.7316
5.4	.7324	.7332	.7340	.7348	.7356	.7364	.7372	.7380	.7388	.7396
N	0	1	2	3	4	5	6	7	8	9

N	0	1	2	3	4	5	6	7	8	9
5.5	.7404	.7412	.7419	.7427	.7435	.7443	.7451	.7459	.7466	.7474
5.6	.7482	.7490	.7497	.7505	.7513	.7520	.7528	.7536	.7543	.7551
5.7	.7559	.7566	.7574	.7582	.7589	.7597	.7604	.7612	.7619	.7627
5.8	.7634	.7642	.7649	.7657	.7664	.7672	.7679	.7686	.7694	.7701
5.9	.7709	.7716	.7723	.7731	.7738	.7745	.7752	.7760	.7767	.7774
6.0	.7782	.7789	.7796	.7803	.7810	.7818	.7825	.7832	.7839	.7846
6.1	.7853	.7860	.7868	.7875	.7882	.7889	.7896	.7903	.7910	.7917
6.2	.7924	.7931	.7938	.7945	.7952	.7959	.7966	.7973	.7980	.7987
6.3	.7993	.8000	.8007	.8014	.8021	.8028	.8035	.8041	.8048	.8055
6.4	.8062	.8069	.8075	.8082	.8089	.8096	.8102	.8109	.8116	.8122
6.5	.8129	.8136	.8142	.8149	.8156	.8162	.8169	.8176	.8182	.8189
6.6	.8195	.8202	.8209	.8215	.8222	.8228	.8235	.8241	.8248	.8254
6.7	.8261	.8267	.8274	.8280	.8287	.8293	.8299	.8306	.8312	.8319
6.8	.8325	.8331	.8338	.8344	.8351	.8357	.8363	.8370	.8376	.8382
6.9	.8388	.8395	.8401	.8407	.8414	.8420	.8426	.8432	.8439	.8445
7.0	.8451	.8457	.8463	.8470	.8476	.8482	.8488	.8494	.8500	.8506
7.1	.8513	.8519	.8525	.8531	.8537	.8543	.8549	.8555	.8561	.8567
7.2	.8573	.8579	.8585	.8591	.8597	.8603	.8609	.8615	.8621	.8627
7.3	.8633	.8639	.8645	.8651	.8657	.8663	.8669	.8675	.8681	.8686
7.4	.8692	.8698	.8704	.8710	.8716	.8722	.8727	.8733	.8739	.8745
7.5	.8751	.8756	.8762	.8768	.8774	.8779	.8785	.8791	.8797	.8802
7.6	.8808	.8814	.8820	.8825	.8831	.8837	.8842	.8848	.8854	.8859
7.7	.8865	.8871	.8876	.8882	.8887	.8893	.8899	.8904	.8910	.8915
7.8	.8921	.8927	.8932	.8938	.8943	.8949	.8954	.8960	.8965	.8971
7.9	.8976	.8982	.8987	.8993	.8998	.9004	.9009	.9015	.9020	.9025
8.0	.9031	.9036	.9042	.9047	.9053	.9058	.9063	.9069	.9074	.9079
8.1	.9085	.9090	.9096	.9101	.9106	.9112	.9117	.9122	.9128	.9133
8.2	.9138	.9143	.9149	.9154	.9159	.9165	.9170	.9175	.9180	.9186
8.3	.9191	.9196	.9201	.9206	.9212	.9217	.9222	.9227	.9232	.9238
8.4	.9243	.9248	.9253	.9258	.9263	.9269	.9274	.9279	.9284	.9289
8.5	.9294	.9299	.9304	.9309	.9315	.9320	.9325	.9330	.9335	.9340
8.6	.9345	.9350	.9355	.9360	.9365	.9370	.9375	.9380	.9385	.9390
8.7	.9395	.9400	.9405	.9410	.9415	.9420	.9425	.9430	.9435	.9440
8.8	.9445	.9450	.9455	.9460	.9465	.9469	.9474	.9479	.9484	.9489
8.9	.9494	.9499	.9504	.9509	.9513	.9518	.9523	.9528	.9533	.9538
9.0	.9542	.9547	.9552	.9557	.9562	.9566	.9571	.9576	.9581	.9586
9.1	.9590	.9595	.9600	.9605	.9609	.9614	.9619	.9624	.9628	.9633
9.2	.9638	.9643	.9647	.9652	.9657	.9661	.9666	.9671	.9675	.9680
9.3	.9685	.9689	.9694	.9699	.9703	.9708	.9713	.9717	.9722	.9727
9.4	.9731	.9736	.9741	.9745	.9750	.9754	.9759	.9763	.9768	.9773
9.5	.9777	.9782	.9786	.9791	.9795	.9800	.9805	.9809	.9814	.9818
9.6	.9823	.9827	.9832	.9836	.9841	.9845	.9850	.9854	.9859	.9863
9.7	.9868	.9872	.9877	.9881	.9886	.9890	.9894	.9899	.9903	.9908
9.8	.9912	.9917	.9921	.9926	.9930	.9934	.9939	.9943	.9948	.9952
9.9	.9956	.9961	.9965	.9969	.9974	.9978	.9983	.9987	.9991	.9996
N	0	1	2	3	4	5	6	7	8	9

APPENDIX C Answers to Selected Problems

Chapter 1

Problem Set 1.1, Pages 7–9

1. False **3.** True **5.** True **7.** True **9.** True **11.** True **13.** True **15.** False **17.** False
19. [number line with points marked at $-.333\cdots$, 0, $.777\cdots$, π, 3.1416] **21.** 17 **23.** 23 **25.** $\pi - 2$ **27.** $\pi - 6$ **29.** $\sqrt{20} - 4$ **31.** $\sqrt{30} - 5$
33. $2\pi - 5$ **35.** $x + 3$ **37.** $5 - y$ **39.** $2s - 5$ **41.** 18 **43.** 20 **45.** $\pi - 3$ **47.** $\sqrt{5} - 2$
49. Distributive **51.** Commutative for addition **53.** Associative for addition **55.** Distributive
57. Commutative for addition **59.** Associative for addition **61.** Inverse for addition **63.** Inverse for multiplication
65. $B = 12$ **67.** All except additive identity, multiplicative and additive inverses **69.** All

Problem Set 1.2, Pages 16–18

1. B **3.** A **5.** H **7.** C **9.** E **11.** U **13.** T **15.** M **17.** R **19.** Y **21.** 13 **23.** 11
25. 36 **27.** -25 **29.** -49 **31.** 33 **33.** -8 **35.** -6 **37.** 15 **39.** -9 **41.** -4 **43.** -16
45. -9 **47.** 19 **49.** 0 **51.** 53 **53.** -71 **55. a.** $z^2 + 4z - 12$ **b.** $s^2 + s - 20$
57. a. $a^2 - 8a + 15$ **b.** $b^2 - b - 12$ **59. a.** $2x^2 - x - 1$ **b.** $2x^2 - 5x + 3$
61. a. $6a^2 + 5a - 6$ **b.** $6a^2 + 13a + 6$ **63. a.** $x^2 - y^2$ **b.** $a^2 - b^2$ **65. a.** $a^2 + 4a + 4$ **b.** $b^2 - 4b + 4$
67. a. $s^2 + 2st + t^2$ **b.** $u^2 - 2uv + v^2$ **69.** $x^3 - x^2 - 4x - 2$ **71.** $3x^2 - 10x + 1$ **73.** $15x^3 - 22x^2 + 5x + 2$
75. $-4x^3 + 25x^2 - 19x + 22$ **77.** $5x^4 - 9x^3 + 13x^2 - 22x - 5$ **79.** $6x^3 + 17x^2 - 4x - 3$ **81.** $x^3 - 3x - 2$
83. $2x^4 - 11x^3 + 19x^2 - 19x + 21$ **85.** $4x^2 + 9y^2 + z^2 + 12xy - 4xz - 6yz$
87. $x^3 - y^3 - 8z^3 - 3x^2y - 6x^2z + 3xy^2 - 6y^2z + 12xz^2 - 12yz^2 + 12xyz$
89. $9w^6 + 12w^5x + 34w^4x^2 + 26w^3x^3 + 29w^2x^4 + 10wx^5 + x^6$
91. $18x^7 - 57x^6y + 50x^5y^2 + 37x^4y^3 - 106x^3y^4 + 81x^2y^5 - 26xy^6 + 3y^7$

Problem Set 1.3, Pages 21–22

1. $4x(5y - 3)$ **3.** $2(3m - 1)$ **5.** $x(y + z^2 + 3)$ **7.** $(a - b)(a + b)$ **9.** $(a - b)(a^2 + ab + b^2)$ **11.** $(s + t)^2$
13. Not factorable **15.** $xy(x + y)$ **17.** $(x + 5)(x - 7)$ **19.** $(x - 2)(3x + 1)$ **21.** $(2x + 3)(x - 5)$
23. $(4x + 3)(x - 6)$ **25.** $(a + b)(x + y)$ **27.** $(x - 2)(x + 2)(3x + 1)$ **29.** $2(s + 3)(s - 8)$
31. $b(2a + 3)(4a - 1)$ **33.** $y(4y^2 + y - 21)$ **35.** $(x - y - 1)(x - y + 1)$ **37.** $5(a - 1)(5a + 1)$
39. $\frac{1}{4}(x - 2)(x + 2)(x^2 + 2x + 4)(x^2 - 2x + 4)$ **41.** $-3(2m - 1)$ **43.** $x^2(x^a + x^b)$
45. $(x^n - y^n)(x^{2n} + x^ny^n + y^{2n})$ **47.** $(x^n - y^n)^2$ **49.** $x(x - 5)$ **51.** $(x - 2)(x + 2)(2x - 1)(2x + 1)$
53. $(z - 2)(z + 2)(z^2 + 2z + 4)(z^2 - 2z + 4)$ **55.** $(z - 2)^2(z + 2)(z^2 + 2z + 4)$
57. $\left(x + \frac{1}{3}\right)\left(x - \frac{1}{2}\right)\left(x^2 - \frac{1}{3}x + \frac{1}{9}\right)\left(x^2 + \frac{1}{2}x + \frac{1}{4}\right)$ or $\frac{1}{216}(3x + 1)(2x - 1)(9x^2 - 3x + 1)(4x^2 + 2x + 1)$
59. $(x - 2)(x^2 + 2x + 4)(x^3 + 2)$ **61.** $(x - y - a - b)(x - y + a + b)$ **63.** $(x - y - a + b)(x - y + a - b)$
65. $4y(x + 2z)$ **67.** $(2x + 2y + a + b)(x + y - 3a - 3b)$ **69.** $(2x - 1)(xy + z)$ **71.** $-\frac{1}{9}(x + 3y)(5x + 3y)$
73. $3(1 - x)^2(4 - 3x)$ **75.** $(x - 2)^4(x^2 + 1)^2(11x^2 - 12x + 5)$

Problem Set 1.4, Pages 30–32

1. a. 4 **b.** -4 **c.** $\frac{1}{4}$ **d.** $\frac{-1}{4}$ **e.** $\frac{1}{4}$ **3. a.** 9 **b.** -9 **c.** $\frac{1}{9}$ **d.** $-\frac{1}{9}$ **e.** $\frac{1}{9}$ **5. a.** 10 **b.** 6×10^8 or 600,000,000

7. 4×10^{-6} **9.** 5.28×10^{14} **11.** 1.5043×10^{14} **13.** $4t^{10}$ **15.** $\frac{1}{4t^{10}}$ **17.** $\frac{81x^4z^4}{y^8}$ **19.** $-\frac{x^8}{8y}$ **21.** $\frac{y^3z^2}{x^2}$ **23.** $\frac{c^{16}}{a^4b^{12}}$ **25.** $\frac{243m^{35}p^{10}}{n^{50}}$ **27.** $\frac{5}{6}$ **29.** $\frac{x+y}{xy}$ **31.** $\frac{3y-2x}{xy}$ **33.** $\frac{2y-3y}{6xy}$ **35.** $\frac{3-2x-2y}{x+y}$ **37.** $\frac{x+y}{3}$ **39.** $\frac{y+6}{y+3}$ **41.** $\frac{5}{2(x-y)}$ **43.** 1 **45.** $(y-3)^2$ **47.** $\frac{3}{y-x}$ **49.** $\frac{x^2+2}{x^2}$ **51.** $\frac{x^5+2x^4y^3+y^2}{x^3y^2}$ **53.** $\frac{2x^4y^3+x^2y^3+1}{x^3y^2}$ **55.** $\frac{(x+y)^2}{2x^2y}$ **57.** $\frac{3s^2+3st-t^2}{(s+t)^3}$ **59.** $\frac{-3}{y+5}$ **61.** $\frac{x^4+x^2+1}{x^2+1}$ **63.** $\frac{x^3+2x^2-x-2}{4x-1}$ or $\frac{(x-1)(x+1)(x+2)}{4x-1}$ **65.** $(x-1)^2$ **67.** $\frac{-(x+3)(x-3)}{3(x^2-3)^2}$ **69.** 1 **71.** $\frac{2x(3x-10)}{(x-3)(2x+1)}$ **73.** $\frac{m(m+n)}{m-mn^2+n}$ **75.** $\frac{3x^3-x^2-1}{x^3}$

Problem Set 1.5, Pages 37–38

1. 3 **3.** Undefined **5.** -3 **7.** 2 **9.** -2 **11.** 0 **13.** 9 **15.** -4 **17.** 10^{-3} or $\frac{1}{1000}$ or .001 **19.** 16 **21.** 8 **23.** 2 **25.** 3 **27.** 3 **29.** 32 **31.** 3125
33. a. 4 **b.** 4 **c.** Undefined **d.** 4 **e.** 4 **f.** Undefined **35. a.** 9 **b.** 9 **c.** Undefined **d.** 9 **e.** 9 **f.** Undefined
37. $x^{5/6}$ **39.** $\frac{x^2}{y}$ **41.** $\frac{1}{64x^6}$ **43.** $\frac{1}{25x^4}$ **45.** $x^{1/6}$ **47.** m^5n **49.** x^5 **51.** $x^{5/6}$ **53.** $x + 2x^{1/2}y^{1/2} + y$
55. $x + y$ **57.** $x + 1 + (x+1)^{5/3}$ **59.** The theorem $(b^q)^p = b^{pq}$ applies only for positive b. **61.** $|x|$ **63.** $2x$
65. $4|x|$ **67.** $4x^2|y|^3$ **69.** $|x-2|$ **71.** $(x+3)^2$ **73.** $12^{1/2} + 18^{1/2} - 30^{1/2}$

Problem Set 1.6, Pages 43–45

1. 11 **3.** -2 **5.** -8 **7.** 8 **9.** -4 **11.** $20\sqrt{3}$ **13.** $-2xyz^2\sqrt[3]{6xy^2}$ **15.** 6 **17.** $\sqrt[4]{2}$ **19.** $\frac{\sqrt[3]{2}}{2}$
21. $\frac{x^5}{4}$ **23.** $-2-3\sqrt{2}$ **25.** $\sqrt[3]{4}$ **27.** $15 + 2\sqrt{10}$ **29.** 13 **31.** $xy - w$ **33.** $y + 2\sqrt{y} + 1$ **35.** -6
37. 2 **39.** -12 **41.** -8 **43.** $\frac{11\sqrt{15}}{5}$ **45.** $2+\sqrt{3}$ **47.** $\frac{\sqrt{x(x+4)}}{2x}$ **49.** $\frac{\sqrt{x+9}}{x+9}$ **51.** $\frac{(1-\sqrt{y})^2}{1-y}$
53. $\frac{\sqrt{3}-1}{2}$ **55.** 9 **57.** 10 **59.** 6 **61.** 1 **63.** $\frac{(\sqrt{x}+2)(\sqrt{x}+1)}{x-1}$ **65.** $\frac{(\sqrt{y}+3)(\sqrt{y}+1)}{y-1}$
67. $\frac{2\sqrt{x}-1}{x-1}$ **69.** $\frac{(m+n)\sqrt{n+2}}{3(n+2)}$ **71.** $\frac{\sqrt{x}-\sqrt{y}}{x-y}$ **73.** $\frac{x-\sqrt{y+z}}{x^2-y-z}$ **75.** $\frac{\sqrt{x+y}-z}{x+y-z^2}$ **77.** $\sqrt[3]{4}+\sqrt[3]{14}+\sqrt[3]{49}$
79. $|x|$ **81.** $x+y$ **83.** $\frac{|x|\sqrt[3]{2x^2}}{2}$ **85.** $\frac{|y|\sqrt[4]{2}}{2}$ **87.** $\frac{7\sqrt[3]{5x}}{5}$ **89. a.** 13

b.

$$X = \sqrt{169 + 15\sqrt{77}} - \sqrt{(5\sqrt{11} - 13\sqrt{2})(3\sqrt{7} - 13\sqrt{2})}$$

Expand and reorder the second radicand.

$$= \sqrt{169 + 15\sqrt{77}} - \sqrt{(169 + 15\sqrt{77}) - 26\left(\frac{5\sqrt{22} + 3\sqrt{14}}{2}\right) + 169}$$

And since $\left(\frac{5\sqrt{22} + 3\sqrt{14}}{2}\right)^2 = 169 + 15\sqrt{77}$,

$$= \sqrt{169 + 15\sqrt{77}} - \sqrt{(\sqrt{169 + 15\sqrt{77}} - 13)^2}$$
$$= \sqrt{169 + 15\sqrt{77}} - \sqrt{169 + 15\sqrt{77}} + 13$$
$$= 13$$

Problem Set 1.7, Pages 49–50

1. $5i$ **3.** $4i$ **5.** $2i\sqrt{2}$ **7.** $24i$ **9.** $-36i$ **11.** $3i$ **13.** $(\sqrt{2}-2)+i\sqrt{2}$ **15.** $(8+\sqrt{3})-3i$ **17.** $-5+3i$ **19.** $-35+14i$ **21.** $10+20i$ **23.** $9+2i$ **25.** 68 **27.** $-1-5i$ **29.** 1 **31.** $-i$ **33.** $-8-6i$ **35.** $-7+24i$ **37.** $-4-6i\sqrt{5}$ **39.** $2-10i$ **41.** $(4+\sqrt{6})+(2\sqrt{2}-2\sqrt{3})i$ **43.** $-1+i$ **45.** $2-3i$ **47.** $2+i$ **49.** $\frac{5}{17}+\frac{14}{17}i$ **51.** $-3i$ **53.** $-i$ **55.** $1-3i$ **57.** $\frac{1}{2}+2\sqrt{3}i$

59. Given $x^2+3=0$, let $x=i\sqrt{3}$:
$(i\sqrt{3})^2+3=-3+3=0$

61. Given $2x^2-4x+5=0$, let $x=1+\dfrac{i\sqrt{6}}{2}$:

$$2\left(1+\frac{i\sqrt{6}}{2}\right)^2-4\left(1+\frac{i\sqrt{6}}{2}\right)+5$$
$$=-1+2i\sqrt{6}-4-2i\sqrt{6}+5$$
$$=0$$

63. $25-8\sqrt{2}+4\sqrt{3}$

65. Given $x^3=15x+4$, let $x=4$:
$4^3=15(4)+4$
$64=60+4$
$64=64$

Given $x^3=15x+4$, let $x=\sqrt[3]{2+11i}+\sqrt[3]{2-11i}$:

$$(\sqrt[3]{2+11i}+\sqrt[3]{2+11i})^3$$
$$=2+11i+3\sqrt[3]{125(2+11i)}+3\sqrt[3]{125(2-11i)}+2-11i$$
$$=15\sqrt[3]{2+11i}+15\sqrt[3]{2-11i}+4$$
$$=15(\sqrt[3]{2+11i}+\sqrt[3]{2-11i})+4$$

67. Answer is lengthy; all properties are satisfied.

Problem Set 1.8, Pages 51–55

1. rational **2.** a; $-a$ **3.** $a>b$; $a<b$ **4.** $|x_2-x_1|$ **5.** base; exponent; power **6. a.** Carry out all operations within parentheses. **b.** Do exponents next. **c.** Complete multiplications and divisions, working from left to right. **d.** do additions and subtractions, working from left to right. **7. a.** $a+b$ **b.** a^2+ab+b^2 **c.** a^2-ab+b^2

8. a. b^{m+n} **b.** a^mb^n **c.** b^{m-n} **d.** a^{nm} **e.** $\dfrac{a^n}{b^n}$ **9.** real number; nonzero real number; rational number; rational number

10. in the form $\dfrac{p}{q}$, where p and q are polynomials and $q\neq 0$

11. a. $PS=QR$ **b.** $\dfrac{P}{Q}$ **c.** $\dfrac{PS+QR}{QS}$ **d.** $\dfrac{PS-QR}{QS}$ **e.** $\dfrac{PR}{QS}$ **f.** $\dfrac{PS}{QR}$ **12. a.** x **b.** x **c.** $|x|$

13. a. There is no factor raised to a power greater than or equal to the index. **b.** No radical appears in the denominator. **c.** No fraction (or negative exponent) appears within a radical. **d.** the power of the radicand and the index of the radical

14. a. $i^2=-1$ **b.** $\sqrt{-a}=i\sqrt{a}$; $a>0$

15. a number of the form $a+bi$, where a and b are real numbers and i is the imaginary unit

16. a. $a=c$ and $b=d$ **b.** $(a+c)+(b+d)i$ **c.** $(a-c)+(b-d)i$ **d.** $(ac-bd)+(ad+bc)i$
e. $\dfrac{ac+bd}{c^2+d^2}+\dfrac{bc-ad}{c^2+d^2}i$ **17.** $-\sqrt{5}$ **18.** $3-2\sqrt{2}$ **19.** $8-2\pi$ **20.** x if $x\geq 0$; $-x$ if $x<0$ **21.** x^2+5

22. $5t-3$ **23.** 11 **24.** 26 **25.** $9-\sqrt{6}$ **26.** $\sqrt{5}-2$ **27.** Commutative for addition **28.** Commutative for multiplication **29.** Distributive **30.** Commutative for addition **31.** Associative for addition **32.** Commutative for addition **33.** Identity for multiplication **34.** Distributive **35.** Commutative for addition **36.** Distributive **37.** Inverse for multiplication **38. a.** -64 **b.** 64 **c.** -64 **d.** -1 **39.** 3 **40.** 16 **41.** $2x+5$ **42.** $-2x-3y$ **43.** $15x^2-x-6$ **44.** $8y^2+4y-4$ **45.** $4x^2-4x+1$ **46.** $9x^2-12x+4$ **47.** x^3-3x^2+3x-1 **48.** $2x^3+7x^2+4x-4$ **49.** $6(6x-1)$ **50.** $xy(x+3y)$ **51.** $(m-n)(m+n)$ **52.** $2(p-2)(p+4)$ **53.** $(9x-1)(4x-3)$ **54.** $(6x+1)(x-4)$ **55.** $(3x-1)(x-1)(x+1)$ **56.** $(x+1)^3(2+x)(4-2x+x^2)$ **57.** $9x(2x^2+x-4)$ **58.** $2(5+3x)(1-3x)$ **59.** $(s^x-t^x)(s^x+t^x)$

60. $5(x - 5)(x^2 + 5x + 25)$ **61.** $(m - n - 1)(m - n + 1)$ **62.** $\frac{1}{16}(k - 4h)(k + 4h)$

63. $(u^t + v^t)(u^{2t} - u^t v^t + v^{2t})$ **64.** $\frac{-x^2}{9}$ **65.** $\frac{-3}{x^4}$ **66.** $\frac{1}{9x^4}$ **67.** $\frac{y^3z^4}{x^2}$ **68.** $\frac{m^9n^6}{q^{12}}$ **69.** $\frac{16s^6}{t^6}$ **70.** $\frac{x + 6y}{3xy}$

71. $\frac{3 - 2x - 2y}{x + y}$ **72.** $\frac{1}{x - y}$ **73.** $2x - 1$ **74.** $\frac{41 - 5x}{(2x - 1)(3x + 4)}$ **75.** $\frac{2(x + 1)}{x - 2}$ **76.** $\frac{y(x + y)^2}{x^2}$

77. $x^2 + x + 1$ **78.** $2x^2 + 3x + 4$ **79.** 4 **80.** -4 **81.** Undefined **82.** $-\frac{1}{4}$ **83.** x^5 **84.** $x^{5/6}$

85. $x^{1/6}$ **86.** $y^{1/6}$ **87.** xy^2 **88.** $\frac{y^2}{x}$ **89.** $x + 2x^{1/2}y^{1/2} + y$ **90.** $x - y$ **91.** $5\sqrt{2}$ **92.** $2\sqrt[3]{9}$

93. $2xy\sqrt{2y^2}$ **94.** $\sqrt[3]{2}$ **95.** $\frac{2x}{y^2}\sqrt[4]{2y^3}$ **96.** $-1 - 3\sqrt{3}$ **97.** $\frac{2 - 3\sqrt{5}}{5}$ **98.** $18 - 2\sqrt{6}$ **99.** $16 - 7\sqrt{5}$

100. $x + 2\sqrt{xy} + y$ **101.** $\frac{12}{5}\sqrt{5}$ **102.** $\frac{\sqrt{y^2 + 9y}}{3y}$ **103.** $\frac{4 - \sqrt{3}}{13}$ **104.** $\frac{-8\sqrt{3} - 4}{11}$ **105.** $3\sqrt{5} + 6$

106. $\frac{2x - x\sqrt{x}}{4 - x}$ **107.** $\frac{x\sqrt{2 + x}}{2 + x}$ **108.** $\frac{y\sqrt{3y - 1}}{3y - 1}$ **109.** $11i$ **110.** $-i$ **111.** $3i$ **112.** $-i$ **113.** $-6i$

114. $2i$ **115.** $\frac{2 - \sqrt{3}i}{7}$ **116.** $7 - i\sqrt{5}$ **117.** $1 + 2i\sqrt{2}$ **118.** $-2 - i$ **119.** $1 - i$ **120.** $-\frac{23}{26} - \frac{15}{26}i$

Chapter 2

Problem Set 2.1, Pages 60–62

1. $x = 2$ **3.** $x = -5$ **5.** $x = -3$ **7.** $x = -3$ **9.** $x = -2$ **11.** $x = 2$ **13.** $x = 2$ **15.** $x = 2$

17. $x = 3$ **19.** $x = 1$

21. $x \leq 3$ **23.** $x < -5$ **25.** $x \geq -4$ **27.** $x > -3$

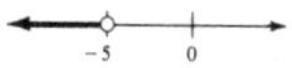

29. $x < -8$ **31.** $m = 3$ **33.** $x < -1$ **35.** $a \leq 10$

37. $h = -3$ **39.** $\varnothing$ **41.** $x < -2$ **43.** $a \leq 1$

45. $x = 1$ **47.** $m > 7$ **49.** $x \geq -7$  **51.** $5 < x < 9$

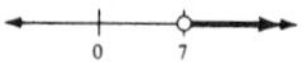

53. $-1 < x < 4$ **55.** $-2 \leq x \leq 5$ **57.** $-\frac{1}{2} \leq x < \frac{5}{2}$ **59.** $-7 < x < -4$

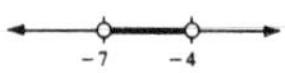

61. $a = -\frac{5}{7}$ **63.** All real numbers **65.** $z = \frac{1}{4}$ **67.** $u = \frac{13}{2}$

69. Arithmetic mean (A.M.) $= \frac{a + b}{2}$, Harmonic mean (H.M.) $= \frac{2ab}{a + b}$, $a, b > 0, a \neq b$

1. If a and b are unequal, then $(a - b)^2 > 0$
2. $a^2 - 2ab + b^2 > 0$
3. $a^2 + 2ab + b^2 > 4ab$
4. $(a + b)^2 > 4ab$
5. $a + b > \frac{4ab}{a + b}$ Since $a + b > 0$, $\mathbf{a \neq b}$
6. $\frac{a + b}{2} > \frac{2ab}{a + b}$
7. $\therefore$ A.M. > H.M.

Problem Set 2.2, Pages 66–69

1. The numbers are 62 and 63. **3.** The first odd integer is 31. **5.** The second integer is 15.
7. The larger rectangle has dimensions 5 ft by 11 ft. **9.** The smaller triangle is 3 cm by 4 cm.
11. True for any such rectangle **13.** \$600 invested at 10% **15.** \$2500 invested at 9%
17. \$600 invested at $9\frac{1}{2}$%; \$900 invested at 14% **19.** \$8500 invested at the lower rate **21.** 288 mi
23. 4 hr spent in the car **25.** $2\frac{1}{2}$ mi by bus **27.** 6 mph; 8 mph **29.** 24 mph
31. 50 gallons of milk; 100 gallons of cream **33.** 2.5 liters of 72% solution; 2 liters of 45% solution **35.** 40 cc of water
37. 12 oz of water **39.** 20 oz of 15% alcohol lotion **41.** $2\frac{2}{3}$ quarts pure antifreeze **43.** 2240 kWh used
45. 2 hr; one boy walks 7 mi and the other 8 mi **47.** 1 hr

Problem Set 2.3, Pages 75–77

1. $x = \pm 4$ **3.** $x = -1$ or 5 **5.** $a = 4$ or -6 **7.** $c = -3$ or 4 **9.** $e = -\frac{3}{2}$ or 3 **11.** $h = -\frac{11}{3}$ or 3

13. $-2 < y < 2$ **15.** $x \leq -5$ or $x \geq 5$ **17.** $-1 < x < 1$ **19.** $-\frac{3}{2} < z < \frac{3}{2}$

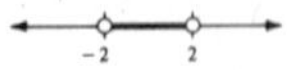

21. $-\frac{2}{5} \leq z \leq \frac{2}{5}$  **23.** $d < -4$ or $d > 10$ 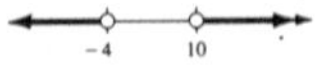 **25.** $-3 \leq k \leq 17$  **27.** $n \leq -7$ or $n \geq 1$

29. $-3 \leq x \leq 10$  **31.** $3 < p < 8$ **33.** $-2 < n < 15$ **35.** $y < -\frac{3}{5}$ or $y > -\frac{1}{5}$

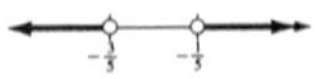

37. $h \leq -25$ or $h \geq -6$  **39.** $-29 \leq m \leq 4$

41. $|x - 17| = 5$ **43.** $|r - 11| \geq 4$ **45.** $|5 - 40| = d$ **47.** $|h + 1| > 9$ **49.** $|x - p| \leq t$
51. $|m - 3.7| \leq .04$ **53.** $|m - 9.15| < .15$ **55.** $|t - 120| \leq 30$ **57.** $|p - .11| \leq .01$ **59.** $|s - 10| \leq 3$
61. $x = \frac{11}{2}$ or $\frac{13}{2}$ **63.** $x = 5$ or 7  **65.** $-4 < x < 12$ **67.** $-\frac{7}{6} \leq x \leq -\frac{1}{6}$

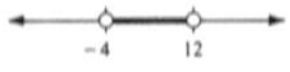

69. $x \neq \frac{5}{6}$

Problem Set 2.4, Pages 82–84

1. 41; 2 **3.** -8; 0 **5.** 0; 1 **7.** 9; 2 **9.** $\frac{1}{4} + 8\sqrt{2}$; 2 **11.** $a = -1$ or 2 **13.** $b = 2$ or 3
15. $c = -3$ or $\frac{1}{2}$ **17.** $x = \frac{1}{2}$ or $\frac{2}{3}$ **19.** $y = 3$ **21.** $a = -3$ or 1 **23.** $b = -2$ or 3 **25.** $c = -1$ or $\frac{4}{3}$
27. $h = 1$ or $\frac{3}{2}$ **29.** $k = -\frac{3}{2}$ or $-\frac{5}{3}$ **31.** $w = 2 \pm \sqrt{3}$ **33.** $y = -1 \pm \sqrt{5}$ **35.** $p = \dfrac{1 \pm \sqrt{3}}{3}$
37. $r = 2 \pm 3i$ **39.** $s = \dfrac{1 \pm i}{2}$ **41.** $t = \dfrac{1 \pm \sqrt{10}}{2}$ **43.** $v = \dfrac{3 \pm 2\sqrt{2}}{3}$ **45.** $x = \dfrac{3 \pm 5i}{2}$ **47.** $m = \dfrac{-3 \pm \sqrt{41}}{4}$
49. $p = \dfrac{-4 \pm i\sqrt{3}}{2}$ **51.** 6 cm by 19 cm **53.** Width is 5 in.; length is 17 in. **55.** 18 sides **57.** 18 mm
59. 6 ft **61.** $8 + 4\sqrt{15}$, $15 + 4\sqrt{15}$, and $23 + 4\sqrt{15}$ or approximately 23 cm, 30 cm, and 38 cm **63.** {2.02, −1.98}
65. {.67, −.90} **67.** {4.87, −2.87} **69.** $k = 2$

Problem Set 2.5, Pages 89–90

1. $\{2 \pm \sqrt{13}\}$ **3.** $\left\{\frac{4}{3}\right\}; y \neq 0$ **5.** $\{4\}; y \neq 0, 1$ **7.** $\{1, 6\}; z \neq 0$ **9.** $\{-8\}; x \neq \pm 2$
11. $\{2\}; x \neq 0, \frac{1}{3}; x = 0$ is extraneous **13.** $\left\{5, -\frac{1}{5}\right\}; a \neq \pm 1$ **15.** $\left\{1, -\frac{9}{5}\right\}; b \neq \pm 3$ **17.** $\{1\}$ **19.** $\{7\}$ **21.** $\{1\}$
23. $\{4\}; m = 1$ is extraneous **25.** $\{3\}; r = 18$ is extraneous **27.** 0 **29.** $\{1, -7\}$ **31.** $\left\{\frac{3}{2}\right\}; x \neq \pm\frac{1}{2}$
33. $\{10\}; r \neq 1$ **35.** $\{1, 5\}; x \neq 2, -3$ **37.** $\left\{\frac{5}{3}, -2\right\}; x \neq 2, -\frac{1}{2}$ **39.** $\{4, 7\}; x \neq -2, -4, -5$
41. $\{1\}; b = 5$ is extraneous **43.** $\{0\}$ **45.** $\{12\}; x = 0$ is extraneous **47.** $\{0, -1\}$ **49.** $\{3\}; a = -\frac{33}{49}$ is extraneous
51. $\{4\}; w \neq 0; w = -2$ is extraneous **53.** $\left\{1, \frac{3}{2}\right\}$ **55.** $1\frac{1}{2}$ feet

57.
$$\begin{aligned}
d^2 + r^2 &= (r + h)^2 \\
d^2 &= 2rh + h^2 \\
d^2 &\approx 2rh \\
d^2 &\approx 2 \cdot 4000 \cdot \frac{h}{5280} \\
d^2 &\approx \frac{8000}{5280}h \\
d^2 &= 1.\overline{51}h \\
d^2 &\approx \tfrac{3}{2}h \\
d &\approx \sqrt{\tfrac{3}{2}h}
\end{aligned}$$

59. $a \neq b; a > 0, b > 0$
$$\begin{aligned}
a &> b \\
a - b &> 0 \\
(a - b)^2 &> 0 \\
a^2 - 2ab + b^2 &> 0 \\
a^2 + 2ab + b^2 &> 4ab \\
(a + b)^2 &> 4ab \\
a + b &> 2\sqrt{ab} \\
\frac{a + b}{2} &> \sqrt{ab}
\end{aligned}$$
Arithmetic mean > Geometric mean

Problem Set 2.6, Pages 95–96

1. $-1 < x < 0$
3. $x \leq 2$ or $x \geq 5$
5. $-7 < x < 3$
7. $-2 \leq y \leq \frac{1}{2}$
9. $z < -\frac{2}{3}$ or $z > 3$
11. $x \leq -2$ or $x \geq 3$
13. $x < \frac{1}{3}$ or $x > 4$
15. $y \leq -4$ or $0 \leq y \leq 3$
17. $-3 \leq y \leq 2$ or $y \geq 4$
19. $z < -\frac{5}{2}$ or $-1 < z < \frac{7}{3}$
21. $-2 < x < 0$
23. $x < 0$ or $x > 3$
25. $-3 < x \leq 2$
27. $y < 0$ or $\frac{1}{2} < y < 3$
29. $z < -2$ or $0 < z < 3$
31. $x < 1$ or $x > 2$
33. $-1 \leq y \leq 2$
35. Always true
37. $-\frac{1}{5} < r < 1$
39. $t \leq 1$ or $t \geq 5$
41. $-4 < u < \frac{3}{2}$
43. $\frac{2}{3} \leq w \leq \frac{3}{2}$
45. $y \leq \frac{2}{5}$ or $y \geq \frac{5}{2}$
47. $-1 < x < 0$ or $x > 2$

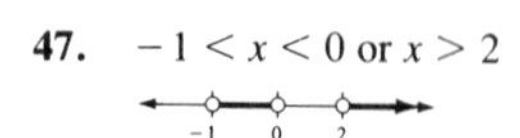

49. $x \leq -2$ or $0 < x \leq 2$
51. $x > 2$
53. $x < 2$ or $3 < x \leq 4$

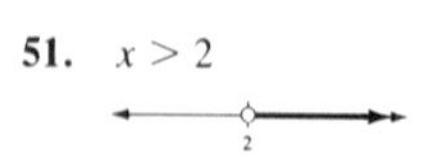

55. $x < -4$ or $-2 < x < 2$ or $x > 4$
57. $y < -3$ or $-2 \leq y \leq 0$ or $2 \leq y < 5$

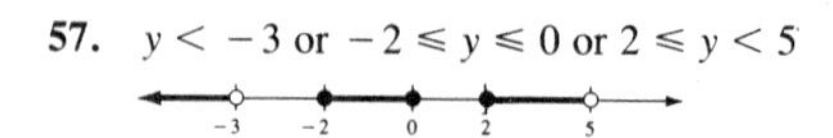

59. $-3 \le y < -1$ or $0 < y < 2$ or $y \ge 3$ **61.** All numbers except those between -17 and 20

63. All numbers except those between -3 and 0, inclusive **65.** Width must be greater than 3 and length must be greater than 6

67. $a \ne b; a > 0, b > 0$

$$a > b$$
$$a^3 > b^3$$
$$a^3(a-b) > b^3(a-b)$$
$$a^4 - a^3b > ab^3 - b^4$$
$$a^4 + b^4 > a^3b + ab^3$$
$$a^4 + \mathbf{2a^2b^2} + b^4 > a^3b + \mathbf{2a^2b^2} + ab^3 \qquad \text{Add } 2a^2b^2 \text{ to complete the square.}$$
$$(a^2+b^2)^2 > ab(a+b)^2$$
$$a^2 + b^2 > \sqrt{ab}\,(a+b)$$
$$a^2 + \mathbf{2ab} + b^2 > \mathbf{2ab} + \sqrt{ab}(a+b) \qquad \text{Add } 2ab.$$
$$(a+b)^2 > 2ab + \sqrt{ab}(a+b)$$
$$a + b > \frac{2ab}{a+b} + \sqrt{ab} \qquad \text{Divide by } a+b$$
$$\frac{a+b}{2} > \frac{\frac{2ab}{a+b} + \sqrt{ab}}{2} \qquad \text{Divide by 2.}$$
$$\text{Arithmetic mean} > \frac{\text{Harmonic mean} + \text{Geometric mean}}{2}$$

Problem Set 2.7, Pages 97–99

1. equation **2.** roots; solutions **3.** identity **4.** contradiction **5.** equivalent **6.** added; subtracted
7. nonzero; equivalent **8.** conditional **9.** multiplication; division; negative quantity; sense (or order); inequality
10. number; opposite (or additive inverse) **11.** $-a$; x; a **12.** distance **13.** b; a **14.** x; -5; 2
15. second-degree **16.** zero **17.** $\dfrac{-b \pm \sqrt{b^2-4ac}}{2a}$ **18.** $b^2 - 4ac$ **19.** real **20.** integer; subset
21. extraneous; checked **22.** restrictions **23.** LCD **24.** critical values **25.** signs; intervals **26.** $x = 2$
27. $x = -3$ **28.** $x = \frac{2}{3}$ **29.** $y = 3$ **30.** $z \le 2$ **31.** $z < 1$

32. $-2 < x < 3$ **33.** $-6 \le x \le -3$ **34.** $x = \frac{37}{5}$ **35.** $x = 1$ **36.** $y = \frac{5}{9}$ **37.** $y = -2$

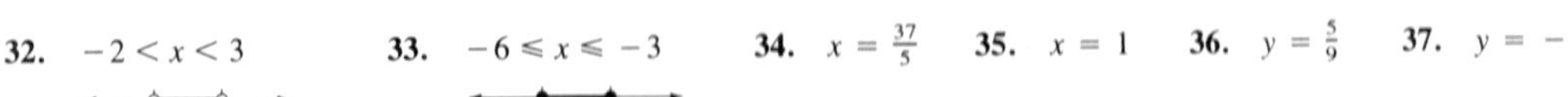

38. $\varnothing$ **39.** $z < 7$ **40.** All real numbers **41.** $x = \pm 5$ **42.** $x = \pm 11$

43. $-2 < x < 2$ **44.** $x < -13$ or $x > 13$ **45.** $y = \pm\frac{3}{2}$ **46.** $-\frac{2}{3} < y < \frac{2}{3}$ **47.** $y = 4, -2$

48. $y = 4, 2$ **49.** $x = 2, -4$ **50.** $x = -2, -4$ **51.** $-11 \le y \le 3$ **52.** $y < -16$ or $y > 6$

53. $z \leqslant 2$ or $z \geqslant 10$ **54.** $-2 < z < 10$ **55.** $z = 2, -\frac{10}{3}$ **56.** $z = 1, -4$ **57.** $z = 1, -\frac{1}{5}$

58. $-\frac{1}{3} < z < 1$ **59.** $z > 1$ or $z < \frac{1}{5}$ **60.** $z < -2$ or $z > \frac{1}{2}$ **61.** $x = 5, -9$ **62.** $\frac{1}{3}, -\frac{5}{2}$

63. $y = 3, -1$ **64.** $y = 1, -\frac{1}{3}$ **65.** $x = 3 \pm \sqrt{2}$ **66.** $x = -\frac{1}{2}, \frac{5}{2}$ **67.** $y = \frac{3}{2}, -3$ **68.** $y = \frac{3}{2}, -\frac{5}{3}$

69. $5 < x < 7$ **70.** $x < -\frac{1}{2}$ or $x > \frac{3}{2}$ **71.** $y \leqslant -5$ or $y \geqslant \frac{1}{2}$ **72.** $y \leqslant \frac{2}{5}$ or $y \geqslant 2$

73. $0 < z < \frac{2}{3}$ or $z > 2$ **74.** $-3 < z < -1$ or $z > 2$ **75.** $x \leqslant -2$ or $x \geqslant 6$ **76.** $y < -\frac{2}{3}$ or $y > \frac{5}{2}$

77. $-2 < y < -1$ or $y \geqslant 0$ **78.** $-1 \leqslant y < 1$ or $y \geqslant 3$ **79.** $z \leqslant -\frac{7}{5}$ or $z \geqslant -1$ **80.** $z \geqslant -\frac{3}{2}$

81. $x = 3$ **82.** $x = 3$ **83.** $y = 2$ **84.** $y = -\frac{2}{3}, 2$ **85.** $x = \frac{5}{3}$ **86.** $x = 9, 10$
87. $x = 14$ **88.** $x = -\frac{1}{3}$ **89.** $z = 3$ **90.** $x = 2$ **91.** $x = 4$ **92.** $x = 25$ **93.** $z = -2$
94. No solution **95.** 12 m by 20 m **96.** 6 cm by 14 cm **97.** \$1200 invested at 8%
98. 5 gallons of milk; 10 gallons of cream **99.** 108 miles **100.** 4 hr

Chapter 3

Problem Set 3.1, Pages 107–108

1.
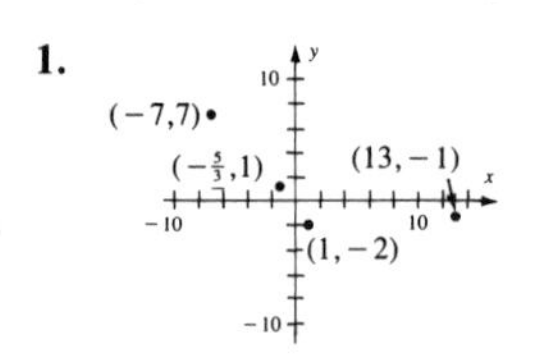

3.
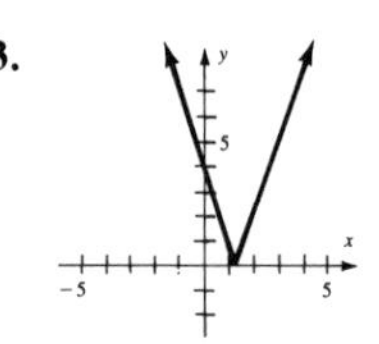

5.
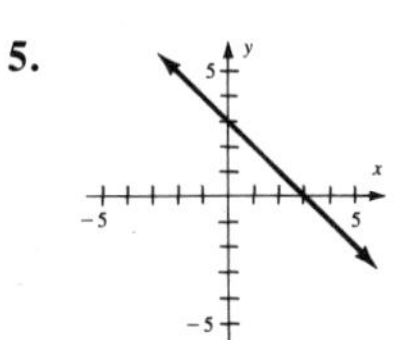

7.
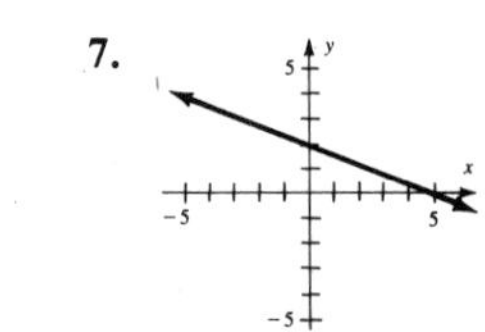

9.
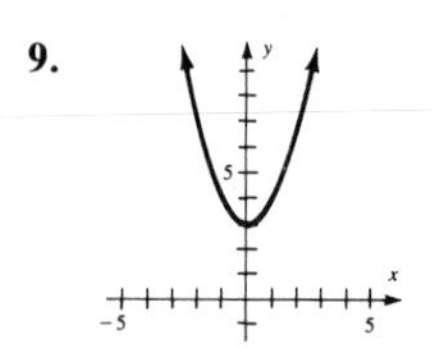

11.
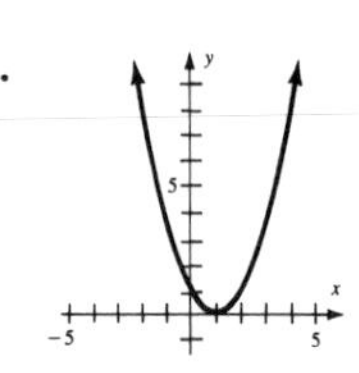

13. 5 **15.** 17 **17.** $12\sqrt{2}$ **19.** $\sqrt{109}$ **21.** 3 **23.** $\sqrt{2}$ **25.** $\sqrt{a^2 + b^2}$ **27.** $\left(6, \frac{15}{2}\right)$ **29.** $(1, 4)$

31. $\left(-2, \frac{7}{2}\right)$ **33.** $\left(0, -\frac{3}{2}\right)$ **35.** $\left(\frac{r+t}{2}, s\right)$ **37.** $\left(\frac{2a+b}{2}, \frac{2b-a}{2}\right)$ **39.** $\frac{3}{2}$ **41.** $-\frac{2}{3}$ **43.** -1

45. $-\sqrt{3}$ **47.** $\frac{q-s}{p-r}$ or $\frac{s-q}{r-p}$ **49.** 1

51. 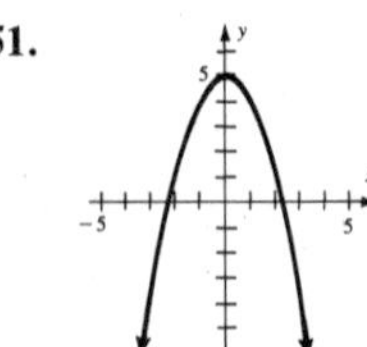

53. **55.** 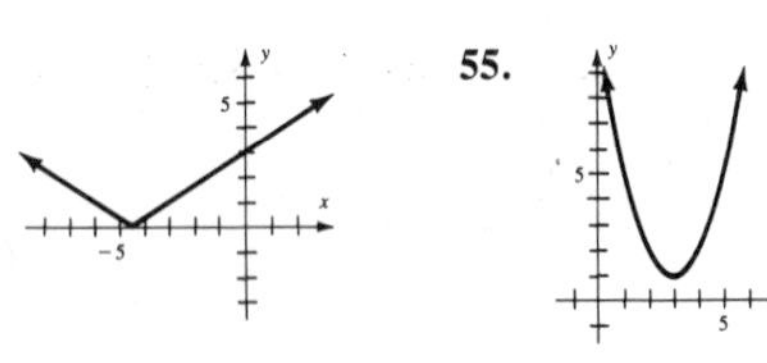

57. 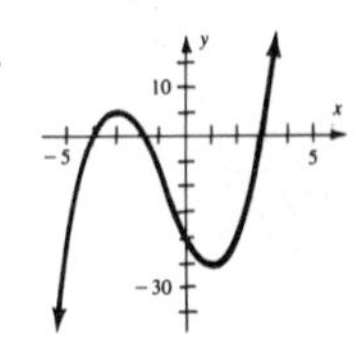

59. 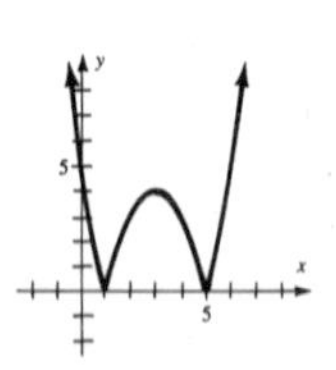

61. $d_{PA} = 10$; $d_{AT} = 5\sqrt{5}$; $d_{PT} = 5$; is a right triangle

63. $\overline{AB} = \overline{BC} = \overline{CD} = \overline{DA} = 2\sqrt{5}$; $m_{AB} = 2$, $m_{BC} = -\frac{1}{2}$; since the sides are equal and perpendicular, *ABCD* is a square
65. $\overline{TR} = \overline{RI} = \overline{IT} = 2$; thus $\triangle TRI$ is equilateral
67. $m_{BC} = m_{DA} = \frac{1}{2}$, $m_{AB} = -2$; since two sides are parallel and perpendicular to a third side, *ABCD* is a trapezoid with two right angles **69.** $m_{IJ} = m_{KL} = 0$, $m_{JK} = m_{LI} = \frac{b}{a}$; since opposite sides are parallel, *IJKL* is a parallelogram

Problem Set 3.2, Pages 116–118

1. $2x - y - 1 = 0$ **3.** $7x + 2y - 22 = 0$ **5.** $8x - y + 6 = 0$ **7.** $x + 3y - 12 = 0$ **9.** $y + 2 = 0$
11. $y - 5 = 0$ **13.** $2x - 5y + 10 = 0$ **15.** $2x + y + 1 = 0$

17. 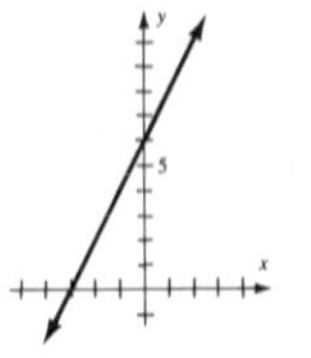

19. 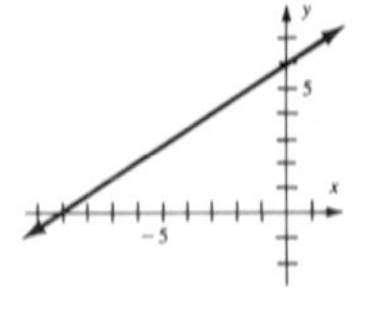

21. 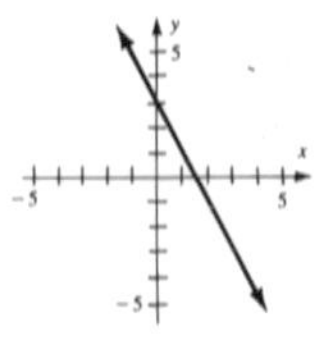

23. 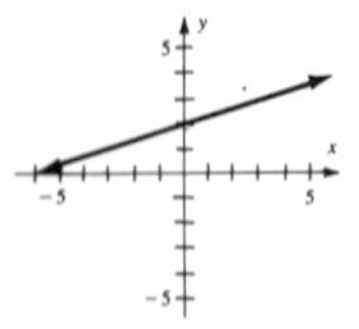

25. 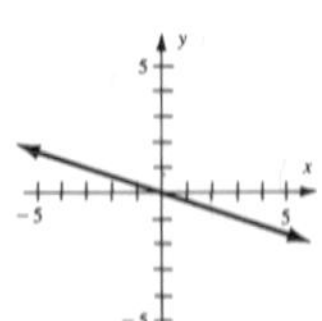

27. 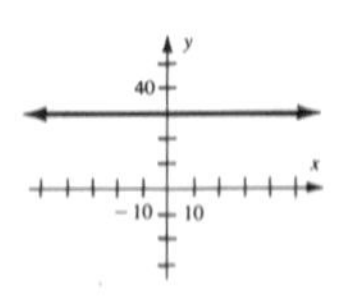

29. 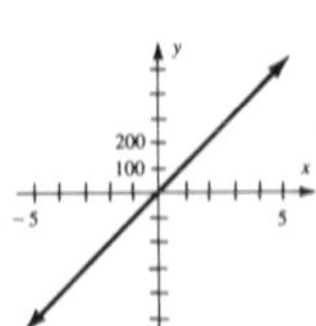

31. 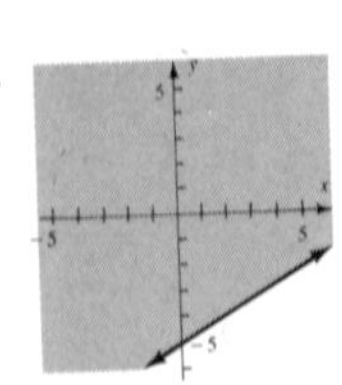

33 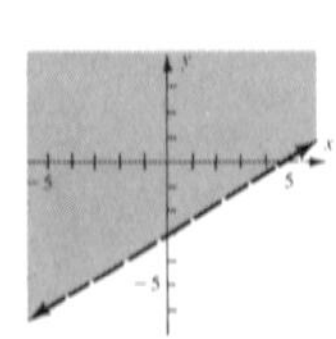

35. 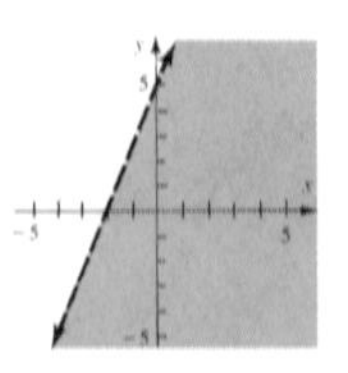

37. 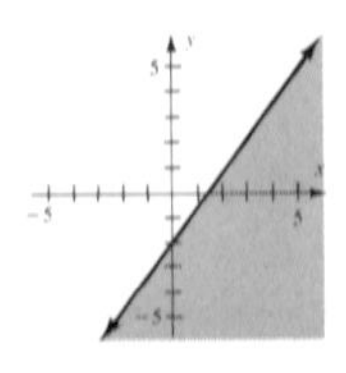

39. 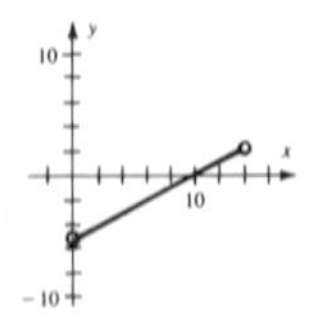

41. 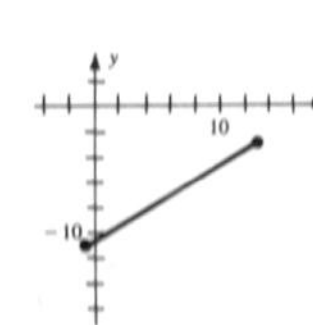

43. 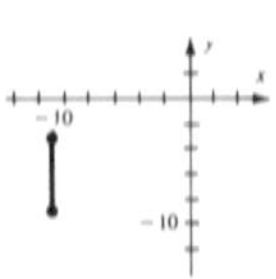

45. 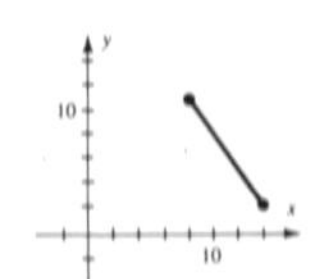

47. 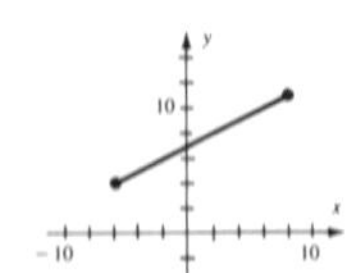

49. 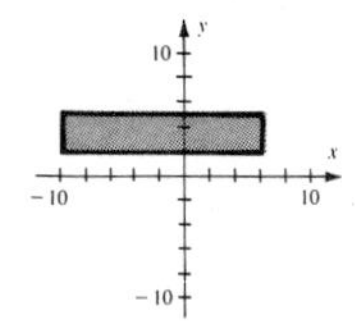

51. 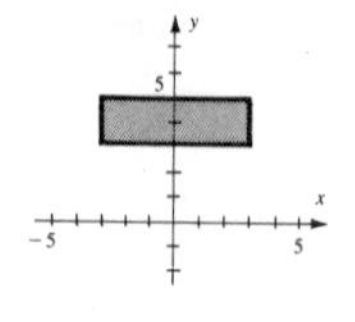

53. 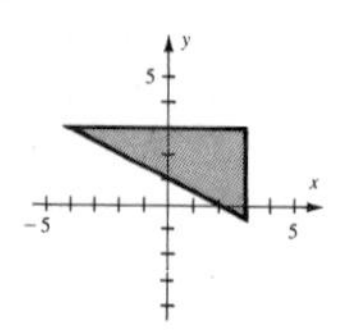

55. 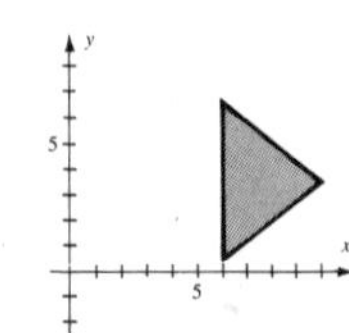

57. 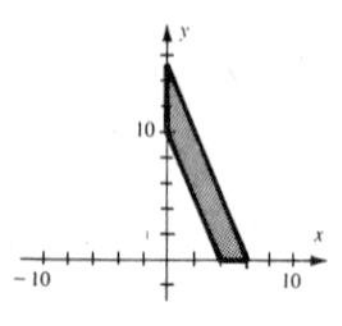

59. 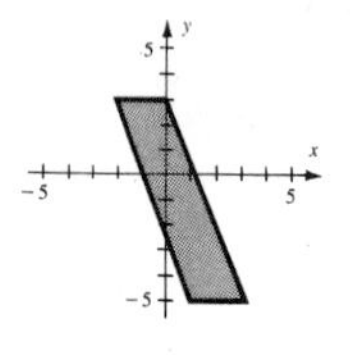

61. 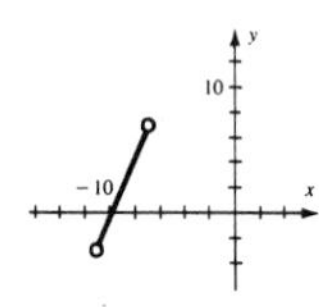

63. 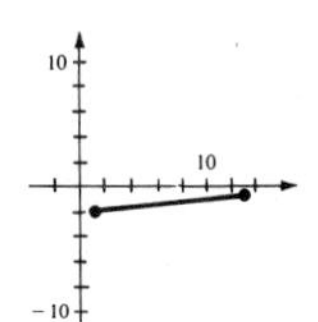

65. 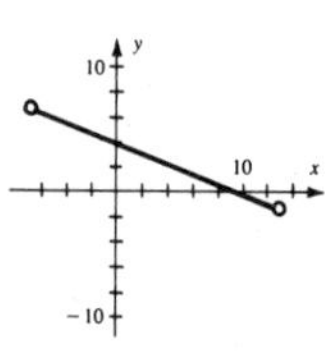

67. 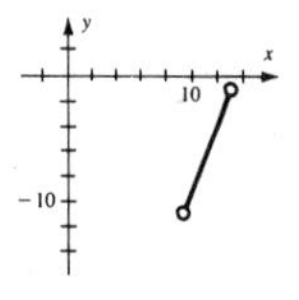

Problem Set 3.3, Pages 123–125

1. $y - 3 = (x + 2)^2$; $(-2, 3)$; up **3.** $x + 10 = (y - 3)^2$; $(-10, 3)$; right **5.** $y + 15 = (x + 5)^2$; $(-5, -15)$; up
7. $y - \frac{1}{2} = (x + 3)^2$; $\left(-3, \frac{1}{2}\right)$; up **9.** $y + \frac{4}{3} = (x + 2)^2$; $\left(-2, -\frac{4}{3}\right)$; up **11.** $x - 2 = 2(y - 5)^2$; $(2, 5)$; right
13. $y - 7 = -3(x - 3)^2$; $(3, 7)$; down **15.** $y - 1 = \frac{2}{3}\left(x + \frac{1}{2}\right)^2$; $\left(-\frac{1}{2}, 1\right)$; up **17.** $x - 5 = -\frac{3}{2}(y + 2)^2$; $(5, -2)$; left
19. $y - 1 = \frac{2}{3}(x + 2)^2$; $(-2, 1)$; up

21. 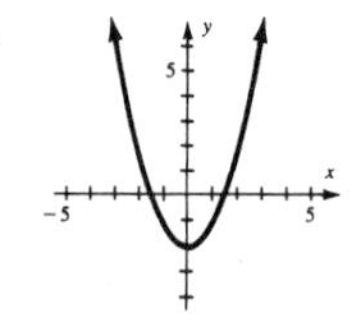

23. 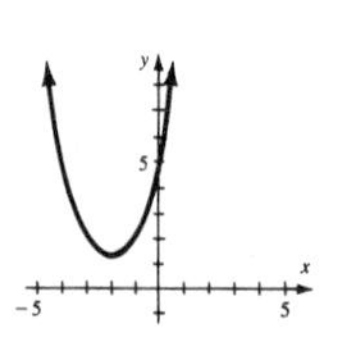

25. 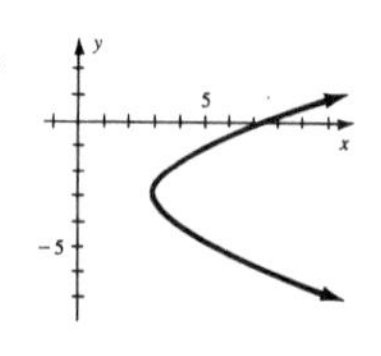

27. 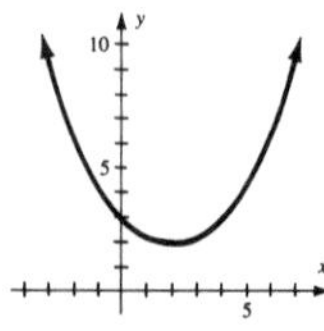

29. 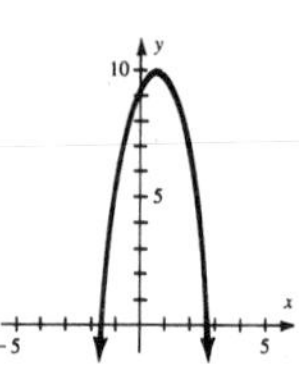

31. 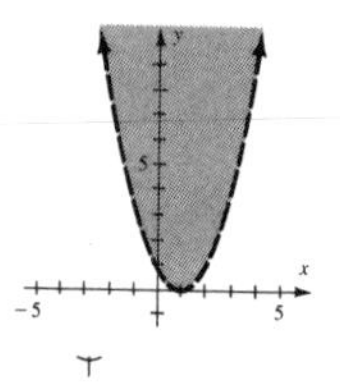

33. 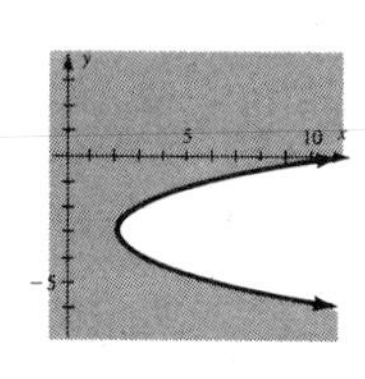

35. 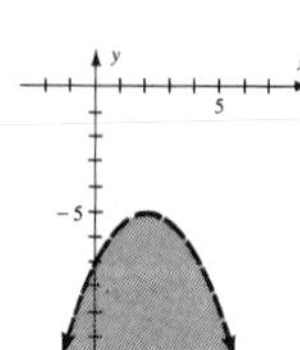

37. 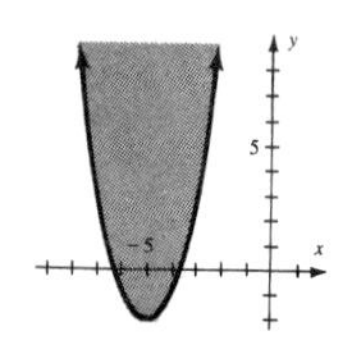

39. 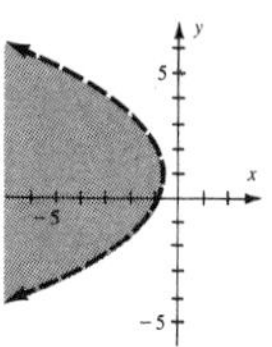

41. 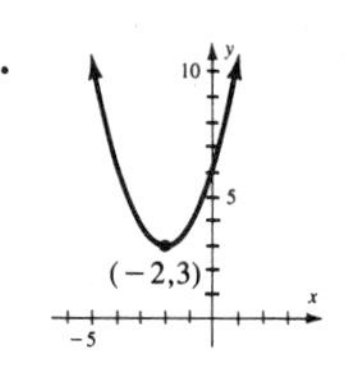

43. 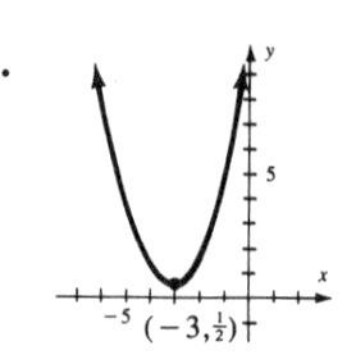

45. 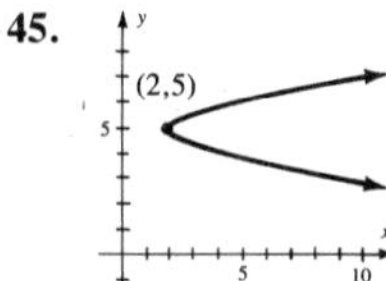

47. 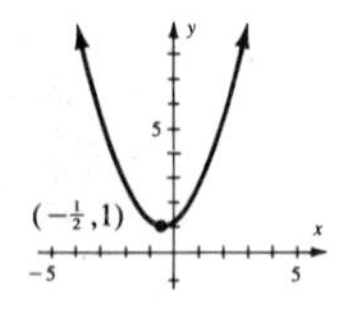

49. 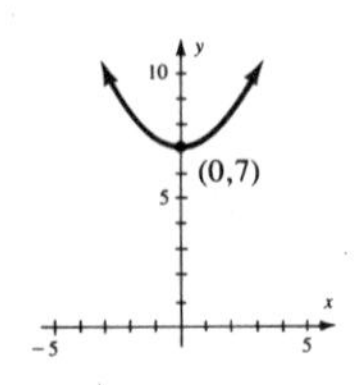

51. 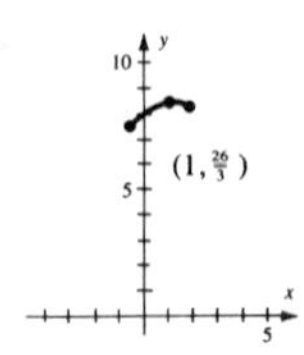

53. 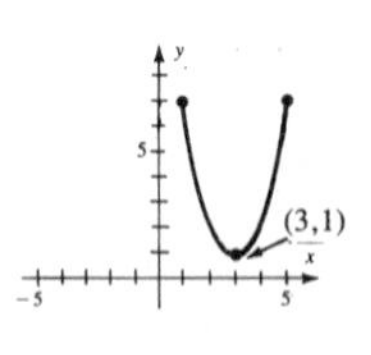

55. 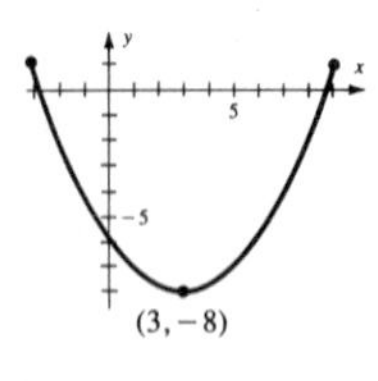

57. The maximum value is when $x = 9$. Thus, the sides are 9 ft and 18 ft, and the maximum area is 162 ft^2. **59.** 16 ft tall

61. 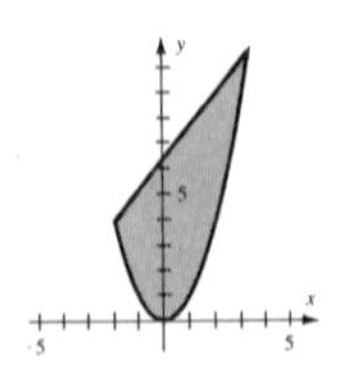

63. 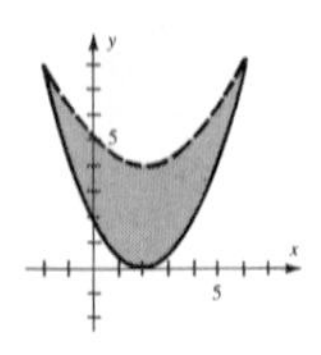

65. 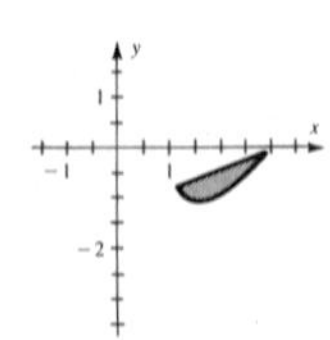

67. 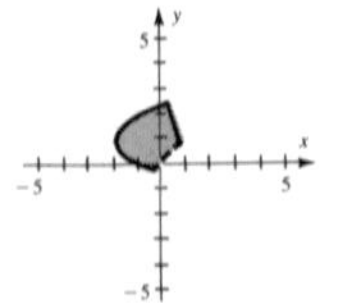

69. Let $F(0, c)$ and let $P(x, y)$ be any point on the parabola. Draw PD so that D is a point on the directrix and PD is perpendicular to the directrix. Then, from the definition,

$$\overline{FP} = \overline{PD}$$
$$\sqrt{x^2 + (y - c)^2} = \sqrt{(y + c)^2}$$
$$x^2 + y^2 - 2cy + c^2 = y^2 + 2cy + c^2$$
$$y = \frac{1}{4c}x^2$$
$$y = ax^2 \quad \text{where } a = \frac{1}{4c}$$

71. Let ℓ be the length, w the width, and p a constant, where $\ell + w = p$.

$$\text{Area} = a = \ell w = (p - w)w$$
$$a = -(w^2 - wp)$$
$$a - \frac{p^2}{4} = -\left(w^2 - wp + \frac{p^2}{4}\right)$$
$$a - \frac{p^2}{4} = -\left(w - \frac{p}{2}\right)^2$$

Vertex is at $\left(\frac{p}{2}, \frac{p^2}{4}\right) = (w, a);\ w = \frac{p}{2}$ and $\ell + w = p$

$$\ell + w = p$$
$$\ell + \frac{p}{2} = p$$
$$\ell = \frac{p}{2} = w$$

73. It follows from the definition of symmetry with respect to the origin that if a point (x, y) is on the graph, then $(-x, -y)$ is also on the graph. Furthermore, if both the points (x, y) and $(-x, -y)$ are on the graph, then it is symmetric with respect to the origin. Therefore, the coordinates of the point $(-x, -y)$ as well as (x, y) must satisfy an equation of the graph. Thus, the graph of an equation in x and y is symmetric with respect to the origin if and only if an equivalent equation is obtained when x is replaced by $-x$ and y by $-y$ in the equation.
75. Examples vary.

Problem Set 3.4, Pages 128–130

1. $(x - 2)^2 + (y - 5)^2 = 1; (2, 5); r = 1$ **3.** $(x + 1)^2 + \left(y - \frac{1}{2}\right)^2 = \frac{1}{4}; \left(-1, \frac{1}{2}\right); r = \frac{1}{2}$
5. $\left(x - \frac{1}{3}\right)^2 + (y + 2)^2 = \frac{1}{9}; \left(\frac{1}{3}, -2\right); r = \frac{1}{3}$ **7.** $(x - 2)^2 + (y - 0)^2 = 9; (2, 0); r = 3$
9. $(x + 3)^2 + (y + 4)^2 = 25; (-3, -4); r = 5$

11.
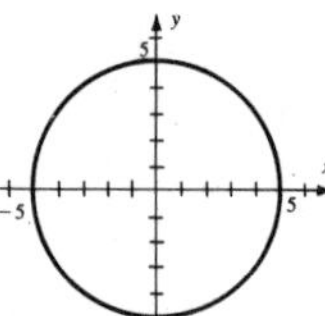
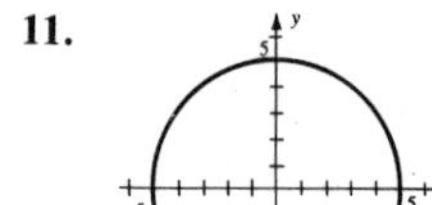

13.
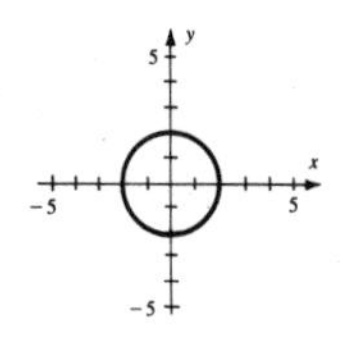

15.
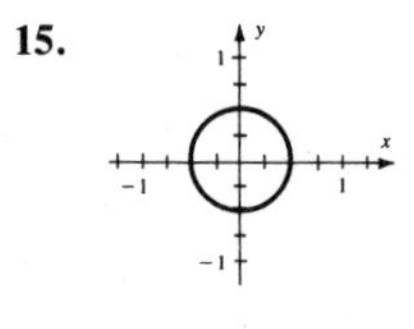

17.
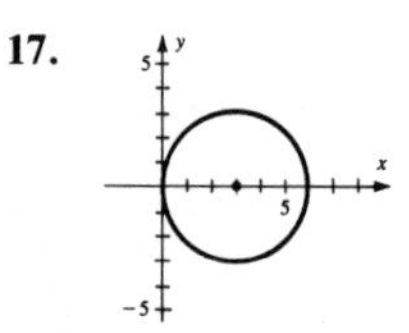

19.
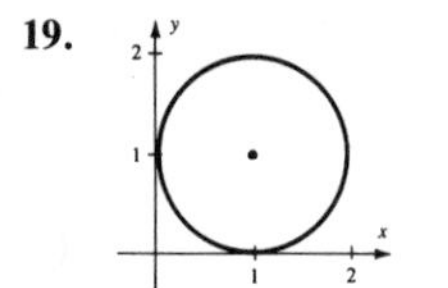

21.
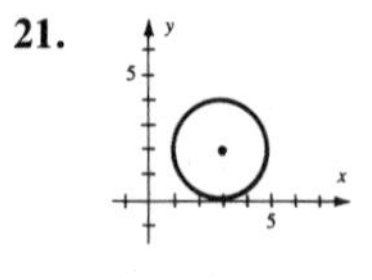

23.
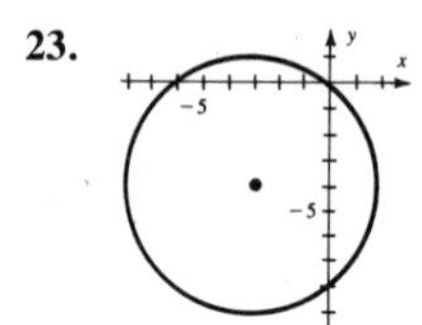

25.
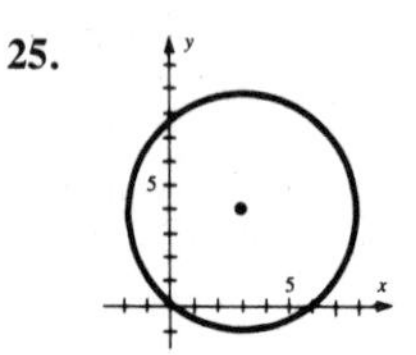

27.
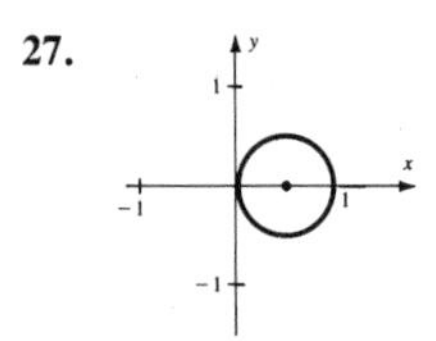

29.
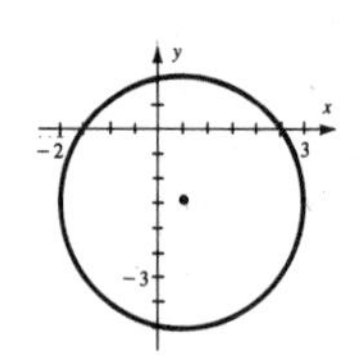

31. $(x - 3)^2 + (y + 4)^2 = 25$ **33.** $(x + 4)^2 + (y - 5)^2 = 25$ **35.** $(x + 3)^2 + (y - 6)^2 = 16$
37. $(x - 4)^2 + (y - 1)^2 = 169$ **39.** $(x + 1)^2 + (y + 2)^2 = 32$

41.
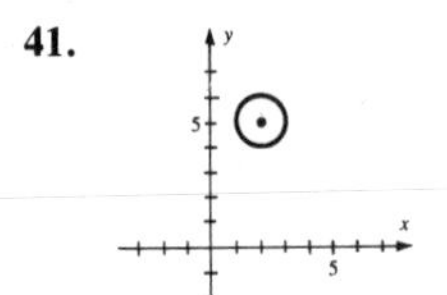

43.
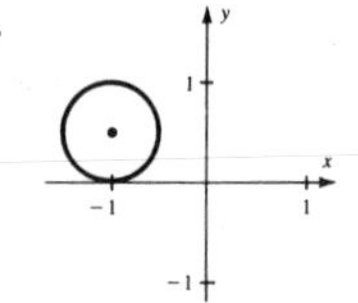

45.
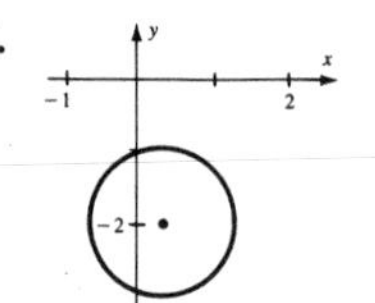

47.
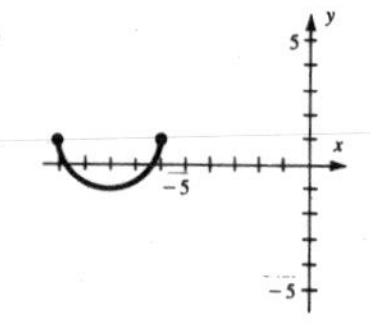

49.
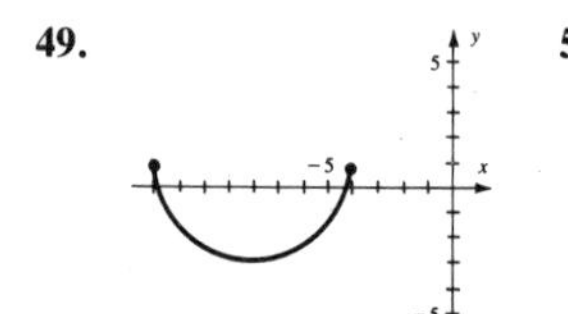

51.
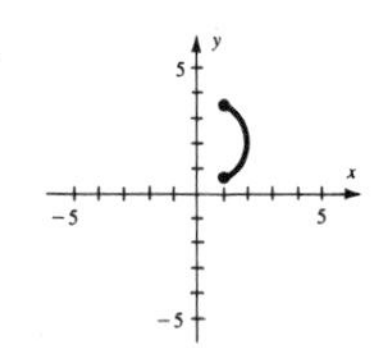

53. $\varnothing$

55.
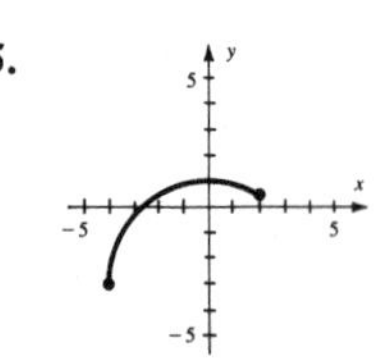

57. 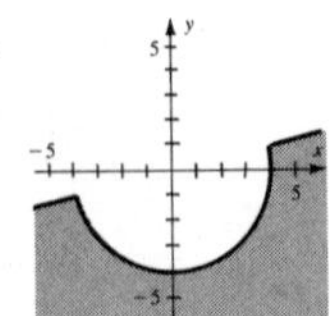

59. 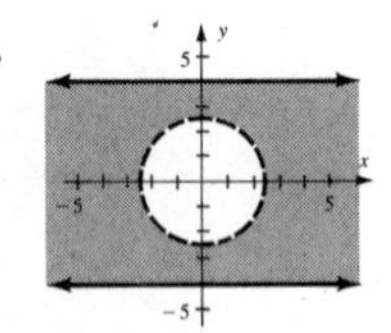

61. 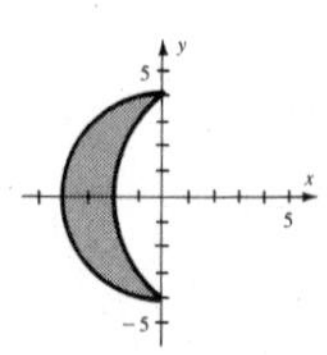

63. 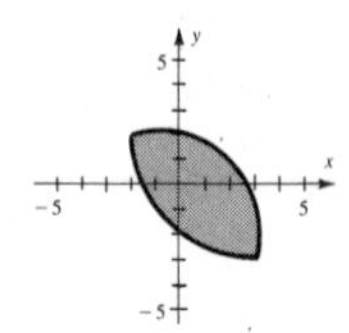

65. 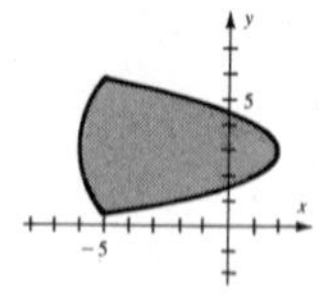

67. If $A(a, 0)$, $B(0, b)$, and $C(0, 0)$, then the midpoint M of the hypotenuse is $M\left(\frac{a}{2}, \frac{b}{2}\right)$. Since $\overline{MA} = \overline{MB} = \overline{MC}$; thus, M is the center of the circumscribed circle.

Problem Set 3.5, Pages 135–137

1. Circle **3.** Hyperbola **5.** No graph **7.** Parabola **9.** Ellipse **11.** Line **13.** Hyperbola **15.** Ellipse **17.** Hyperbola **19.** Two intersecting lines

21. 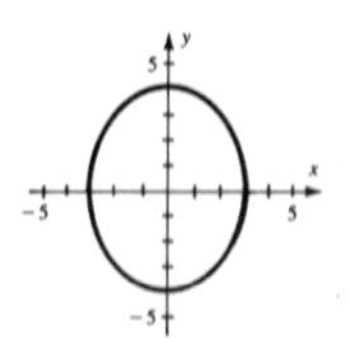

23. 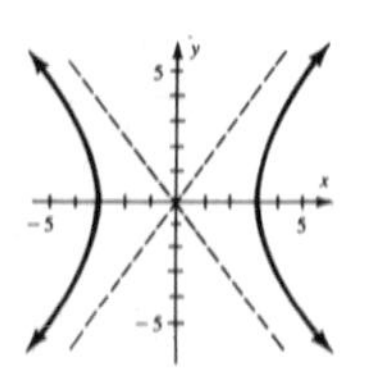

25. 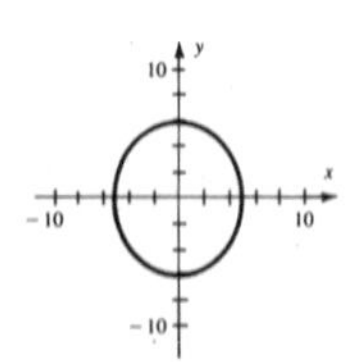

27. 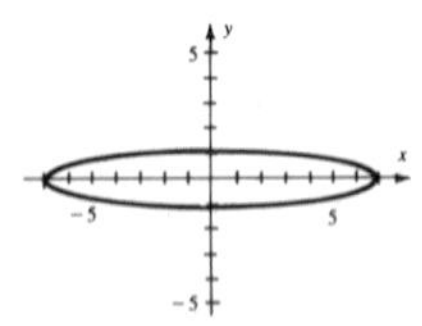

29. 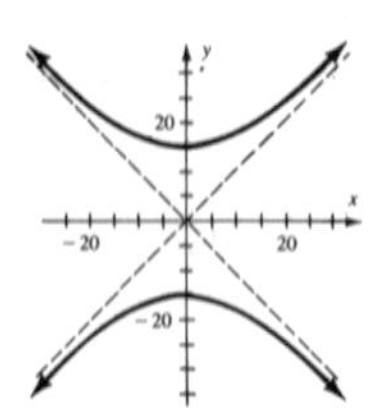

31. 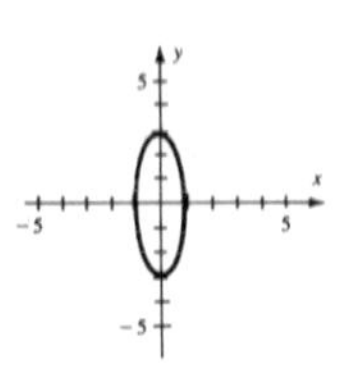

33. 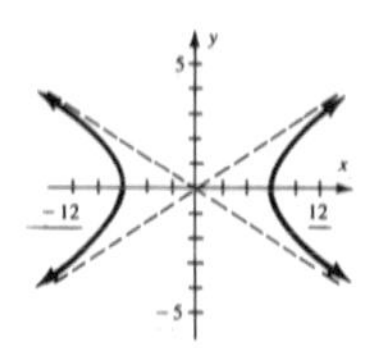

35. No graph

37. 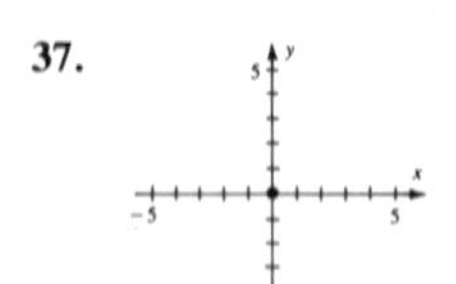

39. 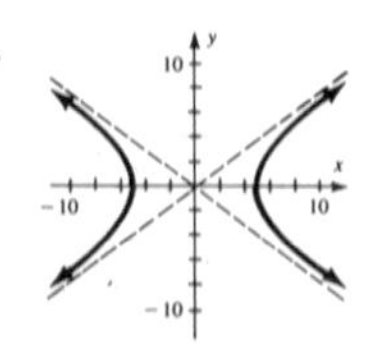

41. 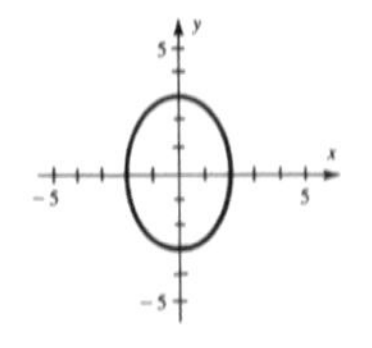

43. 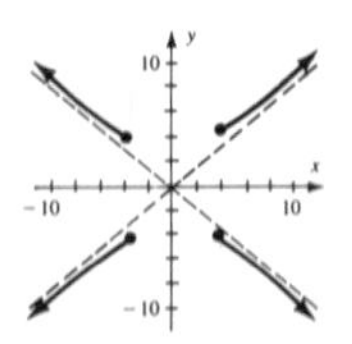

45. 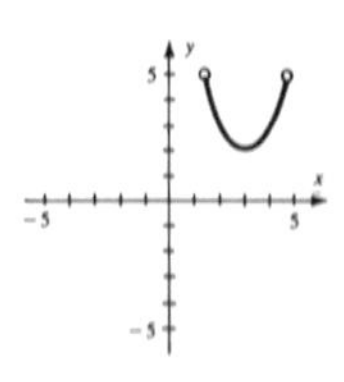

47. $\frac{x^2}{25} + \frac{y^2}{16} = 1$

49. 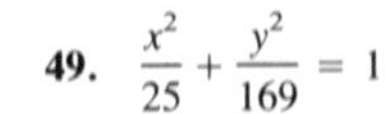
$\frac{x^2}{25} + \frac{y^2}{169} = 1$

51. Apogee is 94,500,000 miles; perigee is 91,500,000 miles.

53. $e = \dfrac{10}{189} \approx .053$

55. If $e = 0$, then

$$\sqrt{1 - \frac{b^2}{a^2}} = 0$$

$$1 - \frac{b^2}{a^2} = 0$$

$$\frac{b^2}{a^2} = 1$$

$$b^2 = a^2$$

$$b = a \quad \text{Since } a \text{ and } b \text{ represent distances}$$

Thus, the ellipse is a circle.

57. The asymptotes are $bx - ay = 0$ and $bx + ay = 0$; thus,

$$\left(\frac{|bx - ay|}{\sqrt{a^2 + b^2}}\right)\left(\frac{|bx + ay|}{\sqrt{a^2 + b^2}}\right) = \frac{|b^2x^2 - a^2y^2|}{a^2 + b^2} = \frac{a^2b^2}{a^2 + b^2} \text{ is a constant}$$

Problem Set 3.6, Pages 142–143

1.

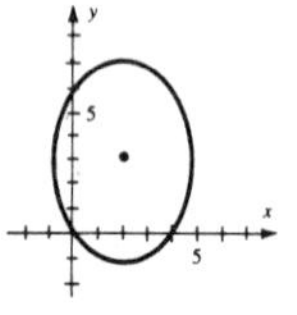

3.

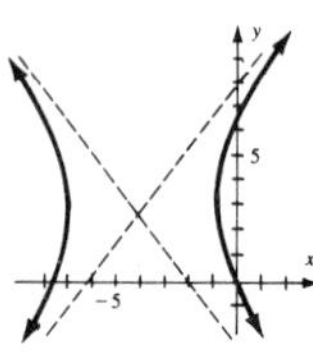

5.

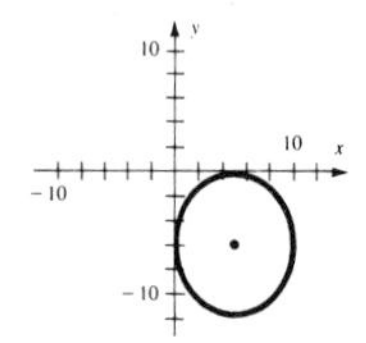

7.

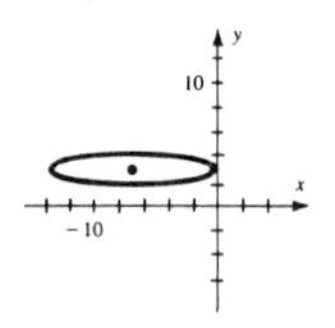

9.

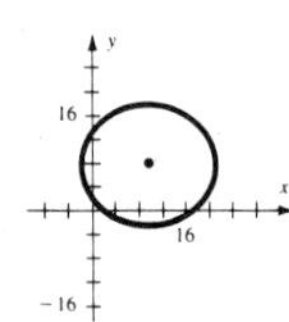

	Equation	Curve	Center	a	b
11.	$\dfrac{(x+5)^2}{16} + \dfrac{(y-3)^2}{9} = 1$;	ellipse;	$(-5, 3)$;	4;	3
13.	$\dfrac{(y-4)^2}{4} - \dfrac{(x+6)^2}{9} = 1$;	hyperbola;	$(-6, 4)$;	2;	3
15.	$\dfrac{(x-3)^2}{49} - \dfrac{(y+5)^2}{49} = 1$;	hyperbola;	$(3, -5)$;	7;	7
17.	$\dfrac{x^2}{4} + \dfrac{(y-2)^2}{25} = 1$;	ellipse;	$(0, 2)$;	2;	5
19.	$\dfrac{(x-1)^2}{9} - \dfrac{y^2}{4} = 1$;	hyperbola;	$(1, 0)$;	3;	2
21.	$\dfrac{(x+1)^2}{1} - \dfrac{(y+1)^2}{4} = 1$;	hyperbola;	$(-1, -1)$;	1;	2
23.	$\dfrac{(x-5)^2}{9} - \dfrac{(y+7)^2}{9} = 1$;	hyperbola;	$(5, -7)$;	3;	3
25.	$\dfrac{(x+3)^2}{4} - \dfrac{(y-2)^2}{16} = 1$;	hyperbola;	$(-3, 2)$;	2;	4

27. $\dfrac{(x-2)^2}{11} + \dfrac{(y-1)^2}{\frac{44}{9}} = 1$; ellipse; $(2, 1)$; $\sqrt{11}$; $\dfrac{2\sqrt{11}}{3}$

29. $\dfrac{(x+3)^2}{81} + \dfrac{(y+1)^2}{4} = 1$; ellipse; $(-3, -1)$; 9; 2

31. Let (x, y) be any point on the curve. Then, the slopes of the lines are

$$\frac{y}{x+m} \quad \text{and} \quad \frac{y}{x-m}$$

$$\left(\frac{y}{x+m}\right)\left(\frac{y}{x-m}\right) = -n^2$$

$$\frac{y^2}{x^2-m^2} = -n^2$$

$$y^2 = -n^2x^2 + m^2n^2$$

$$n^2x^2 + y^2 = m^2n^2$$

$$\frac{x^2}{m^2} + \frac{y^2}{m^2n^2} = 1$$

This is the equation of an ellipse centered at $(0, 0)$ with $a = m$ and $b = mn$.

33. If $G > F$, then $-F + G > 0$, so the curve is an ellipse since A and C are both positive.

35. If $G = F$, then $G - F = 0$ and A and C have the same signs, so the curve is a single point (a degenerate ellipse).

37. If $G < F$, then $-F + G < 0$ and A and C are both positive. Thus, it is a degenerate ellipse with no solution points.

Problem Set 3.7, Pages 144–147

1. ordered **2.** quadrants **3.** domain **4.** satisfy; coordinates **5.** horizontal **6.** distance; $x_2 - x_1$; $y_2 - y_1$
7. $\left(\dfrac{x_1+x_2}{2}, \dfrac{y_1+y_2}{2}\right)$ **8.** $\dfrac{\Delta y}{\Delta x}$ or $\dfrac{y_2-y_1}{x_2-x_1}$ **9.** equal **10.** perpendicular **11.** (h, k); slope **12.** x-axis
13. slope–intercept **14.** standard **15.** half-planes **16.** intersection **17.** parabola **18.** vertex; down; up
19. maximum **20.** $x - h = a(y-k)^2$; $a > 0$; $a < 0$ **21.** x-axis **22.** (y, x) **23.** $(x-h)^2 + (y-k)^2 = r^2$
24. plane; cone **25.** sum **26.** center **27.** $(a, 0)$; $(-a, 0)$ **28.** difference; constant **29.** x; $(a, 0)$; $(-a, 0)$
30. asymptotes **31.** 10 **32.** 10 **33.** 13 **34.** 17 **35.** 5 **36.** 13 **37.** 5 **38.** 17 **39.** $4\sqrt{17}$
40. 13 **41.** $(4, 4)$ **42.** $(4, 6)$ **43.** $(3, 2)$ **44.** $(4, -1)$ **45.** $(-2, 2)$ **46.** $(2, -2)$ **47.** $(0, \frac{1}{2})$
48. $(\frac{3}{2}, 1)$ **49.** $(-1, -1)$ **50.** $(2, -\frac{5}{2})$ **51.** 1 **52.** $\frac{1}{2}$ **53.** $-\frac{2}{3}$ **54.** $-\frac{2}{3}$ **55.** $-\frac{5}{6}$ **56.** $\frac{5}{4}$ **57.** $\frac{3}{4}$
58. $\frac{8}{15}$ **59.** $-\frac{1}{4}$ **60.** $-\frac{5}{12}$ **61.** $2x + 3y + 15 = 0$ **62.** $3x - 4y - 12 = 0$ **63.** $3x - 5y + 41 = 0$
64. $7x + 2y - 4 = 0$ **65.** $x - 6 = 0$ **66.** $y + 3 = 0$ **67.** $2x + 3y - 7 = 0$ **68.** $2x + 7y + 3 = 0$
69. $2x - 3y + 7 = 0$ **70.** $y - 1 = 0$

71.
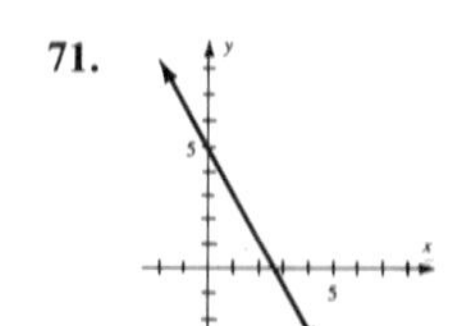

72.
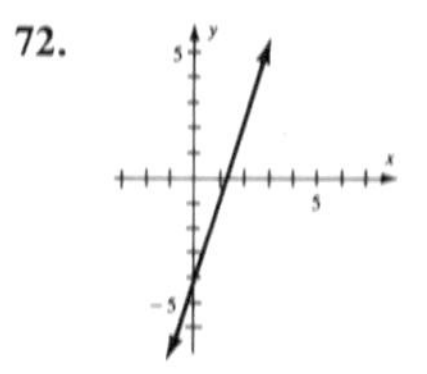

73.
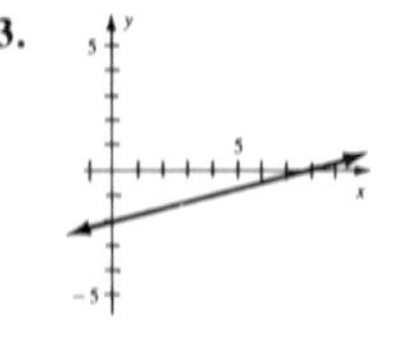

74.
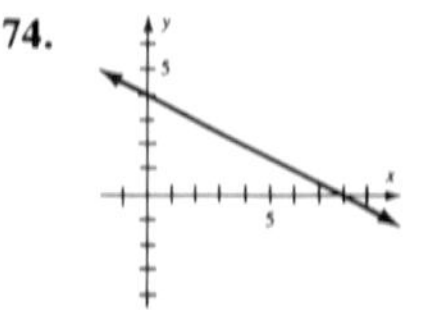

75.
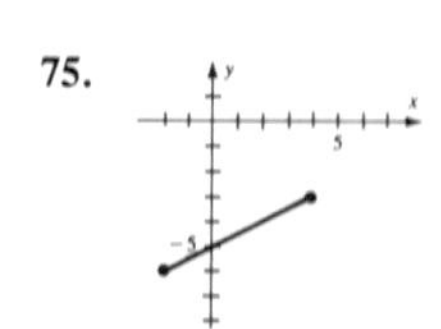

76.
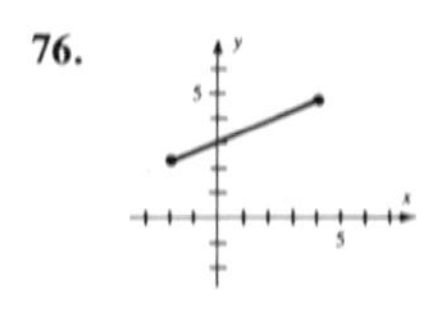

77.
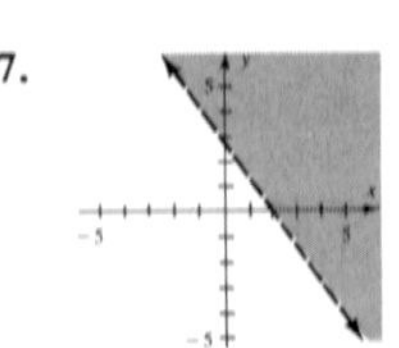

78.
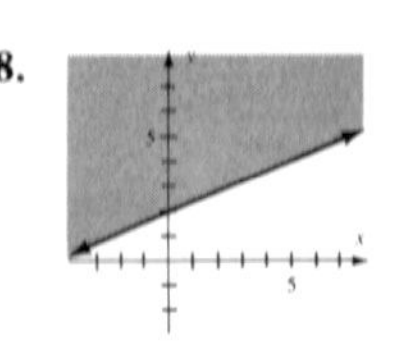

79. 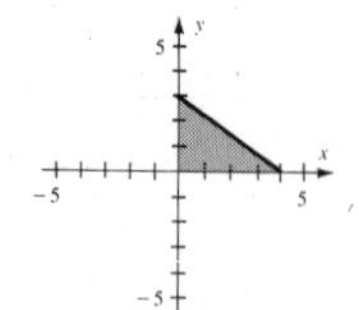

80. 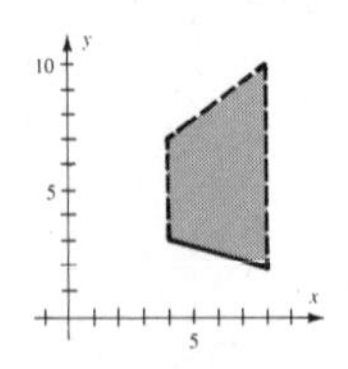

81. 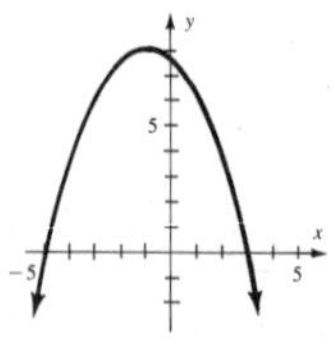

82. 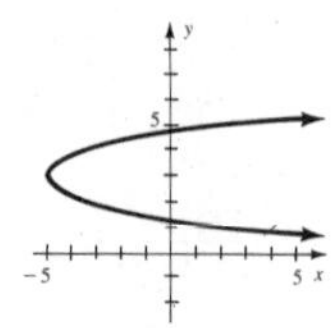

83. 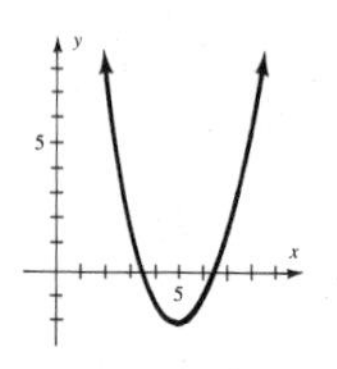

84. 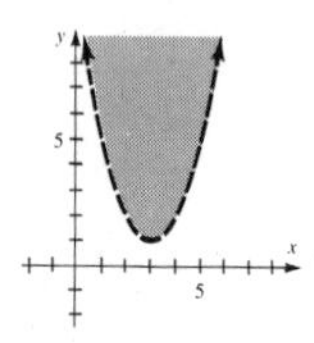

85. 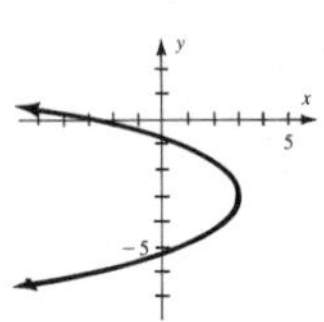

86. 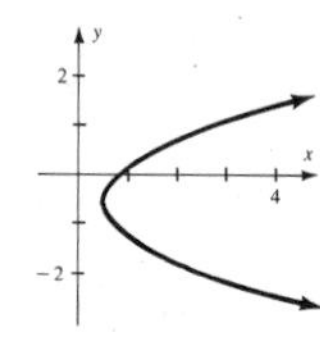

87. 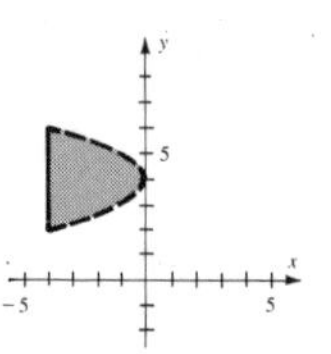

88. 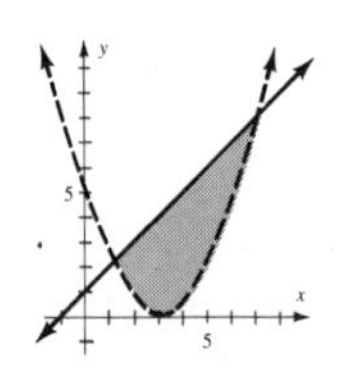

89. 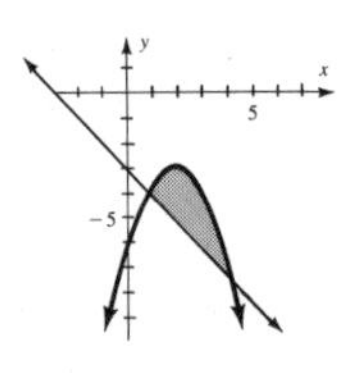

90. 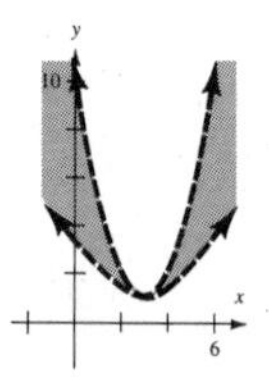

91. 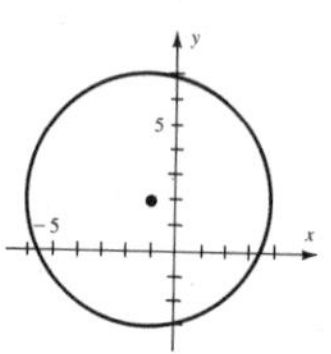

92. 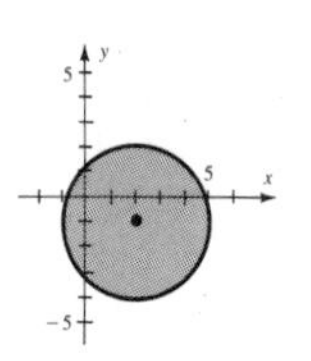

93. 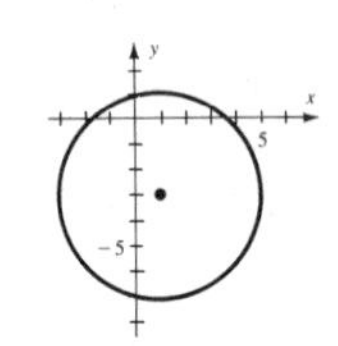

94. 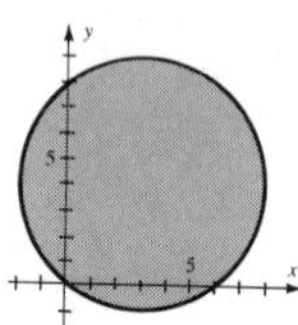

95. 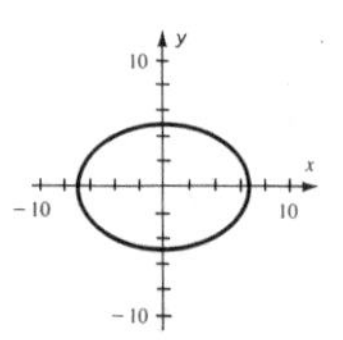

96. 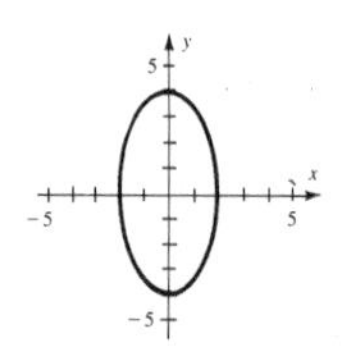

97. 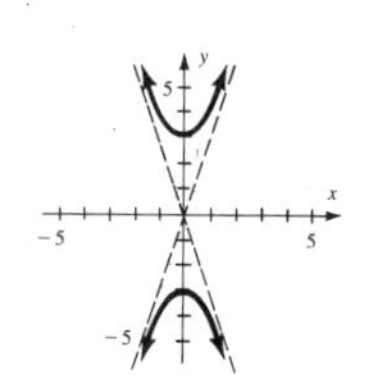

98. 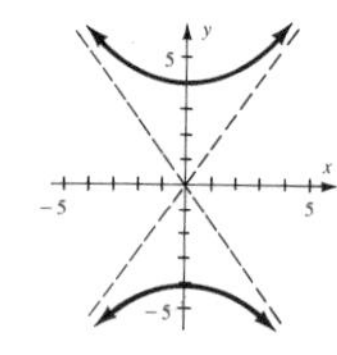

99. 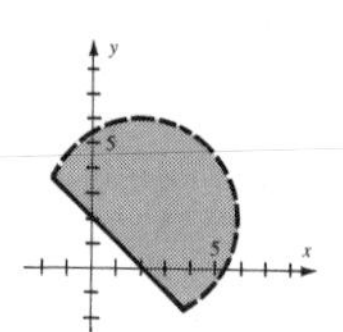

100. 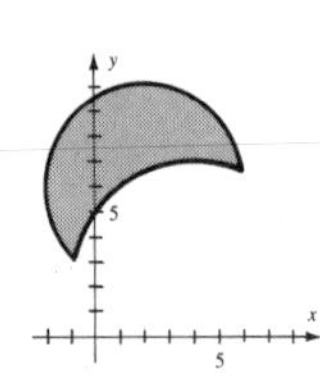

101. 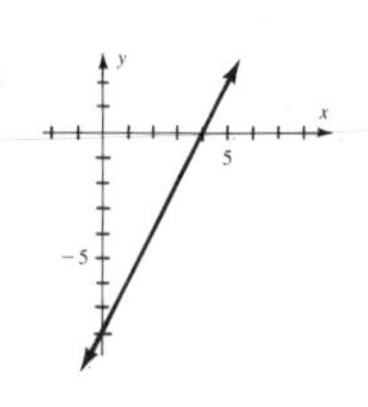

102. 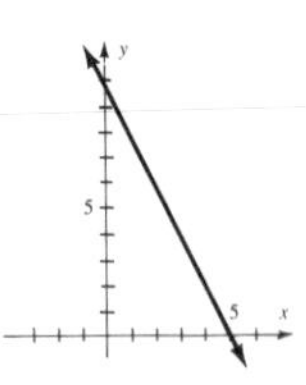

103. 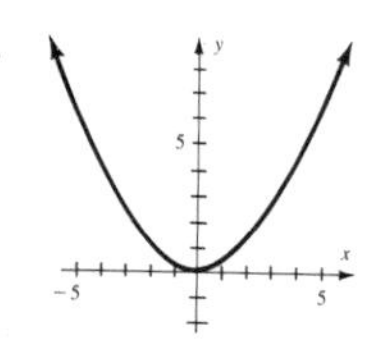

104. 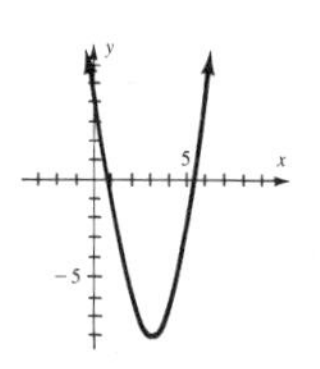

105. 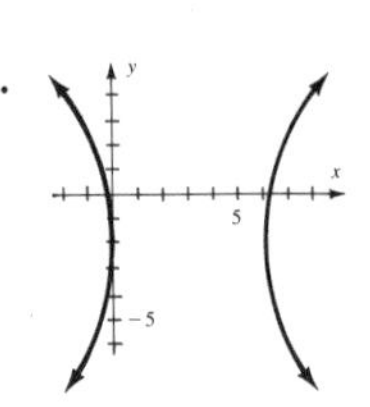

106.

107. 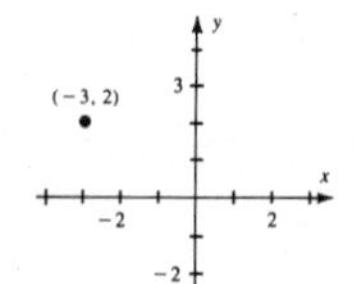

108. 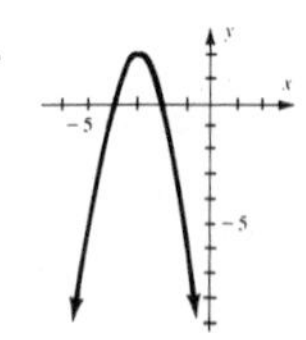

109. 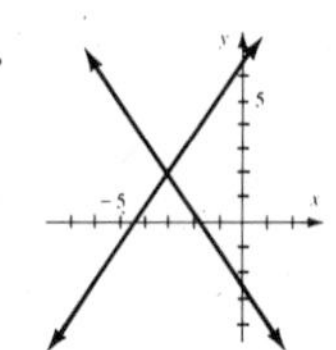

110. 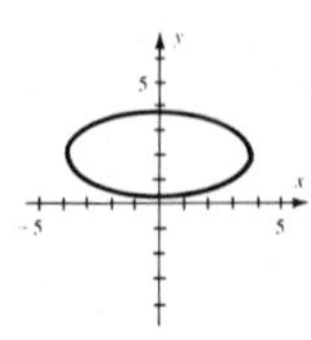

111. 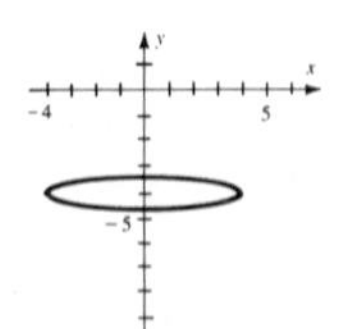

112. 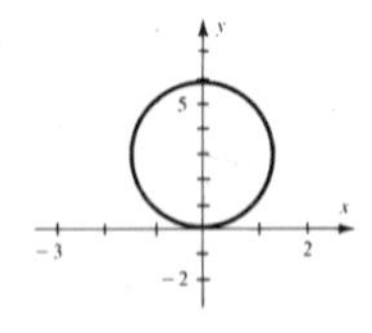

113. 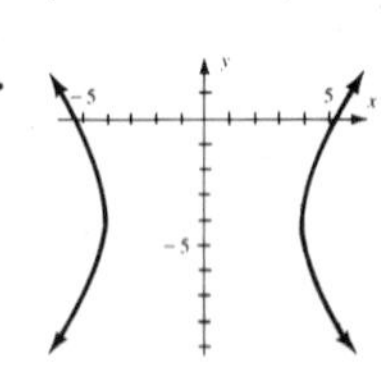

114. 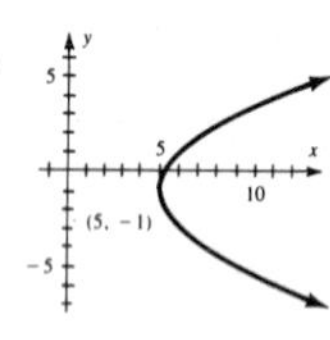

115. 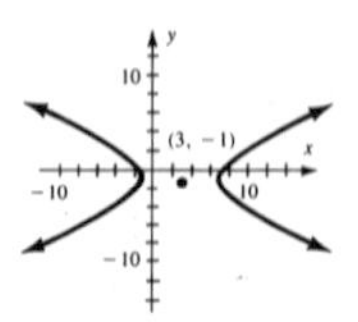

116. 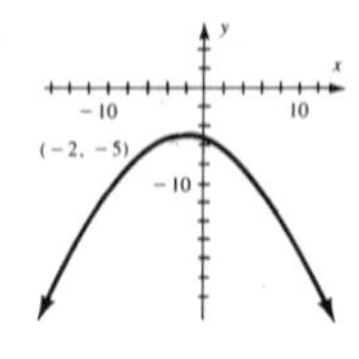

117. 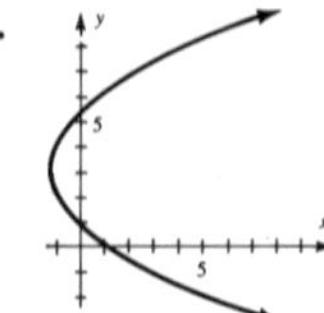

118. 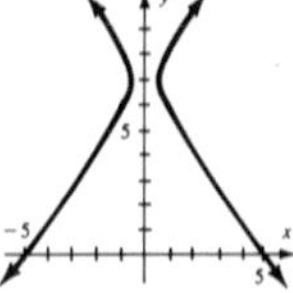
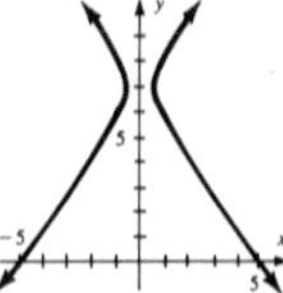

119. 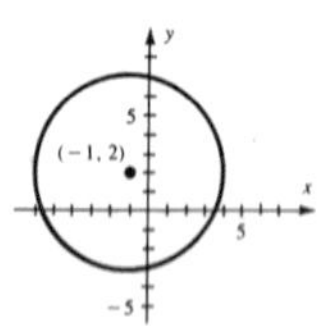

120. 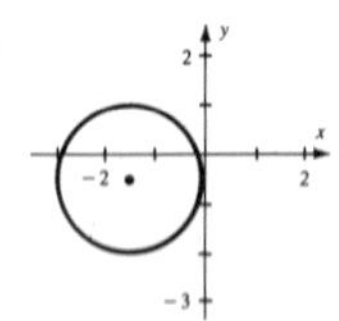

Chapter 4

Problem Set 4.1, Pages 153–155

1. **a.** 1 **b.** 5 **c.** -5 **d.** $2\sqrt{5}+1$ **e.** $2\pi+1$ **3.** **a.** $2w+1$ **b.** $2w^2-1$ **c.** $2t^2-1$ **d.** $2v^2-1$ **e.** $2m+1$
5. **a.** $3+2\sqrt{2}$ **b.** $5+4\sqrt{2}$ **c.** $2t^2+12t+17$ **d.** $2t^2+4t+3$ **e.** $2m^2-4m+1$ **7.** $4x+10$
9. $-6t^2-12t-9$ **11.** $-8x^2-26x-13$ **13.** $4t^2-4t+4$ **15.** 2 **17.** 2 **19.** 2 **21.** Function
23. Not a function **25.** Function **27.** Function **29.** Function **31.** Not a function **33.** -1 **35.** -2
37. 4 **39.** 15 **41.** 5 **43.** $10x+5h$ **45.** $2x+h$ **47.** $2x-1+h$ **49.** $\dfrac{-1}{x(x+h)}$
51. **a.** 1.07 **b.** .45 **c.** .21 **d.** 1.31 **e.** 1.49 **53.** **a.** $1.49-.34=1.15$ **b.** $1.29-.09=1.20$
55. **a.** $1.15-.64=.51$ **b.** $e(1984)-e(1944)$ **57.** $\dfrac{1.52-.21}{40}\approx .03$
59. The average yearly change in the price of gasoline between 1944 and 1984 **61.** 512 **63.** 80 **65.** $32hx+16h^2$

Problem Set 4.2, Pages 160–161

1. **a.** 1 **b.** 1 **c.** 1 **d.** -1 **e.** -1 **3.** **a.** 2 **b.** 28.09 **c.** π^2 **d.** -1 **e.** 5.3
5. **a.** 1 **b.** 5 **c.** 3 **d.** -1 **e.** -6 **7.** **a.** 2 **b.** 6.3 **c.** $\pi+1$ **d.** $\frac{3}{2}$ **e.** 6.3
9. **a.** 2 **b.** 10.6 **c.** 2π **d.** 0 **e.** 0

11. 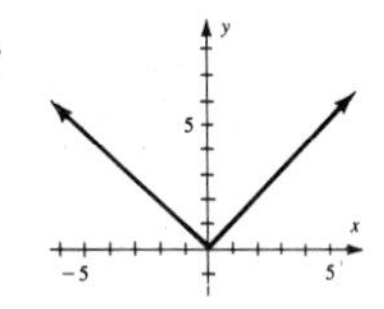
13. 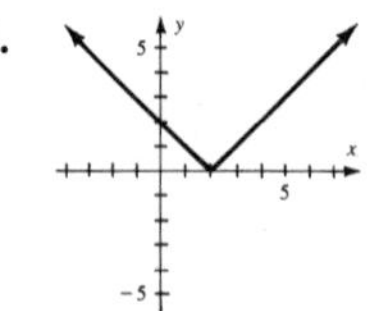
15. 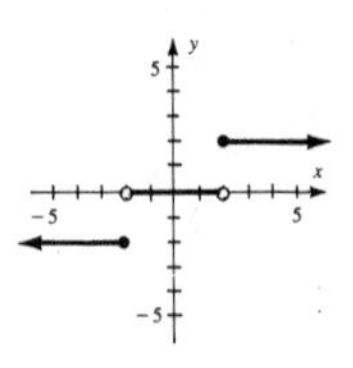
17. 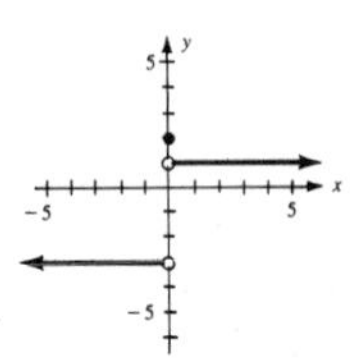

19. 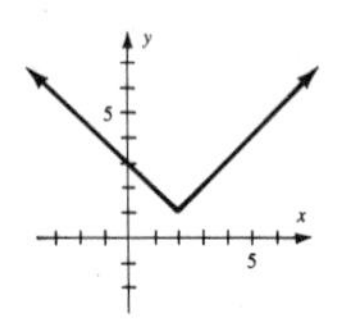
21. 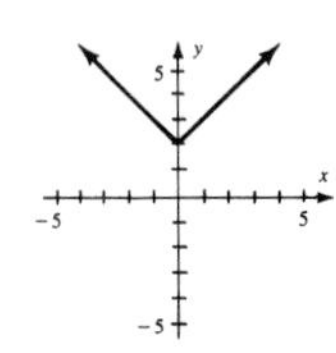
23. 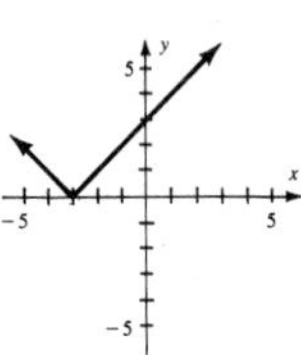
25. 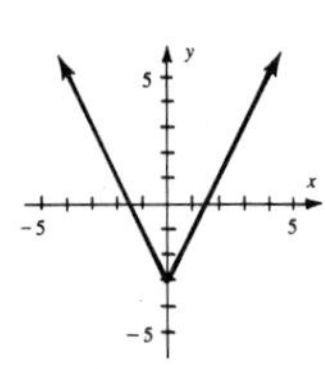

27. 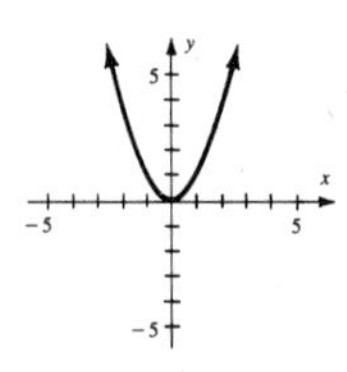
29. 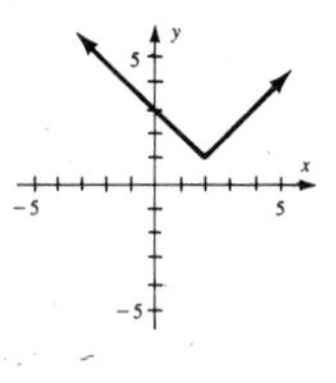
31. 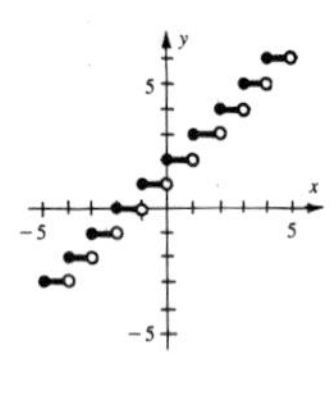
33. 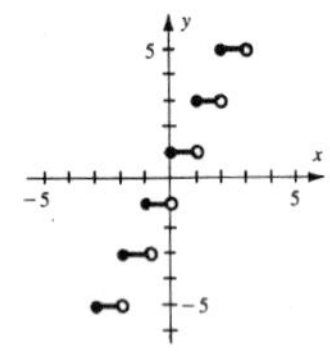

35. 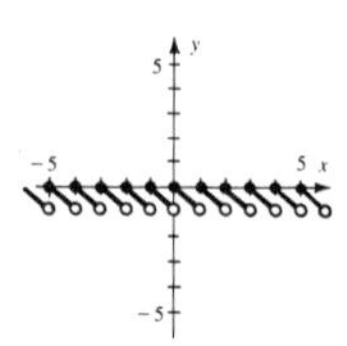
37. 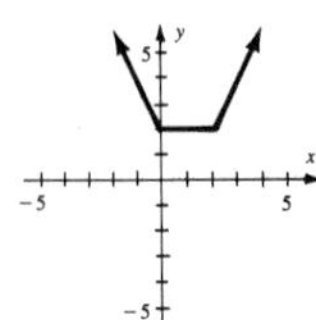
39. 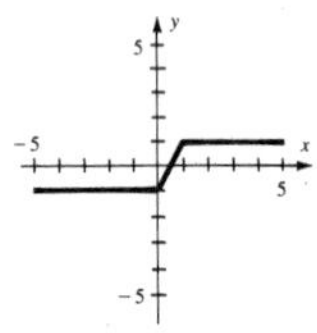

41.–45. Answers should vary. In our sample answers, $[\![\ ]\!]$ is the greatest integer function.

41. $c(x) = \begin{cases} .75 & \text{if } 0 < x \leq 3 \\ .75 - .25[\![3 - x]\!] & \text{if } x > 3 \end{cases}$

43. $m(x) = \begin{cases} .20 & \text{if } 0 < x \leq 1 \\ .20 - .17[\![1 - x]\!] & \text{if } x > 1 \end{cases}$

45. $t(x) = .80 - .10[\![-10x]\!]$

Problem Set 4.3, Pages 165–167

1. 33 **3.** 5 **5.** 7 **7.** $-\frac{35}{2}$ **9.** $\frac{1}{10{,}000}$ **11.** 0 **13.** 1500 **15.** $-\frac{7}{4}$ **17.** 7 **19.** $-\frac{1}{3}$ **21.** 7
23. 24 **25.** 3 **27.** 5 **29.** $\frac{49}{9}$ **31.** $x^2 + 2x - 2$ **33.** $2x^3 - 3x^2 + 2x - 3$ **35.** $\dfrac{x^3 - 2x - 4}{x + 1}$
37. $x^2 - 4x + 4$ **39.** $\dfrac{(x + 1)(x - 1)^2}{x - 2}$ **41.** $(2x - 3)(x + 1)^2$ **43.** $x^3 + 4x + 4$ **45.** $4x^4 + x^3 + 12x + 3$
47. $x^3 + x - 2$ **49.** $x^4 - x^3 - x + 1$
51. $f \cdot g = \{(0, 2), (1, 0), (2, 3), (3, 1)\}$; $g \cdot f = \{(0, 2), (1, 0), (2, 3), (3, 1)\}$
53. $(f \cdot g)(x) = x = (g \cdot f)(x)$ **55.** $(f \cdot g)(x) = x = (g \cdot f)(x)$
57. $(f \cdot g)(x) = 2x^2 - 1$; $(g \cdot f)(x) = 4x^2 - 12x + 10$
59. $(f \cdot g)(x) = x^4 - 2x^3 - 3x^2 + 4x + 4$; $(g \cdot f)(x) = x^4 - x^2 - 2$
61. **a.** $\dfrac{16\pi}{3}$ **b.** $\dfrac{2\pi t^3}{3}$ **c.** $0 < t \leq 3$ **63.** Examples vary **65.** Examples vary **67.** 1

Problem Set 4.4, Pages 171–173

1. One-to-one **3.** Not one-to-one **5.** One-to-one **7.** Not one-to-one **9.** One-to-one **11.** One-to-one
13. Not one-to-one **15.** Not one-to-one **17.** Not one-to-one **19.** Inverses **21.** Inverses **23.** Inverses
25. Not inverses **27.** Inverses **29.** Inverses **31.** $y = x - 2$ **33.** $y = \frac{1}{5}x + \frac{2}{5}$ **35.** $y = -\frac{5}{3}x + 5$
37. $y = \frac{3}{2}x - \frac{7}{2}$ **39.** $y = \sqrt[3]{x}$ **41.** $y = \sqrt[3]{x - 1}$

43.

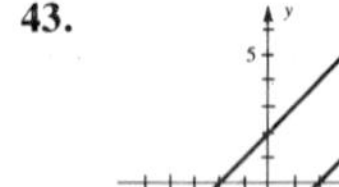

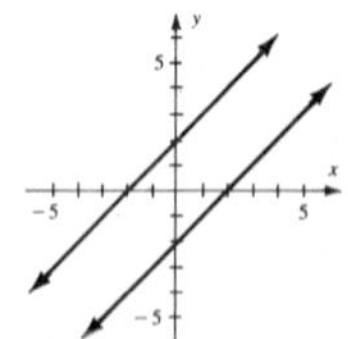

45.

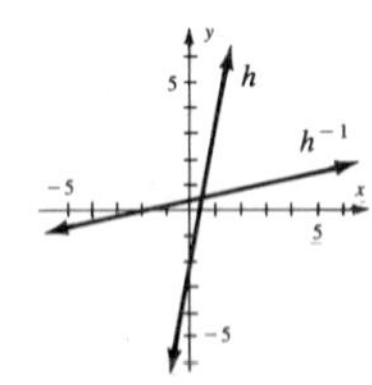

47.

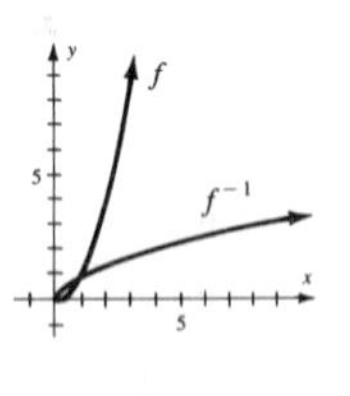

49.

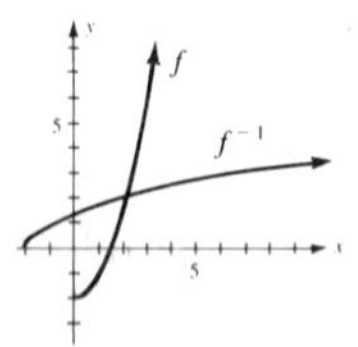

51.

53.

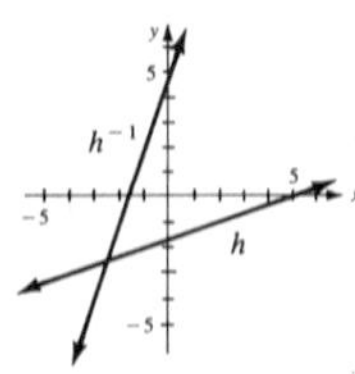

55.

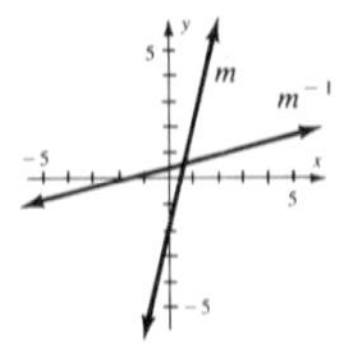

57. **a.** $A(x_0, f(x_0)); B(x_0 + \Delta x, f(x_0 + \Delta x))$ **b.** $\dfrac{f(x_0 + \Delta x) - f(x_0)}{\Delta x}$

59. **a.** 4 **b.** 3 **c.** 2.5 **d.** 2.25 **e.** 2.1 **f.** 2.01 **g.** 2.001
61. **a.** 3 **b.** 2.5 **c.** 2.25 **d.** 2.1 **e.** 2.01 **f.** 2.001 **g.** 2.0001

Problem Set 4.5, Pages 175–176

1. $x = ky$ **3.** $s = kt^2$ **5.** $A = k\ell w$ **7.** $V = kr^3$ **9.** $C = \dfrac{kt}{r}$ **11.** $V = kr^3$ **13.** $I = \dfrac{k}{R}$ **15.** $i = kPr$

17. $F = \dfrac{k}{d^2}$ **19.** $V = \dfrac{kT}{P}$ **21.** $E = kmv^2$ **23.** $A = ktRI^2$ **25.** $w = 21$ **27.** $p = \frac{5}{4}$ **29.** $A = 60$

31. 288π cm^3 **33.** 20 amps **35.** 1200 lb/ft^2 **37.** 3 liters

39. $\dfrac{f(x_1)}{f(x_2)} = \dfrac{x_2}{x_1}$

$x_1 f(x_1) = x_2 f(x_2)$ Thus, $x\,f(x) = k$, where k is a constant; $f(x) = \dfrac{k}{x}$; and $f(x)$ varies inversely as x.

Problem Set 4.6, Pages 177–180

1. a rule that assigns a single number $f(x)$ to each x in the domain of the function
2. a function; a member of the range; a member of the domain **3.** $f(x) = a$ **4.** $g(x) = mx + b$
5. $h(x) = ax^2 + bx + c, a \neq 0$ **6.** $F(x) = |x|$ **7.** $n \leqslant x < n + 1$; greatest integer function **8.** $f(x) + g(x)$
9. $f(x) - g(x)$ **10.** $f(x) \cdot g(x)$ **11.** $\dfrac{f(x)}{g(x)}$ **12.** $f[g(x)]$ **13.** $(f \cdot g)(x) = x$; $(g \cdot f)(x) = x$ **14.** reversing
15. directly as x **16.** inversely as x **17.** p varies directly as the cube of x and inversely as the square of y **18.** 12
19. 9 **20.** 47 **21.** 23 **22.** $5t - 3$ **23.** $2t^2 - 9$ **24.** $2t^2 + 4t - 7$ **25.** $5t + 2$ **26.** $5x + 5h - 3$
27. $2x^2 + 4xh + 2h^2 - 9$ **28.** $2h^2 - 12h + 9$ **29.** $22 - 5h$ **30.** $15x - 9$ **31.** $15x - 3$ **32.** $10w - 1$
33. $10w + 44$ **34.** $-5h$ **35.** $-4xh + 2h^2$ **36.** $2th + h^2$ **37.** h **38.** 5 **39.** $4x + 2h$

40. 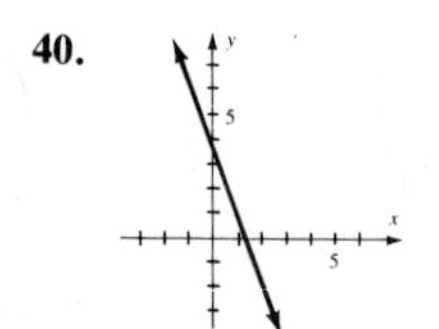 **41.** 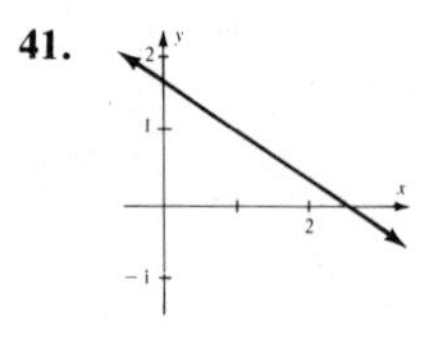 **42.** 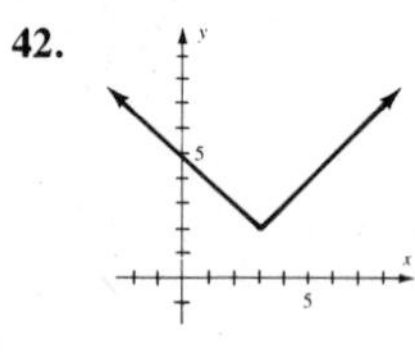 **43.** 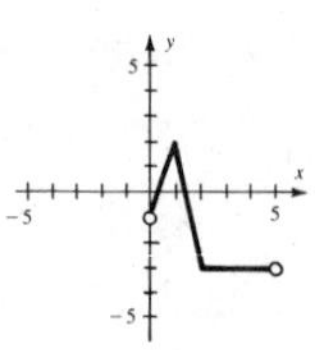

44. 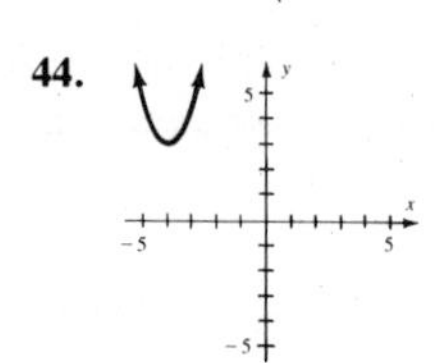 **45.** 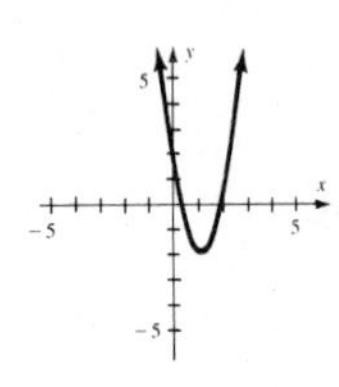 **46.** 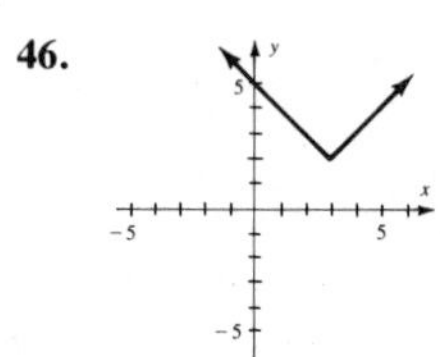 **47.** 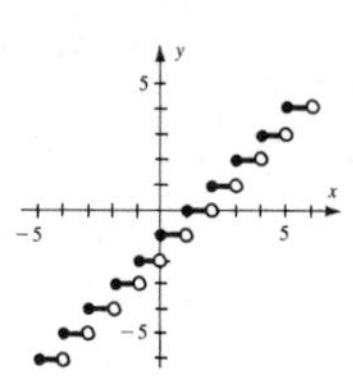

48. $m(x) = \begin{cases} .05x & \text{if } 0 < x < 30 \\ 1.50 & \text{if } x \geq 30 \end{cases}$

49. 26 **50.** 6 **51.** -12 **52.** $\frac{11}{2}$ **53.** 53 **54.** 59 **55.** $2x^2 + 3x - 1$ **56.** $2x^2 - 3x + 7$

57. $6x^3 - 8x^2 + 9x - 12$ **58.** $\dfrac{2x^2 + 3}{3x - 4}$ **59.** $18x^2 - 48x + 35$ **60.** $6x^2 + 5$ **61.** 9 **62.** 3 **63.** 18

64. $\frac{1}{2}$ **65.** 2 **66.** 5 **67.** 11 **68.** $\dfrac{11x - 4}{2}$ **69.** $\dfrac{9x - 12}{2}$ **70.** $\dfrac{5x^2 + 16x - 16}{2}$ **71.** $\dfrac{10x - 8}{x + 4}$

72. $\dfrac{x + 4}{10x - 8}$ **73.** $\dfrac{5x}{2}$ **74.** $25x - 24$ **75.** 5 **76.** $\frac{1}{2}$ **77.** Not inverses **78.** Inverses **79.** Not inverses

80. Inverses **81.** Not inverses **82.** Not inverses **83.** Inverses **84.** Not inverses **85.** Not inverses

86. Inverses **87.** $f^{-1}(x) = \frac{1}{2}x + \frac{5}{2}$ **88.** $f^{-1}(x) = \frac{1}{3}x - \frac{2}{3}$ **89.** $g^{-1}(x) = \dfrac{3}{x} - 1$ **90.** $g^{-1}(x) = \dfrac{7}{x} - 3$

91. $h^{-1}(x) = 6 - 2x$ **92.** $h^{-1}(x) = 6 - 3x$ **93.** $x = ky^2$ **94.** $z = \dfrac{k}{w}$ **95.** $x = \dfrac{ky}{z}$ **96.** $P = kt^3$

97. $T = \dfrac{k}{s^2}$ **98.** $B = \dfrac{kx^2}{y^3}$ **99.** 80 **100.** 10 **101.** $\frac{12}{5}$ **102.** 125 **103.** $\frac{20}{9}$ **104.** 600 kg/cm^2

Chapter 5

Problem Set 5.1, Pages 186–188

1. $3x^2 + 7x + 25$ **3.** $x^3 - 7x^2 + 8x - 8$ **5.** $2x^3 - 6x^2 + 3x - 1$ **7.** $4x^4 + x^3 - 4x^2 - 4x - 4$ **9.** $3x^2 + 4x + 12$
11. $x^2 + 3x + 2$ **13.** $x^2 - 1$ **15.** $x^2 - x - 2$ **17.** $x^2 - 2x - 15$ **19.** $x^2 + x - 6$ **21.** $x^2 - 5x - 14$

23. $2x^2 + x + 6 + \dfrac{2}{x - 2}$ **25.** $x^3 - 3x^2 - 4x - 9 + \dfrac{-15}{x - 2}$ **27.** $2x^3 + x^2 + 5 + \dfrac{-8}{x + 2}$

29. $x^3 + 5x^2 + 5x + 15 + \dfrac{25}{x - 5}$ **31.** $6x^2 - x - 2$ **33.** $6x^2 + 14x - 12$ **35.** $6x^2 + 21x + 9$ **37.** $4x^3 - 3x - 1$

39. $4x^3 - 6x^2 + 2$ **41.** $6x^3 - 17x^2 - 4x + 3$ **43.** $6x^3 + 19x^2 + 2x - 3$ **45.** $6x^3 - 2x^2 - 54x + 18$

47. $6x^3 + 3x^2 - 54x - 27$ **49.** $6x^3 + 3x^2 - 6x - 3$ **51.** $\dfrac{(2x + 1)(x + 1)}{x + 2}$ **53.** $\dfrac{(2x + 1)(x - 3)}{2x - 3}$

55. $\dfrac{2x^3 + 11x^2 + 17x + 6}{x + 4}$ **57.** $\dfrac{(3x - 2)(x + 3)}{x + 2}$ **59.** $\dfrac{6x^3 - 13x^2 + x + 2}{x - 3}$ **61.** $k = -4$ **63.** $m = 6$ **65.** $k = -69$

67. $m = -2$ or 4

69.

	1	$-m$	2	$-8k$
1	1	$1 - m$	$3 - m$	$3 - m - 8k$
-2	1	$-2 - m$	$6 + 2m$	$-12 - 4m - 8k$

yields the system: $\begin{cases} 8k + m - 3 = 0 \\ 8k + 4m + 12 = 0 \end{cases}$ Since each remainder must be zero

Thus, $m = -5$, $k = 1$

Problem Set 5.2, Page 193

1. $f(1) = -5$; $f(2) = -3$; $f(-1) = -9$; $f(-3) = -53$; $f(10) = 805$ **3.** $h(-1) = 12$; $h(-2) = 103$; $h\left(\frac{1}{2}\right) = 3$
5. $m(2) = 74$; $m\left(-\frac{2}{3}\right) = 2$; $m\left(-\frac{1}{2}\right) = 1.5$ **7.** $f(2) = 0$; $f(-3) = 0$; $f\left(\frac{5}{2}\right) = 0$; $f(1) = 12$; $f(-2) = 36$
9. $h(-1) = 0$; $h(1) = 12$; $h(-2) = 0$; $h(2) = 60$; $h\left(\frac{1}{3}\right) = 0$

11.
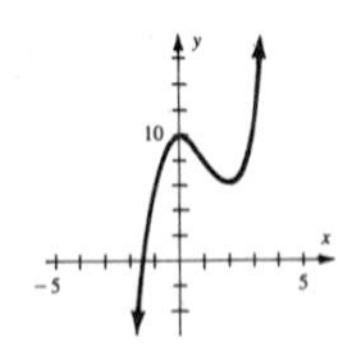

13.
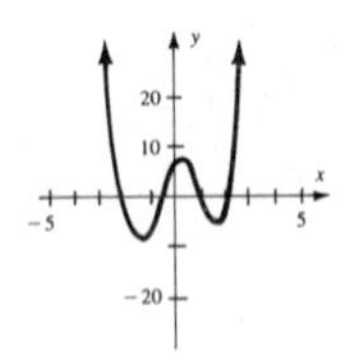

15.
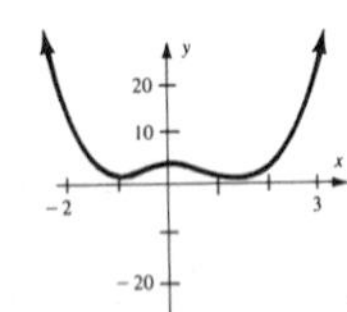

17.
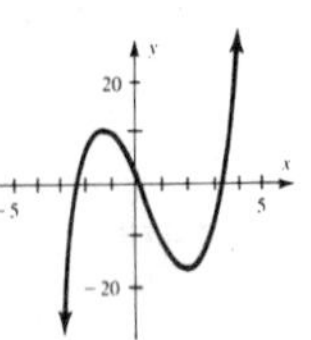

19.
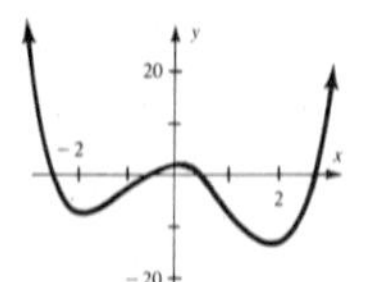

21.
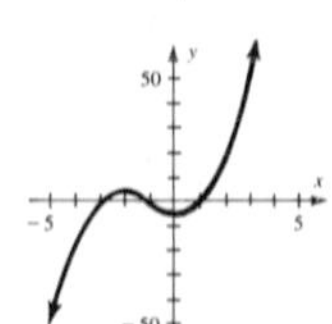

23.
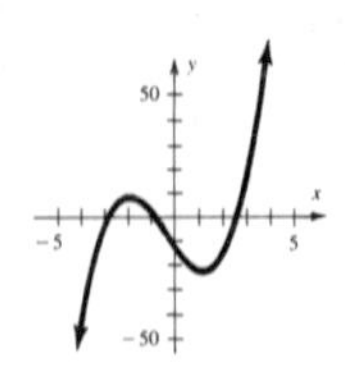

25.
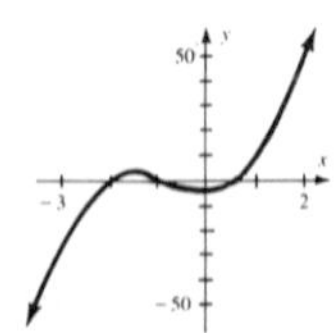

27.
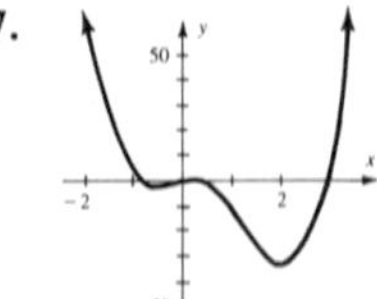

29.
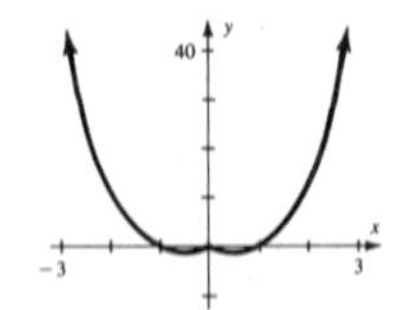

31.
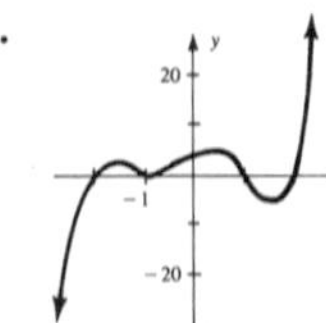

33.
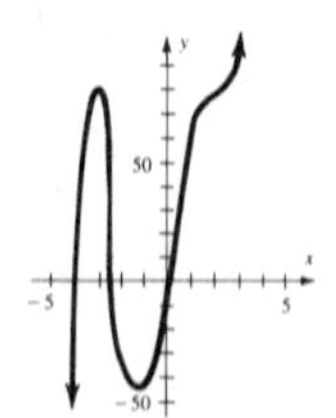

Problem Set 5.3, Pages 198–199

1. $\left\{-1, \frac{1}{2}, -\frac{1}{3}\right\}$ **3.** $\{-2, 3, -1$ (multiplicity 2)$\}$ **5.** $\{1, -2, 3, -4\}$ **7.** $\{1, -1, 2 \pm \sqrt{3}\}$ **9.** $\{1$ (multiplicity 4)$\}$
11. $\{1, -2, 3, -4\}$ **13.** $a < -\frac{5}{4}$ or $\frac{1}{2} < a < 2$ **15.** $\left\{3, \dfrac{1 + i\sqrt{23}}{12}\right\}$ **17.** $p \leq -1$ or $\frac{3}{2} \leq p \leq 4$
19. $\{1, -1, 2, -2, -3\}$ **21.** $\left\{-2, -\frac{2}{3}, -1 \pm \sqrt{3}\right\}$ **23.** $\left\{-1, \frac{1}{2}, \dfrac{1 \pm i\sqrt{3}}{2}\right\}$ **25.** $-\frac{1}{2} < x < \frac{1}{3}$ or $x > 6$ **27.** $\left\{\frac{1}{2}, \frac{2}{3}, -\frac{3}{2}\right\}$
29. $x < -1$ or $-\frac{3}{5} < x < \frac{1}{2}$ **31.** All possibilities ± 1, ± 2, ± 4 fail. **33.** All possibilities ± 1, ± 3 fail.
35. 7 ft × 8 ft × 9 ft **37.** 3 in. or 6 in. **39.** 8 cm, 15 cm, 17 cm **41.** $\left\{\frac{1}{2}, -\frac{1}{3}\text{(multiplicity 2)}, \pm i\sqrt{2}\right\}$
43. $x < -\frac{3}{4}$ or $x < \frac{1}{4}$ **45.** $\left\{2, -3, \frac{1}{2}, -\frac{1}{2}, \frac{2}{3}\right\}$

Problem Set 5.4, Pages 203–205

1. 1 positive; 0 negative **3.** 3 or 1 positive; 0 negative **5.** 2 or 0 positive; 1 negative **7.** $-24 + 4i$ **9.** $24 - 14\sqrt{2}$
11. $12 + 16i$ **13.** $-30 + 6i$ **15.** $62 - 38\sqrt{3}$ **17.** 0 **19.** Not a root **21.** $1 + 2i$ **23.** $\{3i, -3i, 2i, -2i\}$
25. $\{2 + i, 2 - i, \sqrt{2}, -\sqrt{2}\}$
27. $x^n - 1$ has one change of sign and thus, 1 positive real root; $(-x)^n - 1$ has no change if n is odd and 1 if n is even; hence, 1 real root if n is odd and 2 if n is even
29. $f(-x)$ has two changes in sign and thus, there are 2 or 0 negative real roots; $f(0) = 2$, $f(-1) = -3$, thus, by the Location Theorem there is at least 1 root between -1 and 0; hence, there are exactly 2 negative real roots
31. Descartes' Rule shows 1 positive real root and 1 negative real root; the Fundamental Theorem ensures 6 roots, so the remaining 4 must be complex and nonreal. **33.** $\left\{-1, \frac{3}{2}, \pm\sqrt{5}\right\}$ **35.** $\left\{\frac{3}{2}, -3 \pm \sqrt{2}, \pm i\sqrt{2}\right\}$ **37.** $\{-2, \pm i, 2 \pm 3i\}$
39. $\{1 \pm 2i$ (multiplicity 2)$\}$
41. No; $f(3 + \sqrt{2}) = 0$; $f(3 - \sqrt{2}) = -6\sqrt{2}$; the Conjugate Pair Theorem requires rational coefficients

43. Case-by-case, Descartes' Rule shows that there is at least 1 negative real root.
45. For $k \leqslant n$, if there are k changes in sign in $f(x)$, then there are at most k positive real roots (and at most $n - k$ negative real roots). Thus, there can be at most $n - k$ changes in sign in $f(-x)$. If all n roots are real, then there must be exactly k positive real roots.

Problem Set 5.5, Pages 210–211

1. 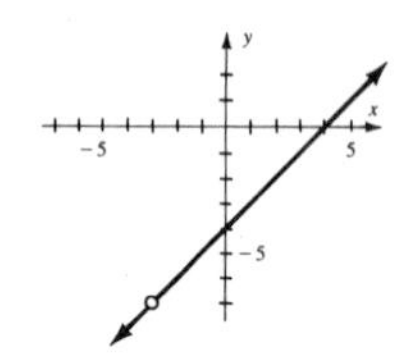

3. 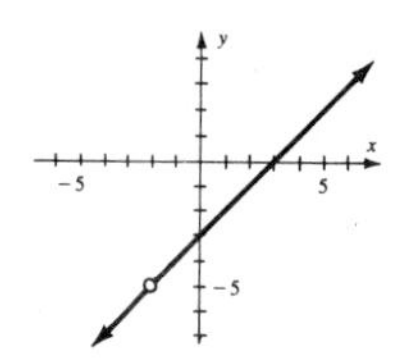

5. 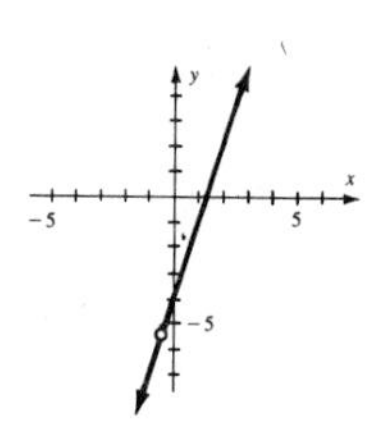

7. 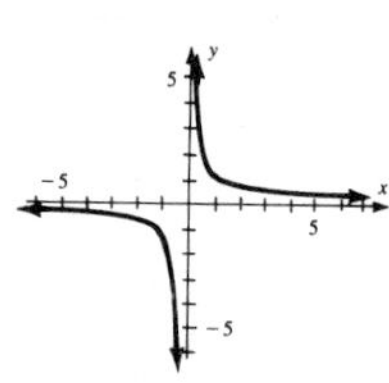

9. 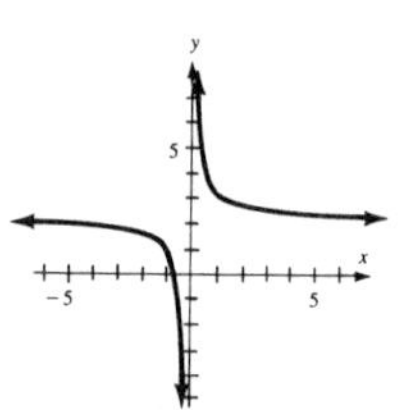

11. 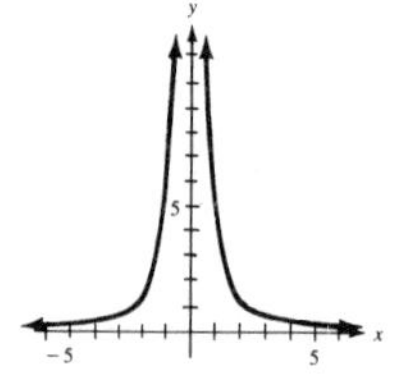

13. 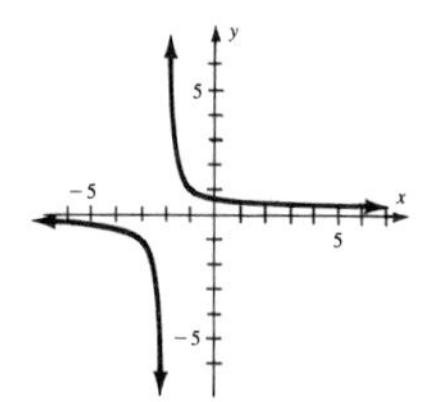

15. 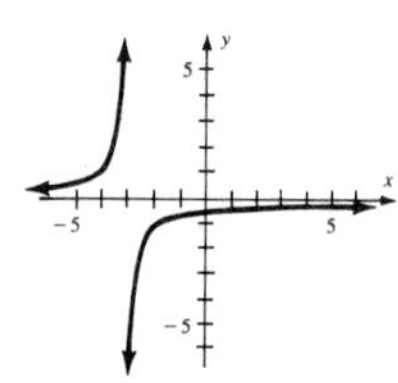

17. 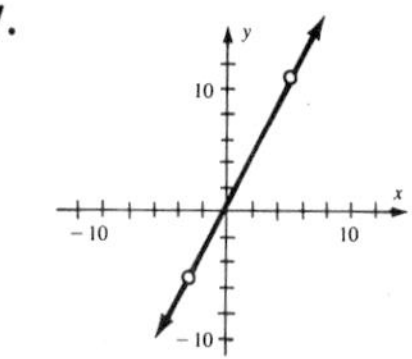

19. 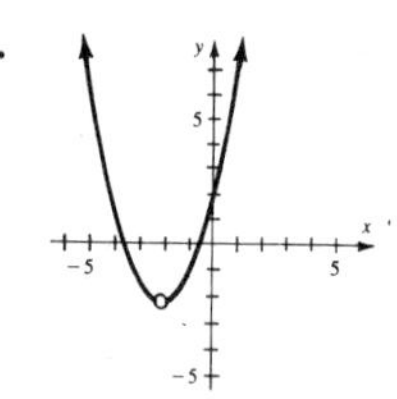

21. 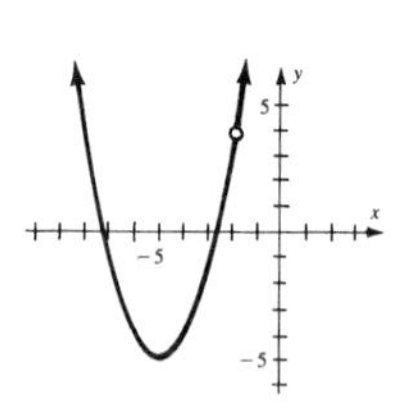

23. 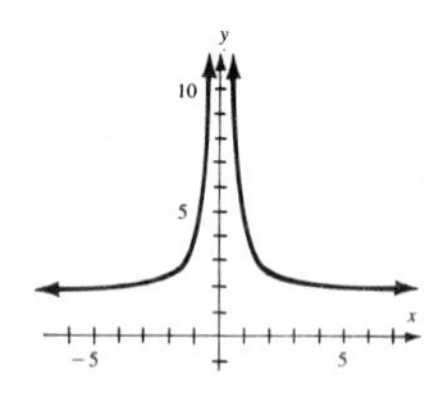

25. 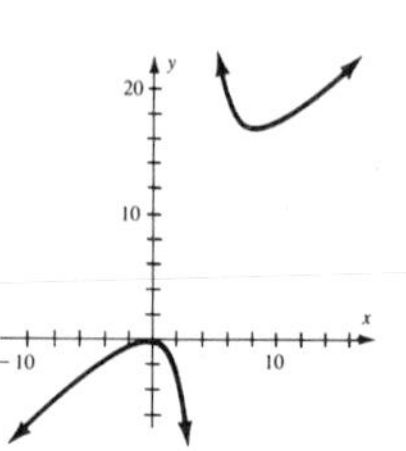

27. 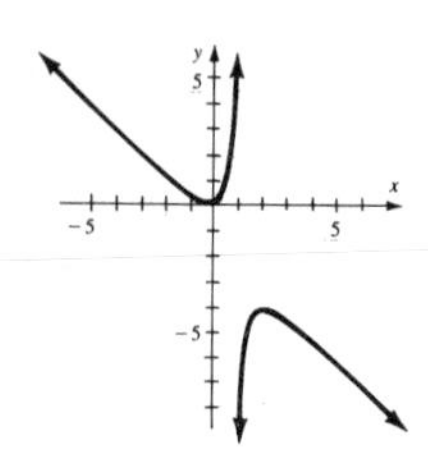

29. 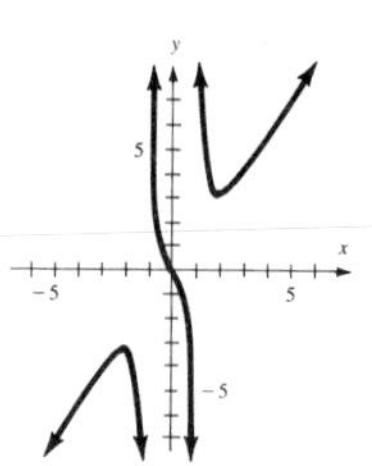

31. 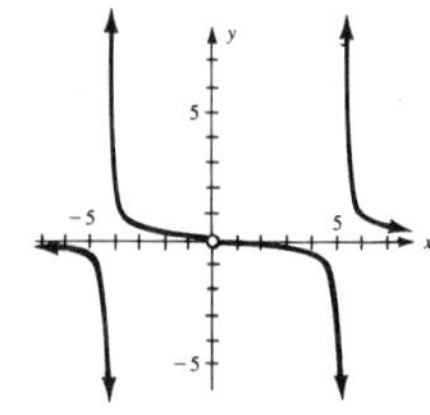

33. 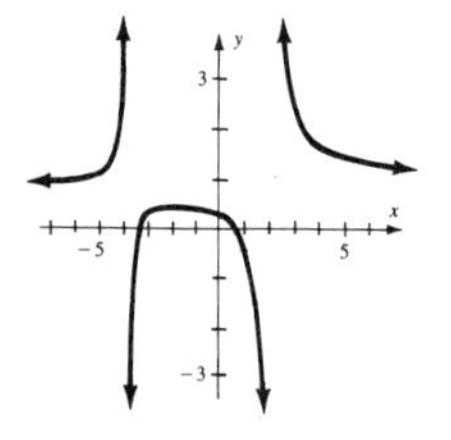

Problem Set 5.6, Page 215

1. $\frac{1}{x} + \frac{2}{x^2} + \frac{5}{x^3}$ **3.** $\frac{2}{x} - \frac{5}{x^2} + \frac{4}{x^3}$ **5.** $\frac{1}{x+4} - \frac{1}{x+5}$ **7.** $\frac{4}{x-2} + \frac{3}{x-1}$ **9.** $\frac{2}{x+2} + \frac{5}{x-4}$ **11.** $\frac{4}{x+3} - \frac{2}{x-2}$
13. $\frac{1}{x-2} + \frac{3}{x+2}$ **15.** $\frac{3}{x-4} - \frac{2}{x-5}$ **17.** $\frac{3}{x} - \frac{3}{x-1} + \frac{4}{x+1}$ **19.** $\frac{2}{x-2} + \frac{3}{(x-2)^2}$ **21.** $\frac{1}{x} + \frac{3}{(x+1)^2}$
23. $\frac{2}{x+1} + \frac{3}{(x+1)^2} - \frac{3}{(x+1)^3}$ **25.** $\frac{5}{6(x+5)} + \frac{1}{6(x-1)}$ **27.** $\frac{13}{3(x-2)} + \frac{8}{3(x+1)}$ **29.** $\frac{1}{x} + \frac{3}{x+3} - \frac{4}{x-2}$
31. $\frac{3}{x-1} + \frac{2x-4}{x^2+1}$ **33.** $x + 2 + \frac{3}{x-1} + \frac{1}{(x-1)^2}$ **35.** $\frac{1}{1-x} - \frac{3}{1+x} + \frac{2x+1}{1+x^2}$

Problem Set 5.7, Pages 216–219

1. positive integer or exponent; quadratic function; linear function; constant function **2.** depressed equation **3.** factor
4. $P(x) = R$ **5.** a real root between a and b **6.** a_0; a_n
7. the sums are the same sign in the synthetic division of $P(x)$ by $x - a$; $a \geq 0$
8. the sums alternate in sign in the synthetic division of $P(x)$ by $x - b$; $b \leq 0$ **9.** a root of multiplicity 3
10. equal to the number of variations in sign in $P(x)$ or is less than that number by an even counting number
11. equal to the number of variations in sign in $P(-x)$ or is less than that number by an even counting number **12.** n
13. $x - 2 - \sqrt{5}$ **14.** rational **15.** the quotient of polynomial functions $P(x)$ and $D(x)$ so that $f(x) = \frac{P(x)}{D(x)}$, $D(x) \neq 0$
16. $m < n$ **17.** the denominator is 0 **18.** horizontal; $m = n$ **19.** slant; $m = n + 1$ **20.** linear; quadratic
21. $x^2 + 2x + 5$ **22.** $x^3 - x^2 + 2x - 5$ **23.** $x^2 + 2x + 3$ **24.** $x^2 - 3x + 2$ **25.** $4x^2 - 2x + 6$
26. $9x^2 + 6x + 12$ **27.** $3x^2 + x - 2 + \frac{2}{x+2}$ **28.** $4x^2 + 2x - 2 + \frac{1}{x-2}$ **29.** $5x^2 - 3x + 4 + \frac{2}{x-3}$
30. $3x^3 + 2x^2 - 3x - 4 + \frac{-12}{x-2}$ **31.** 6 **32.** 0 **33.** -60 **34.** 0 **35.** 0 **36.** 0 **37.** -8 **38.** $-8 + 4\sqrt{2}$
39. $-8 - 4\sqrt{2}$ **40.** $-5 + 4i$ **41.** $16 + 8i$ **42.** $16 - 8i$

43.
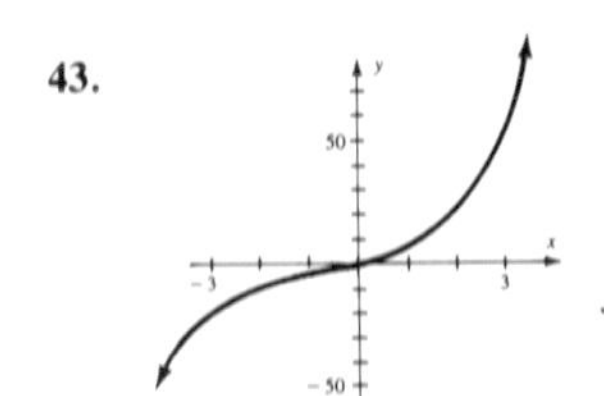

44.
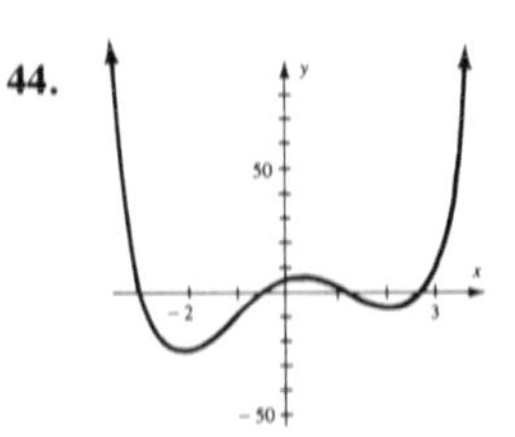

45.
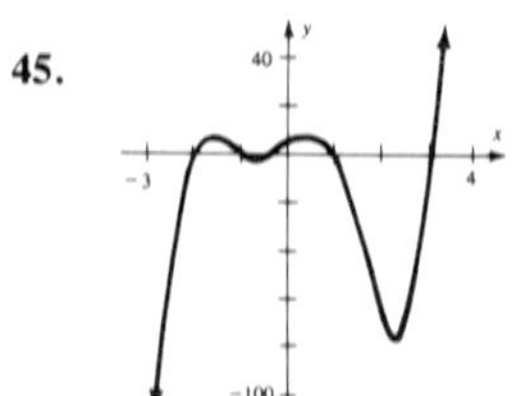

46.
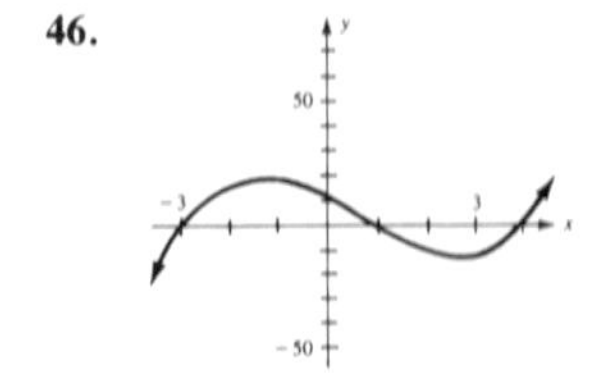

47.
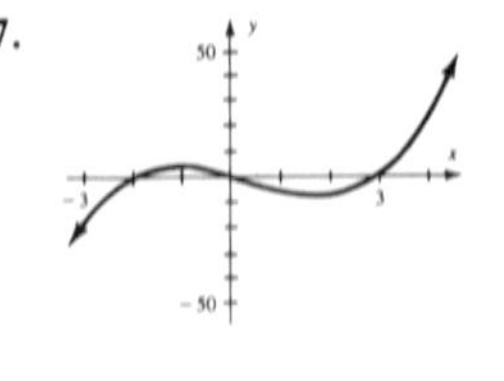

48.
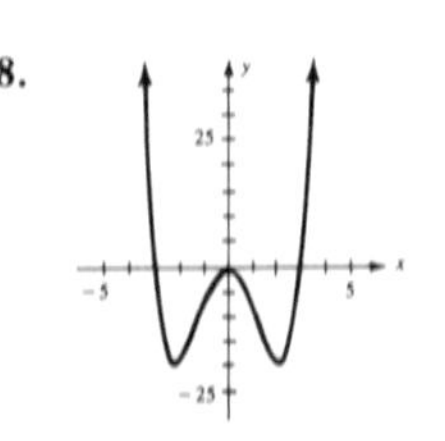

49. $(x - 1)(x + 1)(x - 2) = 0$; $\{-1, 1, 2\}$ **50.** $(x + 1)(x + 3)(x - 3) = 0$; $\{-3, -1, 3\}$

51. $(x + 2)(x - 3)(x + 5) = 0; \{-5, -2, 3\}$ **52.** $(x + 2)(x + 3)(x - 4) = 0; \{-3, -2, 4\}$
53. $(x - 2)(3x - 2)(2x + 5) = 0; \left\{-\frac{5}{2}, \frac{2}{3}, 2\right\}$ **54.** $(x - 3)(2x - 3)(3x + 4) = 0; \left\{-\frac{4}{3}, \frac{3}{2}, 3\right\}$
55. $(x + 2)^3(x + 1) = 0; \{-2, -1\}$ **56.** $(x - 2)(2x - 1)(2x + 1)^2 = 0; \left\{-\frac{1}{2}, \frac{1}{2}, 2\right\}$
57. $(x - 2)(x + 3)(x^2 - 4x + 1) = 0; \{-3, 2, 2 \pm \sqrt{3}\}$ **58.** $(x - 1)(x + 4)(x^2 - 6x + 7) = 0; \{-4, 1, 3 \pm \sqrt{2}\}$
59. $(x - 2)(x + 2)(x^2 + 9) = 0; \{-2, 2, \pm 3i\}$ **60.** $(x + 3)(x - 2)(x^2 - 4x + 13) = 0; \{-3, 2, 2 \pm 3i\}$
61. $-4 \leq x \leq -2$ or $x \geq \frac{2}{3}$ **62.** $x \leq -\frac{1}{2}$ or $\frac{3}{5} \leq x \leq 3$ **63.** $x < 5$ or $0 < x < 7$ **64.** $x < -4$ or $x > 3$
65. $-1 < x < 0$ **66.** $x \leq 0$ or $x \geq 5$ **67.** $(x - 1)(x + 1)(x - 2) \geq 0; -1 \leq x \leq 1$ or $x \geq 2$
68. $(x + 1)(x + 3)(x - 3) \leq 0; x \leq -3$ or $-1 \leq x \leq 3$ **69.** $(x + 2)(x + 3)(x - 5) \leq 0; x \leq -3$ or $-2 \leq x \leq 5$
70. $(x + 2)(x + 3)(x - 4) \geq 0; -3 \leq x \leq -2$ or $x \geq 4$ **71.** $(x - 2)(2x + 5)(3x - 2) \geq 0; -\frac{5}{2} \leq x \leq \frac{2}{3}$ or $x \geq 2$
72. $(x + 3)(3x - 2)(5x + 1) > 0; -3 < x < -\frac{1}{5}$ or $x > \frac{2}{3}$ **73.** $(x - 1)^2(x + 4)(x - 3) < 0; -4 < x < 1$ or $1 < x < 3$
74. $(x - 2)(2x - 3)(x^2 + 4x - 4) < 0; \frac{3}{2} < x < 2$ **75.** 2 in.
76. Descartes' Rule shows that there are 2 or 0 positive real roots and 1 negative real root; the Fundamental Theorem ensures that there are 9 roots, so there are at least 6 complex roots that are not real. **77.** $\{\pm\sqrt{2}, -3 \pm 2\sqrt{2}\}$ **78.** $k = -10$

79.
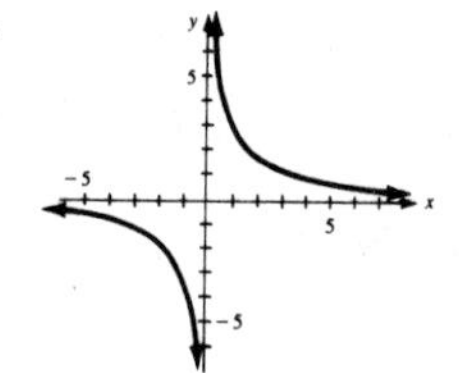

80.
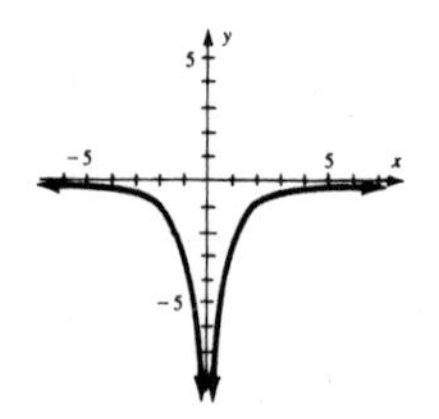

81.
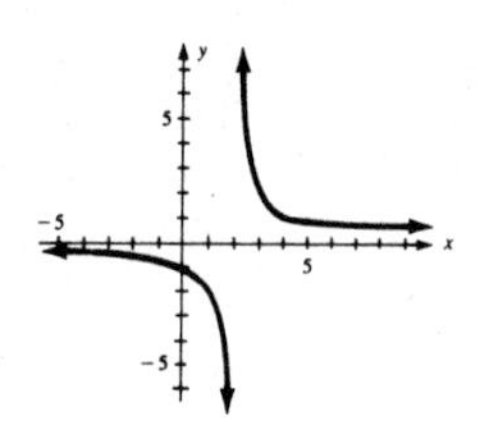

82.
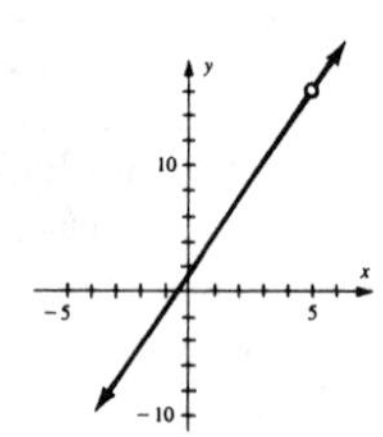

83.
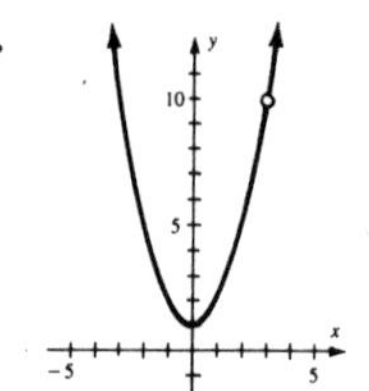

84.
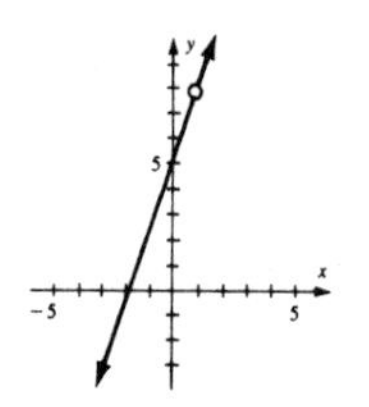

85.
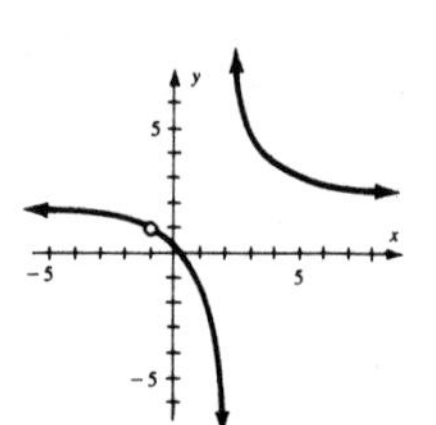

86.
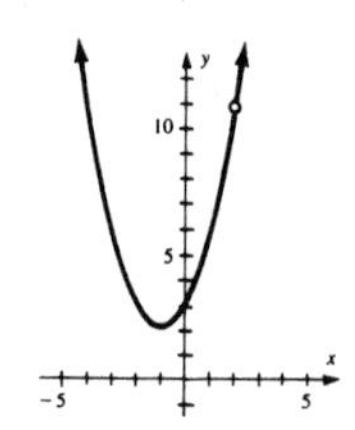

87.
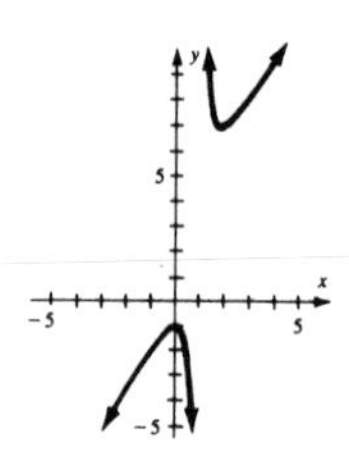

88.
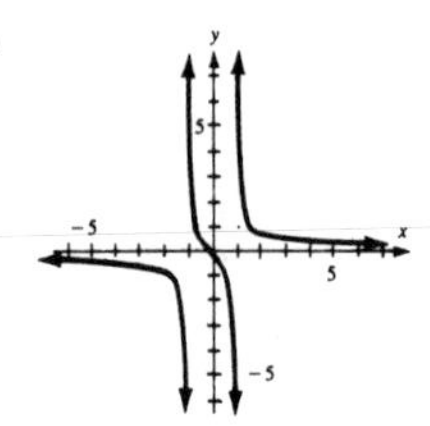

89. $\frac{3}{x} + \frac{5}{x^2} - \frac{2}{x^3}$ **90.** $2x - 3$ **91.** $\frac{4}{x - 3} - \frac{2}{x + 2}$ **92.** $\frac{5}{x + 1} - \frac{3}{x - 3}$ **93.** $2 + \frac{6}{x - 3} - \frac{1}{x - 2}$
94. $2 + \frac{3}{x - 5} - \frac{2}{x + 3}$ **95.** $\frac{3}{x - 1} + \frac{5}{(x - 1)^2}$ **96.** $\frac{5}{x - 2} - \frac{3}{(x - 2)^2}$ **97.** $5 + \frac{2}{x} - \frac{11}{x + 2}$ **98.** $2 - \frac{2}{x} + \frac{7}{x - 3}$
99. $2x + 3 + \frac{1}{x} + \frac{2}{x - 1}$ **100.** $3x - 1 + \frac{2}{x} - \frac{1}{x - 2}$

Chapter 6

Problem Set 6.1, Pages 224–226

1.

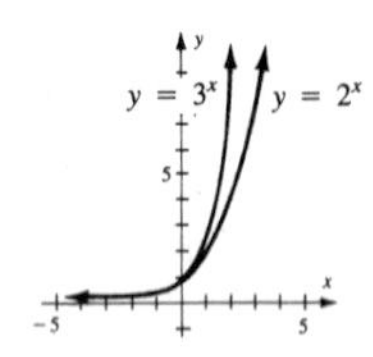

3.

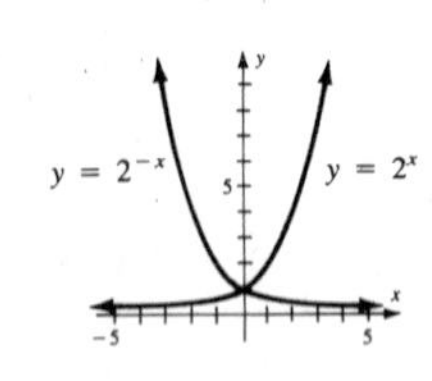

5.

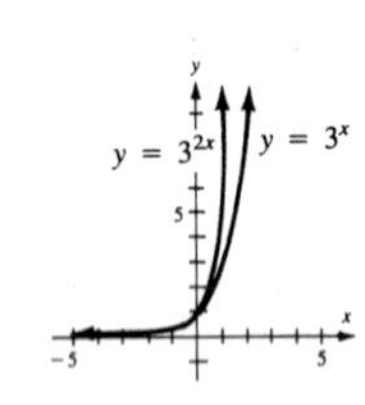

7.

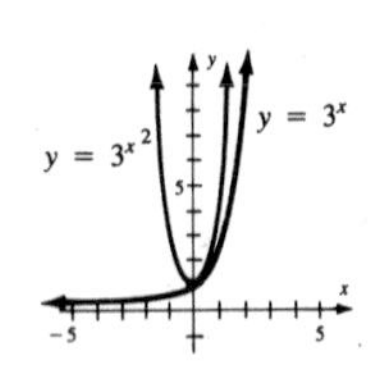

9.

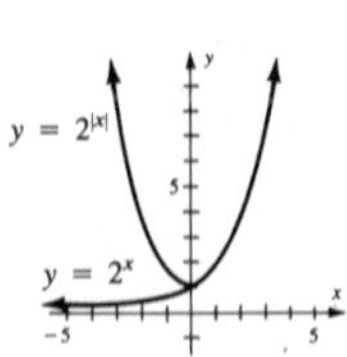

11. $x = 3$ **13.** $x = 4$ **15.** $x = \frac{1}{2}$ **17.** $x = 3$ **19.** $x = \frac{2}{3}$ **21.** $x = \frac{3}{4}$ **23.** $b = 7$

25. $a = 11$ **27.** $x = -\frac{1}{2}$ **29.** $x = \frac{1}{3}$ **31.** $x = -3$ **33.** $x = -\frac{1}{2}$ **35.** $y = \frac{1}{4}$ **37.** $z = -5$ **39.** $x = -1, \frac{1}{2}$

41.

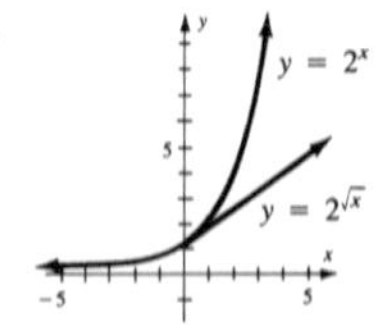

43.

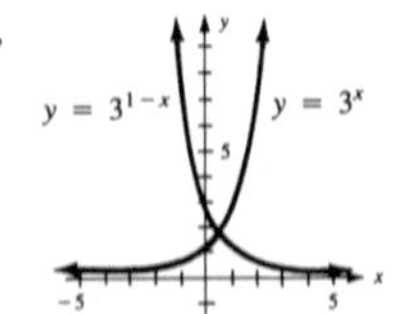

45.

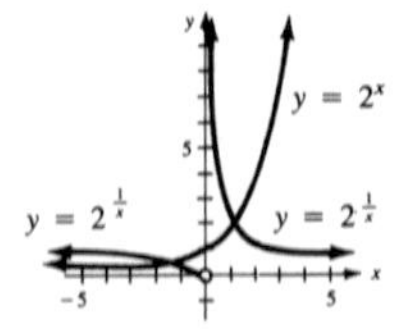

47.

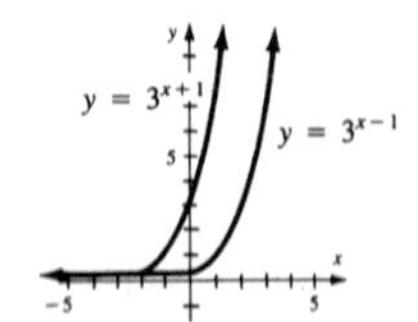

49.

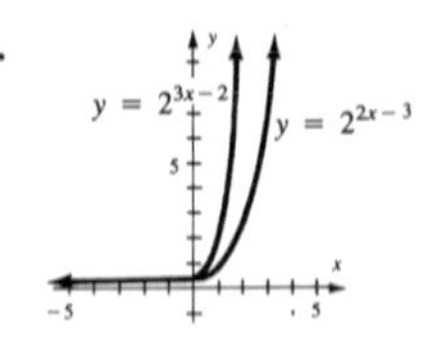

51. 4.72880439 **53.** 77.8802336 **55.** 37.3358408 **57.** 1.82263466 **59.** 22.3552653

61.

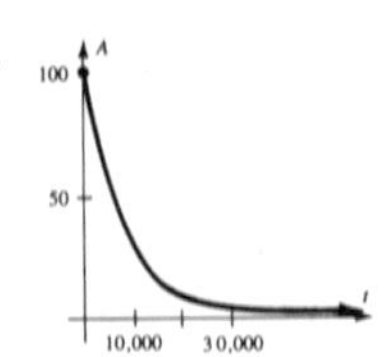

63.

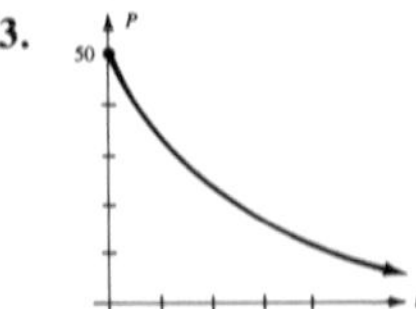

Problem Set 6.2, Pages 231–235

1. 97.3 g **3.** .45 g **5.** .0098 g **7.** About 626,000 in 1990 **9.** About 174,000 in 1985 **11.** \$2001.60
13. \$2288.98 **15.** \$13,742.19 **17.** \$2,013.62; \$12.02 more than in Problem 11; \$3.96 more than in Problem 12
19. \$101.68 **21.** About 14,050,000 in 1980 **23.** about 9,740,000 in 1980
25. 2,960,000 in 1980; 3,390,000 in 1990; 3,890,000 in 2000 **27.** 6.55 psi **29.** 221 mg remain
31. For $A_0 = 1$, $A = .5488116361$ at $t = 6$ and $A = .4955853038$ at $t = 7$; thus, it takes nearly 7 days.
33. About 22 words per minute

35.

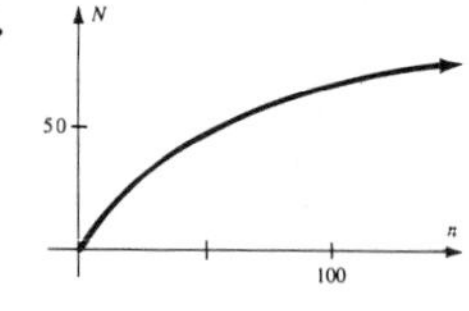

37.

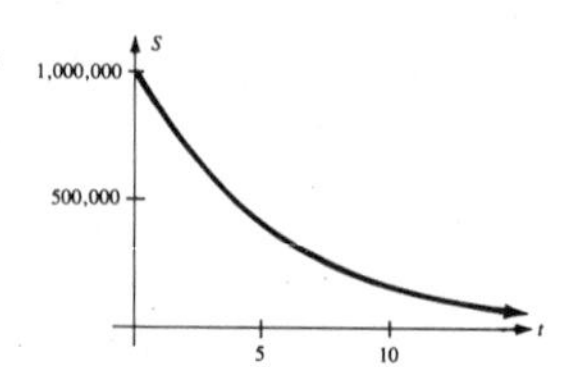

39. The two graphs are virtually identical. The actual function slows a little quicker at the top of its path and arrives back on the ground a little before the parabola, but otherwise the paths are difficult to tell apart.

Problem Set 6.3, Pages 240–241

1. $3 = \log_5 125$ **3.** $-2 = \log_3 \frac{1}{9}$ **5.** $\frac{3}{2} = \log_4 8$ **7.** $-2 = \log_{3/2} \frac{4}{9}$ **9.** $c = \log_b a$ **11.** $2^3 = 8$ **13.** $10^{-2} = .01$
15. $4^{.5} = 2$ **17.** $\left(\frac{1}{3}\right)^{-2} = 9$ **19.** $k^c = h$ **21.** $x = 3$ **23.** $x = 2$ **25.** $x = \frac{2}{3}$ **27.** $x = 9$ **29.** $b = 2$
31. $y = 2$ **33.** $y = 3$ **35.** $b > 0, b \neq 1$ **37.** $x = e$ **39.** $y = 2$

41.

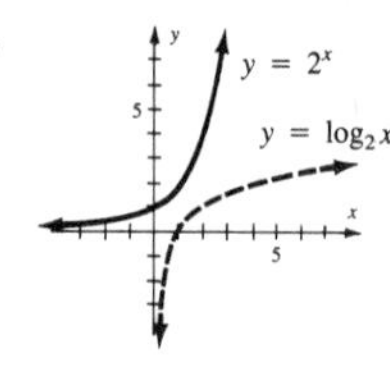

43.

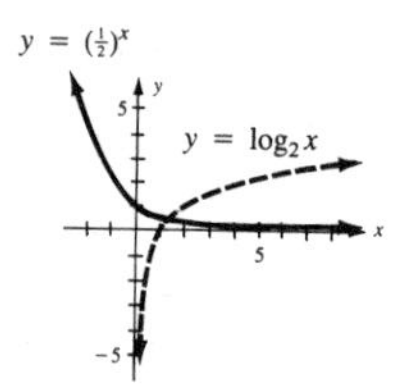

45.

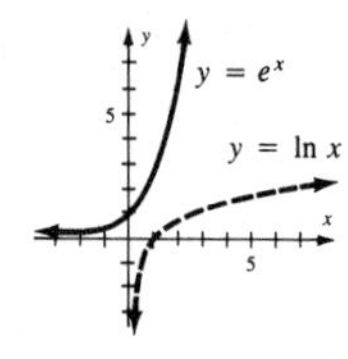

47.

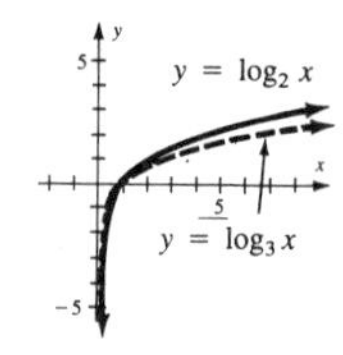

49.

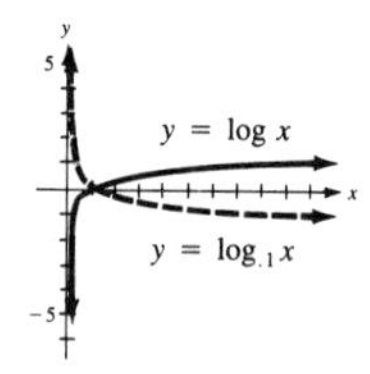

51. $\log_b(4ac) = \log_b 4 + \log_b a + \log_b c$ **53.** $\log(\pi r^2) = \log \pi + 2 \log r$ **55.** $\log_b(P\sqrt{Q}) = \log_b P + \frac{1}{2} \log_b Q$

57. $\log_b \dfrac{rs^2}{\sqrt[3]{t}} = \log_b r + 2 \log_b s - \frac{1}{3} \log_b t$ **59.** $\ln(Pe^{-rt}) = \ln P - rt$ **61.** $t = \dfrac{1}{r} \ln \dfrac{P}{P_0}$ **63.** $a = -\dfrac{1}{.21} \ln \dfrac{P}{14.7}$

65. $h = \dfrac{t \log \frac{1}{2}}{\log\left(\dfrac{A}{A_0}\right)}$ **67.** $n = 1 - \dfrac{1}{.11} \ln \dfrac{P_n}{P_1}$ **69.** $t = \dfrac{-L}{R} \ln\left(1 - \dfrac{IR}{E}\right)$

71. $\log \frac{1}{2} < 0$; thus, both sides of the inequality are multiplied by a negative number and the sense of the inequality should have been reversed when multiplied by $\log \frac{1}{2}$.

73. Let $A = b^x$; then $A^n = (b^x)^n = b^{nx}$. Thus,

$\log_b A^n = nx$	Definition of logarithm
$= n \log_b A$	Substitution

75.

$\log_b \dfrac{1}{A} = \log_b 1 - \log_b A$	Property II
$= 0 - \log_b A$	Property IV
$= -\log_b A$	

Problem Set 6.4, Pages 247–250

1. 1.658964843 **3.** -2.750435918 **5.** 4.29666519 **7.** -1.148853505 **9.** -9.500312916 **11.** 19.94019395

13.

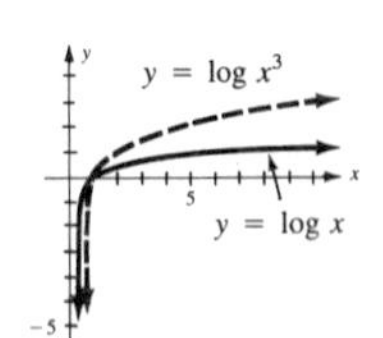

15.

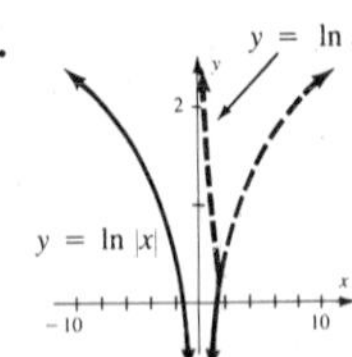

17.

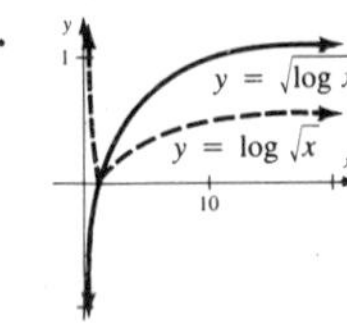

19.

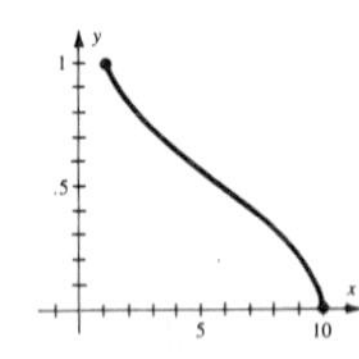

21. Approximately 2.9 **23.** 6.2 **25.** 10.5 **27.** 6.4 **29.** 3.6 **31.** .873568527 **33.** .910239226 **35.** 37.16700973 **37.** -21.60461574 **39.** -46.63343313 **41.** $x = 2 \cdot 5^4\sqrt{3} = 1250\sqrt{3}$ **43.** $x = 25$ **45.** $x = 1$ **47.** $x = \sqrt{2e}$ **49.** $x = 10^8$ **51.** $x = 10$ **53.** $x = e^2$ **55.** Approximately 30 decibels **57.** 107 decibels **59.** $7.53672175 \approx 7\frac{3}{4}$ years **61.** $4.607438359 \approx 4$ years 8 months **63.** $.14105969 \approx 14.1\%$ **65.** $89.666805 \approx 90°F$ **67.** $90.710678 \approx 91°F$ **69.** $139.31568 \approx 140$ minutes **71.** $1.78 \cdot 10^{24}$ ergs
73. Multiplying two numbers is performed by adding the lengths that represent the logarithms of the numbers. Division is performed similarly by subtracting logarithms. **75.** Sketch of $y = x$ is a line through the origin with slope 1.
77. Sketch of $y = \log x$ is a line through the origin with slope 1 for $x > 0$.

Problem Set 6.5, Pages 254–257

1. Approximately \$4714 **3.** Approximately 16.6% per year **5.** $2760.7954 \approx \$2761$
7. $.036389732 \approx 3.64\%$; $.010709402 \approx 1.07\%$; $.01801694 \approx 1.80\%$ **9.** $573{,}353.68 \approx 573{,}000$
11. $1{,}158{,}452.6 \approx 1{,}158{,}000$ **13.** $544{,}701.93 \approx 545{,}000$ **15.** \$362.08 **17.** $231.92375 \approx \$231.92$
19. $24.778650 \approx \$24.78$ **21.** $.0086643397 \approx .9\%$ **23.** 1988; 1998; 2007; 2014
25. Answers may vary. Using the daily increase of 195,000, an annual rate of 1.7793312%, the figure would be 6,240,000,000.
27. a. Sketching the data on a graph should "establish" that the relationship is exponential. **b.** Rate has not been constant. Like population growth, certain periods reflect social conditions. The rate for 1930–1935 was only 2.5%, whereas the 1955–1960 rate was 9.7%. The 1945–1960, 1960–1965, and 1965–1970 rates were all approximately 7.5%, but more recent rates are much lower.
c. Using the most recent rate of 4.8% (based on 1975–1980): 1985, 3050 million barrels; 1990, 3880 million barrels; 2000, 6270 million barrels **d.** A world almanac is a ready source of data. **e.** Scary!

Problem Set 6.6, Pages 258–261

1. exponential **2.** base; exponent **3.** $0 < b < 1$ **4.** asymptote **5.** $x = y$ **6.** bases **7.** e; 2.718281828
8. annual rate r; n times per year; t years **9.** $A = Pe^{rt}$ **10.** $s^r = t$; $r = \log_s t$ (These two answers can be reversed.)
11. $\log_b x$ **12.** $x \leq 0$ **13.** asymptote **14.** sum; logs **15.** 1; 0 **16.** $x = y$ **17.** common **18.** P_0; P; t; k
19. number; b **20.** P; i; n

21.

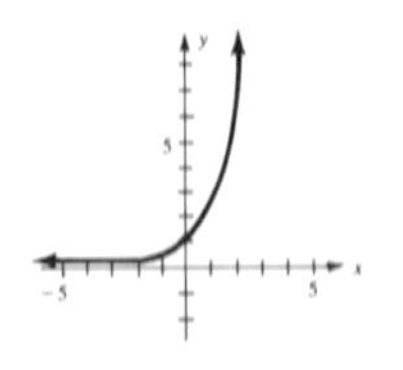

22.

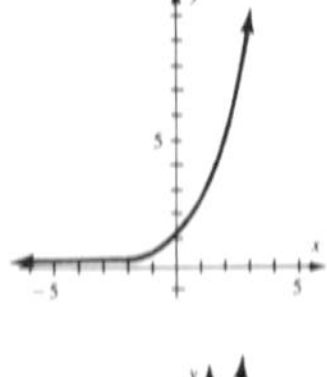

23.

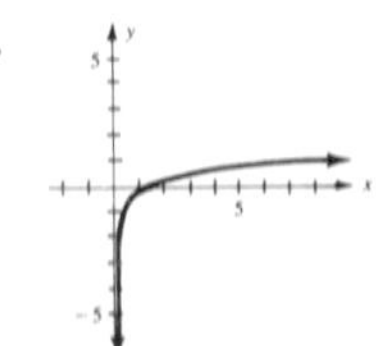

24.

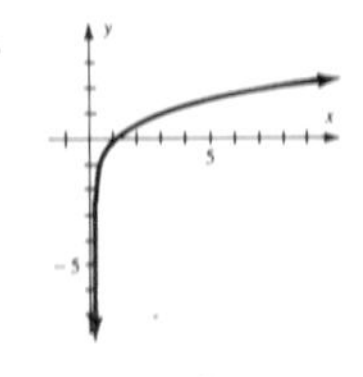

25.

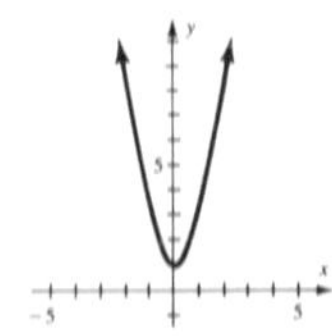

26.

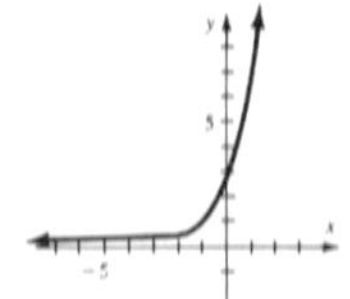

27.

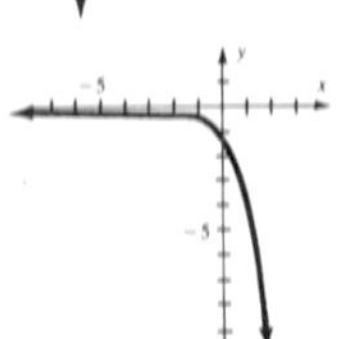

28.

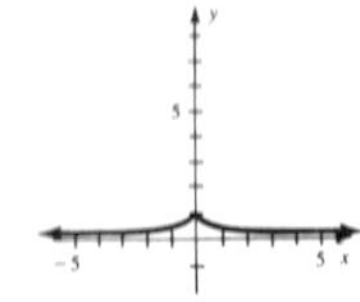

29. 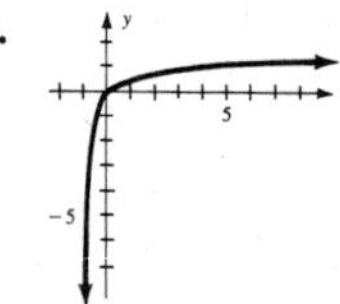

30. 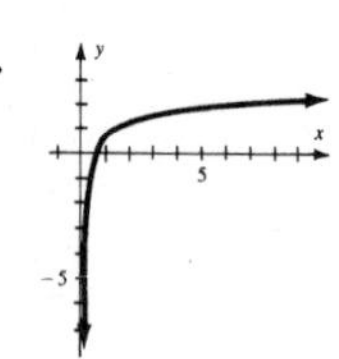

31. $a = 2$ **32.** $a = 3$ **33.** $a = \frac{3}{2}$ **34.** $b = 3^{\frac{16}{3}}$ **35.** $c = -\frac{2}{3}$ **36.** $x = -\frac{3}{4}$ **37.** $x = -\frac{1}{2}$ **38.** $b = \frac{1331}{1000}$
39. $x = -\frac{1}{3}$ **40.** $\log_2 16 = 4$ **41.** $\log_3 81 = 4$ **42.** $\log_2 64 = 6$ **43.** $\log_{11} 14{,}641 = 4$ **44.** $\log_7 343 = 3$
45. $\log_{13} 2197 = 3$ **46.** $\log_b y = x$ **47.** $\log_h p = k$ **48.** $\log_s r = t$ **49.** $3^3 = 27$ **50.** $2^4 = 16$ **51.** $5^5 = 3125$
52. $10^{-2} = .01$ **53.** $e^0 = 1$ **54.** $10^3 = 1000$ **55.** $b^y = x$ **56.** $h^s = r$ **57.** $e^m = n$ **58.** $a \approx 1.5010593$
59. $b \approx -2.1655793$ **60.** $c \approx 284.18424$ **61.** $d \approx .048790165$ **62.** $x \approx .8735685$ **63.** $f \approx .47369631$ **64.** $g = 8$
65. $h = 2^{-16}$ **66.** $y = 0$ **67.** $\log a + \log b$ **68.** $\ln k - kt$ **69.** $\log 5 + 3 \log 6 - \frac{1}{2} \log 160$ **70.** $\log_b c + \frac{1}{2} \log_b d$
71. $10 \log I - 10 \log I_0$ **72.** $\ln A_0 - \frac{t}{h} \ln 2$ **73.** $t = h \log_{1/2} \frac{A}{A_0}$ **74.** $r = \frac{1}{t} \ln \frac{P}{P_0}$ **75.** $N = \frac{10}{\log (I/I_0)}$
76. $L = \frac{-Rt}{\ln(1 - RI/E)}$ **77.** $B = \frac{T - A(1 - 10^{-kt})}{10^{-kt}} = 10^{kt}(T - A) + A$ **78.** $x = \log 18$ **79.** $x = e^2$ **80.** $x = 100$
81. 70.7 g **82.** 7.8 years **83.** 2.4 hours **84.** 25,360 **85.** 18.7 years **86.** $n \approx 18$; 28.1 years **87.** \$10,391.74
88. \$10,449.98 **89.** \$10,478.70 **90.** $7\frac{3}{4}$ years **91.** \$10,479.68 **92.**

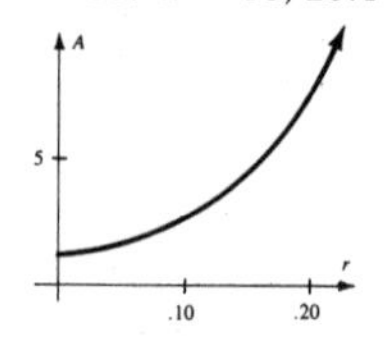

93. $.057762265 \approx 5.78\%$ **94.** 3.5 **95.** $1.58 \cdot 10^{-12}$ **96.** 120 decibels **97.** $I = \frac{1}{100}$ or 10^{-2} watt per cm^2 **98.** .7%
99. 602,000 **100.** \$631.32

Chapter 7

Problem Set 7.1, Pages 270–272

1. (3, 1)

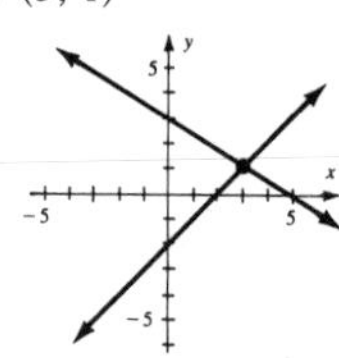

3. (−2, −5)

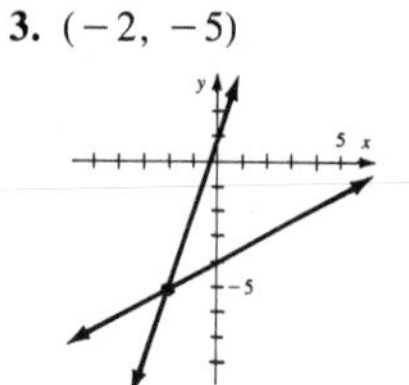

5. (3, − 3)

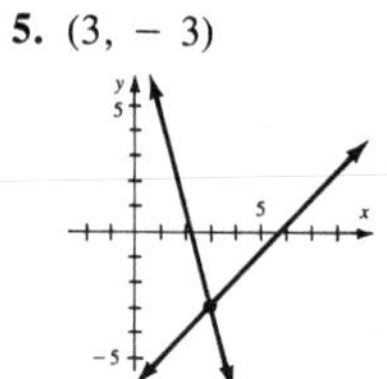

7. (−1, −1)

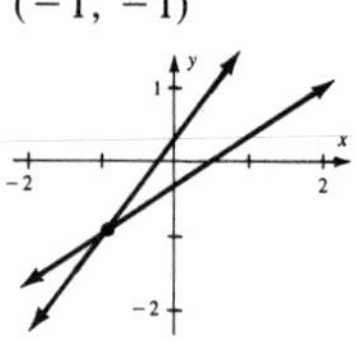

9. (1, 2) and (7, 8)  **11.** (13, 3) **13.** Dependent **15.** $(a_1, a_2) = (-7, 3)$ **17.** (2, 2) and (−1, −4)

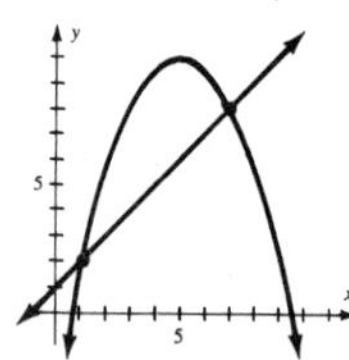

19. (3, − 1) and (5, − 4) **21.** (23, −43) **23.** Dependent **25.** Inconsistent **27.** (0, − 2) and $\left(-\frac{3}{4}, 1\right)$
29. $(\sqrt{13}, 1)$, $(-\sqrt{13}, 1)$, (5, 2), and (−5, 2) **31.** (3, 4) **33.** (4, 3) **35.** (−2, −5) **37.** (3, 1) **39.** $\left(\frac{3}{2}, -3\right)$
41. (2, 200) **43.** (q, d) = (63, 84) **45.** (3, 4) and (0, − 2) **47.** (4, 3), (4, −3), (−4, 3), and (−4, −3)
49. (2, 1) and (−2, −1) **51.** No real solutions **53.** The numbers are −2 and −7, or 7 and 2.
55. The numbers are 3 and 1, or 4 and $\frac{1}{2}$. **57.** 8 liters of 40% solution and 2 liters of 90% solution
59. The legs are 8 mm and 15 mm long.

Problem Set 7.2, Pages 280–284

1. $\begin{bmatrix} 9 & 5 & 6 \\ 6 & 4 & 9 \\ 2 & 10 & 7 \end{bmatrix}$ **3.** $\begin{bmatrix} -6 & 7 & -6 \\ 3 & -7 & -3 \\ -8 & -7 & 8 \end{bmatrix}$ **5.** $\begin{bmatrix} 16 & 19 & 10 \\ 29 & 16 & 15 \\ 35 & 9 & 31 \end{bmatrix}$ **7.** $\begin{bmatrix} 13 & 25 & 20 \\ 25 & 31 & 23 \\ 42 & 24 & 33 \end{bmatrix}$ **9.** $\begin{bmatrix} 34 & 117 & 44 \\ 45 & 143 & 97 \\ 109 & 100 & 151 \end{bmatrix}$ **11.** $\begin{bmatrix} 2 & 8 \\ 16 & 8 \\ -7 & 7 \\ 7 & 7 \end{bmatrix}$

13. $\begin{bmatrix} 14 & 14 \\ -7 & 7 \end{bmatrix}$ **15.** $\begin{bmatrix} 14 & 4 & 8 & 30 \\ -7 & -1 & -9 & 3 \end{bmatrix}$ **17.** Nonconformable **19.** $\begin{bmatrix} 1 & 0 & 0 & 0 \\ 0 & 1 & 0 & 0 \\ 0 & 0 & 1 & 0 \\ 0 & 0 & 0 & 1 \end{bmatrix}$ **21.** Inverses **23.** Inverses

25. Inverses **27.** Not inverses **29.** Inverses **31.** $S_{84} = \begin{bmatrix} 335 & 280 & 100 \\ 250 & 120 & 50 \end{bmatrix}$; $S_{85} = \begin{bmatrix} 350 & 270 & 130 \\ 290 & 270 & 90 \end{bmatrix}$

33. $S_{85} - S_{84} = \begin{bmatrix} 15 & -10 & 30 \\ 40 & 50 & 40 \end{bmatrix}$; change in sales from 1984 to 1985

35. $(.08)S_{84} = \begin{bmatrix} 26.8 & 22.4 & 8 \\ 20 & 9.6 & 4 \end{bmatrix}$; might be amounts payable for state sales tax (answers vary) **37.** [51.8 34.4 13]

39. [108.8 83 32.4]

41. $S_{84} = \begin{bmatrix} 3370 & 1510 & 740 & 720 & 430 \\ 3620 & 1850 & 430 & 640 & 520 \end{bmatrix}$; $S_{85} = \begin{bmatrix} 3000 & 1200 & 700 & 640 & 370 \\ 3550 & 1670 & 480 & 740 & 430 \end{bmatrix}$

43. $\begin{bmatrix} -370 & -310 & -40 & -80 & -60 \\ -70 & -180 & 50 & 100 & -90 \end{bmatrix}$; change in sales from 1984 to 1985

45. $\begin{bmatrix} 505.5 & 226.5 & 111 & 108 & 64.5 \\ 543 & 277.5 & 64.5 & 96 & 78 \end{bmatrix}$; answers vary **47.** [1229.5 596.5 197 236 168.5]

49. [2389.5 1110.5 398 480 310] **51.** X = [10 −6 −7] **53.** $Z = \begin{bmatrix} 34 & -17 & -7 \\ 13 & -12 & -7 \\ -9 & -10 & 7 \end{bmatrix}$ **55.** Yes; examples vary

57. AJ and JA interchange first and third rows of A, thus JAJ leaves A unchanged. **59.** $AB = BA = \begin{bmatrix} vx - wy & vy + wx \\ -wx - vy & vx - wy \end{bmatrix}$

61. [94.1 89.9 86.3 81.6 91.4] **63.** $BC = \begin{bmatrix} 330 \\ 560 \\ 535 \end{bmatrix}$; cost of raw materials for each of the three products

65. $3^2 + 4^2 = 9 + 16 = 25 = 5^2$; XP = [20 21 29] and $20^2 + 21^2 = 400 + 441 = 841 = 29^2$

67. $XP = [2a + b + 2x \quad a + 2b + 2c \quad 2a + 2b + 3c]$;

$$(2a + b + 2c)^2 + (a + 2b + 2c)^2 \stackrel{?}{=} (2a + 2b + 3c)^2$$
$$5a^2 + 5b^2 + 8c^2 + 8ab + 12ac + 12bc \stackrel{?}{=} 4a^2 + 4b^2 + 9c^2 + 8ab + 12ac + 12bc$$
$$5(a^2 + b^2) + 8c^2 + 8ab + 12ac + 12bc \stackrel{?}{=} 4(a^2 + b^2) + 9c^2 + 8ab + 12ac + 12bc$$

But if $[a \quad b \quad c]$ is a Pythagorean triple, then $a^2 + b^2 = c^2$, and

$$5c^2 + 8c^2 + 8ab + 12ac + 12bc \stackrel{?}{=} 4c^2 + 9c^2 + 8ab + 12ac + 12bc$$
$$13c^2 + 8ab + 12ac + 12bc = 13c^2 + 8ab + 12ac + 12bc$$

Thus, XP is also a Pythagorean triple.

Problem Set 7.3, Pages 289–291

1. (3, 4) **3.** (4, 3) **5.** (1, 1) **7.** (3, 1) **9.** $\left(\frac{3}{2}, -3\right)$ **11.** (2, −1, −3) **13.** $\left(\frac{3}{2}, \frac{1}{2}, -\frac{1}{2}\right)$ **15.** (−1, −2, 2)
17. (−1, 2, −3) **19.** (1 −3, 5) **21.** (0, 1, 0) **23.** (3, 2, 5) **25.** (4, −1, 2) **27.** (8, 5, 0) **29.** (−2, −1, 3)
31. (4, 2, 1) **33.** $\left(\frac{1}{2}, \frac{7}{2}, \frac{7}{2}\right)$ **35.** $(w, x, y, z) = (4, 3, 2, 1)$ **37.** $(w, x, y, z) = (-4, 3, -1, 2)$
39. $(w, x, y, z) = (-1, 10, -5, -4)$ **41.** $y = x^2 - 6x + 5$ **43.** $y = x^2 - 2x - 2$ **45.** $y = 2x^2 - 2x + 1$

47. $\left[\begin{array}{rrr|r} 1 & 2 & -1 & 3 \\ -3 & -6 & 1 & 1 \\ 2 & 4 & -1 & 2 \end{array}\right] \sim \left[\begin{array}{rrr|r} 1 & 2 & -1 & 3 \\ 0 & 0 & -2 & 10 \\ 0 & 0 & 1 & -4 \end{array}\right] \begin{array}{l} R_1 \\ R_2 + 3R_1 \\ R_3 - 2R_1 \end{array} \sim \left[\begin{array}{rrr|r} 1 & 2 & -1 & 3 \\ 0 & 0 & 0 & 2 \\ 0 & 0 & 1 & -4 \end{array}\right] \begin{array}{l} R_1 \\ R_2 + 2R_3 \\ R_3 \end{array}$ Row 2 implies 0 = 2

49. $\left(1 - 2y,\ y,\ \dfrac{3 - 4y}{3}\right)$ **51.** $\left(\dfrac{18 - 7y}{7},\ y,\ \dfrac{7y + 1}{7}\right)$

Problem Set 7.4, Pages 294–296

1. $\begin{bmatrix} 2 & 7 \\ 1 & 4 \end{bmatrix}$ **3.** $\frac{1}{5}\begin{bmatrix} 3 & 1 \\ -2 & 1 \end{bmatrix}$ **5.** $\frac{1}{5}\begin{bmatrix} -2 & -1 \\ -1 & -3 \end{bmatrix}$ **7.** $\frac{1}{22}\begin{bmatrix} 2 & -3 \\ 1 & 4 \end{bmatrix}$ **9.** $\begin{bmatrix} 2 & 3 \\ 2 & 1 \end{bmatrix}$ **11.** $\begin{bmatrix} -3 & 2 & -1 \\ -4 & 2 & -1 \\ 2 & -1 & 1 \end{bmatrix}$

13. $\begin{bmatrix} 9 & -4 & -2 \\ -18 & 9 & 4 \\ -4 & 2 & 1 \end{bmatrix}$ **15.** $\frac{1}{12}\begin{bmatrix} -2 & 2 & 2 \\ 8 & -8 & 4 \\ 7 & -1 & -1 \end{bmatrix}$ **17.** $\frac{1}{2}\begin{bmatrix} 1 & 1 & 1 \\ -1 & 1 & 1 \\ -1 & 1 & -1 \end{bmatrix}$ **19.** $\begin{bmatrix} 1 & -1 & 0 \\ -2 & 3 & -1 \\ 1 & -1 & 1 \end{bmatrix}$ **21.** (4, 3)

23. (3, 1) **25.** (−2, −5) **27.** $\left(-\frac{1}{2}, \frac{1}{2}\right)$ **29.** (8, 4) **31.** (2, 7, 3) **33.** (3, 2, 5) **35.** $\left(1, 2, -\frac{1}{2}\right)$ **37.** (1, 0, 0)
39. (10, −15, 12) **41.** (1, 1) **43.** (4, 2, 1) **45.** (−1, −2, 2) **47.** $(w, x, y, z) = (4, 3, 2, 1)$ **49.** Dependent
51. 3 parts (21 gal) of spray I, 4 parts (36 gal) of spray II **53.** 4 parts (52 kg) of I, 5 parts (35 kg) of II, 6 parts (60 kg) of III

55.
$$A^{-1}A = I$$
$$(A^{-1})^{-1}A^{-1}A = (A^{-1})^{-1}I$$
$$IA = (A^{-1})^{-1}$$
$$\therefore\ A = (A^{-1})^{-1}$$

57.
$$AB = AC$$
$$A^{-1}AB = A^{-1}AC$$
$$IB = IC$$
$$\therefore\ B = C$$

Problem Set 7.5, Pages 301–303

1. False **3.** False **5.** False **7.** True **9.** True **11.** 1 **13.** −15 **15.** −2 **17.** 1 **19.** 7 **21.** −12
23. 0 **25.** 0 **27.** −25 **29.** 0 **31.** 5 **33.** 63 **35.** 557 **37.** 90 **39.** −10 **41.** $-x + 2y - z$
43. $abc - a - b - c + 2$
45. The determinant equation is equivalent to the quadratic $x^2 - 2Lx + L^2 - d^2 = 0$, with solutions $L \pm d$; and the absolute value equation has the same solution set. (Discussion and examples will vary.)

Problem Set 7.6, Pages 308–311

1. 1 **3.** 9 **5.** −5 **7.** 0 **9.** −25 **11.** 7 **13.** 1 **15.** 3 **17.** 252 **19.** $x = 0$ **21.** $2abc$ **23.** $x = -8$
25. $x = 1$ **27.** $x = 4$ **29.** $x = -3$ **31.** $2x + 3y + 6 = 0$ **33.** $3x + 4y - 5 = 0$ **35.** $2x - 3y + 13 = 0$
37. $x = 2$ or -2 **39.** $x = 1$ or -1 **41.** −132 **43.** 0 **45.** 0 **47.** 59 **49.** 1008 **51.** 26 sq. units

53. Verifies **55.** Verifies **57.** 220 sq. units **59.** $13\frac{1}{2}$ sq. units

61. $\begin{vmatrix} x & y & 1 \\ x_1 & y_1 & 1 \\ x_2 & y_2 & 1 \end{vmatrix} = \begin{vmatrix} x_1 & y_1 \\ x_2 & y_2 \end{vmatrix} - \begin{vmatrix} x & y \\ x_2 & y_2 \end{vmatrix} + \begin{vmatrix} x & y \\ x_1 & y_1 \end{vmatrix} = 0$

$$x(y_2 - y_1) - y(x_2 - x_1) + (x_2y_1 - x_1y_2) = 0$$

which is a linear equation in x and y; (x_1, y_1) and (x_2, y_2) satisfy the determinant equation; thus, it is the linear equation of the line containing (x_1, y_1) and (x_2, y_2).

63. If a row (or a column) has all zero entries, then expand the determinant about that row (or column) and each minor will be multiplied by zero; thus, the determinant will be zero.

65. $$\begin{vmatrix} a & b & b & b \\ a & b & a & a \\ b & b & a & b \\ a & a & a & b \end{vmatrix} = \begin{vmatrix} a & b & b & b \\ 0 & 0 & a-b & a-b \\ b & b & a & b \\ a-b & a-b & 0 & 0 \end{vmatrix} = \begin{vmatrix} a & b & b & b \\ 0 & 0 & 0 & a-b \\ b & b & a-b & b \\ a-b & a-b & 0 & 0 \end{vmatrix}$$

$$= \begin{vmatrix} a-b & b & b & b \\ 0 & 0 & 0 & a-b \\ 0 & b & a-b & b \\ 0 & a-b & 0 & 0 \end{vmatrix} = (a-b)\begin{vmatrix} 0 & 0 & a-b \\ b & a-b & b \\ a-b & 0 & 0 \end{vmatrix}$$

$$= (a-b)^2 \begin{vmatrix} b & a-b \\ a-b & 0 \end{vmatrix} = (a-b)^2[0 - (a-b)]^2 = (a-b)^4$$

Problem Set 7.7, Page 315–317

1. (2, 3) **3.** $\left(\frac{2}{3}, \frac{1}{3}\right)$ **5.** $\left(\frac{24}{13}, \frac{-5}{13}\right)$ **7.** $(-10, 13)$ **9.** $(-1, 3)$ **11.** $(1, -3, 1)$ **13.** $\left(\frac{1}{2}, -\frac{1}{3}, \frac{1}{6}\right)$ **15.** $\left(-\frac{1}{2}, -\frac{2}{3}, -\frac{3}{4}\right)$
15. $\left(-\frac{1}{2}, -\frac{2}{3}, -\frac{3}{4}\right)$ **17.** $\left(\frac{3}{2}, 4, -3\right)$ **19.** $\left(\frac{3}{2}, -2, \frac{1}{2}\right)$ **21.** $x^2 + y^2 - 6x - 6y + 8 = 0$
23. $x^2 + y^2 + 6x - 10y + 9 = 0$ **25.** $x^2 + y^2 + 8x + 4y - 5 = 0$ **27.** $(-2, -4)$ **29.** $\left(-\frac{1}{2}, \frac{1}{3}\right)$
31. $\left(\frac{2}{3}, 2, -2\right)$ **33.** $\left(\frac{1}{2}, \frac{1}{3}, -\frac{1}{5}\right)$ **35.** $(2, -3, 6)$ **37.** $(w, x, y, z) = \left(-\frac{6}{5}, \frac{3}{5}, \frac{3}{5}, -\frac{9}{5}\right)$
39. $(w, x, y, z) = (-4, -1, 3, 2)$
41. If $r_1 = 1$, $r_2 = i$, $r_3 = -1$, and $r_4 = -i$, then

$$\begin{vmatrix} 1 & i & -1 & -i \\ i & -1 & -i & 1 \\ -1 & -i & 1 & i \\ -i & 1 & i & -1 \end{vmatrix} = \begin{vmatrix} 1 & i & -1 & -i \\ i & -1 & -i & 1 \\ 0 & 0 & 0 & 0 \\ -i & 1 & i & -1 \end{vmatrix} = 0$$

Problem Set 7.8, Pages 318–321

1. intersection **2.** opposites **3.** substitution **4.** inconsistent **5.** dependent **6.** m; n **7.** square
8. order; corresponding **9.** conformable **10.** entry **11.** conformable; p **12.** identity **13.** inverse; A^{-1}
14. elementary row **15.** Gauss–Jordan **16.** identity matrix **17.** ones; zeros **18.** nonsingular **19.** A^{-1}
20. real numbers **21.** rows; columns **22.** minor **23.** cofactor; minor **24.** identical **25.** zero
26. corresponding entries of another row; unchanged **27.** a system of linear equations **28.** determinant of the system
29. replacing the column of coefficients of x by the column of the constant terms **30.** nonzero **31.** (3, 4) **32.** (0, 1)
33. $(-2, 3), (3, -2)$ **34.** $(2, 0), (6, 16)$ **35.** $(3, -4), (4, 3)$ **36.** $(2, 2), (4, 2)$ **37.** $\begin{bmatrix} 4 & -5 & 0 \\ 7 & -1 & 2 \\ -1 & -2 & 4 \end{bmatrix}$

38. $\begin{bmatrix} 5 & -2 \\ 5 & -1 \end{bmatrix}$ **39.** $\begin{bmatrix} 4 & 4 \\ 3 & 3 \end{bmatrix}$

40. $\begin{bmatrix} 3 & -2 \\ 5 & -5 \end{bmatrix}$ **41.** Nonconformable **42.** $\begin{bmatrix} 4 & -2 & 0 \\ 8 & -5 & -2 \\ -3 & 1 & -1 \end{bmatrix}$ **43.** $(-2, 5)$ **44.** $(-1, 3)$ **45.** $\left(-\frac{3}{8}, \frac{7}{12}\right)$

46. Dependent **47.** Inconsistent **48.** $(-1, 2, 1)$ **49.** $(w, x, y, z) = (4, -2, 5, 3)$ **50.** $(w, x, y, z) = (1, 0, -1, 1)$

51. $y = x^2 - 4x + 4$ **52.** $\frac{1}{2}\begin{bmatrix} -4 & 2 \\ 3 & -1 \end{bmatrix}$ **53.** $\begin{bmatrix} 1 & 2 \\ 2 & 3 \end{bmatrix}$ **54.** $\begin{bmatrix} -1 & -2 \\ 3 & 4 \end{bmatrix}$ **55.** $\frac{1}{2}\begin{bmatrix} 2 & -1 & -1 \\ 0 & 1 & 1 \\ 0 & -1 & 1 \end{bmatrix}$ **56.** $\begin{bmatrix} 1 & 3 & 3 \\ 1 & 4 & 3 \\ 1 & 3 & 4 \end{bmatrix}$

57. $\frac{1}{2}\begin{bmatrix} -5 & 4 & -1 \\ 2 & -2 & 2 \\ 1 & 0 & -1 \end{bmatrix}$ **58.** $(-3, 3)$ **59.** $(9, 14)$ **60.** $(-5, 13)$ **61.** $\left(-\frac{1}{2}, -\frac{1}{2}, -\frac{3}{2}\right)$ **62.** $(11, 12, 15)$

63. $\left(\frac{1}{2}, 1, -\frac{1}{2}\right)$ **64.** Mix 20 g of I with 40 g of II.
65. They can manufacture 100 of product I, 150 of product II, and 250 of product III. **66.** 43 **67.** -42 **68.** 3 **69.** .1
70. 25 **71.** $\frac{1}{4}$ **72.** 8 **73.** -9 **74.** -1 **75.** 8 **76.** 16 **77.** -5 **78.** -19 **79.** 185
80. 6 sq. units **81.** Area is 0. **82.** Since the area of $\triangle NIL = 0$ (see Problem 81), the points must be collinear.
83. 34 sq. units **84.** $y = \frac{3}{2}x - \frac{7}{2}$ **85.** $(-2, 3)$ **86.** $\left(-\frac{3}{8}, \frac{7}{12}\right)$ **87.** $(-3, 3)$ **88.** $(-3, 4, 7)$ **89.** $(-1, 2, 1)$
90. $(x_1, x_2, x_3, x_4) = (-1, 1, -4, 1)$

Chapter 8

Problem Set 8.1, Pages 328–330

1. a. Arithmetic **b.** $d = 2$ **c.** 8 **3. a.** Arithmetic **b.** $d = 10$ **c.** 35 **5. a.** Geometric **b.** $r = \frac{1}{3}$ **c.** $\frac{1}{3}$
7. a. Neither **b.** Alternating terms form arithmetic sequences: 3, 4, 2, 5, 1, 6, 0, 7, . . . **c.** 6
9. a. Neither **b.** One more subtracted than on previous subtraction; that is, subtract 1, subtract 2, . . . **c.** 6
11. a. Arithmetic **b.** $d = 3$ **c.** 17 **13. a.** Geometric **b.** $r = -3$ **c.** -1215
15. a. Arithmetic or geometric **b.** $d = 0$ or $r = 1$ **c.** -8
17. a. Neither **b.** After first two terms, add previous two terms **c.** 16 **19. a.** Arithmetic **b.** $d = -11$ **c.** 53
21. a. Neither **b.** Perfect cubes of consecutive counting numbers **c.** 216 **23. a.** Geometric **b.** $r = 3^3$ **c.** 3^{14}
25. a. Neither **b.** 3 followed by 1, 2, 3, . . . sevens **c.** 3
27. a. Neither **b.** 1 followed by all not previously listed reduced common fractions; halves, thirds, etc. **c.** $\frac{5}{6}$
29. a. Geometric **b.** $r = \frac{3}{2}$ **c.** $6\frac{3}{4}$ or $\frac{27}{4}$ **31. a.** 1, 5, 9 **b.** Arithmetic; $d = 4$ **33. a.** 10, 20, 30 **b.** Arithmetic; $d = 10$
35. a. 4, 1, -2 **b.** Arithmetic; $d = -3$ **37. a.** 2, 1, $\frac{2}{3}$ **b.** Neither **39. a.** 0, $\frac{1}{3}$, $\frac{1}{2}$ **b.** Neither
41. a. 1, 5, 14 **b.** Neither **43. a.** 6, 12, 24 **b.** Geometric; $r = 2$ **45. a.** 3, $\frac{3}{2}$, $\frac{3}{4}$ **b.** Geometric; $r = \frac{1}{2}$
47. a. -2, 3, -4 **b.** Neither **49. a.** -5, -5, -5 **b.** Both; $d = 0$, $r = 1$ **51.** 57 **53.** 303 **55.** $\frac{3}{128}$ or .0234375
57. $\frac{48}{49}$ **59.** 625 **61.** 2, 6, 18, 54, 162 **63.** 1, 1, 2, 3, 5 **65.** 47, add previous two terms
67. $1\frac{2}{3}$, 8, $14\frac{1}{3}$, $20\frac{2}{3}$, 27, $33\frac{1}{3}$, . . . **69.** Answers vary. One possibility is $\frac{3}{2}$, 8, $\frac{19}{2}$, $\frac{35}{2}$, 27, $\frac{89}{2}$, . . .; $s_n = s_{n-1} + s_{n-2}$.

Another possibility is 27, 8, 27, 8, . . .; $s_n = \begin{cases} 27 & \text{if } n \text{ odd} \\ 8 & \text{if } n \text{ even} \end{cases}$

Problem Set 8.2, Pages 335–337

1. 5, 9, 13, 17; $4n + 1$ **3.** -15, -7, 1, 9; $8n - 23$ **5.** 100, 95, 90, 85; $105 - 5n$ **7.** $\frac{1}{2}$, 2, $\frac{7}{2}$, 5; $\frac{3}{2}n - 1$
9. 5, $5 + x$, $5 + 2x$, $5 + 3x$; $5 + (n - 1)x$ **11.** $a_1 = 3$; $d = 4$ **13.** $a_1 = 6$; $d = 5$ **15.** $a_1 = -8$; $d = 7$
17. $a_1 = 151$; $d = -9$ **19.** $a_1 = x$; $d = x$ **21.** $4n - 1$ **23.** $5n + 1$ **25.** $7n - 15$ **27.** $160 - 9n$ **29.** nx
31. 101 **33.** 14 **35.** -2 **37.** 1 **39.** 845 **41.** -190 **43.** 480 **45.** 145 **47.** 400 **49.** 2375
51. 11,500 **53.** 2030 **55.** 4048 **57.** $n(n + 1)$ **59.** 3828 blocks **61.** $\frac{9}{2}$ **63.** -1 **65.** 48 **67.** $\frac{19}{2}$ **69.** -6
71. Harmonic **73.** Harmonic **75.** Not harmonic **77.** Option *B*
79. *A*: \$21,000; *B*: \$21,360, \$360 more than *A*; *C*: \$21,540, \$540 more than *A*, \$180 more than *B*; *D*: \$21,660, \$660 more than *A*; \$300 more than *B*; \$120 more than *C*

81. a. $21{,}000 + 22{,}440 + 23{,}880 + \cdots; a_1 = 21{,}000, n = 10, d = 1440$ **b.** $10{,}500 + 10{,}860 + 11{,}220 + 11{,}580 + \cdots;$ $a_1 = 10{,}500, n = 20, d = 360$

Problem Set 8.3, Pages 342–344

1. $5, 15, 45, 135; (5)3^{n-1}$ **3.** $1, -2, 4, -8; (-2)^{n-1}$ **5.** $-15, -3, -\frac{3}{5}, -\frac{3}{25}; -15(\frac{1}{5})^{n-1}$ or $(-3)5^{2-n}$ **7.** $54, -36, 24, -16; 54(-\frac{2}{3})^{n-1}$ or $-2^n(-3)^{4-n}$ **9.** $8, 8x, 8x^2, 8x^3; 8x^{n-1}$ **11.** $g_1 = 3; r = 2$ **13.** $g_1 = 1; r = \frac{1}{2}$ **15.** $g_1 = 3125; r = -\frac{1}{5}$ **17.** $g_1 = 100; r = \frac{3}{2}$ **19.** $g_1 = x; r = x$ **21.** $(3)2^{n-1}$ **23.** $(\frac{1}{2})^{n-1}$ or 2^{1-n} **25.** $3125(-\frac{1}{5})^{n-1}$ or $-(-5)^{6-n}$ **27.** $100(\frac{3}{2})^{n-1}$ or $2^{3-n}(3^{n-1})(5^2)$ **29.** $x(x^{n-1})$ or x^n **31.** 486 **33.** 4 **35.** $\frac{1}{27}$ **37.** $\frac{9}{32}$ **39.** 726 **41.** 2044 **43.** 25,220 **45.** $\dfrac{61}{243} \approx .2510288066$ **47.** $-\dfrac{21}{8} = -2.625$ **49.** $\dfrac{6560}{729} \approx 8.9986283$ **51.** $r = 3, n = 7; G_7 = 5465$ **53.** $r = \frac{2}{3}, n = 9; G_9 = 19{,}171$ **55.** $r = \frac{4}{5}, n = 6; G_6 = 11{,}529$ **57.** $r = -\frac{3}{4}, n = 7; G_7 = 2653$ **59.** $r = x, n = 11; G_{11} = \dfrac{8(1 - x^{11})}{1 - x}$ **61.** 9330 letters mailed **63.** 11 mailings (435,356,465) would be sufficient. **65.** 1,024,000,000 present after 10 days **67.** $2\sqrt{2}$ **69.** $-\sqrt{15}$ **71.** $4\sqrt{5}$

Problem Set 8.4, Pages 349–352

1. 2 **3.** 4 **5.** 2000 **7.** $-\frac{40}{3}$ **9.** $\frac{1}{2}$ **11.** $\frac{1}{4}$ **13.** 256 **15.** 512 **17.** $\frac{3}{2}$ **19.** $\frac{9}{20}$ **21.** $\frac{1}{3}$ **23.** $\frac{2}{3}$ **25.** $\frac{2}{11}$ **27.** $\frac{923}{999}$ **29.** $\frac{133}{99}$ **31.** $\frac{1}{6}$ **33.** $\frac{25}{111}$ **35.** $\frac{250}{111}$ **37.** $\frac{44}{333}$ **39.** $\frac{440}{333}$ **41.** Does not have a sum; $r = \frac{3}{2}$ **43.** $\frac{81}{160}$ **45.** $\frac{1}{3}$ **47.** $\frac{27}{2}$ or 13.5 **49.** $\frac{1132}{225}$ **51.** $\frac{4}{7}$ **53.** $4 + 2\sqrt{2}$ **55.** $\frac{81}{7}\sqrt{2} - \frac{54}{7}$ **57.** $2 + \frac{3}{2}\sqrt{2}$ **59.** $\dfrac{\sqrt{2}}{2}$ **61.** The pendulum swings 200 cm. **63.** The flywheel makes 1500 revolutions. **65.** The superball travels 190 ft. **67.** The total depreciation is \$10,000. **69.** $g_1 = \frac{50}{3}, r = \frac{1}{6}$ **71.** $\dfrac{4a^2}{3}$

73. a. Perimeters: $3a, 4a, \dfrac{16a}{3}, \ldots, 3a\left(\dfrac{4}{3}\right)^{n-1}$, which increases without limit since $\frac{4}{3} > 1$.

b.
$$\text{AREA} = \frac{a^2\sqrt{3}}{4} + \frac{a^2\sqrt{3}}{12} + \frac{a^2\sqrt{3}}{12}\left(\frac{4}{9}\right) + \frac{a^2\sqrt{3}}{12}\left(\frac{4}{9}\right)^2 + \cdots$$
$$= \frac{a^2\sqrt{3}}{4} + \frac{3a^2\sqrt{3}}{20}$$
$$= \frac{2a^2\sqrt{3}}{5}$$
Thus the snowflake curve has finite area and infinite perimeter.

75. $\left(1 + \dfrac{.0806}{4}\right)^{4(20)} \approx 4.9331$ **77.** $\left(1 + \dfrac{.0806}{365}\right)^{365(20)} \approx 5.0119$ **79.** It appears that \$5.01 is the maximum amount present.

Problem Set 8.5, Pages 358–360

1. 40 **3.** 49 **5.** 20 **7.** 30 **9.** 2184 **11.** $\frac{155}{32}$ **13.** 362,880 **15.** 39,916,800 **17.** 696 **19.** 2 **21.** 72 **23.** 7920 **25.** 36 **27.** 220 **29.** 1 **31.** 5 **33.** 270,725 **35.** 56 **37.** 70 **39.** 1 **41.** $\sum_{k=1}^{7}\left(\frac{1}{2}\right)^k$ or $\sum_{k=1}^{7} 2^{-k}$ **43.** $\sum_{k=1}^{50} 2k$ **45.** $\sum_{k=1}^{5} 6^{k-1}$ **47.** $\sum_{k=1}^{11} (10k - 9)$ **49.** $\sum_{k=1}^{7} \dfrac{2k+1}{6}$ **51.** $\sum_{k=1}^{7} 3^{-k}$ **53.** $\sum_{k=1}^{7}\left(\dfrac{-2}{3}\right)^{k-1}$ **55.** $\sum_{k=1}^{9} 8\left(-\dfrac{\sqrt{2}}{2}\right)^{k-1}$ or $\sum_{k=1}^{9} (-1)^{k-1}2^{(7-k)/2}$ **57.** $1 + 2 + 3 + \cdots + r$ **59.** $a_1b_1 + a_2b_2 + \cdots + a_rb_r$ **61.** $x = 70$

63. $x = 1, y = -5$ **65.** $\sum_{j=1}^{r} ka_j = k \sum_{j=1}^{r} a_j$

$$= k \sum_{j=1}^{r} 1$$

$$= k\underbrace{(1 + 1 + 1 + \cdots + 1)}_{r \text{ ones}} = kr$$

67. a. 100 **b.** 200

69. $11^0 = 1$; $11^1 = 11$; $11^2 = 121$; $11^3 = 1331$; the rows of Pascal's Triangle are the powers of 11; however, for 11^5 or higher powers look at the fifth row: 1, 5, 10, 10, 5, 1 and note that in order to obtain powers of 11 when two digits are in the same column, the tens digit must be carried into the next column, just as in addition: $11^5 = 161051$

71. $\binom{n}{r} = \frac{n!}{r!(n-r)!} = \frac{n!}{[n-(n-r)](n-r)!} = \binom{n}{n-r}$

73. Not true in general. For example, ab could be greater than n, as in $\binom{4}{3}\binom{4}{2}$ and $\binom{4}{6}\binom{4}{1}$, where $\binom{4}{6}$ has no meaning. More importantly, within the proper parameters, counterexamples exist, as in $\binom{7}{3}\binom{7}{2} = 7^2 \cdot 5 \cdot 3$ and $\binom{7}{6}\binom{7}{1} = 7^2$.

Problem Set 8.6, Pages 362–363

1. $x^4 + 4x^3 + 6x^2 + 4x + 1$ **3.** $x^5 - 5x^4 + 10x^3 - 10x^2 + 5x - 1$
5. $x^6 - 6x^5y + 15x^4y^2 - 20x^3y^3 + 15x^2y^4 - 6xy^5 + y^6$ **7.** $x^5 + 10x^4 + 40x^3 + 80x^2 + 80x + 32$
9. $x^5 - 15x^4 + 90x^3 - 270x^2 + 405x - 243$ **11.** 462 **13.** 1001 **15.** 1820 **17.** 2016 **19.** -16
21. $x^{15} - 15x^{14}y + 105x^{13}y^2 - 455x^{12}y^3$ **23.** $x^8 + 8\sqrt{2}\,x^7 + 56x^6 + 112\sqrt{2}\,x^5$ **25.** $x^{10} - 30x^9y + 405x^8y^2 - 3240x^7y^3$
27. $a^{15}b^{15} - 30a^{14}b^{15} + 420a^{13}b^{15} - 3640a^{12}b^{15}$ **29.** $z^{22} + 55z^{20}k + 1375z^{18}k^2 + 20{,}625z^{16}k^3$
31. $105x^{4/3}y^{13/3} + 15x^{2/3}y^{14/3} + y^5$ **33.** $190(2^{18})\,z^{3/2}x^{9/2} - 20(2^{19})\,z^{3/4}x^{19/4} + (2^{20})x^5$ **35.** $1792pq^3 - 1024q^3\sqrt{pq} + 256q^4$
37. $\frac{16}{243}m^{-16} + \frac{2}{729}m^{-17} + \frac{1}{19{,}683}m^{-18}$ **39.** $60q^{-4}r^{-4} - 24\sqrt{2}\,q^{-2}r^{-5} + 8r^{-6}$

41. $2^n = (1 + 1)^n$

$$2^n = \binom{n}{0}1^n + \binom{n}{1}1^{n-1}1^1 + \binom{n}{2}1^{n-2}1^2 + \cdots + \binom{n}{n}1^n$$

$$= \binom{n}{0} + \binom{n}{1} + \binom{n}{2} + \cdots + \binom{n}{n}$$

43. $3^n = (1 + 2)^n$

$$= \sum_{j=0}^{n}\binom{n}{j}1^{n-j}(2^j) = \sum_{j=0}^{n} 2^j\binom{n}{j}$$

Problem Set 8.7, Pages 363–366

1. finite sequence **2.** infinite sequence **3.** difference **4.** general term **5.** series **6.** $n\left(\frac{a_1 + a_n}{2}\right)$ **7.** g_1r^{n-1}
8. $\frac{g_1(1 - r^n)}{1 - r}$ **9.** multiplying **10.** limit **11.** $|r| \geq 1$ **12.** $\frac{g_1}{1 - r}$; $|r| < 1$ **13.** sum; 1; n **14.** factorial **15.** nth; rth
16. $n(n - 1)(n - 2) \cdot \cdots \cdot 3 \cdot 2 \cdot 1$ **17.** $\frac{n!}{r!(n - r)!}$ **18.** binomial coefficient **19.** $\sum_{k=0}^{n}\binom{n}{k}a^{n-k}b^k$ **20.** Pascal's Triangle
21. Geometric; 11^{n-1} **22.** Neither; 11111, 111111 **23.** Arithmetic; $4 + 3n$ **24.** Geometric; $2 \cdot 3^{4-n}$
25. Neither; 25, 36 **26.** Arithmetic; $10n - 9$ **27.** Geometric; -2^{-n} **28.** Neither; $\frac{9}{25}, \frac{11}{36}$ **29.** Neither; $\frac{4}{5}, \frac{5}{6}$
30. Arithmetic; $\frac{n + 1}{4}$ **31.** 5 **32.** 37 **33.** 200 **34.** 5 **35.** 14 **36.** 23 **37.** 2 **38.** .1 **39.** 110
40. $\frac{9}{2}$ or 4.5 **41.** 2560 **42.** $\frac{5}{512}$ **43.** 155 **44.** 410 **45.** $\frac{16}{5}$ **46.** -2000 **47.** $\frac{1}{2}$ **48.** $\frac{3}{4}$ **49.** 1023
50. 1050 **51.** 5, 8, 11, 14 **52.** 3, 8, 13, 18 **53.** $\frac{1}{10}, \frac{1}{5}, \frac{2}{5}, \frac{4}{5}$
54. 10^5, $5 \cdot 10^4$, $25 \cdot 10^3$, $125 \cdot 10^2$ or 100,000, 50,000, 25,000, 12,500 **55.** $-1, \frac{1}{2}, -\frac{1}{3}, \frac{1}{4}$ **56.** 1, -2, 3, -4
57. $0, \frac{1}{3}, \frac{1}{2}, \frac{3}{5}$ **58.** $-1, \frac{1}{2}, -\frac{1}{6}, \frac{1}{24}$ **59.** 1, 2, 3, 5 **60.** 1, -5, -4, -9 **61.** 65,792 **62.** 1022 **63.** 1330

64. 5187 **65.** $\frac{19{,}171}{6912}$ **66.** $\frac{2}{5}$ **67.** 200 **68.** 125 **69.** No sum; $r = \frac{3}{2}$ **70.** 432 **71.** 40,296 **72.** 24 **73.** 1680
74. 2 **75.** 70 **76.** 35 **77.** 1 **78.** 15 **79.** $\frac{p!}{q!(p-q)!}$ **80.** $\frac{(r+1)!}{(s-1)!(r-s+2)!}$
81. $1+2+3+4+5+6+7+8+9+10$ **82.** $-1-2-3-4-5-6$
83. $-1+1+3+5+7+9+11+13$ **84.** $7+5+3+1-1-3-5-7-9$ **85.** $2+6+18+54+162$
86. $3+6+12+24+48$ **87.** $2+6+12+20+30$ **88.** $\frac{1}{2}+\frac{2}{3}+\frac{3}{4}+\frac{4}{5}$ **89.** $\binom{n}{0}+\binom{n}{1}+\binom{n}{2}+\cdots+\binom{n}{n}$
90. $a^n + a^{n-1}b + a^{n-2}b^2 + \cdots + ab^{n-1} + b^n$
91. $\sum_{k=1}^{5} 2k$ **92.** $\sum_{k=1}^{9} 2^{9-k}$ **93.** $\sum_{k=1}^{n} a_k$ **94.** $\sum_{k=1}^{n} g_1 r^k$ **95.** $\sum_{k=1}^{n} \binom{n}{k} x^{n-k} y^k$ **96.** $\frac{24}{11}$
97. **a.** $x^5 - 5x^4y + 10x^3y^2 - 10x^2y^3 + 5xy^4 - y^5$ **b.** $32x^5 + 80x^4y + 80x^3y^2 + 40x^2y^3 + 10xy^4 + y^5$ **c.** 7920
98. 2046 people **99.** There will be 72 divisions in 24 hours; $g_{73} = 2^{10} \cdot 2^{72} = 2^{82} \approx 4.835703249 \cdot 10^{24}$.
100. The bob will travel 625 cm.

Chapter 9

Problem Set 9.1, Pages 373–374

1. 9 **3.** 504 **5.** 1 **7.** 132,600 **9.** 24 **11.** 95,040 **13.** 1680 **15.** $\frac{g!}{(g-h)!}$ **17.** 52. **19.** 3360 **21.** 210
23. $\frac{50!}{2}$ or $25 \cdot 49!$ **25.** $\frac{m!}{(m-3)!}$ or $m(m-1)(m-2)$ **27.** 1 **29.** 7920 **31.** 7! or 5040 **33.** 360 **35.** 34,650
37. 4,989,600 **39.** 831,600 **41.** 120 **43.** 720 **45.** 16 **47.** 336 **49.** 6,400,000 **51.** 900,000,000
53. 2,600,000 **55.** 786,240 **57.** 7776 **59.** 17,576,000(TOTAL) − 15,600,000(NONREPEATING) = 1,976,000(REPEATING)
61. 41,472 possibilities (of which only one is a solution) **63.** 48

Problem Set 9.2, Pages 379–381

1. 9 **3.** 84 **5.** 1 **7.** 22,100 **9.** 1 **11.** 35 **13.** 1225 **15.** $\frac{g!}{h!(g-h)!}$ **17.** 1 **19.** 792 **21.** 1 **23.** 4
25. $\frac{n!}{5!(n-5)!}$ **27.** 21 **29.** 66 **31.** 6 **33.** 28 **35.** 28 **37.** 75,287,520 **39.** 6 **41.** 1287 **43.** 24
45. Explanations and examples will vary. **47.** Permutation; ${}_8P_5$; 6720 **49.** Permutation; ${}_5P_5$; 120
51. Combination; ${}_{100}C_6$; 1,192,052,400 **53.** Permutation; ${}_6P_6$; 720 **55.** Neither; $3 \cdot 3 \cdot 2 \cdot 2 \cdot 1 \cdot 1$; 36
57. Neither; $4 \cdot 3 \cdot 2 \cdot 1$; 2^{100} **59.** Permutation; ${}_6P_5$; 720 **61.** Combination; ${}_{12}C_{10}$; 66 **63.** Permutation; ${}_7P_7$; 5040
65. Neither; $2 \cdot 2 \cdot 2 \cdot \cdots \cdot 2 = 2^{10}$; 1024 **67.** Neither; $5 \cdot 5 \cdot 5 \cdot \cdots \cdot 5 = 5^{10}$; 9,765,625
69. Neither; ${}_{17}P_2 \cdot {}_{19}P_2 \cdot {}_{32}C_6$; approximately $8.4298 \cdot 10^{10}$
71. ${}_nC_r = \frac{n!}{r!(n-r)!}$; also, ${}_nC_{n-r} = \frac{n!}{(n-r)!(n-n+r)!} = \frac{n!}{(n-r)!r!} = {}_nC_r$

Problem Set 9.3, Pages 386–388

1. $\frac{1}{13}$ **3.** $\frac{4}{13}$ **5.** $\frac{1}{13}$ **7.** $\frac{2}{13}$ **9.** $\frac{1}{52}$ **11.** $\frac{10}{13}$ **13.** $\frac{1}{6}$ **15.** $\frac{1}{2}$ **17.** 0 **19.** $\frac{2}{3}$ **21.** $\frac{5}{6}$ **23.** $\frac{6}{13}$ **25.** $\frac{3}{13}$ **27.** $\frac{10}{13}$
29. .01 **31.** .95 **33.** .20 **35.** $\frac{5}{36}$ **37.** $\frac{13}{18}$ **39.** $\frac{4}{9}$ **41.** $\frac{2}{9}$ **43.** $\frac{5}{8}$ **45.** $\frac{3}{4}$ **47.** $\frac{1}{6} \approx .17$ **49.** 0 **51.** .21
53. .1 **55.** .04 **57.** .15 **59.** .07 **61.** .00 **63.** .06 **65.** .25 **67.** $\frac{1}{108{,}290} \approx .00000923$

Problem Set 9.4, Pages 394–396

1. $\frac{1}{2}$ **3.** $\frac{5}{6}$ **5.** $\frac{1}{12}$ **7.** $\frac{2}{3}$ **9.** $\frac{4}{9}$ **11.** $\frac{11}{12}$ **13.** $\frac{1}{3}$ **15.** $\frac{5}{9}$ **17.** $\frac{3}{7}$ **19.** $\frac{3}{7}$ **21.** $\frac{3}{7}$ **23.** 1 **25.** $\frac{4}{11}$ **27.** $\frac{2}{5}$
29. $\frac{4}{15}$ **31.** $\frac{1}{4}$ **33.** 1 **35.** $\frac{2}{3}$ **37.** $\frac{1}{6}$ **39.** $\frac{5}{18}$ **41.** $\frac{11}{12}$ **43.** $\frac{7}{12}$ **45.** $\frac{1}{9}$ **47.** $\frac{1}{6}$ **49.** $\frac{1}{12}$ **51.** $\frac{1}{6}$ **53.** $\frac{3}{4}$

55. $\frac{1}{2}$ **57.** $\frac{1}{3}$ **59.** $\frac{1}{2}$ **61.** $\frac{1}{2}$; past events do not influence future independent events **63.** 60% **65.** $\frac{25}{64}$ **67.** $\frac{15}{64}$ **69.** 1
71. $\frac{5}{14}$ **73.** $\frac{15}{28}$ **75.** 1 **77.** $1 - P(\text{No aces or kings}) = 1 - \dfrac{_{88}C_8}{_{104}C_8} \approx \dfrac{3}{4}$

79.
$$\begin{aligned} P(\overline{A} \cap \overline{B}) &= P(\overline{A \cup B}) \\ &= 1 - P(A \cap B) \\ &= 1 - [P(A) + P(B) - P(A \cap B)] \\ &= 1 - P(A) - P(B) + P(A \cap B) \\ &= [1 - P(A)][1 - P(B)] \\ &= P(\overline{A}) \cdot P(\overline{B}) \end{aligned}$$

Problem Set 9.5, Pages 396–398

1. an ordered arrangement of those q elements selected without repetitions
2. the number of permutations of m elements selected from a set of q elements **3.** $\dfrac{q!}{(q-m)!}$
4. If task A can be performed in m ways, and after task A is performed a second task B can be performed in n ways, then task A followed by task B can be performed in $m \cdot n$ ways. **5.** $\dfrac{m!}{q_1!q_2! \cdot \cdots \cdot q_k!}$ **6.** a subset of S that contains q distinct elements
7. $_mC_q$ **8.** $\dfrac{q!}{r!(q-r)!}$ **9.** permutation; combination
10. If an experiment can occur in any of n, $n \geqslant 1$, mutually exclusive and equally likely ways and if s of these ways are considered favorable, then the probability of the event E, denoted by $P(E)$, is $P(E) = \dfrac{s}{n}$. **11.** 1 **12.** 0 **13.** $1 - P(F)$ **14.** $\dfrac{P(E \cap F)}{P(F)}$
15. $P(E) \cdot P(F)$; E and F are independent **16.** $P(E) + P(F) - P(E \cap F)$ **17.** 0 **18.** $P(E) + P(F)$
19. that the result is noted and then replaced before the next event
20. that the result is noted and then the second event occurs without replacing the card **21.** 60 **22.** $8! = 40{,}320$
23. 4,989,600 **24.** 360 **25.** 151,200 **26.** 7,560 **27.** 60 **28.** 2520 **29.** 19,958,400 **30.** $n(n-1)(n-2)$
31. $\dfrac{j!}{(j-h)!}$ **32.** 10 **33.** 21 **34.** 495 **35.** $\dfrac{n!}{3!(n-3)!}$ **36.** $\dfrac{j!}{h!(j-h)!}$ **37.** 12 **38.** 1 **39.** 7
40. $8! = 40{,}320$ **41.** 1 **42.** 11,880 **43.** 495 **44.** 1 **45.** 63 **46.** 42 **47.** 35 **48.** 0 **49.** $\frac{1}{13}$ **50.** $\frac{1}{4}$
51. $\frac{3}{13}$ **52.** $\frac{4}{13}$ **53.** $\frac{1}{52}$ **54.** 0 **55.** .005 **56.** .059 **57.** .090 **58.** .000 (.0003698225) **59.** .003 **60.** $\frac{2}{5}$
61. 1 **62.** $\frac{1}{3}$ **63.** $\frac{1}{6}$ **64.** $\frac{1}{2}$ **65.** $\frac{1}{3}$ **66.** $\frac{1}{6}$ **67.** $\frac{1}{2}$ **68.** $\frac{5}{9}$ **69.** $\frac{5}{12}$ **70.** $\frac{17}{18}$ **71.** $\frac{5}{6}$ **72.** $\frac{2}{3}$ **73.** $\frac{1}{12}$
74. $\frac{36}{121}$ **75.** $\frac{25}{121}$ **76.** $\frac{30}{121}$ **77.** $\frac{30}{121}$ **78.** $\frac{91}{121}$ **79.** $\frac{91}{121}$ **80.** $\frac{60}{121}$ **81.** $\frac{3}{11}$ **82.** $\frac{2}{11}$ **83.** $\frac{3}{11}$ **84.** $\frac{3}{11}$ **85.** $\frac{8}{11}$
86. $\frac{8}{11}$ **87.** $\frac{6}{11}$

Chapter 10

Problem Set 10.1, Pages 404–405

1.–3. Answers will vary; see discussion in text. **5.** If there is time enough, then everything happens to everybody sooner or later.
7. If life is useless, then it is an early death. **9.** If only you can find it, then everything has a moral. **11.** Answers vary
13. Commutative (F3) **15.** Inverse (F9) **17.** Closure (F2) **19.** Transitive (E3) or substitution (E4) **21.** Identity (F8)
23. 4 is divisible by 1 and 4, but is not prime **25.** 1 is not prime **27.** If $a = 0$, then x is undefined.
29. $\dfrac{2^3 - 1^3}{2 - 1} = \dfrac{7}{1} \neq 4 - 1$ **31.** $\dfrac{6^1}{2^1} \neq 3^{1-1}$ **33.** I become lazy. **35.** $ABCD$ has equal diagonals.
37. I will attend college in the fall.

Problem Set 10.2, Pages 410–415

1. Converse: If you go to jail, then you will break the law.
3. If the sun shines, then you are happy. Converse: If you are happy, then the sun shines.

5. Converse: If you do not have any friends, then your car is not air-conditioned.
7. If it is a cockroach, then it is ugly. Converse: If it is ugly, then it is a cockroach.
9. If we do not stop the pollution of our environment, then your grandchildren will not survive. Converse: If your grandchildren do not survive, then we will not stop polluting our environment. **11.** If I eat that piece of pie, then I will be unhealthy.
13. All snarks are ugly. **15.** If the World Health Organization sends pesticide to Borneo, then there is a real danger of plague.
17. p: the judge will sentence you; q: the prosecution can prove your guilt; p iff q **19.** p: $x^2 = 16$; q: $x = 4$; $p \leftarrow q$ or $q \rightarrow p$
21. p: $\triangle ABC$ is a right triangle; q: $a^2 + b^2 = c^2$; p iff q **23.** Distributive (F11) **25.** Commutative (F4) **27.** Transitive (E3)
29. Inverse (F9) **31.** $a + 0 = 0 + (-b)$ **33.** $a, b \in \mathbf{R}, a, b \neq 0$ **35.** $b = 1/a$ **37.** Inverse (F10)
39. Commutative (F4) **41.** $b(1) = 1(1/a)$ **43.** Reflexive (E1) **45.** $= a + [(-b) + b]$ **47.** $= a$ **49.** $a, b \in \mathbf{R}$
51. Reflexive (E1) **53.** Commutative (F3) **55.** $= 3a + 2a + 0$ **57.** Distributive (F11) **59.** $(ab)\frac{1}{a} \cdot \frac{1}{b} = 1$
61. Commutative (F4) **63.** $= 1 \cdot 1$ **65.** $= 1$ **67.** Transitive (E3) **69.** $b + 2a = 0$ **71.** Given
73. $(2a)(\frac{1}{2}) = (-b)(\frac{1}{2})$ **75.** Associative (F6) **77.** $a = -b/2$
79. $a, b \in \mathbf{R}$; $a, b \neq 0$; $a = 1/b$ TO PROVE: $b = 1/a$

Statements	Reasons
1. $a, b \in \mathbf{R}$; $a, b \neq 0$; $a = \frac{1}{b}$	1. Given
2. $ab = \frac{1}{b}(b)$	2. Multiplication Law of Equality (Theorem 2)
3. $ab = 1$	3. Inverse (F10)
∴ 4. $b = \frac{1}{a}$	4. Reciprocal of a real number is unique (Theorem 10.2.3)

81. GIVEN: $a, b \in \mathbf{R}$ TO PROVE: $(2b - 3a) + (3a - 2b) = 0$

Statements	Reasons
1. $a, b \in \mathbf{R}$	1. Given
2. $(2b - 3a) + (3a - 2b) = (2b - 3a) + (3a - 2b)$	2. Reflexive (E1)
3. $= 2b + (-3a) + 3a + (-2b)$	3. Definition of subtraction
4. $= 2b + [(-3a) + 3a] + (-2b)$	4. Associative (F5)
5. $= 2b + 0 + (-2b)$	5. Inverse (F9)
6. $= 2b + (-2b)$	6. Identity (F7)
7. $= 0$	7. Inverse (F9)
∴ 8. $(2b - 3a) + (3a - 2b) = 0$	8. Transitive (E3)

83. GIVEN: $a, b \in \mathbf{R}$; $b \neq 0$ TO PROVE: $\frac{1}{b}(ba) = a$

Statements	Reasons
1. $a, b \in \mathbf{R}$; $b \neq 0$	1. Given
2. $\frac{1}{b}(ba) = \frac{1}{b}(ba)$	2. Reflexive (E1)
3. $= \left(\frac{1}{b}b\right)a$	3. Associative (F6)
4. $= 1a$	4. Inverse (F10)
5. $= a$	5. Identity (F8)
∴ 6. $\frac{1}{b}(ba) = a$	6. Transitive (E3)

85. *Part 1* If $a - 2b = 0$, then $b = a/2$.
GIVEN: $a, b \in \mathbf{R}; a - 2b = 0$ TO PROVE: $b = a/2$

Statements	Reasons
1. $a - 2b = 0; a, b \in \mathbf{R}$	1. Given
2. $a + (-2b) = 0$	2. Definition of subtraction
3. $a + (-2b) + 2b = 0 + 2b$	3. Addition Law of Equality (Theorem 1)
4. $a + [-2b + 2b] = 0 + 2b$	4. Associative (F5)
5. $a + 0 = 0 + 2b$	5. Inverse (F9)
6. $a = 2b$	6. Identity (F7)
$\therefore$ 7. $b = a/2$	7. Problem 84

Part 2 If $b = a/2$, then $a - 2b = 0$.
GIVEN: $a, b \in \mathbf{R}; b = a/2$ TO PROVE: $a - 2b = 0$

Statements	Reasons
1. $b = \frac{a}{2}; a, b \in \mathbf{R}$	1. Given
2. $2b = a$	2. Problem 84
3. $2b + (-2b) = a + (-2b)$	3. Theorem 1
4. $0 = a + (-2b)$	4. Inverse (F9)
5. $0 = a - 2b$	5. Definition of subtraction
$\therefore$ 6. $a - 2b = 0$	6. Symmetric (E2)

87. GIVEN: $0 \in \mathbf{R}$ TO PROVE: $-0 = 0$

Statements	Reasons
1. $0 \in \mathbf{R}$	1. Given
2. $0 + 0 = 0$	2. Identity (F7)
3. $0 + (-0) = 0$	3. Inverse (F9)
4. $0 + 0 = 0 + (-0)$	4. Substitution (E4)
5. $0 + 0 = -0 + 0$	5. Commutative (F3)
6. $0 = -0$	6. Cancellation Law for Addition (Theorem 4) or Identity (F7)

Problem Set 10.3, Pages 421–425

1. Distributive (F11) (or right-distributive, Theorem 10.2.1) **3.** $0 = ab + (-a)b$ **5.** Commutative (F3) **7.** Opposite of a real number is unique (Theorem 10.2.2) **9.** Substitution (E4) **11.** $(-1)a = -a$ **13.** Given **15.** Commutative (F4) **17.** Theorem 10.3.2 **19.** $a > b; c < 0$ **21.** Given **23.** Theorem 10.3.3 **25.** Order axiom (O1) **27.** $ca - cb < 0$ **29.** $ac < bc$ **31.** $a > b; b > c$ **33.** Given **35.** Order axiom (O2) **37.** Inverse (F9) **39.** Definition of subtraction **41.** $a \in \mathbf{R}$ **43.** $a^2 > 1$ **45.** Definition of $>$ **47.** Definition of $>$ **49.** $a \in \mathbf{R}$ **51.** Definition of negative **53.** Double negative (Theorem 7) **55.** Definition of absolute value **57.** Order axiom (O1) **59.** Substitution (E4) **61.** $|a|^2 = a^2$ **63.** Definition of absolute value **65.** Definition of absolute value **67.** Substitution (E4) **69.** Definition of absolute value **71.** $|a|\,|a| = aa$ **73.** Definition of exponent

75. GIVEN: $a, b, c \in \mathbf{R}$ TO PROVE: $a(b - c) = ab - ac$

Statements	Reasons
1. $a(b - c) = a(b - c)$	1. Reflexive (E1)
2. $a(b - c) = a[b + (-c)]$	2. Definition of subtraction
3. $a(b - c) = ab + a(-c)$	3. Distributive (F11)
4. $a(b - c) = ab + [-(ac)]$	4. Property of Opposites (Theorem 8, part 1)
$\therefore$ 5. $a(b - c) = ab - ac$	5. Definition of subtraction

77. GIVEN: $a, b \in \mathbf{R}$ TO PROVE: $(-a) + (-b) = -(a + b)$

Statements	Reasons
1. $(-a) + (-b) = (-a) + (-b)$	1. Reflexive (E1)
2. $(-a) + (-b) = (-1)a + (-1)b$	2. Property of Opposites (Theorem 8, part 4)
3. $(-a) + (-b) = (-1)(a + b)$	3. Distributive (F11)
$\therefore$ 4. $(-a) + (-b) = -(a + b)$	4. Property of Opposites (Theorem 8, part 4)

79. GIVEN: $a, b \in \mathbf{R}$ TO PROVE: $-(a - b) = b - a$

Statements	Reasons
1. $-(a - b) = -(a - b)$	1. Reflexive (E1)
2. $-(a - b) = (-1)(a - b)$	2. Property of Opposites (Theorem 8, part 4)
3. $-(a - b) = (-1)a - (-1)(b)$	3. Problem 75
4. $-(a - b) = (-1)a + (-[(-1) \cdot b]$	4. Definition of subtraction
5. $-(a - b) = (-1)a + (1 \cdot b)$	5. Double negative (Theorem 7)
6. $-(a - b) = (-1)a + b$	6. Identity (F8)
7. $-(a - b) = b + (-1)a$	7. Commutative (F3)
$\therefore$ 8. $-(a - b) = b - a$	8. Definition of subtraction

81. GIVEN: $a, b \in \mathbf{R}$; $a = \pm b$ TO PROVE: $|a| = |b|$

Statements	Reasons
1. $\lvert a\rvert = \lvert a\rvert$	1. Reflexive (E1)
Case *i*: $a = b$	
$\therefore$ 2. $\lvert a\rvert = \lvert b\rvert$	2. Substitution (E4)
Case *ii*: $a = -b$	
3. $\lvert a\rvert = \lvert -b\rvert$	3. Substitution (E4)
$\therefore$ 4. $\lvert a\rvert = \lvert b\rvert$	4. Theorem 10.3.7

83. GIVEN: $a, b \in \mathbf{R}$; $a - b < 0$ TO PROVE: $|a - b| = |b - a|$

Statements	Reasons
1. $a - b < 0$	1. Given
2. $\lvert a - b\rvert = -(a - b)$	2. Definition of absolute value
3. $\lvert a - b\rvert = b - a$	3. Problem 79
4. $-(a - b) > 0$	4. Order axiom (O1)
5. $b - a > 0$	5. Problem 79
6. $\lvert b - a\rvert = b - a$	6. Definition of absolute value
$\therefore$ 7. $\lvert a - b\rvert = \lvert b - a\rvert$	7. Substitution, Step 6 into Step 3

85. GIVEN: $a, b \in \mathbf{R}$ TO PROVE: $|a - b| = |b - a|$

Statements	Reasons
1. $a - b < 0$, $a - b = 0$, or $a - b > 0$	1. Order axiom (O1)
2. If $a - b < 0$, then $\lvert a - b\rvert = \lvert b - a\rvert$.	2. Problem 83
3. If $a - b = 0$, then $\lvert a - b\rvert = \lvert b - a\rvert$.	3. Problem 84
4. If $a - b > 0$, then $\lvert a - b\rvert = \lvert b - a\rvert$.	4. Problem 82
$\therefore$ 5. $\lvert a - b\rvert = \lvert b - a\rvert$	5. Steps 1–4

87. GIVEN: $a, b \in \mathbf{R}$ TO PROVE: $a^2 - 2ab + b^2 = (a - b)^2$

Statements	Reasons
1. $(a - b)^2 = (a - b)(a - b)$	1. Definition of exponent
2. $= [a + (-b)][a + (-b)]$	2. Definition of subtraction
3. $= [a + (-b)]a + [a + (-b)](-b)$	3. Distributive (F11)
4. $= aa + (-b)a + a(-b) + (-b)(-b)$	4. Distributive (F11)
5. $= aa + (-ab) + (-ab) + bb$	5. Theorem 8
6. $= a^2 + (-ab) + (-ab) + b^2$	6. Definition of exponents
7. $= a^2 + (-1)ab + (-1)ab + b^2$	7. Theorem 8
8. $= a^2 + [(-1) + (-1)]ab + b^2$	8. Distributive (F11)
9. $= a^2 + (-2)ab + b^2$	9. Closure (F1)
10. $= a^2 + [-(2ab)] + b^2$	10. Theorem 8
11. $= a^2 - 2ab + b^2$	11. Definition of subtraction
12. $(a - b)^2 = a^2 - 2ab + b^2$	12. Transitive (E3)
$\therefore$ 13. $a^2 - 2ab + b^2 = (a - b)^2$	13. Symmetric (E2)

89. GIVEN: $a, \in \mathbf{R}$ TO PROVE: $-|a| \leq a \leq |a|$

Case i: $a > 0$

$|a| = a$, so $a \leq |a|$ since $a \leq a$
and $-|a| \leq a$ since $-a \leq a$ if $a > 0$
Thus, $-|a| \leq a \leq |a|$ if $a > 0$.

Case ii: $a = 0$

$|a| = 0$, so $a \leq |a|$ since $0 \leq 0$
and $-|a| \leq a$ since $-0 = 0$ and $0 \leq 0$
Thus, $-|a| \leq a \leq |a|$ if $a = 0$.

Case iii: $a < 0$

$|a| = -a$, so $a \leq |a|$ since $a < 0$ and $|a| \geq 0$
and $-|a| \leq a$ since $-(-a) = a$ and $a \leq a$
Thus, $-|a| \leq a \leq |a|$ if $a < 0$.

Therefore, by trichotomy, $a > 0$, $a = 0$, or $a < 0$, so in all cases $-|a| \leq a \leq |a|$.

91. GIVEN: $a, b \in \mathbf{R}$; $b \neq 0$ TO PROVE: $\left|\frac{a}{b}\right| = \frac{|a|}{|b|}$

Statements	Reasons
1. $\left|\frac{a}{b}\right| = \left|a \cdot \frac{1}{b}\right|$	1. Definition of division
2. $\left|a \cdot \frac{1}{b}\right| = |a| \cdot \left|\frac{1}{b}\right|$	2. Problem 90
3. $\left|\frac{1}{b}\right| = \frac{1}{|b|}$ *case i*: $b > 0$ $\left|\frac{1}{b}\right| = \frac{1}{b}$ is true *case ii*: $b < 0$ $\left|\frac{1}{b}\right| = \frac{1}{-b}$ is true	3. Definition of absolute value
4. $|a| \cdot \left|\frac{1}{b}\right| = |a| \cdot \frac{1}{|b|}$	4. Substitution (E4)

Statements	Reasons
5. $\lvert a\rvert \cdot \frac{1}{\lvert b\rvert} = \frac{\lvert a\rvert}{\lvert b\rvert}$	5. Definition of division
$\therefore$ 6. $\left\lvert \frac{a}{b} \right\rvert = \frac{\lvert a\rvert}{\lvert b\rvert}$	6. Transitive (E3)

Problem Set 10.4, Pages 426–429

1. Direct **3.** Indirect **5.** Indirect **7.** Transitive **9.** Direct **11.** Examples will vary. **13.** Invalid **15.** Direct **17.** Indirect **19.** Indirect **21.** Invalid **23.** Substitution (E4) **25.** Distributive (F11) **27.** Distributive (F11) **29.** Substitution; since a^2 is equal to an odd number **31.** a^2 is odd **33.** a is even

35. Let $(2x + 1)$ and $(2y + 1)$ be any odd integers. Then

$$(2x + 1)(2y + 1) = (2x + 1)(2y) + (2x + 1)\cdot 1 \quad \text{Distributive}$$
$$= (2x)(2y) + 2y + 2x\cdot 1 + 1\cdot 1 \quad \text{Distributive}$$
$$= 2(x\cdot 2y + y + x\cdot 1) + 1\cdot 1 \quad \text{Distributive}$$
$$= 2(x\cdot 2y + y + x\cdot 1) + 1 \quad \text{Identity}$$

$\therefore (2x + 1)(2y + 1)$ is odd

37. Let $p = 3k$ or $p = 3k + 1$ or $p = 3k + 2$, where k is an integer. Then

$$p^2 = 9k^2 \text{ or } \underbrace{p^2 = 9k^2 + 6k + 1 \text{ or } p^2 = 9k^2 + 12k + 4}_{\text{These are not divisible by 3}}$$

So $p^2 = 9k^2 \Rightarrow p = 3k \Rightarrow p$ is divisible by 3.

39. Suppose that $\frac{1}{2}$ is integral (an integer).

Then $\frac{1}{2} = k$ is some integer.

$1 = 2k$ Multiply both sides by 2.

$\therefore 1$ is even Contradiction

$\therefore \frac{1}{2}$ is nonintegral

41. Answers vary (for example, the rationals between 0 and 1)

Problem Set 10.5, Pages 433–435

1. *Step 1* Prove it true for $n = 1$: $1 = \frac{1(1+1)}{2}$ ✓

Step 2 Assume it true for $n = k$:

$$1 + 2 + 3 + \cdots + k = \frac{k(k+1)}{2}$$

Step 3 Prove it true for $n = k + 1$.

TO PROVE: $1 + 2 + 3 + \cdots + (k + 1) = \frac{(k+1)(k+2)}{2}$

Statements	Reasons
1. $1 + 2 + 3 + \cdots + k = \dfrac{k(k+1)}{2}$	1. Hypothesis
2. $1 + 2 + 3 + \cdots + k + (k+1) = \dfrac{k(k+1)}{2} + k + 1$	2. Add $k + 1$ to both sides.
3. $= \dfrac{k(k+1) + 2(k+1)}{2}$	3. Adding fractions
$\therefore$ 4. $= \dfrac{(k+1)(k+2)}{2}$	4. Factoring

Step 4 The proposition is true for all positive integers by PMI.

3. *Step 1* Prove it true for $n = 1$:

$$(2\cdot 1 - 1)^2 \stackrel{?}{=} \frac{1(2\cdot 1 - 1)(2\cdot 1 + 1)}{3}$$

$$1 = \frac{1\cdot 1\cdot 3}{3} \checkmark$$

Step 2 Assume it true for $n = k$:

$$1^2 + 3^2 + 5^2 + \cdots + (2k-1)^2 = \frac{k(2k-1)(2k+1)}{3}$$

Step 3 Prove it true for $n = k + 1$.

TO PROVE: $1^2 + 3^2 + \cdots + (2k+1)^2 = \dfrac{(k+1)(2k+1)(2k+3)}{3}$

BY HYPOTHESIS:

$$1^2 + 3^2 + 5^2 + \cdots + (2k-1)^2 = \frac{k(2k-1)(2k+1)}{3}$$

Add $(2k+1)^2$ to both sides:

$$1^2 + 3^2 + 5^2 + \cdots + (2k-1)^2 + (2k+1)^2 = \frac{k(2k-1)(2k+1)}{3} + (2k+1)^2$$

$$= \frac{(2k-1)(2k+1) + 3(2k+1)^2}{3}$$

$$= \frac{(2k+1)[k(2k-1) + 3(2k+1)]}{3}$$

$$= \frac{(2k+1)(2k^2 + 5k + 3)}{3}$$

$$= \frac{(2k+1)(k+1)(2k+3)}{3}$$

Step 4 The proposition is true for all positive integers by PMI.

5. *Step 1* Prove it true for $n = 1$:

$$2^2 \stackrel{?}{=} \frac{2\cdot 1\cdot 2\cdot 3}{3}$$

$$4 = 4 \checkmark$$

Step 2 Assume it is true for $n = k$:

$$2^2 + 4^2 + 6^2 + \cdots + (2k)^2 = \frac{2k(k+1)(2k+1)}{3}$$

Step 3 Prove it true for $n = k + 1$.

TO PROVE: $2^2 + 4^2 + 6^2 + \cdots + (2k+2)^2 = \dfrac{2(k+1)(k+2)(2k+3)}{3}$

BY HYPOTHESIS:

$$2^2 + 4^2 + 6^2 + \cdots + (2k)^2 = \frac{2k(k+1)(2k+1)}{3}$$

Add $(2k + 2)^2$ to both sides:

$$2^2 + 4^2 + 6^2 + \cdots + (2k)^2 + (2k + 2)^2 = \frac{2k(k + 1)(2k + 1)}{3} + (2k + 2)^2$$
$$= \frac{2k(k + 1)(2k + 1) + 12(k + 1)^2}{3}$$
$$= \frac{2(k + 1)(2k^2 + k + 6k + 6)}{3}$$
$$= \frac{2(k + 1)(k + 2)(2k + 3)}{3}$$

Step 4 The proposition is true for all positive integers by PMI.

7. *Step 1* Prove it true for $n = 1$:

$$1 \cdot 3 \stackrel{?}{=} \frac{1(2)(9)}{6}$$
$$3 = 3 \checkmark$$

Step 2 Assume it true for $n = k$:

$$1 \cdot 3 + 2 \cdot 4 + \cdots + k(k + 2) = \frac{k(k + 1)(2k + 7)}{6}$$

Step 3 Prove it true for $n = k + 1$:

TO PROVE: $1 \cdot 3 + 2 \cdot 4 + \cdots + k(k + 2) = \dfrac{k(k + 1)(2k + 7)}{6}$

Add $(k + 1)(k + 3)$ to both sides:

$$1 \cdot 3 + 2 \cdot 4 + \cdots + k(k + 2) + (k + 1)(k + 3) = \frac{k(k + 1)(2k + 7)}{6} + (k + 1)(k + 3)$$
$$= \frac{k(k + 1)(2k + 7) + 6(k + 1)(k + 3)}{6}$$
$$= \frac{(k + 1)(2k^2 + 7k + 6k + 18)}{6}$$
$$= \frac{(k + 1)(2k^2 + 13k + 18)}{6}$$
$$= \frac{(k + 1)(k + 2)(2k + 9)}{6}$$

Step 4 The proposition is true for all positive integers by PMI.

9. *Step 1* Prove it true for $n = 1$:
$1^5 - 1 = 0$, which is divisible by 5.

Step 2 Assume it true for $n = k$:
$k^5 - k$ is divisible by 5.

Step 3 Prove it true for $n = k + 1$.
TO PROVE: $(k + 1)^5 - (k + 1)$ is divisible by 5.

$$(k + 1)^5 - (k + 1) = k^5 + 5k^4 + 10k^3 + 10k^2 + 5k + 1 - k - 1$$
$$= (k^5 - k) + (5k^4 + 10k^3 + 10k^2 + 5k)$$

$k^5 - k$ is divisible by 5 by hypothesis. Thus,
$5k^4 + 10k^3 + 10k^2 + 5k = 5(k^4 + 2k^3 + 2k^2 + k)$ is divisible by 5.
Therefore, $(k + 1)^5 - (k + 1)$ is divisible by 5.

Step 4 The proposition is true for all positive integers by PMI.

11. *Step 1* Prove it true for $n = 1$:

$$2^2 \geqslant 1 + 1^2$$
$$4 \geqslant 2 \checkmark$$

Step 2 Assume it true for $n = k$:

$$(1 + k)^2 \geqslant 1 + k^2$$

Step 3 Prove it true for $n = k + 1$.

TO PROVE: $(2 + k)^2 \geq 1 + (1 + k)^2$

Clearly, $3 \geq 1$ and $2k + 3 \geq 2k + 1$, and by hypothesis:

$$(1 + k)^2 \geq 1 + k^2$$

Adding corresponding members:

$$(1 + k)^2 + (2k + 3) \geq (1 + k^2) + (2k + 1)$$
$$4 + 4k + k^2 \geq 1 + 1 + 2k + k^2$$
$$(2 + k)^2 \geq 1 + (1 + k)^2$$

Step 4 The proposition is true for all positive integers by PMI.

13. Conjecture: $1 + 4 + 7 + \cdots + (3n - 2) = \dfrac{n(3n - 1)}{2}$

Step 1 Prove it true for $n = 1$:

$$1 = \frac{1(2)}{2} = 1 \checkmark$$

Step 2 Assume it true for $n = k$:

$$1 + 4 + 7 + \cdots + (3k + 2) = \frac{k(3k - 1)}{2}$$

Step 3 Prove it true for $n = k + 1$.

TO PROVE: $1 + 4 + 7 + \cdots + (3k + 1) = \dfrac{(k + 1)(3k + 2)}{2}$

BY HYPOTHESIS:

$$1 + 4 + 7 + \cdots + (3k - 2) = \frac{k(3k - 1)}{2}$$

$$\begin{aligned} 1 + 4 + 7 + \cdots + (3k - 2) + (3k + 1) &= \frac{k(3k - 1)}{2} + (3k + 1) \\ &= \frac{k(3k - 1) + 2(3k + 1)}{2} \\ &= \frac{3k^2 - k + 6k + 2}{2} \\ &= \frac{(k + 1)(3k + 2)}{2} \end{aligned}$$

Step 4 The proposition is true for all positive integers by PMI.

15. *Step 1* Prove it true for $n = 1$:

$(b^m)^1 = b^{m \cdot 1} \checkmark$

Step 2 Assume it true for $n = k$:

$(b^m)^k = b^{mk}$

Step 3 Prove it true for $n = k + 1$.

TO PROVE: $(b^m)^{k+1} = b^{m(k+1)}$

$(b^m)^k = b^{mk}$	By hypothesis
$(b^m)^k b^m = b^{mk} b^m$	Multiply both sides by b^m.
$(b^m)^{k+1} = b^{mk} b^m$	Definition
$= b^{mk+m}$	Problem 14
$= b^{m(k+1)}$	Distributive

Step 4 The proposition is true for all positive integers by PMI.

17. *Step 1* Prove it true for $n = 1$:

$$\left(\frac{a}{b}\right)^1 = \frac{a^1}{b^1} \checkmark$$

Step 2 Assume it true for $n = k$:

$$\left(\frac{a}{b}\right)^k = \frac{a^k}{b^k}$$

Step 3 Prove it true for $n = k + 1$

TO PROVE: $\left(\frac{a}{b}\right)^{k+1} = \frac{a^{k+1}}{b^{k+1}}$

$\left(\frac{a}{b}\right)^k = \frac{a^k}{b^k}$ Hypothesis

$\left(\frac{a}{b}\right)^k\left(\frac{a}{b}\right) = \frac{a^k}{b^k}\left(\frac{a}{b}\right)$ Multiply both sides by $\frac{a}{b}$.

$\left(\frac{a}{b}\right)^{k+1} = \frac{a^k}{b^k} \cdot \frac{a}{b}$ Definition

$= \frac{a^k \cdot a}{b^k \cdot b}$ Multiplication of fraction

$= \frac{a^{k+1}}{b^{k+1}}$ Definition

Step 4 The proposition is true for all positive integers by PMI.

19. *Step 1* Prove it true for $n = 1$:

$2^1 > 1$ ✓

Step 2 Assume it true for $n = k$:

$2^k > k$

Step 3 Prove it true for $n = k + 1$.

TO PROVE: $2^{k+1} > k + 1$

$2^{k+1} = 2^k \cdot 2$

$= 2^k \cdot (1 + 1)$

$= 2^k + 2^k$

$> 2^k + 1$ Since $2^k > 1$ for all k

$> k + 1$ By hypothesis, $2^k > k$

Step 4 The proposition is true for all positive integers by PMI.

21. *Step 1* Prove it true for $n = 1$:

$|a_1| \leq |a_1|$ ✓

Step 2 Assume it true for $n = k$:

$|a_1 + a_2 + \cdots + a_k| \leq |a_1| + |a_2| + \cdots + |a_k|$

Step 3 Prove it true for $n = k + 1$.

$|a_1 + a_2 + \cdots + a_k| \leq |a_1| + |a_2| + \cdots + |a_k|$ By hypothesis

Add $|a_{k+1}|$ to both sides:

$|a_1 + a_2 + \cdots + a_k| + |a_{k+1}| \leq |a_1| + |a_2| + \cdots + |a_k| + |a_{k+1}|$

But

$|(a_1 + a_2 + \cdots + a_k) + a_{k+1}| \leq |a_1 + a_2 + \cdots + a_k| + |a_{k+1}|$

by the triangle inequality. Therefore,

$|a_1 + a_2 + \cdots + a_k + a_{k+1}| \leq |a_1| + |a_2| + \cdots + |a_{k+1}|$

Step 4 The proposition is true for all positive integers by PMI.

23. *Step 1* Prove it true for $n = 1$:

$(a + b)^1 = \sum_{j=0}^{1} \binom{1}{j} a^{1-j}b^j = \binom{1}{0}a + \binom{1}{1}b = a + b$ ✓

Step 2 Assume it is true for $n = k$:

$(a + b)^k = \sum_{j=0}^{k} \binom{k}{j} a^{k-j}b^j$

Step 3 Prove it true for $n = k + 1$.

TO PROVE: $(a + b)^{k+1} = \sum_{j=0}^{k+1} \binom{k+1}{j} a^{k+1-j}b^j$

BY HYPOTHESIS:

$$(a+b)^k = \sum_{j=0}^{k} \binom{k}{j} a^{k-j} b^j$$

$$= a^k + \cdots + \binom{k}{r-1} a^{k-r+1} b^{r-1} + \binom{k}{r} a^{k-r} b^r + \cdots + b^k$$

Multiply both sides by $(a+b)$:

$$(a+b)^k(a+b) = \left[a^k + \cdots + \binom{k}{r-1} a^{k-r+1} b^{r-1} + \binom{k}{r} a^{k-r} b^r + \cdots + b^k \right](a+b)$$

$$(a+b)^{k+1} = \left[a^{k+1} + \cdots + \binom{k}{r-1} a^{k-r+2} b^{r-1} + \binom{k}{r} a^{k-r+1} b^r + \cdots + ab^k \right]$$

$$+ \left[a^k b + \cdots + \binom{k}{r-1} a^{k-r+1} b^r + \binom{k}{r} a^{k-r} b^{r+1} + \cdots + b^{k+1} \right]$$

$$= a^{k+1} + \cdots + \left[\binom{k}{r} + \binom{k}{r-1} \right] a^{k-r+1} b^r + \cdots + b^{k+1}$$

In Problem 22 we proved $\binom{k}{r} + \binom{k}{r-1} = \binom{k+1}{r}$ and the result is proved.

Step 4 The proposition is true for all positive integers by PMI.

Problem Set 10.6, Pages 436–438

1. $a = a$ **2.** If $a = b$, then $b = a$. **3.** If $a = b$ and $b = c$, then $a = c$.
4. If $a = b$, then a may be replaced throughout by b in any statement without changing the truth or falsity of the statement.
5. For every $a, b \in \mathbf{R}$, $(a + b) \in \mathbf{R}$. **6.** For every $a, b \in \mathbf{R}$, $ab \in \mathbf{R}$. **7.** $(ab)c = a(bc)$ **8.** $(a + b) + c = a + (b + c)$
9. $a + b = b + a$ **10.** $ab = ba$ **11.** There exists a number 0 such that $a + 0 = a$.
12. There exists a number 1 such that $a \cdot 1 = a$.
13. For each $a \in \mathbf{R}$, $a \neq 0$, there exists one element $\left(\frac{1}{a}\right) \in \mathbf{R}$ so that $a\left(\frac{1}{a}\right) = 1$.
14. For each $a \in \mathbf{R}$ there exists one element $(-a) \in \mathbf{R}$ so that $a + (-a) = 0$. **15.** $a(b + c) = ab + ac$ **16.** $p \rightarrow q$; p; q
17. $p \rightarrow q$; $q \rightarrow r$; $p \rightarrow r$ **18.** if p then q; if q then p **19.** $a + (-b)$ **20.** $a \cdot \left(\frac{1}{b}\right)$
21. positive numbers; $a = 0$; a is positive; $-a$ is positive **22.** $a - b$ is positive **23.** $a < b$ or $a = b$
24. $|a + b| \leq |a| + |b|$ **25.** $p \rightarrow q$; $\sim q$; $\sim p$
26. If a given proposition $P(n)$ is true for $P(1)$ and if the truth of $P(k)$ implies the truth of the proposition for $P(k + 1)$, then $P(n)$ is true for all positive integers. **27.** If $b = 9$, then $b^2 = 81$. Converse: If $b^2 = 81$, then $b = 9$.
28. If $b > 0$, then $b^{1/2} = \sqrt{b}$. Converse: If $b^{1/2} = \sqrt{b}$, then $b > 0$.
29. If $P^2 = Q$, then $|P| = \sqrt{Q}$. Converse: If $|P| = \sqrt{Q}$, then $P^2 = Q$.
30. If $h < x < k$, then $b^h < b^x < b^k$. Converse: If $b^h < b^x < b^k$, then $h < x < k$.
31. If one is a mathematician, then one is strange. Converse: If one is strange, then one is a mathematician.
32. If $x^2 = 9$, then $x = 3$ or $x = -3$. If $x = 3$ or $x = -3$, then $x^2 = 9$. The converse is the same. **33.** Invalid
34. Transitive **35.** Indirect **36.** Invalid **37.** Direct **38.** Invalid **39.** Indirect **40.** Invalid **41.** Transitive
42. Direct **43.** Invalid **44.** Reflexive (E1) **45.** Definition of subtraction **46.** Property of Opposites (Theorem 8, part 4)
47. Distributive (F11) **48.** Property of Opposites (Theorem 8, part 2) **49.** Commutative (F3) **50.** Associative (F5)
51. Distributive (F11) **52.** Closure (F1) **53.** Identity (F8) **54.** Transitive (E3)

55. GIVEN: $a, b, c, x, y \in \mathbf{R}$ TO PROVE: $y = -\frac{a}{b}x - \frac{c}{b}$

$b \neq 0$
$ax + by + c = 0$

Statements	Reasons
1. $ax + by + c = 0$	1. Given
2. $by + (ax + c) = 0$	2. Commutative (F3), Associative (F5)
3. $by = -(ax + c)$	3. Theorem 10.2.2
4. $by = (-1)(ax + c)$	4. Theorem 8
5. $by = (-1)ax + (-1)c$	5. Distributive (F11)
6. $\frac{1}{b} \in \mathbf{R}$	6. $b \neq 0$; Inverse (F10)
7. $by\left(\frac{1}{b}\right) = [(-1)ax + (-1)c]\frac{1}{b}$	7. Multiplication Law of Equality (Theorem 2)
8. $by\left(\frac{1}{b}\right) = (-1)ax\left(\frac{1}{b}\right) + (-1)c\left(\frac{1}{b}\right)$	8. Distributive (F11)
9. $y\left[b \cdot \frac{1}{b}\right] = (-1)\left[a\left(\frac{1}{b}\right)\right]x + (-1)\left[c\left(\frac{1}{b}\right)\right]$	9. Commutative (F4), Associative (F6)
10. $y\left[b \cdot \frac{1}{b}\right] = (-1)\frac{a}{b}x + (-1)\frac{c}{b}$	10. Definition of division
11. $y\left(b \cdot \frac{1}{b}\right) = (-1)\frac{a}{b}x + \left[-\left(1 \cdot \frac{c}{b}\right)\right]$	11. Theorem 8
12. $y \cdot 1 = (-1)\frac{a}{b}x + \left[-\left(1 \cdot \frac{c}{b}\right)\right]$	12. Inverse (F10)
13. $y = (-1)\frac{a}{b}x + \left[-\left(\frac{c}{b}\right)\right]$	13. Identity (F8)
14. $y = (-1)\frac{a}{b}x - \frac{c}{b}$	14. Definition of subtraction
∴ 15. $y = -\frac{a}{b}x - \frac{c}{b}$	15. Theorem 8

56. GIVEN: $a, b, c \in \mathbf{R}$, $a < b$ TO PROVE: $a + c < b + c$

Statements	Reasons
1. $a < b$	1. Given
2. $a - b < 0$	2. Definition of $<$
3. $(a + 0) - b < 0$	3. Identity (F7)
4. $c + (-c) = 0$	4. Inverse (F9)
5. $a + [c + (-c)] - b < 0$	5. Substitution (E4)
6. $(a + c) + [(-c) - b] < 0$	6. Associative (F5)
7. $(a + c) + [(-c) + (-b)] < 0$	7. Definition of subtraction
8. $(a + c) + [(-1)c + (-1)b] < 0$	8. Theorem 8, part 4
9. $(a + c) + (-1)(c + b) < 0$	9. Distributive (F11)
10. $(a + c) - (c + b) < 0$	10. Definition of subtraction
11. $a + c < c + b$	11. Definition of $<$
∴ 12. $a + c < b + c$	12. Commutative (F3)

57. $x > 0$, $x = 0$, or $x < 0$ from Order axiom

Case i If $x > 0$, $x^2 > 0$ since POS · POS = POS.

$|x^2| = x$ By definition of absolute value

Case ii: If $x = 0$, $|x^2| = 0$, $x^2 = 0$ so $|x^2| = x^2$

Case iii: If $x < 0$, $x^2 > 0$ since NEG · NEG = POS.

$|x^2| = x^2$ By definition of absolute value

58. Let $x = (-1)$. Then $x^3 = -1$ and $|x^3| = |-1| = 1$; since $-1 \neq 1$, $|x^3| \neq x^3$.

59. *Step 1* Prove it true for $n = 1$:

$$1 = \tfrac{1}{2}\cdot 1\cdot(1 + 1)$$
$$= \tfrac{1}{2}\cdot 1\cdot 2$$
$$= 1 \quad \checkmark$$

Step 2 Assume it true for $n = k$:

$$1 + 2 + 3 + \cdots + k = \tfrac{1}{2}k(k + 1)$$

Step 3 Prove it true for $n = k + 1$.

TO PROVE: $1 + 2 + 3 + \cdots + k + (k + 1) = \tfrac{1}{2}(k + 1)(k + 2)$

BY HYPOTHESIS: $1 + 2 + 3 + \cdots + k = \tfrac{1}{2}k(k + 1)$

Add $k + 1$ to each side:

$$1 + 2 + 3 + \cdots + k + (k + 1) = \tfrac{1}{2}k(k + 1) + (k + 1)$$
$$= \tfrac{1}{2}k(k + 1) + \tfrac{1}{2}\cdot 2(k + 1)$$
$$= \tfrac{1}{2}(k + 1)(k + 2)$$

Step 4 The proposition is true for all positive integers n by PMI.

60. All the field axioms are satisfied.

Index

FUNCTIONS

Constant Function

$f(x) = a$

Linear Function

$f(x) = mx + b,\ m \neq 0$

Quadratic Function

$f(x) = ax^2 + bx + c,\ a \neq 0$

Absolute Value Function

$$f(x) = |x| = \begin{cases} x \text{ if } x \geqslant 0 \\ -x \text{ if } x < 0 \end{cases}$$

Exponential Function

$f(x) = b^x,\ b > 0,\ b \neq 1$

Logarithmic Function

$f(x) = \log_b x,\ b > 0,\ b \neq 1$

Arithmetic Sequence

$a_n = a_1 + (n - 1)d$

Arithmetic Series

$$A_n = n\left(\frac{a_1 + a_n}{2}\right)$$

or $\frac{n}{2}[2a_1 + (n - 1)d]$

Geometric Sequence

$g_n = g_1 r^{n-1}$

Geometric Series

$$G_n = \frac{g_1(r^n - 1)}{r - 1}$$

$$G_\infty = \frac{g_1}{1 - r},\ |r| < 1$$

GRAPHS

Line, Slope-intercept form

$y = mx + b$

Line, Point-slope form

$y - k = m(x - k)$

Parabola with vertex at (h, k)

$y - k = a(x - h)^2$

opens up if $a > 0$,
or down if $a < 0$.

$x - h = a(y - k)^2$

opens right if $a > 0$,
or left if $a < 0$.

Circle with center at (h, k)

$(x - h)^2 + (y - k)^2 = r^2$

radius, $r > 0$.

Ellipse with center at (0, 0)

$$\frac{x^2}{a^2} + \frac{y^2}{b^2} = 1$$

x-intercepts $\pm\ a$
y-intercepts $\pm\ b$

Hyperbola with center at (0, 0)

$$\frac{x^2}{a^2} - \frac{y^2}{b^2} = 1 \text{ or } \frac{y^2}{a^2} - \frac{x^2}{b^2} = 1$$

with intercepts $\pm\ a$.